z	.00	.01	.02	.03	.04	.05	.06	.07	.08	.09
0.0	.5000	.5040	.5080	.5120	.5160	.5199	.5239	.5279	.5319	.5359
0.1	.5398	.5438	.5478	.5517	.5557	.5596	.5636	.5675	.5714	.5753
0.2	.5793	.5832	.5871	.5910	.5948	.5987	.6026	.6064	.6103	.6141
0.3	.6179	.6217	.6255	.6293	.6331	.6368	.6406	.6443	.6480	.6517
0.4	.6554	.6591	.6628	.6664	.6700	.6736	.6772	.6808	.6844	.6879
0.5	.6915	.6950	.6985	.7019	.7054	.7088	.7123	.7157	.7190	.7224
0.6	.7257	.7291	.7324	.7357	.7389	.7422	.7454	.7486	.7517	.7549
0.7	.7580	.7611	.7642	.7673	.7704	.7734	.7764	.7794	.7823	.7852
0.8	.7881	.7910	.7939	.7967	.7995	.8023	.8051	.8078	.8106	.8133
0.9	.8159	.8186	.8212	.8238	.8264	.8289	.8315	.8340	.8365	.8389
1.0	.8413	.8438	.8461	.8485	.8508	.8531	.8554	.8577	.8599	.8621
1.1	.8643	.8665	.8686	.8708	.8729	.8749	.8770	.8790	.8810	.8830
1.2	.8849	.8869	.8888	.8907	.8925	.8944	.8962	.8980	.8997	.9015
1.3	.9032	.9049	.9066	.9082	.9099	.9115	.9131	.9147	.9162	.9177
1.4	.9192	.9207	.9222	.9236	.9251	.9265	.9278	.9292	.9306	.9319
1.5	.9332	.9345	.9357	.9370	.9382	.9394	.9406	.9418	.9429	.9441
1.6	.9452	.9463	.9474	.9484	.9495	.9505	.9515	.9525	.9535	.9545
1.7	.9554	.9564	.9573	.9582	.9591	.9599	.9608	.9616	.9625	.9633
1.8	.9641	.9649	.9656	.9664	.9671	.9678	.9686	.9693	.9699	.9706
1.9	.9713	.9719	.9726	.9732	.9738	.9744	.9750	.9756	.9761	.9767
2.0	.9772	.9778	.9783	.9788	.9793	.9798	.9803	.9808	.9812	.9817
2.1	.9821	.9826	.9830	.9834	.9838	.9842	.9846	.9850	.9854	.9857
2.2	.9861	.9864	.9868	.9871	.9875	.9878	.9881	.9884	.9887	.9890
2.3	.9893	.9896	.9898	.9901	.9904	.9906	.9909	.9911	.9913	.9916
2.4	.9918	.9920	.9922	.9925	.9927	.9929	.9931	.9932	.9934	.9936
2.5	.9938	.9940	.9941	.9943	.9945	.9946	.9948	.9949	.9951	.9952
2.6	.9953	.9955	.9956	.9957	.9959	.9960	.9961	.9962	.9963	.9964
2.7	.9965	.9966	.9967	.9968	.9969	.9970	.9971	.9972	.9973	.9974
2.8	.9974	.9975	.9976	.9977	.9977	.9978	.9979	.9979	.9980	.9981
2.9	.9981	.9982	.9982	.9983	.9984	.9984	.9985	.9985	.9986	.9986
3.0	.9987	.9987	.9987	.9988	.9988	.9989	.9989	.9989	.9990	.9990
3.1	.9990	.9991	.9991	.9991	.9992	.9992	.9992	.9992	.9993	.9993
3.2	.9993	.9993	.9994	.9994	.9994	.9994	.9994	.9995	.9995	.9995
3.3	.9995	.9995	.9995	.9996	.9996	.9996	.9996	.9996	.9996	.9997
3.4	.9997	.9997	.9997	.9997	.9997	.9997	.9997	.9997	.9997	.9998

FINITE MATHEMATICS

LAWRENCE E. SPENCE
Illinois State University

CHARLES VANDEN EYNDEN
Illinois State University

DANIEL GALLIN

SCOTT, FORESMAN/LITTLE, BROWN HIGHER EDUCATION
A Division of Scott, Foresman and Company
Glenview, Illinois   London, England

■ ■ ■ ■ ■ ■ ■ ■ ■ **TO THE STUDENT**

If you want further help with this course, you may want to obtain a copy of the *Student's Solutions Manual* that accompanies this textbook. This manual provides detailed step-by-step solutions to the odd-numbered exercises in the textbook and can help you study and understand the course material. Your college bookstore either has this manual or can order it for you.

Library of Congress Cataloging-in-Publication Data

Spence, Lawrence E.
   Finite mathematics / Lawrence E. Spence, Charles Vanden Eynden, Daniel Gallin.
    p. cm.
   ISBN 0-673-38582-5
   1. Mathematics. I. Vanden Eynden, Charles.
II. Gallin, Daniel. III. Title.
QA39.2.S682   1990
510 s—dc20   89-24376
                                                              CIP

Copyright © 1990 Lawrence E. Spence and Charles Vanden Eynden.
All Rights Reserved.
Printed in the United States of America.
Artwork, illustrations, and other materials supplied by the publisher.
Copyright © 1990 Scott, Foresman and Company.
Portions of this book have been adapted from *Finite Mathematics* by Daniel Gallin.

1 2 3 4 5 6 - RRW - 95 94 93 92 91 90

PREFACE

Finite Mathematics is intended for use in a finite mathematics course that emphasizes mathematical applications and models. Such courses have become commonplace at most American colleges and universities during the past fifteen years, a development that reflects the increased use of quantitative concepts and techniques in many disciplines.

The mathematical topics presented here are linear functions and equations, systems of linear equations and matrices, linear programming, the mathematics of finance, sets and counting techniques, probability, statistics, Markov chains, and game theory. These topics have proven to have broad application to the management, life, and social sciences.

Flexibility

This book contains ample material for a four-credit course lasting one semester or two quarters. Consequently, instructors will have considerable flexibility in designing a course to meet the needs of their students.

Algebra Review

Although it is assumed that the reader has completed the equivalent of one and a half years of high school algebra, essential algebraic concepts are reviewed when necessary throughout the text. The need for review, however, has been balanced against the risk of boredom. We prefer to avoid the discussion of algebra for its own sake in favor of treating mathematics that is more readily applicable.

APPROACH

The mathematics presented in this book is developed intuitively, yet without loss of mathematical precision. Above all we have tried to provide students with conceptual understanding rather than a collection of mechanical procedures. By including a wide variety of applications, we have attempted to motivate students to learn mathematics as well as to see its power and utility in other disciplines.

Calculators

We assume that students will have access to a calculator when working exercises. Nevertheless, most exercises are designed so that the necessary computations can be performed by hand. Only in Chapter 4 is a calculator essential; the financial calculations there require a calculator with a key to evaluate powers of a number.

KEY FEATURES

Applications

Special emphasis has been placed on applications in sections called "Mathematics in Action." These sections present realistic uses of mathematics rather than contrived or artificial examples. All of the sections are optional, however, in the sense that no other parts of the text depend on them. The extent to which these sections are covered will depend on the length of the course and the nature of the students being taught.

Examples and Exercises

Finite Mathematics contains over 200 numbered examples that are carefully chosen to demonstrate both the mathematical techniques under consideration and their usefulness in solving real-world problems. A large number of exercises are included in the book. The exercises within each section range from easy to difficult, and many contain realistic applications of the mathematical ideas presented.

Practice Problems

Over 125 practice problems occur throughout the book. These are designed to test a student's understanding of the mathematical content being developed by providing an immediate opportunity to work a problem involving the concepts and techniques being discussed. As a further help to the student, both answers (at the end of each section) and *complete solutions* (at the back of the book) to practice problems are provided.

Chapter Reviews

Each chapter of *Finite Mathematics* concludes with a comprehensive chapter review that lists the new terms and important formulas introduced in that chapter. The chapter reviews also contain exercises that are intended as a practice chapter test for students. Although these exercises may sometimes require more than an hour to complete, they cover all the principal ideas in the chapter and should enable students to test their understanding prior to an examination.

Format

The format of this book is intended to help students recognize important results and techniques. Terminology being defined appears in boldface. Boxes are used to highlight theorems and procedures for emphasis and ease of reference. A second color is also used functionally in the artwork and in the text (for example, to annotate matrix calculations).

COURSE ORGANIZATION

As noted above, *Finite Mathematics* contains ample material for use in a one-semester or two-quarter course, and so considerable flexibility is given an instructor to teach those mathematical topics that are most meaningful for his or her students. Most courses in finite mathematics are built around three major subjects: linear equations and matrices (Chapters 1–2); linear programming (Chapter 3); and probability and statistics (Chapters 6–7). The content of Chapter 5 (sets and counting techniques) is needed for Chapter 6. The material on the mathematics of finance (Chapter 4), Markov chains (Chapter 8), and game theory (Chapter 9) can be included according to the interests of students. The following diagram shows the interdependence among the chapters.

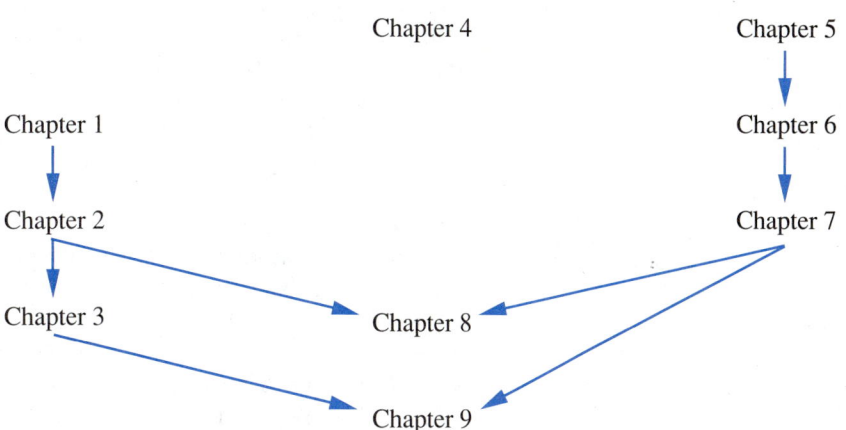

SUPPLEMENTS

Manuals

The following resources were prepared by John I. Hill, Vanessa F. Miller, Linda Ann Spence, and Joan Vanden Eynden of Illinois State University.

The **Instructor's Guide and Solutions Manual** gives a lengthy set of test questions for each chapter, organized by section, plus answers to all the questions. It also provides complete solutions to all of the even-numbered text exercises.

The **Instructor's Answer Manual** gives the answers to every text exercise, collected in one convenient location.

The **Student's Solutions Manual,** available for purchase by students, provides detailed, worked-out solutions to all of the odd-numbered text exercises, plus a chapter test for each chapter. Answers to the chapter test questions are given at the back of the book.

A brief supplement on *Logic* is also available from the publisher for those wishing to cover this topic in their course.

Visuals

A set of overhead transparencies showing charts, figures, and portions of examples is available and can be used to accompany lectures. The transparencies are especially useful in large classroom situations.

Computerized Testing

The **Scott, Foresman/Little, Brown Test Generator for Mathematics** enables instructors to select questions by section or chapter or to use a ready-made test for each chapter. Instructors may generate tests in multiple-choice (IBM^R/$Macintosh^R$ versions) or open-response formats, scramble the order of questions, and produce multiple versions of each test (up to 9 with Apple II^R and up to 25 with IBM^R and $Macintosh^R$). The system features a preview option that allows instructors to view questions before printing, to regenerate variables, and to replace or skip questions.

Software

Computer Applications for Finite Mathematics and Calculus by Donald R. Coscia is a softbound textbook packaged with two diskettes (in Apple II and IBM-PC versions) with programs and exercises keyed to the text. The programs allow students to solve meaningful problems without the difficulties of extensive computation. This book bridges the gap between the text and the computer by providing additional explanations and exercises for solution using a microcomputer.

Matrices, Statistics, and the Mathematical Functions of Finance: An Introduction to the Electronic Spreadsheet by Samuel W. Spero is a workbook for students that will be accompanied by the LOTUS-compatible electronic spreadsheet, VP-Planner Plus for the IBM-PC and compatibles.

The many applications of Finite Mathematics to business and industry are bound up with computing. To learn matrices, statistics, and the mathematical functions of finance effectively students should have available to them a computational tool. This workbook not only provides such a computational tool, but it also introduces the student to the electronic spreadsheet, which is the tool most often used in business and industry for these applications.

The workbook is closely linked to the material in the text by Spence and Vanden Eynden and can be profitably used from Chapter 1 on.

RELATED BOOKS IN THE SERIES

- *Calculus with Applications to the Management, Life, and Social Sciences* by Spence and Vanden Eynden.
- *Applied Mathematics for the Management, Life, and Social Sciences* by Spence and Vanden Eynden. This book is intended for a year-long course on finite mathematics and calculus.

ACKNOWLEDGMENTS

This book is based in part on *Finite Mathematics* by Daniel Gallin and *Finite Mathematics* by Lawrence E. Spence.

We are grateful for the enthusiasm and advice of Pam Carlson, Leslie Borns, George Duda, and Barbara Schneider of Scott, Foresman and Company as this project was developed. We also appreciate the fine work of Carol Leon during the production process.

While preparing this text, we have benefited greatly from the advice of many reviewers. We are especially indebted to those who provided thoughtful comments concerning early drafts:

James Angelos   *Central Michigan University*
Steven E. Blasberg   *West Valley College*
James Daly   *University of Colorado at Colorado Springs*
Frank C. Denney   *Chabot College*
Alan Gorfin   *Western New England College*
Cheryl M. Hawker   *Eastern Illinois University*
Richard W. Marshall   *Eastern Michigan University*
Donald E. Myers   *University of Arizona*
Caroline Woods   *Marquette University*

The following market reviewers provided additional insights which were much appreciated:

Charlotte Lewis   *University of New Orleans*
Thomas J. Miles   *Central Michigan University*
Paul Britt   *Louisiana State University*

We hope to have created a text that instructors can teach from and students can learn from with both success and enjoyment. We welcome and appreciate your feedback and comments.

Lawrence E. Spence
Charles Vanden Eynden

VP-Planner Plus is a registered trademark of Paperback Software International.

CONTENTS

1 LINEAR RELATIONSHIPS 1
- 1.1 Graphing in the Euclidean Plane 2
- 1.2 Linear Functions 10
- 1.3 Intersecting Lines 17
- 1.4 Break-Even Analysis 26
- 1.5 *MATHEMATICS IN ACTION:* Market Equilibrium 35
- Chapter Review 42

2 SYSTEMS OF LINEAR EQUATIONS AND MATRICES 44
- 2.1 Gauss-Jordan Elimination 45
- 2.2 Systems without Unique Solutions 56
- 2.3 Applications of Systems of Linear Equations 67
- 2.4 Matrices 78
- 2.5 Matrix Multiplication 86
- 2.6 Inverse Matrices 92
- 2.7 *MATHEMATICS IN ACTION:* The Leontief Input-Output Model 102
- Chapter Review 111

3 LINEAR PROGRAMMING 114
- 3.1 Optimization with Linear Constraints 115
- 3.2 The Graphical Method 127
- 3.3 The Simplex Method 137
- 3.4 Maximization Problems 148
- 3.5 Minimization Problems and Duality 159
- 3.6 Artificial Variables 170
- Chapter Review 185

4 MATHEMATICS OF FINANCE   188

4.1 Simple Interest   189
4.2 Compound Interest   196
4.3 Annuities and Installment Loans   202
4.4 Sinking Funds and Amortization   211
Chapter Review   219

5 SETS AND COUNTING TECHNIQUES   222

5.1 Sets and Venn Diagrams   223
5.2 Counting with Venn Diagrams and Tree Diagrams   231
5.3 The Multiplication and Addition Principles   242
5.4 Permutations and Combinations   252
5.5 Pascal's Triangle and the Binomial Theorem   263
Chapter Review   267

6 PROBABILITY   271

6.1 Computing Probabilities by Counting   272
6.2 Assigning Probabilities   287
6.3 Basic Laws of Probability   297
6.4 Conditional Probability, the Intersection Rule, and Independent Events   311
6.5 Bayes's Formula   326
6.6 Bernoulli Trials   337
Chapter Review   346

7 STATISTICS   350

7.1 Descriptive Statistics   351
7.2 Grouped Data   361
7.3 Random Variables: Expected Value and Variance   370
7.4 *MATHEMATICS IN ACTION:* Decision Theory   382
7.5 Probability Distributions   396
7.6 The Normal Distribution   407
Chapter Review   421

8 MARKOV CHAINS   425

8.1 Transition Matrices and State Vectors   426
8.2 Regular Markov Chains   436
8.3 Absorbing Markov Chains   445
8.4 *MATHEMATICS IN ACTION:* Genetics   460
Chapter Review   471

9 GAME THEORY 474

9.1 Strictly Determined Games 475
9.2 Games of Strategy 487
9.3 Simplex Solution of $m \times n$ Games 499
Chapter Review 511

TABLES 513
ANSWERS TO SELECTED EXERCISES 517
SOLUTIONS TO PRACTICE PROBLEMS 547
INDEX 587

INDEX OF APPLICATIONS

Management Science and Economics
advertising campaign, 16
annuities, 210–11, 217–18, 221
binomial distribution, 406, 420–21, 424
break-even analysis, 26–32
combinations, 261–62, 269
compound interest, 201, 221
consumer price index, 202
decision theory, 384–85, 389–95
depreciation, 16
expected value, 380
factory equipment, 125
farm land allocation, 123, 124, 158
filling truck, 118
inflation rate, 198, 201
inspections, 16
installment loans, 215–18, 221
insurance planning, 125
interest rate, 195, 201, 221
interest, 195, 201, 221
labor utilization, 22
Leontief input-output model, 102–9
linear programming, Chapter 3
manufacturing decisions, 136, 154, 158, 169, 176–78, 184, 186
marginal cost, 13
market equilibrium, 35–40
market share, 427–29, 431–32, 435–35, 440–41, 444, 446–52, 483–84, 486, 498
Markov chain, 431–32, 434, 440–41, 444, 446–47, 451–52

matrix solution of linear systems, 98
mine operation, 169, 187
oil reserves, 14–15
permutations, 261
probability, 309–11, 335–36
product-mix problems, 67–68
profit comparison, 28–32
service charge allocation, 69–71
simple interest, 195
simplex method, 154
stock charting, 2
taxation, effect on supply and demand, 38–40
travel planning, 124, 129–31, 163–65, 169

Life and Physical Science
albinism, 462–63, 473
alloy composition, 184–85
antigens, 228–29, 240–41, 325
binomial distribution, 407, 420, 423
cholesterol level, 322–23
coronary disease, 322–23
diagnostic testing, 333–34
diet planning, 23, 75, 117–18, 136–37, 168–69, 184, 186
difference equations, 204
disease propagation, 444, 458
ecology, 445–46, 458
family tree, 470
fertilizer composition, 75

genetics, 460–71
Hardy-Weinberg law, 466
health costs, 359
heredity, 460–62, 468
human blood, 228–29, 240–41, 325
inbreeding, 463, 64
Markov chain, 435, 444, 458, 462–64
medical diagnosis, 464–65
mortality rates, 323
pasture ecosystem, 445–46, 458
pollution control, 16, 74, 80, 136, 168
pregnancy test, 337
probability, 322–23, 335–37
snapdragons, 468–69
temperature conversion, 13–14
Venn diagrams, 238
weight gain, 16

Behavioral and Social Science
animal experiments, 136
busing students, 55
decision theory, 394
divorce, 341
labor negotiations, 486
land usage, 442
Markov chain, 434–35, 442, 444, 459
mistakes and temperature, 16
probability distributions, 194
probability, 285, 309, 311, 322–23, 344–46, 348
psychological experimentation, 238
student protests, 239–40
Venn diagrams, 230, 238–41, 324
voter demographics, 322–23

General Interest
binomial distribution, 407, 420
checking accounts, 31–32
expected value, 380
graphology, 403
lawn care, 136
Markov chain, 426–31, 441–42, 452–53, 459, 473
probability, 284–86, 294, 344–45
radio station call letters, 221
school excursion, 187
scissors-paper-stone, 510
statistics, 359, 369
success runs, 454–55
taxi rates, 34
telephone numbers, 221
three-finger morra, 510
tornadoes, 353, 363, 366
traffic flow, 71–74
two-finger morra, 476, 494
Venn diagrams, 238, 241
vitamins, 181–83

LINEAR RELATIONSHIPS

1.1 Graphing in the Euclidean Plane
1.2 Linear Functions
1.3 Intersecting Lines
1.4 Break-Even Analysis
1.5 Mathematics in Action: Market Equilibrium
Chapter Review

1

1.1 Graphing in the Euclidean Plane

When Sarah Marchant's husband died early in 1985 she used some of the insurance money to buy IBM stock, hoping that it would increase in value. After several years she began to wonder how her investment was doing. She was able to find most of the reinvestment reports she had received. These listed the selling price of the stock at various times as follows.

Date	Price
3-31-87	150 1/8
9-30-85	123 7/8
3-29-85	127
6-28-85	123 3/4
9-30-86	134 1/2
12-31-85	155 1/2
6-30-86	146 1/2

These seemed to her to be a jumble of numbers and dates until she arranged the reports in chronological order, then plotted the stock prices against time as in Figure 1.1. From this picture she could see that the general trend of the price was upward for the past two years, although there was a downturn in most of 1986. Even so, the stock was worth approximately 20 percent more than when she bought it. When she called her broker for advice, she already had an overview of the stock's performance since her purchase.

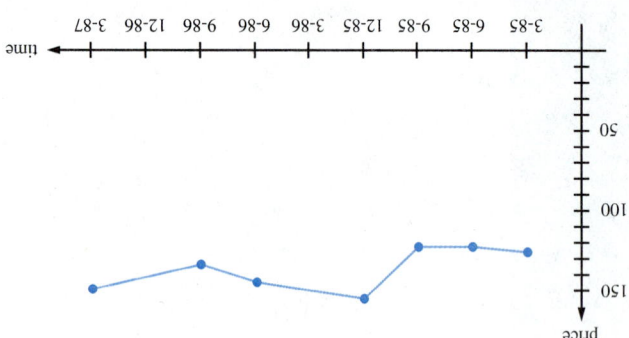

FIGURE 1.1

The simplest possible relationship between two quantities is the linear one; yet this is often a good approximation to reality. In this chapter we investigate linear relationships and how they arise.

1.1 GRAPHING IN THE EUCLIDEAN PLANE   3

THE REAL NUMBER LINE

In this example time is measured horizontally and stock price vertically along **real number lines.** To introduce a coordinate system on a given straight line:

(i) designate a particular point O on the line as the **origin** of the system;
(ii) select one of the two directions on the line to be the **positive direction**; and
(iii) choose a **unit length** on the line for measuring distances. With such a coordinate system established, each real number x can be represented as a unique point on the line. To plot the value x, we start at the origin and move x units along the line, taking into account the sign (positive or negative) of x. As an illustration, several values are plotted as points on the number line of Figure 1.2. The positive direction is chosen (by convention) to the right, as indicated by the arrow. In this way every real number corresponds to exactly one point, and conversely every point corresponds to exactly one number, called its **coordinate**.

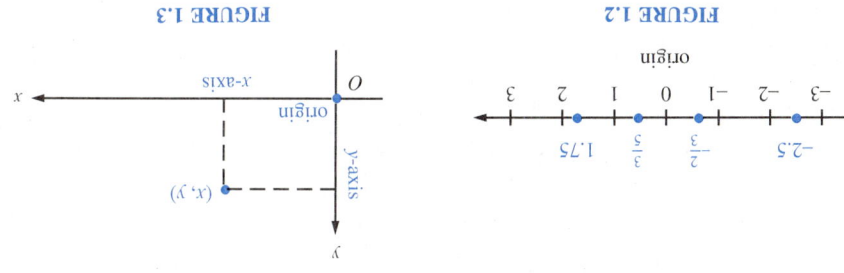

FIGURE 1.2

FIGURE 1.3

RECTANGULAR COORDINATES IN THE PLANE

To establish a coordinate system in the plane we choose two perpendicular lines, or **coordinate axes.** By convention these axes are horizontal and vertical, and are called the **x-axis** and **y-axis** respectively. We introduce a coordinate system on each axis in such a way that the lines share a common **origin** O at their point of intersection. (See Figure 1.3.) The positive direction is usually taken to the right on the x-axis and upward on the y-axis. The unit length may or may not be the same on the two axes.

The location of a point P is now specified by an **ordered pair** (x, y) of real numbers. To plot the ordered pair (x, y), we start at the origin and move x units horizontally and then y units vertically, taking into account the signs of x and y. Figure 1.4 shows an xy-coordinate system in the plane, with several points plotted for illustration.

For example, the ordered pair $(3, 2)$ corresponds to point A in the figure, obtained by starting at the origin and moving 3 units to the right, then 2 units upward. In the same way, every ordered pair (x, y) determines a unique point

4 CHAPTER 1 LINEAR RELATIONSHIPS

GRAPHING MATHEMATICAL STATEMENTS

A mathematical statement involving x and y may be true for certain pairs x, y of numbers and false for others. For example the statement $x + y = 7$ is true for the pair $(x, y) = (3, 4)$, and also for the pair $(-2, 9)$, but false for the pair $(4, 5)$. The **graph** of a statement is the set of all points (x, y) in the plane such that the statement is true for the given x and y. For example, the graph of the statement $x > 0$ and $y > 0$ is the set of all points to the right of the y-axis and above the x-axis, as shown in Figure 1.6(a). Likewise, the graph of the statement $x = 3$ and $1 \leq y \leq 4$ is the vertical line segment shown in Figure 1.6(b).

P in the plane, and conversely. The numbers x and y are called the **coordinates** of the point. For example, point D in Figure 1.4 has x-coordinate -1 and y-coordinate 4; we would therefore describe D as "the point $(-1, 4)$." Notice that points on the x-axis, such as $F(-4, 0)$ and $J(4, 0)$, have y-coordinate zero, while points on the y-axis, such as $C(0, 5)$ and $H(0, -3)$, have x-coordinates zero. The origin O has coordinates $(0, 0)$.

For example, a scientist interested in how temperature affects the distance a football can be kicked finds that a kicking machine propels the ball 60 feet when the outside temperature is 20° Celsius, 55 feet at 10° C, and 40 feet at $-5°$ C. Letting x be the temperature and y the distance leads to the points $(20, 60)$, $(10, 55)$, and $(-5, 40)$ plotted in Figure 1.5. Notice that different scales are used on the two axes.

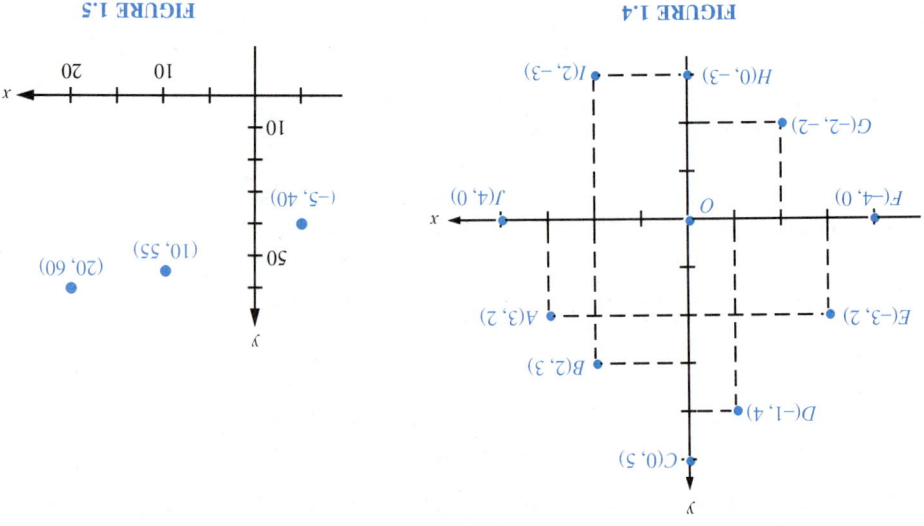

FIGURE 1.4

FIGURE 1.5

1.1 GRAPHING IN THE EUCLIDEAN PLANE 5

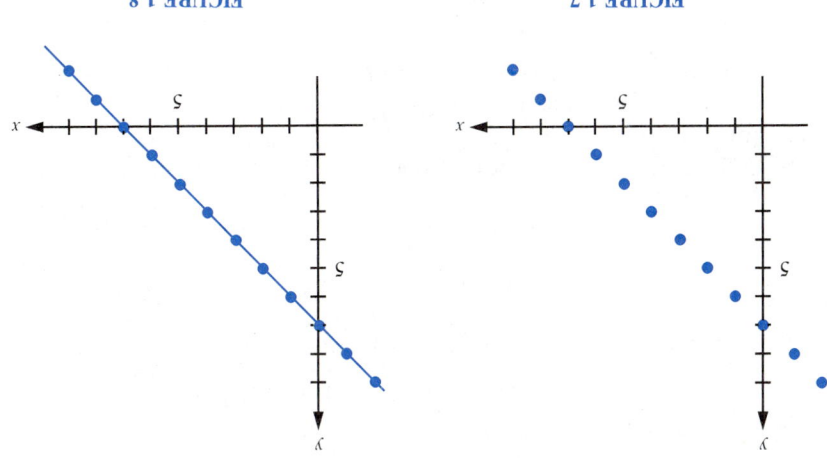

FIGURE 1.6

To get an idea of the graph of the statement $x + y = 7$ we will make a table of various pairs for which it is true, using the fact that for any value of x, the number y satisfies the equation if and only if $y = 7 - x$.

x	1	2	3	4	5	6	7	8	9	0	-1	-2
y	6	5	4	3	2	1	0	-1	-2	7	8	9

The corresponding points are plotted in Figure 1.7.

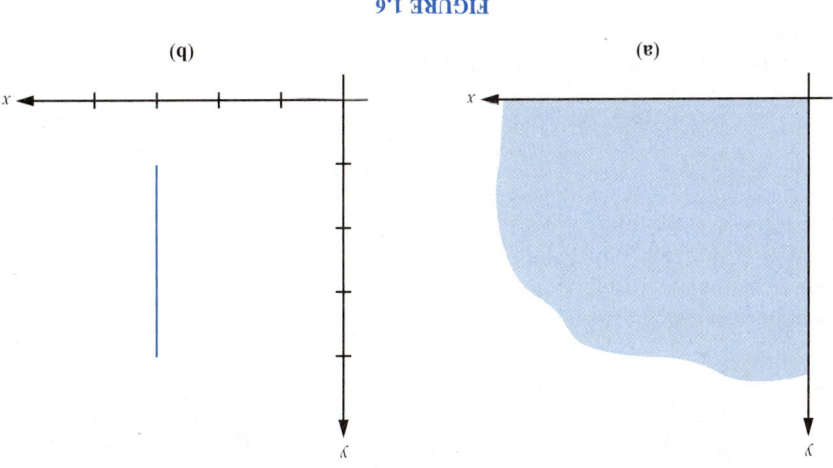

FIGURE 1.7

FIGURE 1.8

Of course there are many other points on the graph of $x + y = 7$, infinitely many, in fact. As one might guess, they are the set of all points on the line shown in Figure 1.8.

both 0, then the graph of the equation

$$Ax + By = C \qquad (1.1)$$

is a straight line. Conversely, any straight line is the graph of an equation of this form. For this reason equations of the form (1.1) are called **linear equations**. For example, the equation $x + y = 7$ is a linear equation with $A = 1$, $B = 1$, and $C = 7$, and $3y = -5/6$ is a linear equation with $A = 0$, $B = 3$, and $C = -5/6$.

GRAPHING BY INTERCEPTS

The line determined by a given equation can be graphed very easily by finding its **intercepts**, the points where it crosses the coordinate axes. The procedure is described below.

Graphing by Intercepts

To graph a linear equation $Ax + By = C$:

1. Set $y = 0$ and solve for x. This determines the **x-intercept** of the line.
2. Set $x = 0$ and solve for y. This determines **y-intercept** of the line.
3. The graph is the straight line through these points.

EXAMPLE 1.1 Graph the equation $2x + 3y = 6$.

Solution We set $y = 0$ in the equation and solve for x.

$$2x + 3(0) = 6$$
$$2x = 6$$
$$x = 3$$

The point (3, 0) thus lies on the graph of $2x + 3y = 6$, and it also lies on the x-axis (since its y-coordinate is zero). We have therefore found the x-intercept of the line, the point where it crosses the x-axis. To find the y-intercept of the line, we set $x = 0$ and solve for y.

$$2(0) + 3y = 6$$
$$3y = 6$$
$$y = 2$$

The line therefore crosses the y-axis at (0, 2). Since two points determine a single straight line, we have only to plot the intercepts and draw the line through them to obtain the graph in Figure 1.9. ■

1.1 GRAPHING IN THE EUCLIDEAN PLANE

EXAMPLE 1.2 Graph the equation $4x - 5y = 0$.

Solution Here we have $Ax + By = C$, where $A = 4$, $B = -5$, and $C = 0$. When we try to find the intercepts we encounter a slight problem: $y = 0$ yields $x = 0$, and vice-versa. This is because *the line goes through the origin* and thus has only one intercept, namely the point $(0, 0)$. To graph the equation we need only find one other point on the line. This can be accomplished by arbitrarily assigning a value to x in the equation, for instance, the value $x = 5$.

$$4(5) - 5y = 0$$
$$20 - 5y = 0$$
$$5y = 20$$
$$y = 4$$

We now have a second point $(5, 4)$ on the line. Joining $(0, 0)$ and $(5, 4)$ we obtain the graph in Figure 1.10. ■

EXAMPLE 1.3 Graph the equation $3x = 4$.

Solution If we write this equation in the form $3x + 0y = 4$, we see that it is of the form $Ax + By = C$, with $A = 3$, $B = 0$, and $C = 4$. Therefore, it must have a straight line as its graph. Setting $y = 0$ we find that $x = 4/3$, so the x-intercept of the line is $(4/3, 0)$. But setting $x = 0$ leads to an absurdity,

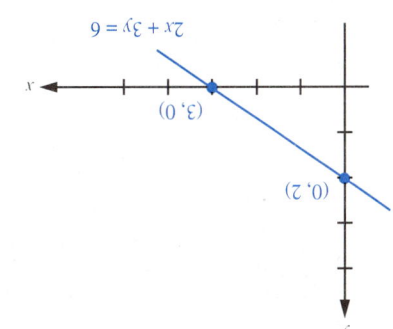

FIGURE 1.9

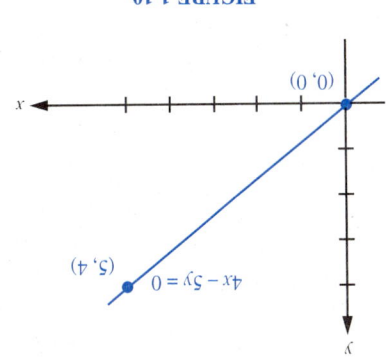

FIGURE 1.10

Practice Problem 1 Find the x-intercept and y-intercept, and graph the equation $3x - 4y = 6$.

If one or two of the constants A, B, C in the linear equation $Ax + By = C$ are zero, then the method of graphing by intercepts must be modified slightly.

8 CHAPTER 1 LINEAR RELATIONSHIPS

$0 = 4$. This tells us that the equation $3x = 4$ *cannot be satisfied* by any point $(0, y)$ on the y-axis. In other words, the line has no y-intercept. From this observation we conclude that the line must be parallel to the y-axis, that is, vertical. (See Figure 1.11(a).) ∎

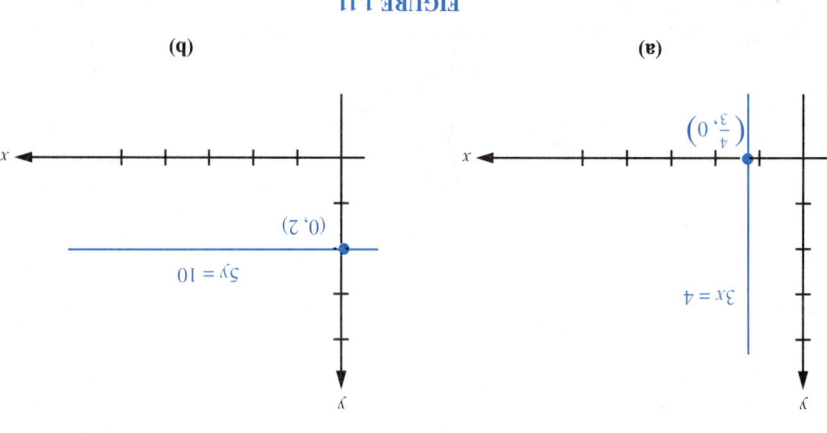

FIGURE 1.11

In a way similar to Example 1.3 we can see that the equation $5y = 10$ describes a line that has y-intercept $(0, 2)$ but no x-intercept. Thus it is parallel to the x-axis, that is, horizontal. (See Figure 1.11(b).)

Notice that the equation $3x = 4$ can be written equivalently as $x = 4/3$, while the equation $5y = 10$ is equivalent to $y = 2$. In general,

■ ■ ■ ■ ■ ■

The equation $x = k$ (k constant) represents a *vertical* line that crosses the x-axis at $(k, 0)$.

The equation $y = k$ (k constant) represents a *horizontal* line that crosses the y-axis at $(0, k)$.

EXERCISES 1.1

In Exercises 1–6, plot the listed points in the xy-plane. Use appropriate units on the axes.

1. $(2, 5)$, $(-1, 3)$, and $(0, -4)$
2. $(-3, -2)$, $(3, 1)$, and $(2, 0)$
3. $(25, 30)$, $(-5, -15)$, and $(20, 0)$
4. $(.05, 0.1)$, $(.03, -.08)$, and $(.11, -.02)$
5. $(350, -400)$, $(0, 800)$, and $(100, 0)$
6. $(4000, 1000)$, $(2000, 0)$, and $(0, -1000)$

In Exercises 7–10, give the ordered pair corresponding to the indicated point in the following figures.

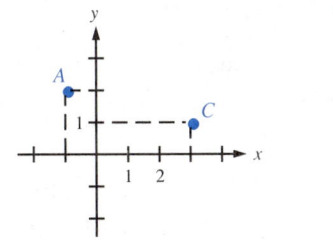
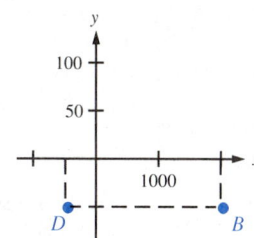

7. Point A in the first graph
8. Point B in the second graph
9. Point D in the second graph
10. Point C in the first graph

11. By going through his old income tax returns, a farmer found he had made $35,000 in 1984, $17,000 in 1985, and $29,000 in 1987, and had lost $7,000 in 1986. Represent this information graphically.

12. A textbook author finds she received royalty checks for $350.75 in January 1986, $202.20 in January 1985, $157.50 in July 1986, and $13.25 in January 1987. Represent this information graphically.

In Exercises 13–18, draw the graph of the given statement.

13. $x \geq 0$ and $y \geq 0$
14. $x \geq 3$ and $y = 2$
15. $1 \leq y$ and $y \leq 3$
16. $0 \leq x$, $x \leq 3$, and $y \geq -1$
17. $2 \leq y$, $y \leq 3$, $-1 \leq x$, and $x \leq 2$
18. $xy = 0$

In Exercises 19–22, write the given equation in the form $Ax + By = C$.

19. $y = 5x + 7$
20. $5x = x + y + 7$
21. $2x + 3y + 4 = x - y$
22. $x - 4y = \dfrac{x}{2} + \dfrac{y}{3} + 4$

In Exercises 23–36, find the x-intercept and y-intercept of the given equation. Then graph the equation.

23. $4x + 6y = 24$
24. $6x + 4y = 24$
25. $4x - 6y = 24$
26. $4x + 6y = -24$
27. $3x - 5y = 10$
28. $4y - 3x = 9$
29. $2x + 3y + 12 = 0$
30. $3x = 5y + 30$
31. $x + 4y + 4 = 3x - y + 24$
32. $3y - 2x = 5x - 4y + 14$
33. $5x + 3y = 450$
34. $4x - 7y = 140$
35. $4x + 5y = 1$
36. $20y - 25x = 2$

Answer to Practice Problem   1. The x-intercept is (2, 0), the y-intercept is (0, −3/2), and the graph is shown below.

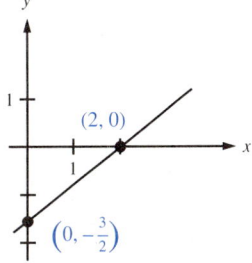

1.2 Linear Functions

THE SLOPE-INTERCEPT FORM

If the quantities x and y satisfy the linear equation $Ax + By = C$ with $B \neq 0$, then we can solve for y, getting an equation of the form

$$y = mx + b,$$

where m and b are constants. In this case we say that y is a **linear function** of x. We usually think of y as depending on x, and call y the **dependent** variable and x the **independent** variable.

EXAMPLE 1.4 Write the equation $2y + 6x = 5$ in the form $y = mx + b$.

Solution We have the equations:

$$2y + 6x = 5$$
$$2y = -6x + 5$$
$$y = -3x + \frac{5}{2}.$$

The last equation is in the form $y = mx + b$ with $m = -3$ and $b = 5/2$. ∎

Notice that if we set $x = 0$ in the equation $y = mx + b$, we get $y = b$. Thus the y-intercept of the graph of this equation is the point $(0, b)$. The constant m also has a geometric interpretation. Consider a pair (x_1, y_1) satisfying the equation $y = mx + b$, so that $y_1 = mx_1 + b$. How does y change if we increase x by 1, to $x_1 + 1$? Then

$$y = m(x_1 + 1) + b = mx_1 + m + b = (mx_1 + b) + m = y_1 + m.$$

We see that increasing x by 1 makes y change by the amount m. This will represent an increase or a decrease, depending on whether m is positive or negative. If $m = 0$, then y does not change at all. Figure 1.12 illustrates the three cases when m is positive, negative, or zero.

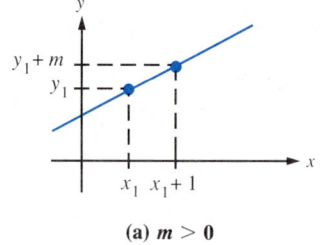
(a) $m > 0$

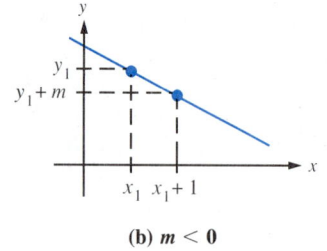
(b) $m < 0$

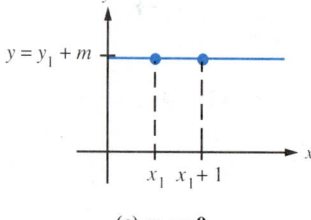
(c) $m = 0$

FIGURE 1.12

The number m is called the **slope** of the line, and the equation
$$y = mx + b$$
is said to be in **slope-intercept form.**

In general, if (x_1, y_1) and (x_2, y_2) are any two points on the line, then
$$y_2 = mx_2 + b \quad \text{and} \quad y_1 = mx_1 + b.$$

Subtracting these equations gives
$$y_2 - y_1 = mx_2 + b - (mx_1 + b) = m(x_2 - x_1),$$
or

Slope Formula

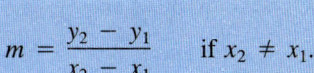

$$m = \frac{y_2 - y_1}{x_2 - x_1} \quad \text{if } x_2 \neq x_1.$$

The slope of a line is a measure of how fast the line rises (or, if $m < 0$, falls) as we move to the right. Some examples are shown in Figure 1.13. The slope of a vertical line is undefined, since the slope formula would entail division by zero.

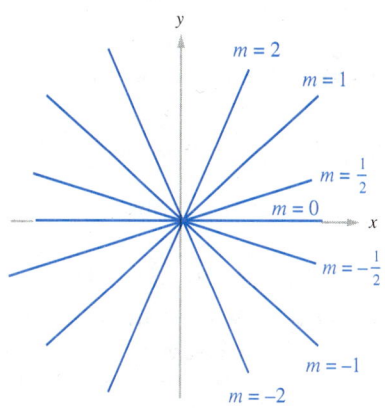

FIGURE 1.13

EXAMPLE 1.5 Find the slope of the line through the points $(2, -5)$ and $(-3, 4)$.

Solution By the slope formula with $(x_1, y_1) = (2, -5)$ and $(x_2, y_2) = (-3, 4)$, we have
$$m = \frac{4 - (-5)}{-3 - 2} = \frac{9}{-5} = -\frac{9}{5}.$$

EXAMPLE 1.6 Find the slope and y-intercept of the line with equation $2x + 3y = 6$.

Solution Solving for y gives

$$3y = -2x + 6$$
$$y = -\frac{2}{3}x + 2.$$

Since this equation is in the form $y = mx + b$, the slope is $-2/3$ and the y-intercept is $(0, 2)$. ∎

THE POINT-SLOPE FORM

Suppose we know the slope m of a line and a point (x_1, y_1) on the line. Then if (x, y) is any other point on the line, from the slope formula we must have

$$m = \frac{y - y_1}{x - x_1},$$

or

Point-Slope Form

$$y - y_1 = m(x - x_1).$$

Practice Problem 2 What is the slope-intercept form of the equation of the line with slope -2 through the point $(5, 3)$?

EXAMPLE 1.7 A manufacturer finds that the cost y of making x light bulbs is a linear function of x. If making 2000 light bulbs costs $900, and if each additional bulb costs $.20 to make, express y as a linear function of x. How much would it cost to make 4000 light bulbs?

Solution We are given that x and y satisfy a linear equation. Since increasing x by 1 increases y by $.20, we have $m = .20$. Also, if $x = 2000$, then $y = 900$, so the point $(x_1, y_1) = (2000, 900)$ is on the corresponding line. Then the point-slope form of the equation is

$$y - 900 = .20(x - 2000).$$

Thus

$$y - 900 = .20x - 400$$
$$y = .20x + 500,$$

which is the required equation. If $x = 4000$, then

$$y = .20(4000) + 500 = 800 + 500 = 1300,$$

so 4000 lightbulbs would cost $1300. ∎

In Example 1.7 the cost of making one more bulb, namely $.20, is called the **marginal cost**. Notice that if the total cost y of making x items is a linear function of x, then the marginal cost is the slope of the corresponding line.

If we are given two points on a line, we can find the equation of the line by first computing its slope, and then using either of the points in the point-slope formula.

EXAMPLE 1.8 Find the equation of the line through the points $(-1, 3)$ and $(4, 2)$. Where does this line cross the coordinate axes?

Solution Let $(x_1, y_1) = (-1, 3)$ and $(x_2, y_2) = (4, 2)$. Then the slope of the line is

$$m = \frac{y_2 - y_1}{x_2 - x_1} = \frac{2 - 3}{4 - (-1)} = -\frac{1}{5}.$$

From the point-slope form its equation is

$$y - 3 = -\frac{1}{5}[x - (-1)]$$

$$= -\frac{1}{5}(x + 1)$$

$$5(y - 3) = -(x + 1)$$

$$5y - 15 = -x - 1$$

$$5y + x = 14.$$

Setting $y = 0$ gives $x = 14$, and setting $x = 0$ gives $y = 14/5$. Thus the intercepts are $(14, 0)$ and $(0, 14/5)$, as shown in Figure 1.14. ∎

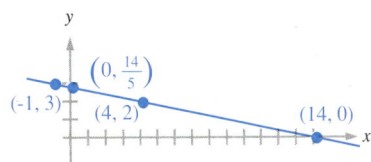

FIGURE 1.14

Practice Problem 3 Find the slope-intercept form of the equation of the line through the points $(1, 3)$ and $(2, -5)$.

EXAMPLE 1.9 It is known that the Fahrenheit temperature y of an object is a linear function of its Celsius temperature x. Because the Celsius scale sets 0 degrees as the freezing point of water and 100 degrees as its boiling point, 0 degrees Celsius corresponds to 32 degrees Fahrenheit and 100 degrees Celsius corresponds to 212 degrees Fahrenheit. What Fahrenheit temperature corresponds to 50 degrees Celsius? What is the Celsius temperature on a hot day when it is 100 degrees Fahrenheit?

14 CHAPTER 1 LINEAR RELATIONSHIPS

Solution   The Fahrenheit temperature is a linear function of the Celsius temperature, and the graph of the corresponding equation is a straight line containing the points (0, 32) and (100, 212). Thus its slope is

$$m = \frac{212 - 32}{100 - 0} = \frac{180}{100} = \frac{9}{5}.$$

Using the point $(x_1, y_1) = (0, 32)$ in the point-slope form we get the equation

$$y - 32 = \frac{9}{5}(x - 0)$$

or

$$y = \frac{9}{5}x + 32.$$

Corresponding to the Celsius temperature $x = 50$ we have

$$y = \frac{9}{5}(50) + 32$$
$$= 90 + 32 = 122.$$

Thus, 50 degrees Celsius corresponds to 122 degrees Fahrenheit. Conversely, if the Fahrenheit temperature is $y = 100$, then we have

$$100 = \frac{9}{5}x + 32$$

$$\frac{9}{5}x = 100 - 32 = 68$$

$$x = \frac{68}{\left(\frac{9}{5}\right)} = 68\left(\frac{5}{9}\right) = \frac{340}{9} = 37\frac{7}{9}.$$

We see that 100 degrees Fahrenheit corresponds to almost 38 degrees Celsius. ∎

EXAMPLE 1.10   A middle-eastern country exports a constant amount of oil each year, so its oil reserves have been decreasing linearly with time. Industry experts estimate that the country's reserves were 350 million barrels in 1975 and 252 million barrels in 1982. If this trend continues, when will the oil run out?

Solution   Letting R denote the remaining oil reserves, in millions of barrels, and t denote the time (in years) since 1975, we have a linear function $R = mt + b$. (Notice that here we are using the more suggestive variable names, t for time and R for reserves, in place of x and y.) The given conditions say that $R = 350$ when $t = 0$ (in 1975), and $R = 252$ when $t = 7$ (in 1982). Thus, the points (0, 350) and (7, 252) lie on the corresponding line, as illustrated in Figure 1.15. By the slope formula the slope of this line is

$$m = \frac{252 - 350}{7 - 0} = \frac{-98}{7} = -14.$$

Since the line crosses the R-axis at $b = 350$, the slope-intercept equation is
$$R = -14t + 350.$$

Note that $m = -14$ represents the rate of change of the function, that is, the amount of oil pumped out each year. According to the above equation, the reserves will reach zero when
$$0 = -14t + 350.$$

Solving this equation, we obtain $14t = 350$, or $t = 25$. In other words, if the linear trend continues, the oil will run out in the year 2000. (See Figure 1.15.) ■

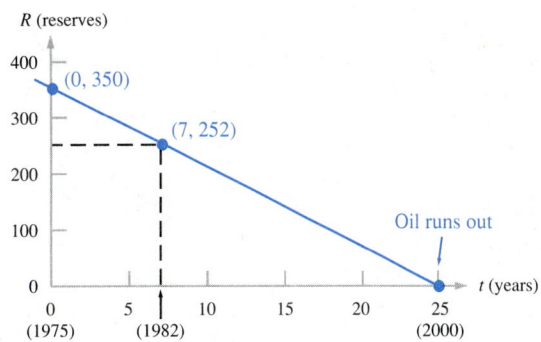

FIGURE 1.15

EXERCISES 1.2

In Exercises 1–6 change the equation given to slope-intercept form.

1. $3x + 4y = 18$ **2.** $5x - 6y = 15$

3. $x = 3y + 5$ **4.** $5x = 6y$

5. $4y + 7 = 0$ **6.** $2x + 3y = -4x + 9$

In Exercises 7–14 give the slope and y-intercept, if they exist, of the line with the given equation.

7. $7x + 8y = 42$ **8.** $-3x - 6y = 8$

9. $3y + 5 = 0$ **10.** $\frac{1}{2}x + \frac{1}{3}y = 5$

11. $2y + 3 = 4x$ **12.** $-3x + 6 = 0$

13. $3(x + 1) = 4$ **14.** $4x - 2y = 5 + x$

In Exercises 15–20 give the slope of the line through the two points, if possible.

15. $(2, 3)$ and $(4, 7)$ **16.** $(5, -7)$ and $(1, -9)$

17. $(-2, 5)$ and $(6, -1)$ **18.** $(-3, 6)$ and $(2, 6)$

19. $(5, 4)$ and $(5, -2)$ **20.** $(1/2, 2)$ and $(5/2, 4/3)$

In Exercises 21–28 give the equation of the line described, in slope-intercept form if possible.

21. slope 4, y-intercept $(0, -3)$
22. slope -3, y-intercept $(0, 5)$
23. slope -3, through $(1, -4)$
24. slope 2, through $(-2, 5)$
25. x-intercept $(2, 0)$, y-intercept $(0, -2)$
26. x-intercept $(-4, 0)$, slope 6
27. slope undefined, through $(7, 5)$
28. horizontal, through $(2, -3)$

In Exercises 29–34 give the equation of the line through the two points, in slope-intercept form if possible.

29. $(3, 4)$ and $(5, 2)$
30. $(6, 1)$ and $(3, 7)$
31. $(-1, 3)$ and $(2, -5)$
32. $(3, 4)$ and $(3, -2)$
33. $(4, 2)$ and $(-1, 2)$
34. $(3, 1/2)$ and $(-1, 5/2)$

35. A motion picture company finds that the number of people who will see one of its pictures the first week it is released is a linear function of the amount of money it spends on pre-release television advertising. If it spends $500,000 on such advertising, 200,000 tickets will be sold the first week, and if it spends $700,000, 250,000 tickets will be sold.
    (a) How many tickets will be sold if $1,000,000 is spent?
    (b) How much advertising is needed to sell 1,000,000 tickets the first week?

36. A psychologist is testing the hypothesis that the number of mistakes per hour a subject will make copying nonsense words is a linear function of the room temperature. The subject makes 13 mistakes per hour at 70° F and 17 mistakes per hour at 80° F. If the hypothesis is correct, how many mistakes per hour can be expected at 95° F?

37. An automobile manufacturer finds that the number of complaints it gets per month is a linear function of the number of inspections given to cars before they are shipped. If there are 40 inspections, there are 60 complaints per month; and if there are 30 inspections, there are 85 complaints per month. How many complaints per month would there be if there were 45 inspections?

38. It is found that the rate at which baby rats gain weight is a linear function of the amount of food they are fed each day. If fed 30 grams per day, they gain 2 ounces per week; and if fed 36 grams per day, they gain 2.5 ounces per week.
    (a) How much would a rat gain per week if fed 40 grams per day?
    (b) What daily ration would produce a gain of 2.8 ounces per week?

39. A machine purchased for $10,000 is depreciated linearly over a 15-year period, at the end of which it has a scrap value of $2500. In other words, the value of the machine decreases linearly from $10,000 to $2500 over fifteen years.
    (a) Let V be the value of the machine t years after its purchase. Express V as a linear function of t.
    (b) What will the machine be worth after three years? Five years?
    (c) How long will it take for the value to reach $4000?

40. The level of oxidant pollution in a certain area has been decreasing linearly since 1980, when an intensive pollution control program began. At that time the oxidant level in the air was measured at .15 parts per million (ppm); by 1990 the figure was .09 ppm.

The number of people who will see this movie the first week is a linear function of the money spent on prerelease advertising.

(a) Let P be the oxidant level (in ppm) and let t be the time (in years) since 1980. Express P as a linear function of t.
(b) What will the oxidant level be in 1995? In 1997?
(c) Air with an oxidant level of .03 ppm or less is considered clean. If the present linear trend continues, when will this be achieved?

Answers to Practice Problems
2. $y = -2x + 13$
3. $y = -8x + 11$

1.3 Intersecting Lines

The geometric problem of finding the point where two given lines intersect corresponds to the algebraic problem of solving two linear equations in two unknowns. In this section we review the methods of substitution and elimination, the generalizations of which will play an important role in Chapter 2.

SYSTEMS OF EQUATIONS

We begin with the fact, familiar from Euclidean geometry, that two nonparallel lines intersect at a point. If we know the equations of the lines, we should be able to find the coordinates of the point.

EXAMPLE 1.11 Find the point where the lines $2x - 3y = 9$ and $x + y = 2$ intersect.

Discussion Since the point of intersection lies on *both* lines, its coordinates (x, y) must satisfy both equations. We must therefore solve the **system of two linear equations in two unknowns**

$$2x - 3y = 9$$
$$x + y = 2.$$

Two different methods of solution are illustrated.

Substitution Method

To solve a system of two linear equations in two unknowns:

1. Use one of the equations to solve for one variable in terms of the other.
2. Substitute this quantity into the remaining equation and solve for the other variable.

In our example we can solve for y in terms of x using the second equation.

$$x + y = 2 \tag{1.2}$$
$$y = 2 - x$$

Substituting $2 - x$ for y in the first equation of the system, we obtain

$$2x - 3(2 - x) = 9$$
$$2x - 6 + 3x = 9$$
$$5x = 15$$
$$x = 3.$$

Finally, letting $x = 3$ in equation (1.2), we obtain y.

$$y = 2 - x = 2 - 3 = -1$$

The point of intersection is therefore $(3, -1)$. We can verify this solution by direct substitution in the original equations.

Check $2x - 3y = 2(3) - 3(-1) = 6 + 3 = 9$
$\phantom{\textbf{Check}\ \ } x + y = 3 + (-1) = 2$

The situation is illustrated in Figure 1.16.

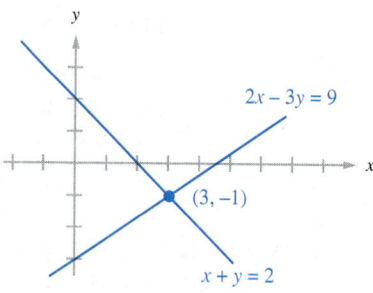

FIGURE 1.16

The second method of solution is called **elimination.**

Elimination Method

To solve a system of two linear equations in two unknowns:

1. Multiply one or both equations by appropriate constants so that one of the variables has coefficient c in one equation and $-c$ in the other.
2. Add the equations together to eliminate this variable.

To apply this method to the system of Example 1.11 we begin by writing down the given equations.

$$2x - 3y = 9$$
$$x + y = 2$$

Multiplying the second equation by 3, we obtain an **equivalent system,** that is, a system having exactly the same solutions, if any.

$$2x - 3y = 9$$
$$3x + 3y = 6$$

If we now add these equations, the term involving y is eliminated.

$$5x = 15$$
$$x = 3$$

Finally, we substitute $x = 3$ into either of the original equations—say, the second—and solve for y.

$$3 + y = 2$$
$$y = -1$$

We thus arrive at the solution $(3, -1)$, as before. ∎

EXAMPLE 1.12 Find the point where the lines $3x + 7y = 27$ and $5x + 4y = 22$ intersect.

Solution We must solve the following system of equations:

$$3x + 7y = 27$$
$$5x + 4y = 22.$$

If we multiply the first equation by 5 and the second by -3, we obtain an equivalent system.

$$15x + 35y = 135$$
$$-15x - 12y = -66$$

When we add these equations the term involving x is eliminated.

$$23y = 69$$
$$y = 3$$

We can now find x by substituting $y = 3$ into the first equation.

$$3x + 7(3) = 27$$
$$3x + 21 = 27$$
$$3x = 6$$
$$x = 2$$

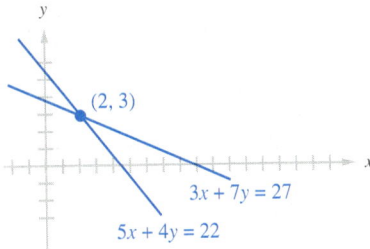

FIGURE 1.17

The lines intersect at the point $(2, 3)$, as shown in Figure 1.17. ∎

Practice Problem 4 Find the point where the lines $x + 3y = -5$ and $3x - 2y = 7$ intersect.

EXAMPLE 1.13 Find the point where the lines $2x - 4y = 7$ and $-3x + 6y = 8$ intersect.

Solution We apply the elimination method to the system
$$2x - 4y = 7$$
$$-3x + 6y = 8.$$

Multiplying the first equation by 3 and the second by 2, we transform the system into
$$6x - 12y = 21$$
$$-6x + 12y = 16.$$

When we add these equations the terms involving both x and y drop out, leaving us with an absurdity:
$$0 = 37.$$

We say that the given equations form an **inconsistent system,** that is, they have no common solution (x, y). This means that the lines they represent are parallel, as in Figure 1.18. ■

Note that both the lines in Example 1.13 have slope $m = 1/2$. It is easy to show that *two nonvertical lines are parallel if and only if they have the same slope*.

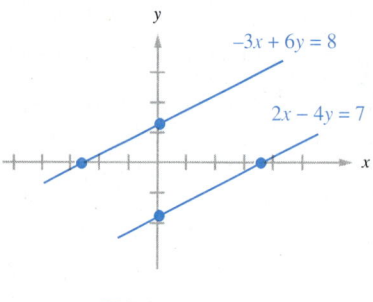

FIGURE 1.18

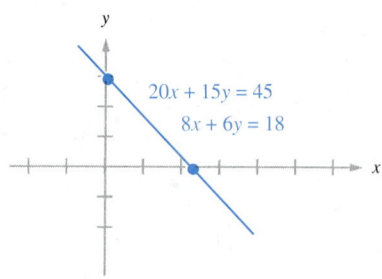

FIGURE 1.19

EXAMPLE 1.14 Find the point where the lines $8x + 6y = 18$ and $20x + 15y = 45$ intersect.

Solution Here we must solve the following system:
$$8x + 6y = 18$$
$$20x + 15y = 45.$$

We multiply the first equation by 5 and the second by -2.
$$40x + 30y = 90$$
$$-40x - 30y = -90$$

When we add these together we obtain

$$0 = 0,$$

which is true but not very informative. The problem here is that the given system is a **dependent system**: the first equation implies the second, as we see by multiplying it by 5/2. Both equations thus have the straight line graph shown in Figure 1.19, and every point on this line is a common solution to the given equations. ■

These examples illustrate all possible cases that can arise in solving two linear equations in two unknowns. The results may be summarized as follows.

Every system

$$a_1 x + b_1 y = c_1, \quad a_1, b_1 \text{ not both } 0,$$
$$a_2 x + b_2 y = c_2, \quad a_2, b_2 \text{ not both } 0,$$

of linear equations leads to exactly one of the following three cases:

Systems with a unique solution: The system has only one solution, and the corresponding lines intersect in one point.

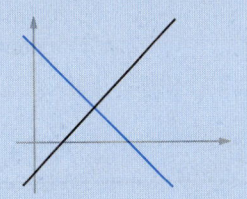

Inconsistent systems: The system has no solution, and the corresponding lines are parallel.

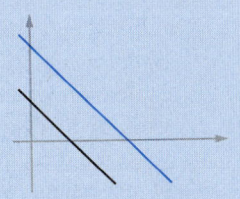

Dependent systems: The system has infinitely many solutions, and the corresponding lines coincide.

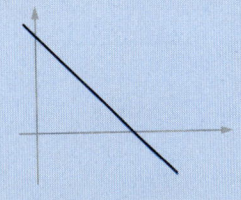

In the next chapter we shall solve systems involving more than two equations in more than two unknowns, using a general technique based on the elimination method. We shall see, however, that even in this more general setting, the only possible cases are the three listed in the box on page 21.

APPLICATIONS

Linear equations play a central role in many practical problems. This is illustrated by the following example.

EXAMPLE 1.15 A furniture manufacturer makes two styles of chair, "Antique" and "Baroque," for sale to retail outlets. Each Antique chair requires two work-hours of construction and three work-hours of finishing, while each Baroque chair requires three work-hours of construction and five work-hours of finishing. The company has a total of 60 work-hours of construction labor and 95 work-hours of finishing labor available each day. Is it possible to use up exactly the available amount of labor?

Solution We begin by introducing two unknowns. Let

x = the number of Antique chairs produced daily,

y = the number of Baroque chairs produced daily.

Next, we summarize the given data in the form of a table, indicating the total available amounts of labor in the last column.

		Style Antique	Style Baroque	Available
	Number of units	x	y	
Labor per unit	Construction	2 work-hours	3 work-hours	60 work-hours
Labor per unit	Finishing	3 work-hours	5 work-hours	95 work-hours

We now make the following simple but crucial observation: *If one Antique chair requires 2 work-hours of construction, then x Antique chairs will require 2x work-hours of construction.* Similarly, y Baroque chairs will require $3y$ work-hours of construction. Hence, in making x Antique and y Baroque chairs we use a total of $2x + 3y$ work-hours of construction labor. By the same kind of reasoning, we arrive at the expression $3x + 5y$ for the total amount of finishing labor used. The condition that exactly the available amount of labor be used up therefore translates into the following pair of equations:

$$2x + 3y = 60$$
$$3x + 5y = 95.$$

We will solve this system by the elimination method. Multiply the first equation by 5 and the second by -3.

$$10x + 15y = 300$$
$$-9x - 15y = -285$$

Add these to eliminate the term involving y.

$$x = 15$$

Finally, substitute $x = 15$ back into the first equation.

$$2(15) + 3y = 60$$
$$30 + 3y = 60$$
$$3y = 30$$
$$y = 10$$

We now have our answer: the company can use up all the available labor by producing 15 Antique chairs and 10 Baroque chairs each day. ∎

Question What if either x or y had come out negative in Example 1.15?

Answer If we had obtained negative values for x or y we would have had to conclude that *the problem has no solution,* since the physical interpretation of x and y makes such values meaningless. For example, if the manufacturer had only 80 work-hours of finishing labor available instead of 95, the corresponding equations would be

$$2x + 3y = 60$$
$$3x + 5y = 80,$$

which have the solution $x = 60$, $y = -20$. In this case it would not be possible to use up exactly the available supply of labor.

Practice Problem 5 A bag of apples weighs 6 pounds and costs $4, while a bag of pears weighs 4 pounds and costs $3. A man bought 112 pounds of fruit costing $79. How many bags of apples and pears did he buy?

EXAMPLE 1.16 A dietitian has available two types of flour: Type 1 with 20% protein and Type 2 with 24% protein. How many ounces of each type must she mix together to get 16 ounces of flour that contains 21% protein?

Solution Suppose she mixes x ounces of Type 1 flour and y ounces of Type 2. We summarize our data in the table below.

	Flour		
	Type 1	Type 2	Desired
Ounces	x	y	16
Protein proportion	.20	.24	.21

Then we must have

$$x + y = 16$$
$$.20x + .24y = .21(16)$$

Multiplying the second equation by 5 yields

$$x + 1.2y = 16.8.$$

We subtract the first equation to get

$$.2y = .8$$
$$y = \frac{.8}{.2} = 4.$$

Then $x = 16 - y = 12$. She should use 12 ounces of Type 1 flour and 4 ounces of Type 2. ■

EXERCISES 1.3

In Exercises 1–20, find the point (if any) where the lines intersect. Identify the cases when there are no solutions or infinitely many.

1. $3x = 6$
   $2x + 3y = -2$

2. $2x + y = 7$
   $-2y = 8$

3. $.5y = 5$
   $2x - 4y = 6$

4. $3x - 2y = 4$
   $\frac{1}{3}x = \frac{2}{3}$

5. $3x + 2y = 8$
   $x - y = 1$

6. $x + 2y = 7$
   $3x + y = 6$

7. $6x + 5y = 21$
   $3x - 4y = 30$

8. $2x - 5y = 14$
   $5x - 2y = 14$

9. $7x + 5y = 41$
   $4x - 3y = 0$

10. $5x + 4y = 0$
    $4x + 5y = 9$

11. $.9x + .8y = 72$
    $.1x + .2y = 10$

12. $1.6x - .8y = 8$
    $-1.2x + .6y = 0$

13. $x + y = 8$
    $x - y = 3$

14. $2x - 3y = 6$
    $3x - 5y = 15$

15. $-\frac{2}{3}x + \frac{1}{2}y = 15$
    $\frac{4}{5}x - \frac{3}{5}y = 12$

16. $\frac{2}{3}x + y = 8$
    $x + \frac{1}{2}y = 8$

17. $2x + 3y = 75$
    $4y = 20$

18. $-x + \frac{2}{5}y = -2$
    $\frac{5}{2}x - y = 5$

19. $x - 3y = 14$
    $4y = 16 - 3x$

20. $3x = 4y + 1$
    $x = y + 6$

21. Each ounce of Food A contains 2 gm of carbohydrate and 2 gm of protein, while each ounce of Food B contains 4 gm of carbohydrate and 2 gm of protein. By combining these foods, is it possible to obtain exactly 60 gm of carbohydrate and 50 gm of protein?

22. An oil company operates two eastern refineries. The Albany facility produces 100 barrels of high-grade oil and 300 barrels of medium-grade oil per day. The Boston facility produces 200 barrels of high-grade and 100 barrels of medium-grade per day. By operating each facility for a certain number of days, can the company produce exactly 12,000 barrels of high-grade and 24,000 barrels of medium-grade oil?

23. The House of Coffee sells two different blends of Brazilian and Colombian beans. Each pound of Blend A contains 1/4 lb of Brazilian and 3/4 lb of Colombian. Each pound of Blend B contains 2/3 lb of Brazilian and 1/3 lb of Colombian. The store would like to use up its entire supply of Brazilian and Colombian beans by producing a certain amount of each blend. Is this possible
    (a) if the store has 1200 lb of Brazilian and 1500 lb of Colombian?
    (b) if the store has 3000 lb of Brazilian and 1400 lb of Colombian?

24. World Travel Service has agreed to transport a group of 1800 football fans from Los Angeles to Chicago for one of the season's big games. The agency can charter two different types of aircraft for the trip. Type A has 20 first-class seats and 80 economy seats. Type B has 40 first-class seats and 60 economy seats. By using a certain number of planes of each type, the agency hopes to accommodate the group's needs exactly, with no seats left empty on any plane.
    (a) Is this possible if 600 fans want first-class seats and 1200 want economy seats?
    (b) Is this possible if 650 fans want first-class seats and 1150 want economy seats?

25. A chemist has available two strengths of acid solution. Solution A is 10% acid (by volume), while Solution B is 20% acid. How many liters of each solution should the chemist mix together to obtain a total of 100 liters of 12% acid solution?

26. A marine biologist uses two different fish nutrients. Nutrient 1 contains 7 units of protein per gm, while Nutrient 2 contains 3 units of protein per gm. How many grams of each should be mixed together, if the biologist needs 300 gm of nutrient containing 4 units of protein per gm?

27. Customs officials suspect a man of smuggling jade rings and bracelets into the U.S., where he sells them for $150 and $250, respectively. Their investigation shows that his latest shipment sold here for a total of $50,000, and officials in Taiwan, where the items cost 40% and 50% less, respectively, have proof that he paid $28,000 there for the merchandise. How many rings and how many bracelets did the shipment contain?

28. During a certain year, Laura had a profit of $5600 from two investments, the first earning 5% and the second 12%. Had she put all her money into the second investment, her profit would have been $8400. How much did she put into each investment?

29. A woman holds 45 shares of stock A and 24 shares of stock B. On the morning of a certain day the value of her holdings was $4800. During the day the price of stock A doubled, while the price of stock B fell by 50%. At day's end her holdings were worth $7800. What was the initial price of each stock?

30. A traveling salesperson represents two companies. The first company pays a weekly salary of $200 plus a 6% commission on gross sales, and the second pays a 10% commission on gross sales, but no salary. One week the salesperson earned a total of $440 on $2800 of gross sales. How much did he sell for each company?

Answers to Practice Problems **4.** $(1, -2)$

5. He bought 10 bags of apples and 13 bags of pears.

1.4 Break-Even Analysis

In any production process there are both **fixed costs** and **variable costs.** Fixed costs include such items as rent, equipment, depreciation, and interest, which remain the same no matter how much output is produced. Variable costs include such items as material and labor, which depend on the amount of output. The **total cost** of an operation is the sum of the fixed and variable costs, which depends on the number of units produced. The **revenue,** or sales income, also depends on the amount produced. The number of items to be produced so that cost exactly equals revenue is called the **break-even point.**

EXAMPLE 1.17 The Zeno Corporation has a division that makes color TV sets. The fixed costs of the operation are $15,000 per day, and the variable cost is $125 for each unit produced. The sets sell for $325 each. How many sets must be produced and sold daily for the operation to break even?

Solution If the firm produces x units per day, the total production cost C (in dollars) will be

$$C = 125x + 15{,}000,$$

and the total revenue R (assuming all the units are sold) will be

$$R = 325x.$$

Note that both cost and revenue are linear functions of x. If they are graphed in a common coordinate system, with y representing both cost and revenue, the **cost** and **revenue lines** in Figure 1.20 are obtained.

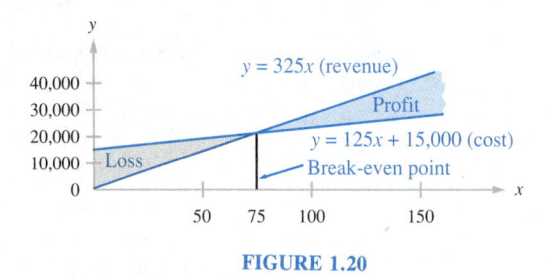

FIGURE 1.20

The break-even point is the value of x for which cost equals revenue.

$$C = R$$
$$125x + 15{,}000 = 325x$$
$$200x = 15{,}000$$
$$x = 75$$

This is the x-coordinate of the point where the cost and revenue lines intersect. For $x = 75$ we have $C = R = \$24{,}375$, so the operation shows neither a profit nor a loss. For $x < 75$ the cost line lies above the revenue line $(C > R)$, so the operation is losing money. For $x > 75$ the cost line lies below the revenue line $(C < R)$, so the operation shows a net profit. ■

In general, **profit** equals the difference between revenue and cost.

Profit

$$P = R - C$$

If revenue and cost are linear functions of x, then so is profit. In the example above,

$$\begin{aligned} P &= R - C \\ &= 325x - (125x + 15{,}000) \\ &= 200x - 15{,}000. \end{aligned}$$

The graph of this function is the **profit line** shown in Figure 1.21. At the break-even point $x = 75$, the profit line crosses the x-axis. For $x < 75$ the profit line lies below the x-axis, indicating a *negative* profit, that is, a net loss. For $x > 75$ the line lies above the x-axis, indicating a *positive* profit, or a net gain.

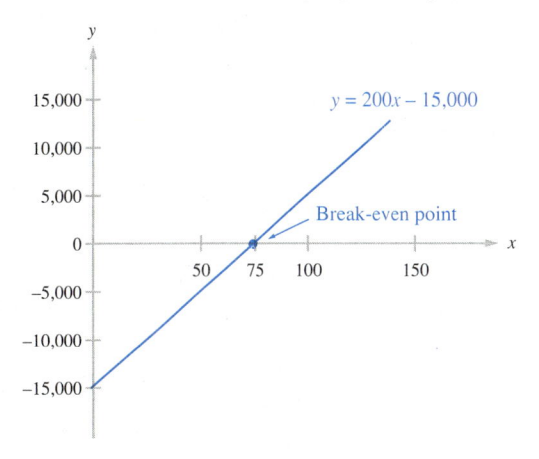

FIGURE 1.21

Using the profit function, we can also answer some other questions of interest.

EXAMPLE 1.17 Continued

How many sets must be produced and sold daily:

(a) to make a profit of $17,000?
(b) to make a profit of $125 on each unit?
(c) to make a profit of 20% on sales?

Solution In each case we begin with the profit function $P = 200x - 15{,}000$.

(a) Setting the profit equal to 17,000, we solve for x.

$$P = 17{,}000$$
$$200x - 15{,}000 = 17{,}000$$
$$200x = 32{,}000$$
$$x = 160$$

Thus, producing and selling 160 units will yield a profit of $17,000.

(b) The profit per unit is P/x. Hence, we must solve the equation

$$\frac{P}{x} = 125.$$
$$P = 125x$$
$$200x - 15{,}000 = 125x$$
$$75x = 15{,}000$$
$$x = 200$$

Thus, producing and selling 200 units will yield a profit of $125 per unit.

(c) The condition that profit is 20% of sales translates into the equation

$$P = .20R.$$
$$200x - 15{,}000 = .20(325x)$$
$$200x - 15{,}000 = 65x$$
$$135x = 15{,}000$$
$$x \approx 111.1$$

(The symbol $\approx$ indicates that the answer has been rounded off.) In practical terms, this means that the company must product at least 112 units per day to achieve a profit margin of at least 20%. ∎

Practice Problem 6 A lemonade stand sells lemonade for 50 cents per glass. The ingredients in a glass of lemonade cost 10 cents, and the fixed costs of running the stand are $44 per day.
(a) How many glasses of lemonade must be made and sold in a day to break even?
(b) Write the profit P in dollars as a linear function of the number x of glasses sold.
(c) How many glasses must be sold in a day to make $100 profit?

COMPARISON OF PROFIT FUNCTIONS

In some applications we are presented with several alternative methods of production. By comparing their associated profit functions we can determine the conditions under which one method is preferable to another.

1.4 BREAK-EVEN ANALYSIS

EXAMPLE 1.18 The firm of Ziferstein & Son manufactures a dental floss dispenser that sells for $1.55. Under the present production method the operation has fixed costs of $200 a day and a variable cost of 75¢ per unit. The junior owner wants to lease some new equipment that will save labor and bring the variable cost down to 35¢ per unit. The senior owner is opposed to the new plan, since payments for the equipment will increase the daily fixed costs by $250. The junior owner argues that the savings in variable costs will more than offset the increase in fixed costs. Is he right?

Discussion Let x be the number of units produced per day. The cost functions for the two production methods are given by

$$C_1 = .75x + 200 \quad \text{(old method)},$$
$$C_2 = .35x + 450 \quad \text{(new method)},$$

while the revenue function is the same for both.

$$R = 1.55x$$

We calculate the profit function for each method.

$$P_1 = R - C_1 = .80x - 200 \quad \text{(old method)}$$
$$P_2 = R - C_2 = 1.20x - 450 \quad \text{(new method)}$$

The two profit lines are shown in Figure 1.22.

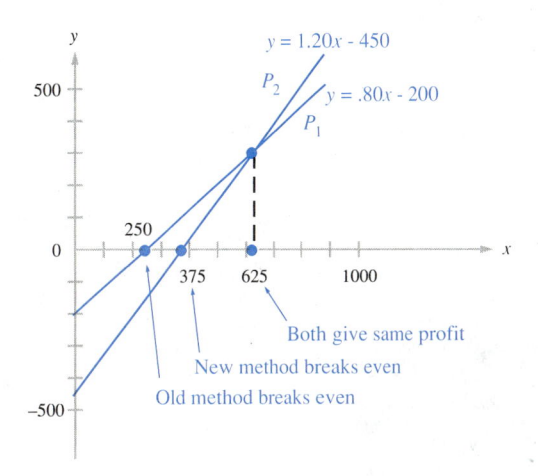

FIGURE 1.22

To find the break-even point under the old method, we set $P_1 = 0$.

$$.80x - 200 = 0$$
$$.80x = 200$$
$$x = 250$$

This is the point where the line giving P_1 crosses the x-axis in Figure 1.22. Similarly, the new method will break even when $P_2 = 0$.

$$1.20x - 450 = 0$$
$$1.20x = 450$$
$$x = 375$$

For values of x in the range $250 < x \le 375$, therefore, only the old method will show a profit. We see from Figure 1.22 that the old method is more profitable (that is, $P_1 > P_2$) up to the point where the two profit lines cross, but beyond that point the new method is more profitable ($P_2 > P_1$). To find this intersection point we set $P_1 = P_2$ and solve for x.

$$.80x - 200 = 1.20x - 450$$
$$.40x = 250$$
$$x = 625$$

When $x = 625$ both methods yield the same profit: $P = \$300$. When $x < 625$ we have $P_1 > P_2$, and when $x > 625$ we have $P_2 > P_1$. Our answer to the problem, therefore, must be conditional: if it is possible to produce more than 625 units per day under the new method, and they can all be sold at the price stated, then the new method is preferable; otherwise the old method is preferable. ■

EXAMPLE 1.19 Sylvia plans to run a private shuttle service between the Astor Hotel and the local airport. Limousine rental, insurance, and license costs total $3600 a year. Each round-trip will average $14 in revenues. Gasoline and maintenance costs average $2 per run. In addition, the city council intends to charge her either a flat franchise fee of $1200 a year or else a franchise tax of 50% of revenues received. Which plan will be more profitable for Sylvia?

Solution Suppose she makes x round-trip runs per year. Under the franchise fee (Plan 1), the fixed yearly costs will be $3600 + $1200 = $4800, and the variable costs will be $2x$ dollars. The total annual cost is therefore given by

$$C_1 = 2x + 4800,$$

while annual revenue is given by

$$R = 14x.$$

The profit function under Plan 1 is therefore

$$P_1 = R - C_1 = 12x - 4800.$$

Under the franchise tax (Plan 2), the fixed cost is reduced to $3600, but $.50R = .50(14x) = 7x$ dollars must be added to the variable cost, for a total annual cost of
$$C_2 = 9x + 3600.$$

Since the revenue function remains the same, the profit function under Plan 2 is given by
$$P_2 = R - C_2 = 5x - 3600.$$

The profit lines are shown in Figure 1.23.

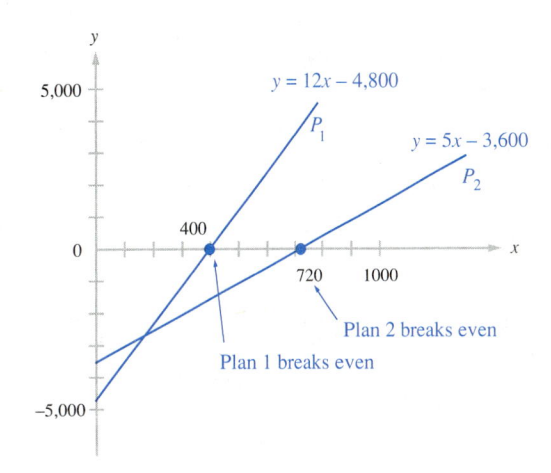

FIGURE 1.23

Under Plan 1, the operation will break even when $P_1 = 0$.
$$12x - 4800 = 0$$
$$12x = 4800$$
$$x = 400$$

Thus, Plan 1 breaks even when $x = 400$. Similarly, Plan 2 breaks even when $P_2 = 0$, or $x = 720$. From the graph, however, we see that Plan 2 never becomes more profitable than Plan 1, since $P_1 > P_2$ for all values of x beyond the first break-even point. Thus Sylvia should favor the franchise fee instead of the franchise tax; and under this plan she will have to make 400 runs per year, or about 34 per month, to break even. ∎

EXAMPLE 1.20 A bank offers its customers three different checking plans. Under Plan 1, you pay a flat monthly service charge of $4.00, regardless of how many checks you write. Under Plan 2 you pay $1.75 a month, plus 5¢ for each check written. Under Plan 3 you pay 12¢ per check, with no fixed service charge. Under what conditions will Plan 2 be the least expensive of the three plans?

Solution Let x be the number of checks written per month. The monthly cost functions for the three plans (in dollars) are as follows:

$$C_1 = 4$$
$$C_2 = .05x + 1.75$$
$$C_3 = .12x.$$

These functions are graphed in Figure 1.24. The cost lines for Plans 2 and 3 cross when $C_2 = C_3$, that is, when

$$.05x + 1.75 = .12x$$
$$.07x = 1.75$$
$$x = 25.$$

Similarly, the cost lines for Plans 1 and 2 cross when $C_1 = C_2$:

$$4 = .05x + 1.75$$
$$.05x = 2.25$$
$$x = 45.$$

From Figure 1.24 we see that the cost line for Plan 2 lies below the other two lines over the interval $25 \leq x \leq 45$. Hence, Plan 2 is the least expensive when the customer writes between 25 and 45 checks per month. ∎

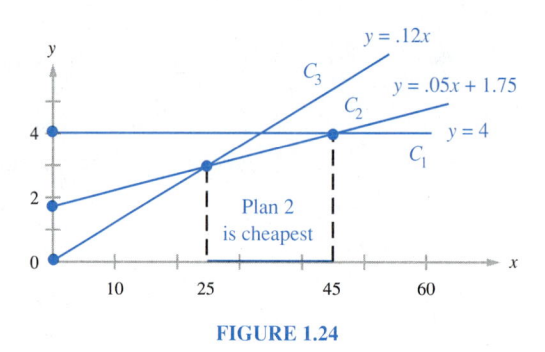

FIGURE 1.24

Remark It is important in these problems to choose the right scale on each axis so your graph will show all the relevant intersection points. If your first attempt is unsuccessful, try a larger unit of measurement. As in the examples above, you may want to use different units of measurement on the x- and y-axes.

Practice Problem 7 A wig manufacturer has fixed costs of \$21,000 per week and variable costs of \$25 per wig. The wigs wholesale for \$60. How many wigs must the company sell per week to break even?

EXERCISES 1.4

1. A manufacturing operation has cost and revenue functions
$$C = 1.25x + 8400$$
$$R = 2x,$$
where x is the number of units of output. How many units must be produced in order for the operation to break even? Draw a graph of the profit function.

2. A power company has fixed costs of $35,000 a day, with a variable cost of 2¢ for each kilowatt-hour (kwh) of electricity produced. The company charges its customers 4¢ per kwh. How many kwh must be sold daily in order for the company to break even? Draw a graph of the profit function.

3. An electronics firm has a plant that produces a single type of silicon chip, which the firm sells for $1.40. If the plant's fixed costs are $1200 a day, and the chips cost 60¢ each to produce, how many must be made daily in order for the plant to break even? Draw a graph of the profit function.

4. An agricultural corporation finds that its asparagus processing division has fixed costs of $14,000 a year, with a variable cost of 15¢ for each pound processed. If the asparagus is sold to a distributor at 35¢/lb, how much must be processed per year in order to break even? Draw a graph of the profit function.

5. In Exercise 1, how many units must be produced
   (a) to make a profit of $6000?
   (b) to make a profit of 25¢ per unit?
   (c) to make a profit of 25% on sales?

6. In Exercise 2, how many kwh must be sold daily
   (a) to make a profit of $25,000 a day?
   (b) to make a profit of 1¢ per kwh?
   (c) to make a profit of 15% on sales?

7. In Exercise 3, how many silicon chips must be produced daily
   (a) to make a profit of $3000 a day?
   (b) to make a profit of 50¢ on each chip?
   (c) to make a profit of 40% on sales?

8. In Exercise 4, how many pounds of asparagus must be processed per year
   (a) to make an annual profit of $100,000?
   (b) to make a profit of 15¢/lb?
   (c) to make a profit of 30% on sales?

9. A manufacturer's profit functions using two different production methods are given by
$$P_1 = .75x - 8400 \quad \text{(method 1)}$$
$$P_2 = .90x - 10{,}800 \quad \text{(method 2)},$$
where x is the number of units of output.
   (a) How many units must be produced in order to break even under method 1? Under method 2?
   (b) Draw a graph of the two profit functions. Which method is better for the manufacturer?

Customers pay the power company 2¢ per kilowatt-hour of electricity.

10. An ecology group is planning a fund-raising dinner to be held in one of the two local hotels. The group must pay $1200 for the speaker and $200 for promotion. In addition, the Astor Hotel wants $400 plus $10 a plate to provide the meal, while the Belmont Hotel is asking $800 plus $7 a plate. Admission to the event is set at $15 a plate.
   (a) How many people must attend for the event to break even, if it is held at the Astor? At the Belmont?
   (b) Draw a graph of the two profit functions involved. Which hotel's offer is better?

11. An auto plant has fixed costs of $60,000 a day and a variable cost of $3000 for each car produced. The cars are sold to the distributor for $4250. A new computerized system could cut the variable cost by 20%, but installing the system would double the present fixed costs.
   (a) How many cars must be produced daily to break even, under the present operation? Under the computerized system?
   (b) Draw a graph of the two profit functions involved. Under what conditions would it pay to install the new system?

12. A corporation's annual operating costs and revenue (in dollars) are given by

$$C = 5x + 90,000$$
$$R = 20x,$$

where x is the number of units of output per year. In addition to its operating costs the company must pay corporate tax. Under Tax Plan 1, the company pays a tax T equal to 15% of revenues received. Under Tax Plan 2, the tax T is equal to 40% of gross profit $(R - C)$. The company's *net* profit P, after taxes, is given by

$$P = R - C - T.$$

   (a) Express the net profits, P_1 and P_2, under the two plans as linear functions of x.
   (b) Compare these functions by graphing.
   (c) What is the company's break-even point under Plan 1? Under Plan 2?
   (d) For what values of x will Plan 1 be better for the company than Plan 2?

13. A private school is planning to produce a desk calendar to be mailed out to the community. Printer 1 charges $1000 for design and layout, plus $2.80 a copy. Printer 2 charges $4000 for design and layout, plus $1.80 a copy. Which offer is better, in terms of the number of copies the school needs? Illustrate with a graph.

14. Three different lawyers are willing to handle a certain lawsuit. Lawyer 1 charges $1000 plus 20% of the judgment to be awarded. Lawyer 2 charges $500 plus 40% of the award. Lawyer 3 charges 60% of the award, with no fixed fee. Let x be the dollar amount awarded to the plaintiff, and let P_1, P_2, and P_3 be his net gain (award minus legal costs) if he accepts lawyer 1, 2, or 3, respectively.
   (a) Express P_1, P_2, and P_3 as linear functions of x.
   (b) Compare these functions by graphing.
   (c) What is the plaintiff's break-even point with each of the three lawyers?
   (d) Assuming that the suit is won, for what values of x will each lawyer's offer be the best?

15. Taxi rates are not regulated in Baker County, and they vary among the three competing companies. Red Cab charges a fixed fee of 65¢ plus a variable fee of 20¢ a mile. White Cab charges 85¢ plus 15¢ a mile, and Blue Cab charges $1.15 plus 12¢ a mile. A sales

representative who uses cabs frequently would like a simple rule to determine which cab is least expensive, given the number x of miles she has to travel. Solve her problem, and illustrate with a graph.

16. A woman is comparing three hospitalization insurance plans. Plan 1 pays all hospital charges for an annual premium of $750. Plan 2 pays 90% of the charges for a premium of $500. Plan 3 pays 60% of the charges for a premium of $200. Let x be the hospital charges the woman might incur during the year. Express her net loss (premium plus uncovered charges) as a linear function of x, under each of the three plans. For what values of x will each plan be best? Illustrate with a graph.

17. Three recording companies are bidding on production rights for a demo featuring a country singer. The first company has offered a flat payment of $1 million. The second has offered $100,000 plus 20% of gross receipts. The third has offered $250,000 plus 10% of gross receipts. Which offer is best, in terms of the expected gross receipts x? Illustrate with a graph.

18. A famous diplomat is trying to decide between two publishers interested in his memoirs. Publisher 1 offers royalties of 90¢ per copy after the first 2000 copies sold in a given year, with no royalties paid on the first 2000 copies. Publisher 2 offers 50¢ per copy after the first 1000 copies sold. Which offer is better, in terms of the number x of copies he expects to be sold per year? Illustrate with a graph.

Answers to Practice Problems 6. (a) 110 (b) $P = .4x - 44$ (c) 360
7. 600

■ ■ ■ ■ ■ ■ ■ ■ MATHEMATICS IN ACTION

1.5 Market Equilibrium

SUPPLY AND DEMAND

As the price of an item goes up, more producers are willing to supply the item, and the amount produced, or **supply** of the item, rises also. For example, when the price of corn increases, so does the supply, because many farmers will grow corn rather than some other crop to take advantage of the better price. Conversely, if the price of corn drops, then farmers turn to other crops and the supply goes down.

The amount of a product consumed, or **demand** for the product, depends on the price of the product in the opposite manner. An increase in price tends to restrict demand. For example, if the price of corn goes up, users of corn products may try to find less expensive substitutes. Specifically, soft drink manufacturers use sugars derived from not only corn, but also sugar cane, beets, and other crops. They will use more or less corn syrup according to whether its price is low or high compared to other sugar sources.

Often supply and demand are approximately linear functions of price, and the techniques of the previous sections may be used to analyze how the market forces determine price.

EXAMPLE 1.21 A company manufacturing kitchen graters estimates that if the price of a grater is x (in dollars), then the monthly demand will be D units, where

$$D = 1200 - 150x.$$

The number of graters the company is willing to make also depends on the price. If the price is low, it will turn to more profitable items. The company is willing to supply S units per month at the price x, where

$$S = 300x - 600.$$

Notice that both supply and demand are linear functions of the price x. The corresponding lines are graphed in Figure 1.25.

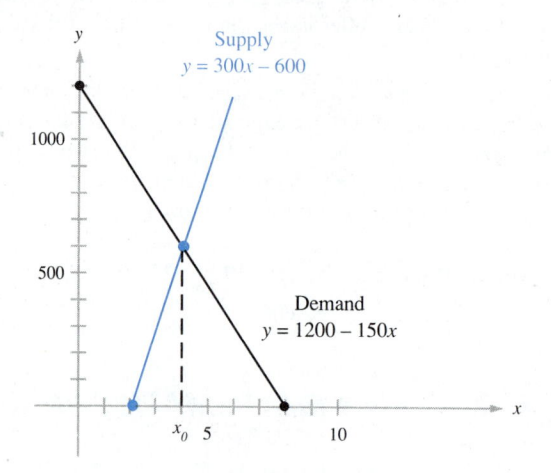

FIGURE 1.25

Discussion The point where the two lines cross has a special significance. To its right supply exceeds demand, and the company will have to lower its price to sell what it is producing. To the left of the intersection point demand exceeds supply, and the company will raise its price to take advantage of the situation. Only at the point where the two lines intersect does the price tend to remain the same, and the value of x at this point is called the **equilibrium price**, denoted x_0 in Figure 1.25. At this price we have $S = D$, so that

$$300x_0 - 600 = 1200 - 150x_0$$
$$450x_0 = 1800$$
$$x_0 = 4.$$

As long as neither the supply nor demand function changes, the price should remain a steady $4 per grater.

1.5 MATHEMATICS IN ACTION: MARKET EQUILIBRIUM

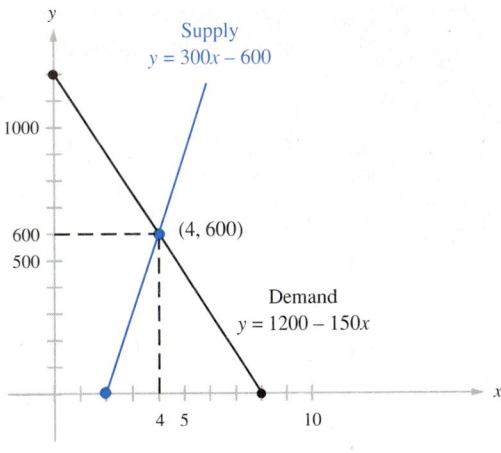

FIGURE 1.26

Supply: $y = 300x - 600$
Demand: $y = 1200 - 150x$
(4, 600)

At the equilibrium price the supply and demand have a common value, called the **equilibrium supply,** or **equilibrium demand.** This can be found by substituting x_0 into either the supply or demand function. In our example at $x_0 = 4$ we have

$$S = 300(4) - 600 = 600.$$

Thus the equilibrium price and equilibrium demand are both 600 units per month. The geometric meaning of these values is shown in Figure 1.26. ∎

EXAMPLE 1.22 A factory manufactures cigars. At a price of x dollars per box it is willing to supply S boxes per week, where

$$S = 120x - 240,$$

and the demand is D boxes per week, where

$$D = -75x + 1125.$$

What is the equilibrium price? How many boxes will be manufactured and sold at this price, and what will the total revenue to the company be?

Solution To find the equilibrium price we set $S = D$.

$$120x - 240 = -75x + 1125$$
$$195x = 1365$$
$$x = \frac{1365}{195} = 7$$

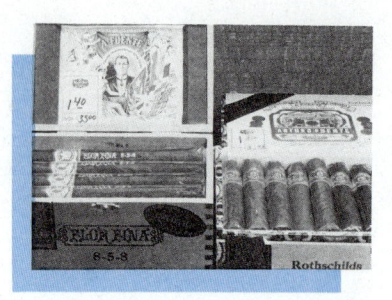

We see that the equilibrium price is $7 per box. At this price the equilibrium supply (which equals the equilibrium demand) is

$$120(7) - 240 = 600.$$

Thus 600 boxes would be sold each week, for a total revenue of $600(7) = \$4200$ per week. The supply and demand functions and equilibrium price are indicated in Figure 1.27. ∎

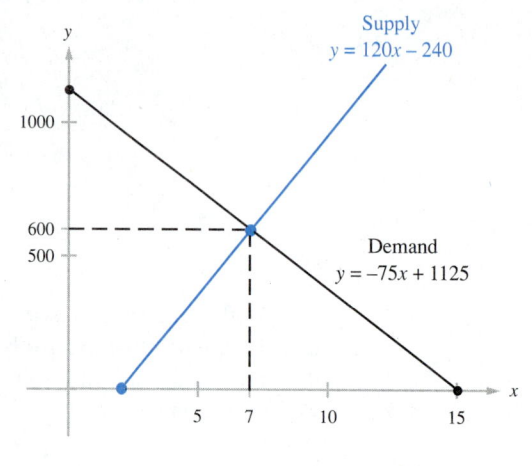

FIGURE 1.27

Practice Problem 8 The supply of a certain computer chip is $S = 10{,}000x - 7000$ at a price of x dollars, and the demand is $D = -8000x + 20{,}000$.
(a) What is the equilibrium price?
(b) How many chips will be supplied at this price?

THE EFFECT OF TAXATION

Suppose that the city containing the cigar factory in Example 1.22 decides to tax the boxes of cigars. It imposes a 20% **sales tax** on each box. Since the factory is selling 600 boxes per week at $7 a box, the city officials expect to take in $.20(7) = \$1.40$ tax on each box, for a total of $600(1.40) = \$840$ per week. They have not taken into account the laws of supply and demand, however.

For simplicity, let x denote the *before-tax* price of each box. Since this is still the price the factory gets for each box, the supply function is the same:

$$S = 120x - 240.$$

1.5 MATHEMATICS IN ACTION: MARKET EQUILIBRIUM

The demand, on the other hand, depends on the price the consumer actually pays for each box, which includes the 20% tax. Thus the new demand function is given by

$$D_1 = -75(x + .20x) + 1125$$
$$D_1 = -90x + 1125.$$

We find the equilibrium price by setting $S = D_1$.

$$120x - 240 = -90x + 1125$$
$$210x = 1365$$
$$x = \frac{1365}{210} = 6.50$$

Because the tax has reduced demand, the factory will be able to get only $6.50 for a box of cigars, and the 20% tax will net the city only $.20(6.50) = \$1.30$ per box instead of the expected $1.40. Notice that the equilibrium supply at $6.50 will be

$$120(6.50) - 240 = 540$$

boxes per week, so that the weekly tax revenue will be 540($1.30) = $702 per week, rather than the expected $840.

Frustrated that the tax revenue on boxes of cigars is not as great as expected, the city changes from the 20% tax to a flat tax of $1.40 per box. Such a flat tax, not depending on the price of the item sold, is called a **stamp tax.** An example of such a tax is the state gasoline tax, which is a fixed amount per gallon, independent of the selling price of gasoline.

Again letting x be the before-tax price of a box of cigars, we still have the supply function

$$S = 120x - 240.$$

Now, however, a consumer is paying $x + 1.40$ per box of cigars, so the demand is

$$D_2 = -75(x + 1.40) + 1125,$$

or

$$D_2 = -75x + 1020.$$

Setting $S = D_2$ gives

$$120x - 240 = -75x + 1020$$
$$195x = 1260$$
$$x = \frac{1260}{195} \approx 6.46.$$

Now the factory gets about $6.46 per box, and will produce

$$120\left(\frac{1260}{195}\right) - 240 \approx 535$$

boxes per week. Since production is reduced, the city will still not get the $840 per week in taxes it originally expected. The three demand functions D, D_1, and D_2 are shown in Figure 1.28.

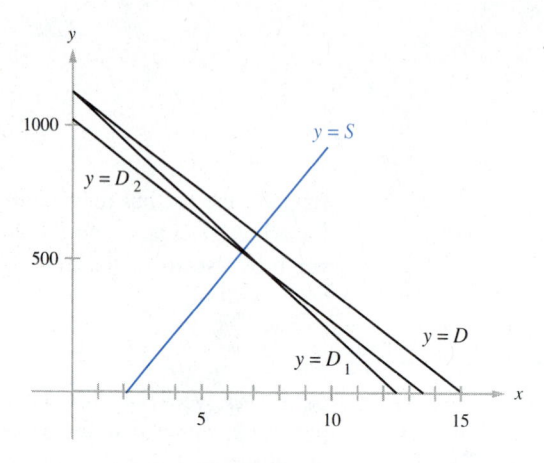

FIGURE 1.28

Practice Problem 9 Example 1.21 considered the production of graters, for which there was a demand

$$D = 1200 - 150x,$$

where x is the price in dollars. Find the demand functions D_1, when a 10% sales tax is imposed, and D_2, when a 20¢ stamp tax is imposed.

EXERCISES 1.5

In Exercises 1–8 find the equilibrium price and demand (supply) for the following demand and supply functions $D(x)$ and $S(x)$, where x is the price in dollars.

1. $D(x) = 4500 - 500x$, $S(x) = 250x$
2. $D(x) = 7000 - 200x$, $S(x) = 300x$
3. $D(x) = 1200 - 60x$, $S(x) = 100x$
4. $D(x) = 9900 - 400x$, $S(x) = 800x$
5. $D(x) = 1500 - 100x$, $S(x) = 400x - 2000$, for $x \geq 6$
6. $D(x) = 15{,}300 - 450x$, $S(x) = 300x - 2700$, for $x \geq 12$
7. $D(x) = 5040 - 280x$, $S(x) = 220x - 1660$, for $x \geq 8$
8. $D(x) = 19{,}000 - 7600x$, $S(x) = 2400x - 1800$, for $x \geq 1$

1.5 MATHEMATICS IN ACTION: MARKET EQUILIBRIUM

In Exercises 9–14 determine how the equilibrium price and demand (supply) change in the given exercise if the government imposes a sales tax as indicated. Find the total tax collected.

9. 10% tax in Exercise 1
10. 20% tax in Exercise 2
11. 25% tax in Exercise 5
12. 15% tax in Exercise 6
13. 30% tax in Exercise 7
14. 8% tax in Exercise 8

In Exercises 15–20 determine how the equilibrium price and demand (supply) change in the given exercise if the government imposes a stamp tax as indicated. Find the total tax collected.

15. $0.75 in Exercise 1
16. $1 in Exercise 2
17. $0.50 in Exercise 5
18. $2 in Exercise 6
19. $0.90 in Exercise 7
20. $0.20 in Exercise 8

21. A manufacturer has found that the demand for one of his products is 1600 units when the price per unit is $8 and is 1000 units when the price per unit is $10. If the manufacturer is willing to supply $S(x) = 200x - 1200$ units at any price x exceeding $7 per unit, determine the equilibrium price and supply if demand is a linear function of x.

22. The manager of a popcorn stand is willing to supply $100x - 2000$ bags of popcorn per week at a price of x cents per bag. She sells 2000 bags per week if the price is 75¢ and 1400 bags per week if the price is 90¢. If the demand is a linear function of price, find the equilibrium price and supply.

23. A part-time barber is willing to give 578 haircuts per year at $8 each, but can sell only 364 at that price. He is willing to give 155 haircuts at $5 each, and can sell 583 at that price. If the supply and demand are both linear functions of price, find the equilibrium price and supply.

24. A manufacturer is willing to supply 800 units of a product at a price per unit of $6 and 1200 units at a price per unit of $7. The manufacturer has found that the demand for the product is 7100 units at a price of $6 per unit and 6800 units at a price of $7 per unit. If the supply and demand functions are both linear, determine the equilibrium price and demand.

25. Suppose in Example 1.22 the city pays the cigar factory a subsidy of $1 per box of cigars it makes and sells. What is the new equilibrium price and demand?

26. In Example 1.22 what flat tax per box would give the city its $840 per week revenue?

Raising the price of popcorn from 75¢ to 90¢ decreases the weekly sales from 2000 bags to 1400 bags.

Answers to Practice Problems
8. (a) $1.50 (b) 8000
9. $D_1 = 1200 - 165x$
$D_2 = 1170 - 150x$

CHAPTER 1 REVIEW

IMPORTANT TERMS

- coordinate axes *(1.1)*
- graph
- linear equation
- ordered pair
- origin
- real number line
- x-axis, y-axis
- x-intercept, y-intercept
- dependent variable *(1.2)*
- independent variable
- linear function
- marginal cost
- point-slope form
- slope
- slope-intercept form
- dependent system *(1.3)*
- elimination method
- equivalent system
- inconsistent system
- substitution method
- system of equations
- break-even point *(1.4)*
- cost and revenue lines
- fixed, variable, and total cost
- profit, profit line
- revenue
- demand *(1.5)*
- equilibrium price
- equilibrium supply, demand
- sales tax
- stamp tax
- supply

IMPORTANT FORMULAS

Slope-intercept form

The equation of the line with slope m and y-intercept $(0, b)$ is $y = mx + b$.

Slope of a line through two points

The slope of the line through the points (x_1, y_1) and (x_2, y_2) is

$$m = \frac{y_2 - y_1}{x_2 - x_1}, \quad \text{if } x_2 \neq x_1.$$

Point-slope form

The equation of the line through the point (x_1, y_1) with slope m is $y - y_1 = m(x - x_1)$.

Profit

If $P = $ profit, $R = $ revenue, and $C = $ cost, then $P = R - C$.

REVIEW EXERCISES

In Exercises 1–6 graph the given statement.

1. $x \geq 0$ and $y \leq 0$
2. $x \geq 1$ and $y \leq 3$
3. $3x + 5y = 600$
4. $x = -2$
5. $2x + 5y = 0$
6. $x + y = 3$ and $x \geq 1$
7. Give the slope, x-intercept, and y-intercept of $3x + 1 = 3 - 2y$.
8. Give the equation of the line through $(2, -3)$ and $(1, 5)$ in slope-intercept form.

9. The number of ice cream cones sold in a day at the Dairy Queen is a linear function of the maximum temperature on that day. If the maximum temperature is 80° F, 55 cones are sold; and if the maximum temperature is 100° F, 95 cones are sold. How many cones are sold when the maximum temperature is 65° F?

10. Solve the system
$$2x + y = 2$$
$$6x + 4y = 3.$$

11. Solve the system
$$2a - b = 0$$
$$12a + 4b = 2.$$

12. In planning a birthday party for 11 children it is decided that each child should have 9 balloons and 4 noisemakers. A Fun-Pac contains 7 balloons and 3 noisemakers, while an Economy Box contains 15 balloons and 8 noisemakers. How many Fun-Pacs and Economy Boxes should be bought to supply exactly the right number of balloons and noisemakers?

13. An investor has $10,000 to invest and wants an 8% return. Investment A pays an 11% return, while investment B, which is safer, pays a 6% return. How should the $10,000 be divided between investments A and B?

14. A company produces toasters which it sells for $25. There are fixed costs of $30,000 each month, and variable costs of $13 per toaster.
    (a) Plot the cost and revenue lines.
    (b) How many toasters must be made and sold to break even?

    How many toasters must be produced and sold monthly
    (c) to make a profit of $12,000?
    (d) to make a profit of $4 on each toaster?
    (e) to make a profit of 24% on sales?

15. A singing telegram business has fixed office expenses of $400 per week, and each telegram delivered costs an average of $7.50 for labor and transportation. It charges $20 for a singing telegram.
    (a) Graph the cost and revenue lines.
    (b) How many telegrams must it deliver per week to break even?

    How many telegrams must it deliver per week
    (c) to make a profit of $175?
    (d) to make a profit of $2.50 on each telegram?
    (e) to make a profit of 22.5% on sales?

16. A photographer is willing to supply $280x - 500$ cheap passport photographs per year at a price of x dollars each. The public will buy $1000 - 120x$ such photographs at x dollars. What is the equilibrium price and equilibrium demand?

17. Work the previous exercise if there is a 10% tax on passport pictures.

18. Work Exercise 16 if there is a stamp tax of 50¢ per passport picture.

SYSTEMS OF LINEAR EQUATIONS AND MATRICES

2.1 Gauss-Jordan Elimination

2.2 Systems without Unique Solutions

2.3 Applications of Systems of Linear Equations

2.4 Matrices

2.5 Matrix Multiplication

2.6 Inverse Matrices

2.7 Mathematics in Action: The Leontief Input-Output Model

Chapter Review

In this chapter we will generalize the methods of Chapter 1 to solve systems of more than two linear equations. It turns out that many applications entail a large number of variables subject to linear constraints. Analyzing such a system can be simplified by organizing and manipulating the numbers involved in a table, called a *matrix* or *tableau*.

2.1 Gauss-Jordan Elimination

In Chapter 1 we presented two different methods for solving a system of two linear equations in two unknowns, substitution and elimination. While the substitution technique can be applied to larger systems, it becomes increasingly cumbersome. The elimination idea, however, can be generalized to give a systematic procedure known as the *Gauss-Jordan elimination* method. In this section this method is used to solve a system of n linear equations in n unknowns.

LINEAR SYSTEMS AND ELEMENTARY OPERATIONS

An equation of the form

$$a_1x_1 + a_2x_2 + \ldots + a_nx_n = c,$$

where $a_1, a_2, \ldots, a_n$, and c are constants and at least one of the coefficients a_i is nonzero, is called a **linear equation** in the unknowns $x_1, x_2, \ldots, x_n$. A system of m linear equations in n unknowns is called an **m by n linear system.** For instance, the systems

$$\begin{array}{lll} 2x + 3y = 9 & x + 2y - z = 4 & 5x + y + z = 8 \\ 3x - 5y = 7 & 3x - y + 5z = 6 & -x + 2y + 2z = 5 \\ & 2x + y - 3z = 1 & \end{array}$$

are 2 by 2, 3 by 3, and 2 by 3, respectively.

When a system has the same number of equations as unknowns (when $m = n$), it will usually have a unique solution $(x_1, x_2, \ldots, x_n)$. For the case $n = 2$ this is reflected in the fact that two lines usually intersect in a unique point. We now take up the general problem of finding the solution of an n by n linear system when it exists.

One approach to solving a system of linear equations is to transform it into a simpler algebraically equivalent system, that is, a simpler system having exactly the same solutions. The types of transformations we shall need, called **elementary operations,** are the following:

Type 1 An equation can be multiplied by a nonzero number.

Type 2 A multiple of one equation can be added to another.

Type 3 Two equations can be interchanged.

CHAPTER 2 SYSTEMS OF LINEAR EQUATIONS

EXAMPLE 2.1 Solve the following 3 by 3 system:

$$2x + 4y + 2z = 6$$
$$-2x + y + 3z = 4 \qquad (2.1)$$
$$3x + 2y + z = 5.$$

Discussion The first task is to eliminate the unknown x from all but the first equation in the system. We multiply the first equation by 1/2 in order to change the coefficient of x to $+1$, so that this variable can be eliminated from the other equations. This Type 1 operation transforms (2.1) into the following system:

$$x + 2y + z = 3$$
$$-2x + y + 3z = 4 \qquad (2.2)$$
$$3x + 2y + z = 5.$$

Note that the x-term in the first equation now has coefficient $+1$. Next, we add 2 times the (new) first equation to the second, in order to eliminate the x-term in the second equation. The scratch work for this Type 2 operation follows.

$$\begin{array}{r} -2x + y + 3z = 4 \\ (+) \quad \underline{2x + 4y + 2z = 6} \\ 5y + 5z = 10 \end{array} \quad \begin{array}{l} \text{(second equation)} \\ \text{(2 times first equation)} \\ \text{(sum)} \end{array}$$

This transforms (2.2) into the equivalent system (2.3).

$$x + 2y + z = 3$$
$$5y + 5z = 10 \qquad (2.3)$$
$$3x + 2y + z = 5$$

Remark Notice that it is the second equation, not the first, which is changed by this operation. We merely *use* the first equation to eliminate x from the second equation.

Working now with the system (2.3), we add -3 times the first equation to the third to eliminate the x-term in the third equation, yielding the system:

$$x + 2y + z = 3$$
$$5y + 5z = 10 \qquad (2.4)$$
$$-4y - 2z = -4.$$

We have achieved our first objective: the unknown x has now been isolated so that it occurs in one and only one equation of the system, and has coefficient $+1$ in that equation. Our next goal is to isolate y in one of the remaining equations; we will use the second. To begin, we simplify the y-coefficient by multiplying the second equation by 1/5.

$$x + 2y + z = 3$$
$$y + z = 2 \qquad (2.5)$$
$$-4y - 2z = -4$$

Next, we use the second equation to eliminate the y-terms in the other two equations. Specifically, we add -2 times the second equation to the first and add 4 times the second equation to the third. These Type 2 operations transform (2.5) into the following system in which both x and y have been isolated.

$$x \quad - z = -1$$
$$y + z = 2 \qquad (2.6)$$
$$2z = 4$$

To isolate z, multiply the third equation by 1/2 to make the z-coefficient 1.

$$x \quad - z = -1$$
$$y + z = 2 \qquad (2.7)$$
$$z = 2$$

Then add (1 times) the third equation to the first and add -1 times the third equation to the second in (2.7). The result is the system

$$x \quad = 1$$
$$y \quad = 0 \qquad (2.8)$$
$$z = 2,$$

which announces its own solution, the ordered triple (1, 0, 2). Since all the systems (2.1) through (2.8) are algebraically equivalent, we conclude that (1, 0, 2) is the unique solution to the original system. This can be checked by direct substitution in (2.1):

$$2(1) + 4(0) + 2(2) = 6$$
$$-2(1) + (0) + 3(2) = 4$$
$$3(1) + 2(0) + (2) = 5. \quad \blacksquare$$

TABLEAU FORM

The elimination procedure can be streamlined by writing the given system in **tableau form,** as follows:

Linear system   *Tableau form*

$$\begin{array}{r} 2x + 4y + 2z = 6 \\ -2x + y + 3z = 4 \\ 3x + 2y + z = 5 \end{array} \qquad \left[\begin{array}{rrr|r} 2 & 4 & 2 & 6 \\ -2 & 1 & 3 & 4 \\ 3 & 2 & 1 & 5 \end{array}\right]$$

This is merely a shorthand device which enables us to avoid rewriting the variables x, y, z and the equals sign after each transformation. The coefficients of the variables are listed to the left of the vertical line. The first column corresponds to x, the second to y, and the third to z. The three rows of the tableau correspond to the three equations of the system.

Our earlier operations on equations correspond to **elementary row operations,** defined as follows:

Elementary Row Operations

Type 1 A row can be multiplied by a nonzero number.
Type 2 A multiple of one row can be added to another.
Type 3 Two rows can be interchanged.

EXAMPLE 2.1 Revisited

Solve the linear system in Example 2.1 using the tableau form.

Solution The sequence of transformations (2.1)-(2.8) in Example 2.1 can be carried out in tableau form, as shown below. (In the annotation to the right R_1 stands for the old first row, etc.) The first tableau represents the original system (2.1).

$$\begin{bmatrix} 2 & 4 & 2 & | & 6 \\ -2 & 1 & 3 & | & 4 \\ 3 & 2 & 1 & | & 5 \end{bmatrix}$$

Multiply the first row by 1/2; this is indicated by the notation $\tfrac{1}{2}R_1$ to the right of this row.

$$\begin{bmatrix} 1 & 2 & 1 & | & 3 \\ -2 & 1 & 3 & | & 4 \\ 3 & 2 & 1 & | & 5 \end{bmatrix} \quad (\tfrac{1}{2}R_1)$$

Add 2 times the first row to the second row. (We indicate this with the notation $R_2 + 2R_1$ to the right of row 2. Note that the row that is changed is written first.)

$$\begin{bmatrix} 1 & 2 & 1 & | & 3 \\ 0 & 5 & 5 & | & 10 \\ 3 & 2 & 1 & | & 5 \end{bmatrix} \quad (R_2 + 2R_1)$$

Add -3 times the first row to the third row.

$$\begin{bmatrix} 1 & 2 & 1 & | & 3 \\ 0 & 5 & 5 & | & 10 \\ 0 & -4 & -2 & | & -4 \end{bmatrix} \quad (R_3 + (-3)R_1)$$

Multiply the second row by 1/5.

$$\begin{bmatrix} 1 & 2 & 1 & | & 3 \\ 0 & 1 & 1 & | & 2 \\ 0 & -4 & -2 & | & -4 \end{bmatrix} \quad (\tfrac{1}{5}R_2)$$

Add -2 times the second row to the first row, and 4 times the second row to the third row.

$$\begin{bmatrix} 1 & 0 & -1 & | & -1 \\ 0 & 1 & 1 & | & 2 \\ 0 & 0 & 2 & | & 4 \end{bmatrix} \quad \begin{array}{l} (R_1 + (-2)R_2) \\ \\ (R_3 + 4R_2) \end{array}$$

Multiply the third row by 1/2.

$$\begin{bmatrix} 1 & 0 & -1 & | & -1 \\ 0 & 1 & 1 & | & 2 \\ 0 & 0 & 1 & | & 2 \end{bmatrix} \quad (\tfrac{1}{2}R_3)$$

Add the third row to the first row, and add -1 times the third row to the second row.

$$\begin{bmatrix} 1 & 0 & 0 & | & 1 \\ 0 & 1 & 0 & | & 0 \\ 0 & 0 & 1 & | & 2 \end{bmatrix} \quad \begin{array}{l} (R_1 + R_3) \\ (R_2 + (-1)R_3) \end{array}$$

This final tableau corresponds to the linear system (2.8). The solution is therefore $x = 1$, $y = 0$, $z = 2$. ∎

THE PIVOT TRANSFORMATION

When a linear system is written in tableau form, the process of isolating its unknowns can be carried out as follows: In each column but the last, we look for a nonzero entry a. Using row operations, we change this entry to $+1$ and change the other entries in the column to 0. This combination of row operations is called a **pivot transformation** on the entry a. If we call the row and column of a the **pivot row** and **pivot column,** respectively, we can describe this transformation as follows:

Pivot Transformation on a Nonzero Entry *a*

Step 1 Multiply the pivot row by $1/a$.
Step 2 For every other nonzero entry b in the pivot column, add $-b$ times the (new) pivot row to the row in which it occurs.

The pivot transformation is the heart of the Gauss-Jordan elimination method. Before going further, we illustrate it with an example.

EXAMPLE 2.2 Perform a pivot transformation on the entry -2 (circled) in the following tableau:

$$\begin{bmatrix} 3 & 1 & -1 & | & 3 \\ \circled{-2} & 4 & 2 & | & 10 \\ 4 & 1 & 0 & | & 6 \end{bmatrix}.$$

Solution Since the pivot entry is in the second row and first column, we designate row 2 as the pivot row and column 1 as the pivot column.

Step 1 The pivot entry is not equal to 1. Hence, we multiply the second row by $-1/2$.

$$\begin{bmatrix} 3 & 1 & -1 & | & 3 \\ ① & -2 & -1 & | & -5 \\ 4 & 1 & 0 & | & 6 \end{bmatrix} \quad (-\tfrac{1}{2}R_2)$$

Notice that this changes the pivot entry to $+1$.

Step 2 There are two other nonzero entries in the pivot column, namely 3 (first row) and 4 (third row). Hence, we add -3 times the second row to the first row, and add -4 times the second row to the third row.

$$\begin{bmatrix} 0 & 7 & 2 & | & 18 \\ 1 & -2 & -1 & | & -5 \\ 0 & 9 & 4 & | & 26 \end{bmatrix} \quad \begin{array}{l} (R_1 + (-3)R_2) \\ \\ (R_3 + (-4)R_2) \end{array}$$

This completes the pivot. Note that the pivot entry is now $+1$ and all other entries in the pivot column are zero. This means that the first unknown is now isolated (with coefficient 1) in the second equation of the system. ■

Practice Problem 1 Perform the pivot transformation using the entry circled below.

$$\begin{bmatrix} 1 & 2 & 5 & | & 0 \\ 0 & ③ & -6 & | & 9 \\ 0 & -1 & -1 & | & 2 \end{bmatrix}$$

Gauss-Jordan Elimination When using Gauss-Jordan elimination to solve an n by n linear system with a unique solution, we perform a sequence of pivots until all the unknowns have been isolated. The procedure is as follows:

Gauss-Jordan Elimination

To solve an n by n system:

1. Write the system in tableau form.

2. Perform a sequence of pivots on nonzero entries, using a different row and column for each pivot.

3. If the system has a unique solution, this process will terminate after at most n pivots. The system is then solved.

Note that the pivot entries can be chosen arbitrarily, subject only to the condition that we use a new row and new column each time. Whenever possible it is desirable to use 1 as a pivot entry, since Step 1 of the pivot transformation then becomes unnecessary.

EXAMPLE 2.3   Solve the linear system

$$2y + 3z = 8$$
$$x + 2y + z = 6$$
$$3x + 2y = 8.$$

Solution   This 3 by 3 system has the following tableau form:

$$\begin{bmatrix} 0 & 2 & 3 & | & 8 \\ ① & 2 & 1 & | & 6 \\ 3 & 2 & 0 & | & 8 \end{bmatrix}.$$

For the first pivot, we choose the entry 1 (circled) in the first column. Step 1, transforming the pivot entry to 1, is unnecessary. We proceed to Step 2, transforming the other nonzero entries in the pivot column to zero. Add -3 times the second row to the third row.

$$\begin{bmatrix} 0 & ② & 3 & | & 8 \\ 1 & 2 & 1 & | & 6 \\ 0 & -4 & -3 & | & -10 \end{bmatrix} \quad (R_3 + (-3)R_2)$$

For the second pivot we must use an entry in a different row and column. For the entry 2 (circled), Step 1 of the pivot is to multiply the first row by 1/2.

$$\begin{bmatrix} 0 & ① & \tfrac{3}{2} & | & 4 \\ 1 & 2 & 1 & | & 6 \\ 0 & -4 & -3 & | & -10 \end{bmatrix} \quad (\tfrac{1}{2}R_1)$$

Step 2 needs two operations. Add -2 times the first row to the second, and add 4 times the first row to the third.

$$\begin{bmatrix} 0 & 1 & \tfrac{3}{2} & | & 4 \\ 1 & 0 & -2 & | & -2 \\ 0 & 0 & ③ & | & 6 \end{bmatrix} \quad \begin{array}{l} (R_2 + (-2)R_1) \\ (R_3 + 4R_1) \end{array}$$

For the last pivot we *must* use the circled entry 3. (Why?) For Step 1, multiply the third row by 1/3.

$$\begin{bmatrix} 0 & 1 & \tfrac{3}{2} & | & 4 \\ 1 & 0 & -2 & | & -2 \\ 0 & 0 & ① & | & 2 \end{bmatrix} \quad (\tfrac{1}{3}R_3)$$

For Step 2, add $-3/2$ times the third row to the first, and add 2 times the third row to the second.

$$\begin{bmatrix} 0 & 1 & 0 & | & 1 \\ 1 & 0 & 0 & | & 2 \\ 0 & 0 & 1 & | & 2 \end{bmatrix} \quad \begin{array}{l} (R_1 + (-\tfrac{3}{2})R_3) \\ (R_2 + 2R_3) \end{array}$$

The system is now solved. However, in order to make interpreting the solution easier we *interchange the first and second rows,* so the pivot entries lie along the diagonal from top-left to bottom-right.

$$\begin{bmatrix} 1 & 0 & 0 & | & 2 \\ 0 & 1 & 0 & | & 1 \\ 0 & 0 & 1 & | & 2 \end{bmatrix} \begin{matrix} (R_2) \\ (R_1) \\ \end{matrix}$$

The tableau corresponds to the linear system

$$x = 2$$
$$y = 1$$
$$z = 2.$$

The reader should check this result by direct substitution in the original equations. ∎

Remark The Gauss-Jordan elimination procedure—and the pivot transformation on which it is based—will be needed for several later topics. For this reason *you should practice the technique until you have thoroughly mastered it.* Here are some suggestions.

1. Annotations are the key to success. Before performing a row operation, *write down its annotation.*
2. Don't try to combine Steps 1 and 2 of the pivot transformation. As in the examples above, *first* do Step 1, *then* do all Step 2 operations.
3. Be especially wary of mistakes involving signs (+ or −) and fractions. Bear in mind that one careless error throws the entire procedure off.
4. Always check your final answer by substitution in the original equations.

Practice Problem 2 Solve the linear system

$$3x + z = 0$$
$$x + y + z = 4$$
$$3y - z = 3.$$

EXAMPLE 2.4 A piggy bank contains 90 coins, each of which is a penny, nickel, or dime. If the total value of the coins is $3.30 and the value of the nickels equals the value of the other two coins, how many coins are there of each type?

Solution Let there be x pennies, y nickels, and z dimes. Then since there are 90 coins with total value $3.30, we have

$$x + y + z = 90$$
$$x + 5y + 10z = 330.$$

2.1 GAUSS-JORDAN ELIMINATION

Because the value of the nickels equals that of the other two coins, we have $5y = x + 10z$, or
$$-x + 5y - 10z = 0.$$

Thus we have the tableau
$$\begin{bmatrix} ① & 1 & 1 & | & 90 \\ 1 & 5 & 10 & | & 330 \\ -1 & 5 & -10 & | & 0 \end{bmatrix}.$$

The computations that follow are self-explanatory.

$$\begin{bmatrix} 1 & 1 & 1 & | & 90 \\ 0 & ④ & 9 & | & 240 \\ 0 & 6 & -9 & | & 90 \end{bmatrix} \quad \begin{array}{l} (R_2 + (-1)R_1) \\ (R_3 + R_1) \end{array}$$

$$\begin{bmatrix} 1 & 1 & 1 & | & 90 \\ 0 & ① & \frac{9}{4} & | & 60 \\ 0 & 6 & -9 & | & 90 \end{bmatrix} \quad (\tfrac{1}{4}R_2)$$

$$\begin{bmatrix} 1 & 0 & -\frac{5}{4} & | & 30 \\ 0 & 1 & \frac{9}{4} & | & 60 \\ 0 & 0 & \tfrac{45}{2} & | & -270 \end{bmatrix} \quad \begin{array}{l} (R_1 + (-1)R_2) \\ (R_3 + (-6)R_2) \end{array}$$

$$\begin{bmatrix} 1 & 0 & -\frac{5}{4} & | & 30 \\ 0 & 1 & \frac{9}{4} & | & 60 \\ 0 & 0 & ① & | & 12 \end{bmatrix} \quad (-\tfrac{2}{45}R_3)$$

$$\begin{bmatrix} 1 & 0 & 0 & | & 45 \\ 0 & 1 & 0 & | & 33 \\ 0 & 0 & 1 & | & 12 \end{bmatrix} \quad \begin{array}{l} (R_1 + (\tfrac{5}{4})R_3) \\ (R_2 + (-\tfrac{9}{4})R_3) \end{array}$$

Since the last tableau corresponds to the system
$$\begin{aligned} x & = 45 \\ y & = 33 \\ z & = 12, \end{aligned}$$

we conclude that 45 pennies, 33 nickels, and 12 dimes are in the bank. This checks, since
$$\begin{aligned} 45 + 33 + 12 &= 90 \\ 45 + 5 \cdot 33 + 10 \cdot 12 &= 330 \\ -45 + 5 \cdot 33 - 10 \cdot 12 &= 0. \end{aligned} \blacksquare$$

Practice Problem 3   A toaster weighs 6 pounds and costs \$20, a blender weighs 5 pounds and costs \$15, and a mixer weighs 3 pounds and costs \$15. If a shipment of 12 of these small appliances weighs 57 pounds and costs \$205, how many of each appliance does it contain?

EXERCISES 2.1

Exercises 1–12 refer to the following three tableaus:

$$A \quad \begin{bmatrix} 2 & 3 & | & 12 \\ 4 & -1 & | & 3 \end{bmatrix} \qquad B \quad \begin{bmatrix} 1 & -2 & \frac{1}{2} & | & 4 \\ 0 & \frac{1}{2} & -1 & | & -5 \\ 2 & 0 & -\frac{2}{3} & | & 6 \end{bmatrix} \qquad C \quad \begin{bmatrix} 2 & -1 & 1 & 0 & | & 6 \\ 0 & \frac{1}{2} & \frac{5}{4} & -1 & | & 2 \\ \frac{3}{4} & -1 & 1 & 2 & | & 2 \\ -1 & 0 & 1 & 1 & | & -5 \end{bmatrix}$$

In Exercises 1–6 perform the indicated row operation on the given tableau.

1. $\frac{1}{4} R_2$, on R_2 of A
2. $R_2 + \left(-\frac{1}{2} R_1\right)$, on R_2 of A
3. $R_3 + (-2)R_1$, on R_3 of B
4. $R_1 + (-1)R_3$, on R_1 of B
5. $2R_2$, on R_2 of C
6. $R_4 + 4R_2$, on R_4 of C

In Exercises 7–12 perform a pivot on the indicated entry in the given tableau.

7. pivot on -1 in row 2, column 2 of A
8. pivot on 3 in row 1, column 2 of A
9. pivot on 1/2 in row 1, column 3 of B
10. pivot on -2 in row 1, column 2 of B
11. pivot on 2 in row 3, column 4 of C
12. pivot on 3/4 in row 3, column 1 of C

In Exercises 13–22 solve the linear system by Gauss-Jordan elimination. Annotate each row operation and indicate pivot entries with a circle.

13. $x + 2y = 18$
    $5x + 4y = 60$

14. $3x + y = 7$
    $-2x + 4y = 7$

15. $.5x + y + .5z = 4$
    $x + 2y - .5z = 2$
    $-.5x + z = 3$

16. $x + y = 1$
    $x + z = 2$
    $y + z = 3$

17. $x + 2y - z = 2$
    $-x + y + 2z = 1$
    $x + y - z = 1$

18. $x + 2y + 3z = 3$
    $3x + y + 2z = 7$
    $2x + 3y + z = 2$

19. $x + y + z = 6$
    $x - y + z = 0$
    $x + y - z = 2$

20. $x + 2y + z = 1$
    $2x + 5y + z = 2$
    $y + 2z = 3$

21. $x + y + z = 0$
    $x + y + w = 1$
    $x + z + w = 2$
    $y + z + w = 3$

22. $-x + y + z + w = 20$
    $x - y + z + w = 40$
    $x + y - z + w = 60$
    $x + y + z - w = 80$

23. Three herbicides are available that contain the chemicals A, B, and C. One gallon of the first herbicide contains 1 unit of chemical A, 1 unit of chemical C, and none of chemical B; one gallon of the second herbicide contains none of chemical A, 2 units of chemical B, and 1 unit of chemical C; and one gallon of the third herbicide contains 1 unit of chemical A, 1 unit of chemical B, and none of chemical C. To control a certain weed, a farmer wants to apply 11 units of chemical A, 12 units of chemical B, and 14 units of chemical C per acre. How many gallons of each herbicide should be used per acre?

24. An investor asked her investment counselor to invest $18,000 for her in a mutual fund, in stocks, and in a speculative land development. She asked only that the counselor invest 5 times as much in the mutual fund and stocks combined as in land. One year later she learned that her $18,000 investment had increased $580 in value, with the mutual fund yielding a 5% return, the stocks yielding a 3% return, and the land losing 1%. How much money had been invested in each way?

25. A caterer is to supply slices of apple pie, at $1 each, strawberry shortcakes, at $2 each, and baked Alaska, at $3 each. There must be a total of 70 slices of pie and baked Alaska, and 40 more shortcakes than pie slices. The total cost of the two more expensive desserts is to be $240. How can this be done?

26. A 1-pound bag of grass seed costs $2, a 2-pound bag costs $3, and a 4-pound bag costs $4. If 35 bags cost a total of $103 and weigh 73 pounds, how many of each type are included?

27. In a certain city there are two high schools. Eastern High School has 1400 white students and 1200 black students, and Western High School has 1600 white students and 800 black students. In order to achieve a racial balance in the two schools some of the students attending Eastern are going to be bused to Western, and some of those now attending Western will be bused to Eastern. The city's school board has suggested the following guidelines:
   (i) The total number of students to be bused should be 20% of the total population of the two schools.
   (ii) The total number of white students who are bused should equal the total number of black students who are bused.
   (iii) The proportion of white to black students who are bused from Western to Eastern should equal the present proportion of white students to black students at Western.
   (iv) After the busing plan is implemented the proportion of black students to white students should be the same at both schools.
   Determine the number of black students and the number of white students now attending each school who should be bused to the other school.

28. A 200-acre farm is used to grow corn and soybeans. For the coming year the owner estimates yields of 100 bushels per acre for corn and 50 bushels per acre for soybeans and selling prices of $2.30 per bushel for corn and $5.00 per bushel for soybeans. The owner wishes to use all of his 550 bags of available fertilizer on the land where the corn and soybeans are planted and intends to apply 4 bags of fertilizer per acre of corn and 3 bags of fertilizer per acre of soybeans. In addition, some land may be left fallow, in which case a federal price support program will pay $85 per acre to the owner. To receive an equitable return on his investment for land and equipment while remaining in a desirable tax bracket, the owner would like a gross return of $42,000 for his farm. How many acres should be planted with corn, how many acres should be planted with soybeans, and how many acres should lie fallow to meet these conditions?

29. If a tableau T is transformed into T' by a Type 1 row operation replacing R_i with aR_i ($a \neq 0$), what row operation will transform T' back into T?

30. If a tableau T is transformed into T' by a Type 2 row operation replacing R_i with $R_i + aR_j$, what row operation will transform T' back into T?

31. Show that the effect of a Type 3 operation interchanging R_i and R_j can be achieved by a sequence of four operations of Types 1 and 2.

Answers to Practice Problems

1. $\begin{bmatrix} 1 & 0 & 9 & | & -6 \\ 0 & 1 & -2 & | & 3 \\ 0 & 0 & -3 & | & 5 \end{bmatrix}$

2. $x = -1, y = 2, z = 3$

3. 5 toasters, 3 blenders, and 4 mixers

2.2 Systems without Unique Solutions

It often happens that a system of linear equations fails to have a unique solution. For an n by n system this is somewhat exceptional, but when the system contains fewer equations than unknowns it is always the case. In this section we consider the general m by n linear system, and show that in all cases—unique solution or not—the Gauss-Jordan elimination procedure tells us everything we might want to know about the system.

INCONSISTENT SYSTEMS

Beneath a healthy-looking system of linear equations there may lurk several types of pathology, as we see in the following examples.

EXAMPLE 2.5 Solve the following system:

$$x - y + z = 1$$
$$x + 2y - z = 4$$
$$3x + 3y - z = 6.$$

Solution We write the system in tableau form and proceed by Gauss-Jordan elimination (pivot entries are circled).

$$\begin{bmatrix} ① & -1 & 1 & | & 1 \\ 1 & 2 & -1 & | & 4 \\ 3 & 3 & -1 & | & 6 \end{bmatrix}$$

We add -1 times the first row to the second row, and -3 times the first row to the third row.

$$\begin{bmatrix} 1 & -1 & 1 & | & 1 \\ 0 & ③ & -2 & | & 3 \\ 0 & 6 & -4 & | & 3 \end{bmatrix} \quad \begin{array}{l} (R_2 + (-1)R_1) \\ (R_3 + (-3)R_1) \end{array}$$

Now we multiply the second row by 1/3.

$$\begin{bmatrix} 1 & -1 & 1 & | & 1 \\ 0 & ① & -\frac{2}{3} & | & 1 \\ 0 & 6 & -4 & | & 3 \end{bmatrix} \quad (\tfrac{1}{3}R_2)$$

2.2 SYSTEMS WITHOUT UNIQUE SOLUTIONS

We add the second row to the first row, and add -6 times the second row to the third row.

$$\begin{bmatrix} 1 & 0 & \tfrac{1}{3} & | & 2 \\ 0 & 1 & -\tfrac{2}{3} & | & 1 \\ 0 & 0 & 0 & | & -3 \end{bmatrix} \quad \begin{array}{l} (R_1 + R_2) \\ \\ (R_3 + (-6)R_2) \end{array}$$

For the final pivot, in the third column, we are now forced to use an entry in the third row, since the first and second rows have already been used as pivot rows. But *zero cannot be used as a pivot entry,* so there is no way to complete the Gauss-Jordan procedure.

If we translate the final tableau back into the language of linear equations, we can see where the problem lies.

$$(1)x + (0)y + (\tfrac{1}{3})z = 2$$
$$(0)x + (1)y + (-\tfrac{2}{3})z = 1$$
$$(0)x + (0)y + (0)z = -3$$

The third equation in this system says $0 = -3$, which is false no matter what values x, y, and z represent. Since this system is equivalent to the one we started with, we conclude that the original system has no solution (x, y, z). ∎

We call a linear system an **inconsistent system** when, as in the last example, it has no solution. The Gauss-Jordan procedure always signals such a system by producing a tableau row of the form $0\ 0\ \ldots\ 0 \mid a$, where $a \neq 0$.

Practice Problem 4   Solve the system

$$x + 2y - z = 3$$
$$2x \quad\quad + z = 2$$
$$4x + 4y - z = 6.$$

A GEOMETRICAL INTERPRETATION OF A 3 BY 3 SYSTEM

We saw in Chapter 1 that equations involving two variables x and y could be graphed using a coordinate system based on two perpendicular axes. In the same way a 3-dimensional coordinate system can be established, based on three mutually perpendicular axes, labeled x, y, and z. Although we will not go into the details of such a coordinate system now, it can be used to give a geometrical meaning to a system of three equations in three unknowns. In such a coordinate system each point corresponds to a triple (x, y, z), as shown in Figure 2.1(a).

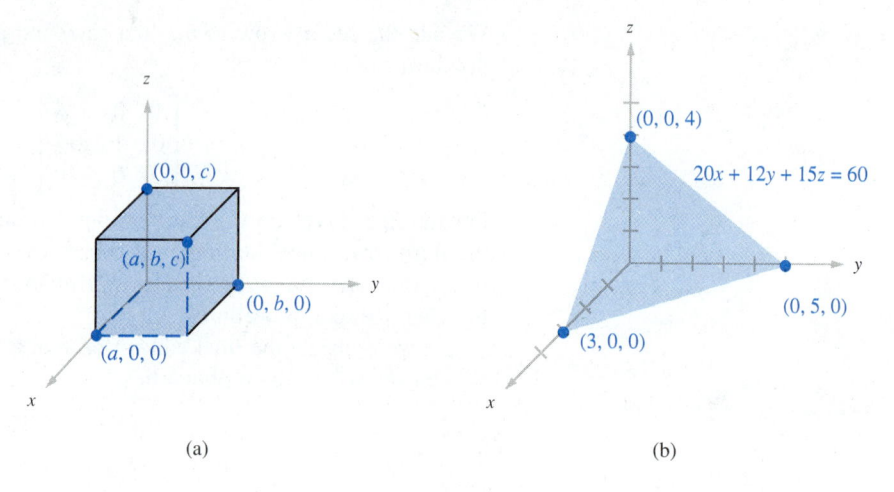

FIGURE 2.1

It turns out that just as the graph of a linear equation in x and y is always a line, the graph of a linear equation in x, y, and z is always a plane. An example is shown in Figure 2.1(b), in which a portion of the plane with equation $20x + 12y + 15z = 60$ is depicted. Notice that each of the points $(3, 0, 0)$, $(0, 5, 0)$, and $(0, 0, 4)$ on the x-axis, y-axis, and z-axis, respectively, satisfies this equation.

Ordinarily two planes intersect in a line, and a third plane will intersect this line in a single point, as in Figure 2.2. This corresponds to the fact that three linear equations in three unknowns usually have a unique solution. It is possible, however, for three planes to have no point in common, even though any two of them intersect in a line. Such a situation is illustrated in Figure 2.3, and corresponds to an inconsistent 3 by 3 system of linear equations.

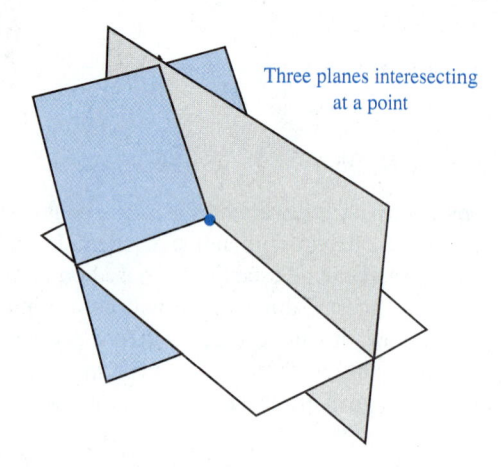

FIGURE 2.2 3 by 3 System with a Unique Solution

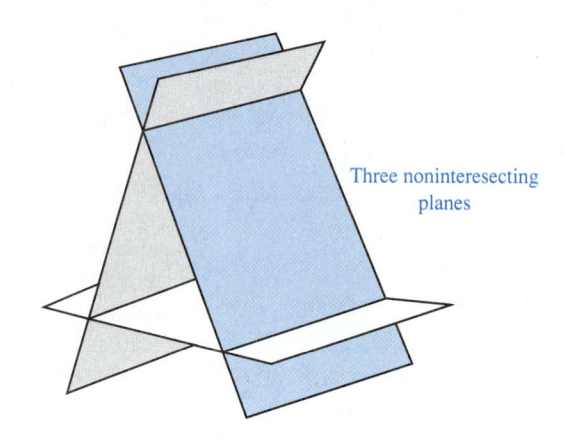

FIGURE 2.3 Inconsistent System

2.2 SYSTEMS WITHOUT UNIQUE SOLUTIONS

DEPENDENT SYSTEMS

The other type of exceptional behavior for an n by n system is *dependency*, which we encounter in the following example.

EXAMPLE 2.6 Solve the system

$$-x + 3y - 2z = 1$$
$$2x - 5y + 2z = 0$$
$$2x - y - 6z = 8.$$

Solution Again, we proceed by Gauss-Jordan elimination.

$$\begin{bmatrix} \boxed{-1} & 3 & -2 & | & 1 \\ 2 & -5 & 2 & | & 0 \\ 2 & -1 & -6 & | & 8 \end{bmatrix}$$

We multiply the first row by -1.

$$\begin{bmatrix} \boxed{1} & -3 & 2 & | & -1 \\ 2 & -5 & 2 & | & 0 \\ 2 & -1 & -6 & | & 8 \end{bmatrix} \quad (-R_1)$$

Now we add -2 times the first row to the second row, and -2 times the first row to the third row.

$$\begin{bmatrix} 1 & -3 & 2 & | & -1 \\ 0 & \boxed{1} & -2 & | & 2 \\ 0 & 5 & -10 & | & 10 \end{bmatrix} \quad \begin{matrix} (R_2 + (-2)R_1) \\ (R_3 + (-2)R_1) \end{matrix}$$

We add 3 times the second row to the first row, and -5 times the second row to the third row.

$$\begin{bmatrix} 1 & 0 & -4 & | & 5 \\ 0 & 1 & -2 & | & 2 \\ 0 & 0 & 0 & | & 0 \end{bmatrix} \quad \begin{matrix} (R_1 + 3R_2) \\ \\ (R_3 + (-5)R_2) \end{matrix}$$

Once again the method breaks down. We cannot use zero as a pivot in the third column, and the other two entries are in rows that have already been used for pivots. This time the third row represents the equation

$$(0)x + (0)y + (0)z = 0,$$

which is true for *every* triple (x, y, z). Since it tells us nothing, we can ignore this row in the final tableau. Its appearance indicates that the system is a **dependent system;** that is, one of the equations is implied by the others. (In our case, it happens that the third equation is equal to 8 times the first plus 5 times the second.)

Ignoring the third row in the final tableau, we conclude that the original system is equivalent to the system

$$x \quad - 4z = 5$$
$$y - 2z = 2$$

of two equations in three unknowns. We can solve these equations for x and y in terms of z, to obtain the general solution:

$$x = 5 + 4z$$
$$y = 2 + 2z$$
$$z = \text{any real number.}$$

Note that z can be assigned any value we please in the general solution to yield a specific solution of the original system. For instance

$z = 0$ gives $x = 5$, $y = 2$, and thus the solution $(5, 2, 0)$;
$z = 2$ gives $x = 13$, $y = 6$, and thus the solution $(13, 6, 2)$;
$z = -1$ gives $x = 1$, $y = 0$, and thus the solution $(1, 0, -1)$.

The original system therefore has *infinitely many solutions* (x, y, z), one for each value of z. We represent the solution we have found by writing

$$(x, y, z) = (5 + 4z, 2 + 2z, z),$$

where it is understood that z can be any real number.

Notice that we can check our solution by plugging it back into the original equations, just as we check a unique solution. Here the original system is

$$-x + 3y - 2z = 1$$
$$2x - 5y + 2z = 0$$
$$2x - y - 6z = 8.$$

Substituting $5 + 4z$ for x and $2 + 2z$ for y yields

$$-(5 + 4z) + 3(2 + 2z) - 2z = 1$$
$$2(5 + 4z) - 5(2 + 2z) + 2z = 0$$
$$2(5 + 4z) - (2 + 2z) - 6z = 8,$$

or

$$-5 - 4z + 6 + 6z - 2z = 1$$
$$10 + 8z - 10 - 10z + 2z = 0$$
$$10 + 8z - 2 - 2z - 6z = 8;$$

and each of these equations is true for all values of z. ∎

Three planes through a common line

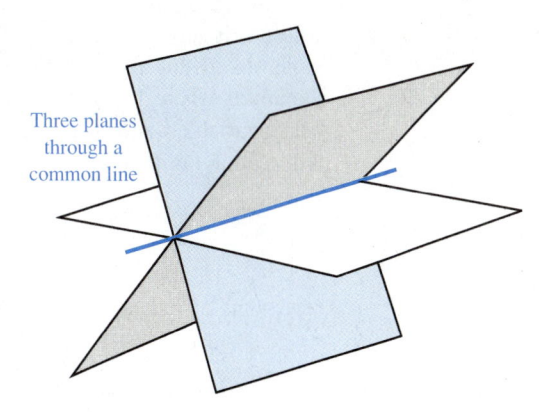

FIGURE 2.4 3 by 3 Dependent System

The geometric meaning of the last example is illustrated in Figure 2.4, which shows three planes intersecting along a common line. Any point on this line lies on all three planes and therefore satisfies all three equations.

BASIC AND FREE VARIABLES

The preceding examples, together with those in the previous section, illustrate the three possibilities that can arise in solving a system of m equations in n unknowns.

■ ■ ■ ■ ■ ■

Every m by n linear system leads to *exactly one* of the following three cases.

System with a unique solution: The system has exactly one solution, and can be solved for all the unknowns.

Inconsistent system: The system has no solutions.

Dependent system: The system has infinitely many solutions, which can be expressed by writing some of the unknowns in terms of the others.

In Example 2.6 we saw that the dependent system

$$-x + 3y - 2z = 1$$
$$2x - 5y + 2z = 0$$
$$2x - y - 6z = 8$$

has the general solution

$$(x, y, z) = (5 + 4z, 2 + 2z, z), \qquad (2.9)$$

where z is any real number. Since z is unrestricted in (2.9), we call z a **free variable** in this form of the solution. The variables x and y are called **basic variables** when the solution is written this way. That x and y turned out to be basic variables and z a free variable depended on various choices we made while applying the Gauss-Jordan procedure. For example it can be shown that

$$(x, y, z) = \left(1 + 2y, y, -1 + \frac{1}{2}y\right) \qquad (2.10)$$

represents exactly the same set of ordered triples as (2.9). In particular,

$y = 2$ gives $x = 5$, $z = 0$, and thus the solution $(5, 2, 0)$;
$y = 6$ gives $x = 13$, $z = 2$, and thus the solution $(13, 6, 2)$;
$y = 0$ gives $x = 1$, $z = -1$, and thus the solution $(1, 0, -1)$;

and these are the same three particular solutions we computed in Example 2.6. When the solution is expressed as in (2.10), y is the free variable and x and z are basic variables.

Practice Problem 5 Solve the system

$$x + 2y + z = 3$$
$$y - z = -2.$$

In our examples and solutions to exercises, we will always choose our pivot columns as far to the left as possible so that solutions can be easily compared. For example, given the system

$$x + y + z \phantom{{}+w} = 3$$
$$x + y \phantom{{}+z} + w = 4,$$

with tableau

$$\begin{bmatrix} 1 & 1 & 1 & 0 & | & 3 \\ 1 & 1 & 0 & 1 & | & 4 \end{bmatrix},$$

we first pivot on the 1 in the first row and column, which produces

$$\begin{bmatrix} 1 & 1 & 1 & 0 & | & 3 \\ 0 & 0 & -1 & 1 & | & 1 \end{bmatrix}. \qquad (R_2 + (-1)R_1)$$

Since we have already pivoted using an element in the first row, we cannot pivot on the 1 in the first row and second column. Instead we pivot on the -1 in row 2 and column 3.

$$\begin{bmatrix} 1 & 1 & 1 & 0 & | & 3 \\ 0 & 0 & 1 & -1 & | & -1 \end{bmatrix} \qquad (-R_2)$$

$$\begin{bmatrix} 1 & 1 & 0 & 1 & | & 4 \\ 0 & 0 & 1 & -1 & | & -1 \end{bmatrix} \qquad (R_1 + (-1)R_2)$$

Since we pivoted in columns 1 and 3 we take x and z as the basic variables, using the equivalent system

$$x + y + w = 4$$
$$z - w = -1.$$

Thus $x = -y - w + 4$, $z = w - 1$, and we have the general solution

$$(x, y, z, w) = (-y - w + 4, y, w - 1, w),$$

with y and w free variables.

Although pivoting from left to right allows us to compare solutions, it may not be the quickest path to an answer. For example in the system

$$x + y + z = 3$$
$$x + y + w = 4$$

just treated, we can see immediately that $z = 3 - x - y$ and $w = 4 - x - y$, giving the general solution

$$(x, y, z, w) = (x, y, 3 - x - y, 4 - x - y),$$

with x and y free variables.

SUMMARY OF GAUSS-JORDAN ELIMINATION

Gauss-Jordan Elimination

To solve an m by n linear system:

1. Write the system in tableau form.

2. Perform a sequence of pivots on nonzero entries, using a different row and column for each pivot. Eventually this process will terminate.

3. Look for rows of the form 0 0 . . . 0 | a, where $a \neq 0$. If any such row appears, *the system has no solution*.

4. Look for rows of the form 0 0 . . . 0 | 0. Ignore all such rows in the final tableau.

5. If the left side of the tableau now has the same number of rows and columns, *the system has a unique solution*, which can be read from the final tableau.

6. If the left side of the tableau has fewer rows than columns, *the system has infinitely many solutions*. Use the final tableau to solve for the basic variables in terms of the free variables.

EXAMPLE 2.7

Solve the following 3 by 5 system:

$$x - 2y + z = 1$$
$$y - 2z + u = 2$$
$$z - 2u + v = 3.$$

Solution In tableau form, the system becomes

$$\begin{array}{ccccc} x & y & z & u & v \end{array}$$
$$\begin{bmatrix} 1 & -2 & 1 & 0 & 0 & | & 1 \\ 0 & 1 & -2 & 1 & 0 & | & 2 \\ 0 & 0 & 1 & -2 & 1 & | & 3 \end{bmatrix}.$$

Since we already have a 1 and all zeros in the x-column, we proceed to the y-column, using the circled 1 as pivot. We add twice the second row to the first row.

$$\begin{bmatrix} 1 & 0 & -3 & 2 & 0 & | & 5 \\ 0 & 1 & -2 & 1 & 0 & | & 2 \\ 0 & 0 & ① & -2 & 1 & | & 3 \end{bmatrix} \quad (R_1 + 2R_2)$$

Now we add 3 times the third row to the first row, and add 2 times the third row to the second row.

$$\begin{bmatrix} 1 & 0 & 0 & -4 & 3 & | & 14 \\ 0 & 1 & 0 & -3 & 2 & | & 8 \\ 0 & 0 & 1 & -2 & 1 & | & 3 \end{bmatrix} \quad \begin{array}{l}(R_1 + 3R_3) \\ (R_2 + 2R_3)\end{array}$$

This tableau corresponds to the equations

$$x - 4u + 3v = 14$$
$$y - 3u + 2v = 8$$
$$z - 2u + v = 3,$$

or

$$x = 4u - 3v + 14$$
$$y = 3u - 2v + 8$$
$$z = 2u - v + 3.$$

We see that the general solution is

$$(x, y, z, u, v) = (4u - 3v + 14, 3u - 2v + 8, 2u - v + 3, u, v),$$

where u and v are free variables. ∎

EXAMPLE 2.8

Chocolate kisses come in 2-pound bags costing $3, jelly beans come in pound boxes costing $2, and mints come in 3-pound cans costing $5. Thirty-two pounds of candy costing $53 are needed for a school party. Express the number of bags of kisses and boxes of jelly beans in terms of the number of cans of mints bought.

2.2 SYSTEMS WITHOUT UNIQUE SOLUTIONS

Solution Let x bags of kisses, y boxes of jelly beans, and z cans of mints be bought. Then
$$2x + y + 3z = 32$$
$$3x + 2y + 5z = 53.$$

The corresponding tableau is
$$\begin{bmatrix} ② & 1 & 3 & | & 32 \\ 3 & 2 & 5 & | & 53 \end{bmatrix}.$$

First we pivot on the first row and column.

$$\begin{bmatrix} ① & \frac{1}{2} & \frac{3}{2} & | & 16 \\ 3 & 2 & 5 & | & 53 \end{bmatrix} \quad (\tfrac{1}{2}R_1)$$

$$\begin{bmatrix} 1 & \frac{1}{2} & \frac{3}{2} & | & 16 \\ 0 & ②\!\!\tfrac{1}{2} & \frac{1}{2} & | & 5 \end{bmatrix} \quad (R_2 + (-3)R_1)$$

Next we pivot on the second row and column.

$$\begin{bmatrix} 1 & \frac{1}{2} & \frac{3}{2} & | & 16 \\ 0 & ① & 1 & | & 10 \end{bmatrix} \quad (2R_2)$$

$$\begin{bmatrix} 1 & 0 & 1 & | & 11 \\ 0 & 1 & 1 & | & 10 \end{bmatrix} \quad (R_1 + (-\tfrac{1}{2})R_2)$$

The corresponding equations are
$$x + z = 11$$
$$y + z = 10,$$

or $x = 11 - z$, $y = 10 - z$. These give the number of bags of kisses and boxes of jelly beans in terms of z, the number of cans of mints. Note that at most 10 cans of mints may be bought, since otherwise y is negative. ■

Practice Problem 6 Wrapping paper comes in three sizes. The standard roll is 9 square feet, the large is 12 square feet, and the jumbo is 18 square feet. A woman needs 30 rolls of paper with an area of 357 square feet. Express the number of standard and large rolls in terms of the number of jumbo rolls she must buy.

EXERCISES 2.2

In Exercises 1–16 solve the given linear system using the Gauss-Jordan procedure. At each stage choose pivots in the left-most column possible.

1. $2x + y - 3z = 2$
   $x + 2y + 3z = 4$

2. $2x + 4y - 2z = 6$
   $3x + y + 2z = -1$

3. $x + 2y - z = 3$
   $2x + 4y + 3z = 1$

4. $3x - y + 6z = 1$
   $x + y + 2z = 3$

5. $w + 3y + 2z = 1$
   $4w + 2x + 2y - 2z = 2$

6. $2p - 3q + s = 6$
   $-6p + 9q + 2r + 5s = -2$

7. $\begin{aligned} u + 2v &= 5 \\ 2u + 3w &= 11 \\ v - 2w &= -4 \end{aligned}$

8. $\begin{aligned} x - 2y + z &= 2 \\ -2x + 3y &= -1 \\ x - 3y + 3z &= 5 \end{aligned}$

9. $\begin{aligned} 2r + 3s - 4t &= 1 \\ -r + 2s + t &= -2 \\ r + 12s - 5t &= -4 \end{aligned}$

10. $\begin{aligned} r + 2s + 5t &= 2 \\ r + 4s - t &= -3 \\ r - 2s + 17t &= 10 \end{aligned}$

11. $\begin{aligned} a + 5b - c &= 4 \\ -2a + 2b &= -3 \\ -2a + 26b - 4c &= 5 \end{aligned}$

12. $\begin{aligned} 2X + Y - Z &= 4 \\ -X + 2Y + Z &= -4 \\ X + 3Y - 3Z &= -3 \end{aligned}$

13. $\begin{aligned} y + 2z + u + 3v &= 3 \\ x + 2y - z + 2v &= 1 \end{aligned}$

14. $\begin{aligned} -x + y + 2z + 3u + v &= 2 \\ 3x - 3y - 6z - 2u + 4v &= 1 \end{aligned}$

15. $\begin{aligned} x - 2y + 3z &= 1 \\ 2x + y - z &= -2 \\ 5x + z &= -3 \\ 4x - 3y + 5z &= 0 \end{aligned}$

16. $\begin{aligned} 2x - y + 2z &= -1 \\ -3x + 3y - z &= 2 \\ x + y + 3z &= 0 \\ -x + 5y - 7z &= 3 \end{aligned}$

17. Use Gauss-Jordan elimination to find the general solution to the system
$$\begin{aligned} x + 2y + 3z &= 2 \\ 3x + 2y + z &= 10 \end{aligned}$$
(a) with x and y basic, z free
(b) with x and z basic, y free
(c) with y and z basic, x free.

18. Use Gauss-Jordan elimination to find the general solution to the system
$$\begin{aligned} x + 2y + 3z + 4w &= 5 \\ x + 3y + 5z + 7w &= 9 \end{aligned}$$
(a) with x and y basic, z and w free
(b) with x and z basic, y and w free
(c) with z and w basic, x and y free.

19. An airline uses a 200-passenger plane on a certain flight. A first-class ticket costs $360, a coach ticket $180, and a super-saver ticket $90. It is desired to take in $27,000 in fares on the flight. Suppose x first-class tickets, y coach tickets, and z super-saver tickets are sold, and w seats are empty. Express x and y in terms of z and w.

20. A college student has 48 hours available to work at her summer job at a discount store. She gets $3.50 per hour ordinarily, but $5.25 per hour on Sunday. One week she worked x hours at ordinary pay, y hours on Sunday, and did not work z of the 48 hours, earning $140. Express x and y in terms of z.

21. A merchant wants to mix peanuts costing $0.60 per pound, cashews costing $1.50 per pound, and almonds costing $2.00 per pound, to make 40 pounds of mixed nuts costing $1.28 per pound. Express the number of pounds of peanuts and cashews he must use in terms of the number of pounds of almonds.

22. At his next party, a host intends to serve a punch made from three types of mixed fruit juices. Mix A contains 0% apple juice, Mix B contains 20% apple juice, and Mix C contains 40% apple juice. He wants to make 200 fluid ounces of punch containing 15.2% apple juice. Express the number of ounces of Mixes A and B he must use in terms of the number of ounces of Mix C.

2.3 APPLICATIONS OF SYSTEMS OF LINEAR EQUATIONS

Answers to Practice Problems

4. The system is inconsistent, and has no solution.
5. $(x, y, z) = (7 - 3z, z - 2, z)$, where z is any real number.
6. If she buys z jumbo rolls, then she must buy $1 + 2z$ standard rolls and $29 - 3z$ large rolls.

2.3 Applications of Systems of Linear Equations

This section gives some examples of problems that are solved using systems of linear equations. More applications of linear systems will be presented in later sections, and in the next chapter.

PRODUCT-MIX PROBLEMS

There is a large class of applications referred to as **product-mix** problems, since they involve finding a suitable combination of products or available resources in order to satisfy certain given conditions. Some 2 by 2 problems of this type were treated in Chapter 1.

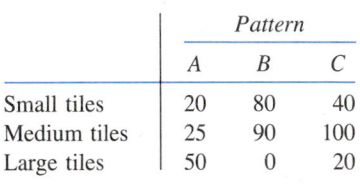

EXAMPLE 2.9 A flooring company makes linoleum in a standard size roll. These large rolls are cut into small, medium, and large tiles according to one of three available cutting patterns.

	Pattern A	Pattern B	Pattern C
Small tiles	20	80	40
Medium tiles	25	90	100
Large tiles	50	0	20

For example, if a roll is cut using pattern A, then 20 small, 25 medium, and 50 large tiles are produced. How many rolls should be cut according to each pattern, if the company wants to fill an order for 2000 small tiles, 3400 medium tiles, and 1400 large tiles, without wasting any linoleum?

Solution The three unknowns in the problem are:

x = number of rolls cut according to Pattern A
y = number of rolls cut according to Pattern B
z = number of rolls cut according to Pattern C.

The condition that we fill the order without any waste—that is, without any tiles left over—translates into a 3 by 3 system of linear equations.

$$20x + 80y + 40z = 2000 \text{ (small tiles)}$$
$$25x + 90y + 100z = 3400 \text{ (medium tiles)}$$
$$50x + 20z = 1400 \text{ (large tiles)}$$

For example, the first equation comes from the fact that 2000 small tiles are needed, while each roll cut to Pattern A yields 20 small tiles, each roll cut to Pattern B yields 80 small tiles, and each roll cut to Pattern C yields 40 small tiles.

We write the system in tableau form and apply Gauss-Jordan elimination. The annotations will explain the row operations done at each step.

$$\begin{bmatrix} \boxed{20} & 80 & 40 & | & 2000 \\ 25 & 90 & 100 & | & 3400 \\ 50 & 0 & 20 & | & 1400 \end{bmatrix}$$

$$\begin{bmatrix} \boxed{1} & 4 & 2 & | & 100 \\ 25 & 90 & 100 & | & 3400 \\ 50 & 0 & 20 & | & 1400 \end{bmatrix} \quad (\tfrac{1}{20}R_1)$$

$$\begin{bmatrix} 1 & 4 & 2 & | & 100 \\ 0 & \boxed{-10} & 50 & | & 900 \\ 0 & -200 & -80 & | & -3600 \end{bmatrix} \quad \begin{array}{l}(R_2 + (-25)R_1) \\ (R_3 + (-50)R_1)\end{array}$$

$$\begin{bmatrix} 1 & 4 & 2 & | & 100 \\ 0 & \boxed{1} & -5 & | & -90 \\ 0 & -200 & -80 & | & -3600 \end{bmatrix} \quad (-\tfrac{1}{10}R_2)$$

$$\begin{bmatrix} 1 & 0 & 22 & | & 460 \\ 0 & 1 & -5 & | & -90 \\ 0 & 0 & \boxed{-1080} & | & -21{,}600 \end{bmatrix} \quad \begin{array}{l}(R_1 + (-4)R_2) \\ (R_3 + 200R_2)\end{array}$$

$$\begin{bmatrix} 1 & 0 & 22 & | & 460 \\ 0 & 1 & -5 & | & -90 \\ 0 & 0 & \boxed{1} & | & 20 \end{bmatrix} \quad (-\tfrac{1}{1080}R_3)$$

$$\begin{bmatrix} 1 & 0 & 0 & | & 20 \\ 0 & 1 & 0 & | & 10 \\ 0 & 0 & 1 & | & 20 \end{bmatrix} \quad \begin{array}{l}(R_1 + (-22)R_3) \\ (R_2 + 5R_3)\end{array}$$

The solution is therefore (20, 10, 20), that is, the company should cut 20 rolls according to Pattern A, 10 according to Pattern B, and 20 according to Pattern C. ■

Practice Problem 7 A slice of bread contains 70 calories and 2 grams of protein, while an egg contains 80 calories and 6 grams of protein. At a certain orphanage the egg and toast breakfasts are to provide 1090 calories and 46 grams of protein per week for each child.

(a) If x slices of bread and y eggs are provided per week, express the above information as a 2 by 2 linear system.

(b) Solve the linear system.

A MODEL FOR ALLOCATING SERVICE CHARGES*

Manufacturing firms typically are composed of two types of departments having entirely different functions. Service departments such as data processing and shipping provide services to other departments within the firm, whereas production departments produce commodities for sale outside the firm. Before the commodities made by the production departments can be priced for sale, the firm must know the total cost for each production department. These costs consist not only of direct costs such as salaries and material costs but also of indirect costs such as charges for services provided by the service departments. Standard accounting procedures require that the direct costs of the service departments be charged to the production departments in accordance with the amount of service used. The allocation of these charges is complicated by the fact that each service department contributes to the expenses of the other service departments, as well as to its own expenses.

EXAMPLE 2.10 An appliance manufacturer has three service departments (maintenance, shipping, and data processing), and two production departments (small appliances and large appliances). The five departments each consume a proportion of the output of the three service departments as given in Table 2.1.

	Provider of Service		
User of services	*Maintenance*	*Shipping*	*Data processing*
Maintenance	0.3	0.3	0.2
Shipping	0.3	0.3	0.2
Data processing	0.1	0.1	0.4
Small appliances	0.1	0.1	0.1
Large appliances	0.2	0.2	0.1

TABLE 2.1

For example, the maintenance department expends 30% of its effort on itself, 30% on the shipping department, 10% on each of data processing and small appliances, and 20% on large appliances. Notice that each column in Table 2.1 adds up to 1, since this sum accounts for the total effort of the corresponding department. The direct costs per week for the five departments are $d_1 = \$1000$, $d_2 = \$800$, $d_3 = \$1200$, $d_4 = \$2000$, and $d_5 = \$3000$, for maintenance, shipping, data processing, small appliances, and large appliances, respectively. These include the salaries, materials, etc. expended directly by each department per week. How should the direct cost

$$d_1 + d_2 + d_3 = \$3000$$

of the service departments be allocated between the two production departments?

---

*For a more detailed explanation of the problems of allocating service charges see Don O. Koehler, "The Cost Accounting Problem," *Mathematics Magazine,* Vol. 53, No. 1 (1980) 3–12.

Solution Let x_4 and x_5 stand for the total costs of the small and large appliance departments, respectively, including both direct and indirect costs. Since all costs are to be allocated to the two production departments, we should have $x_4 + x_5 = d_1 + d_2 + d_3 + d_4 + d_5 = \$8000$. On the other hand, from the last two rows of Table 2.1 we should have

$$x_4 = d_4 + 0.1x_1 + 0.1x_2 + 0.1x_3 \qquad (2.11)$$
$$x_5 = d_5 + 0.2x_1 + 0.2x_2 + 0.1x_3,$$

where x_1, x_2, and x_3 are defined in a similar way from the first three rows of Table 2.1:

$$x_1 = d_1 + 0.3x_1 + 0.3x_2 + 0.2x_3$$
$$x_2 = d_2 + 0.3x_1 + 0.3x_2 + 0.2x_3 \qquad (2.12)$$
$$x_3 = d_3 + 0.1x_1 + 0.1x_2 + 0.4x_3.$$

The meaning of x_1, x_2, and x_3 is somewhat fictitious, since if $x_4 + x_5$ is the total direct cost of all five departments, some of these costs must be counted again in x_1, x_2, and x_3. In fact, if we add up the equations of (2.11) and (2.12), we get

$$x_1 + x_2 + x_3 + x_4 + x_5 = d_1 + d_2 + d_3 + d_4 + d_5 + x_1 + x_2 + x_3,$$

from which we see that

$$x_4 + x_5 = d_1 + d_2 + d_3 + d_4 + d_5,$$

as desired.

Since neither x_4 nor x_5 appears in the three equations of (2.12), we start by solving this 3 by 3 system. Plugging in the values of d_1, d_2, and d_3 and arranging the variables on the left and constants on the right yields

$$0.7x_1 - 0.3x_2 - 0.2x_3 = 1000$$
$$-0.3x_1 + 0.7x_2 - 0.2x_3 = 800$$
$$-0.1x_1 - 0.1x_2 + 0.6x_3 = 1200,$$

and the corresponding tableau

$$\begin{bmatrix} 0.7 & -0.3 & -0.2 & 1000 \\ -0.3 & 0.7 & -0.2 & 800 \\ -0.1 & -0.1 & 0.6 & 1200 \end{bmatrix}.$$

The reader should confirm that Gauss-Jordan elimination leads to the unique solution $x_1 = 4000$, $x_2 = 3800$, and $x_3 = 3300$. Substituting these values and the given $d_4 = 2000$ and $d_5 = 3000$ into (2.11) gives $x_4 = 3110$ and $x_5 = 4890$. Thus of the total of \$8000 of direct costs per week, \$3110 should be allocated to small appliances and \$4890 to large appliances. ∎

2.3 APPLICATIONS OF SYSTEMS OF LINEAR EQUATIONS

Practice Problem 8 A retailer has two service departments, data processing and maintenance, and two production departments, direct sales and mail order. The direct costs and distribution of outputs of the service departments to each department are shown below.

User of Services	Direct cost	Provider of Service: Data processing	Provider of Service: Maintenance
Data processing	$400	.2	.1
Maintenance	$210	.2	.2
Direct sales	$500	.2	.4
Mail order	$600	.4	.3

(a) Write a system of equations for this problem corresponding to (2.12).

(b) How should the total costs of $1710 be allocated to direct sales and mail order?

TRAFFIC FLOW THROUGH A ROAD NETWORK

Network analysis has been used for many years in physics and engineering to study the flow of currents through electric circuits. More recently, this subject has proved useful in other fields as well. The next application involves a technique for analyzing traffic flow through a road network.

EXAMPLE 2.11 A one-way drive circles a reservoir, with two one-way streets coming into it and two leaving it, as shown in Figure 2.5. During rush hour 350 cars enter the drive per hour from North Avenue. Likewise 250, 200, and 300 cars per

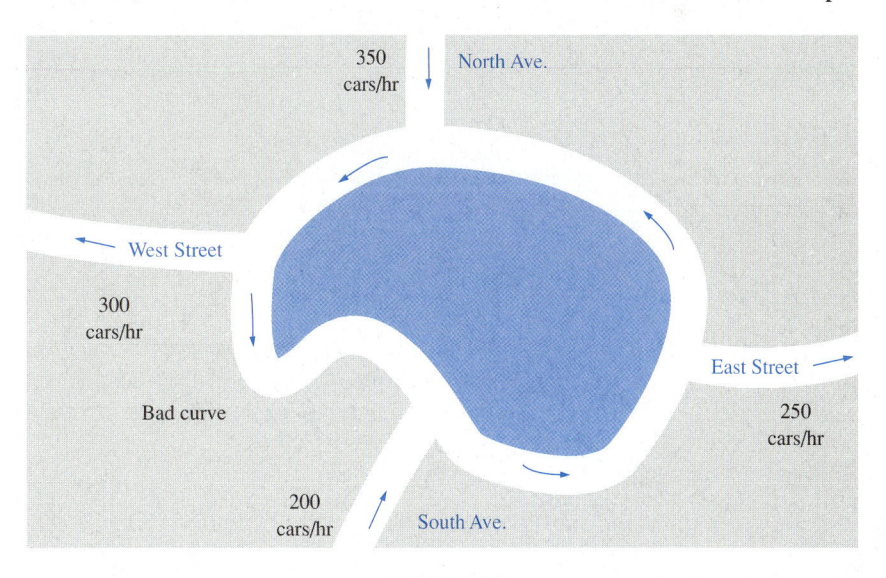

FIGURE 2.5

hour use East Street, South Avenue, and West Street, respectively. There is a bad curve between West Street and South Avenue that should be straightened out, but the construction should not impede rush hour traffic. How many cars per hour must be able to take the portion of the drive between West Street and South Avenue in order to maintain this traffic flow?

Solution Let

w = number of cars per hour from North Avenue to West Street
x = number of cars per hour from East Street to North Avenue
y = number of cars per hour from South Avenue to East Street
z = number of cars per hour from West Street to South Avenue,

and let A, B, C, and D be the intersections shown in Figure 2.6.

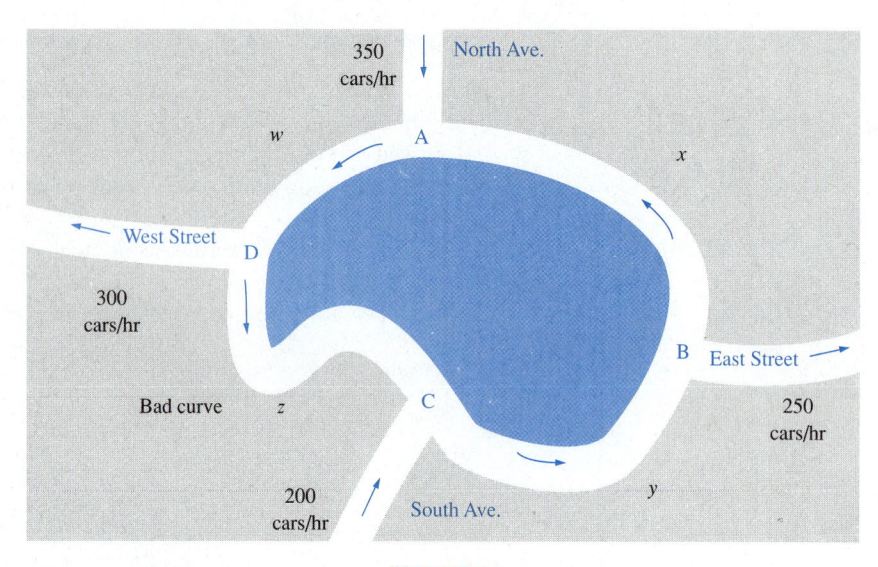

FIGURE 2.6

Now at intersection A there are x cars entering from the east and 350 cars entering from the north, while w cars leave the intersection to the west. Since the total number of cars entering an intersection must equal the number leaving it, we have

$$x + 350 = w.$$

Applying a similar analysis to intersections B, C, and D leads to the system

$$x + 350 = w \quad \text{(intersection A)}$$
$$y = x + 250 \quad \text{(intersection B)}$$
$$z + 200 = y \quad \text{(intersection C)}$$
$$w = z + 300 \quad \text{(intersection D)}.$$

If we arrange the variables on the left and constants on the right, we get

$$\begin{aligned} -w + x & = -350 \\ -x + y & = 250 \\ -y + z & = -200 \\ w \quad\quad\quad - z & = 300, \end{aligned}$$

which translates into the following tableau.

$$\begin{bmatrix} -1 & 1 & 0 & 0 & -350 \\ 0 & -1 & 1 & 0 & 250 \\ 0 & 0 & -1 & 1 & -200 \\ 1 & 0 & 0 & -1 & 300 \end{bmatrix}$$

The reader should confirm that Gauss-Jordan elimination transforms the tableau into the following.

$$\begin{bmatrix} 1 & 0 & 0 & -1 & 300 \\ 0 & 1 & 0 & -1 & -50 \\ 0 & 0 & 1 & -1 & 200 \\ 0 & 0 & 0 & 0 & 0 \end{bmatrix}$$

Thus taking z as a free variable gives

$$\begin{aligned} w &= z + 300 \\ x &= z - 50 \\ y &= z + 200. \end{aligned} \quad (2.13)$$

Of course w, x, y, and z must all be nonnegative since they represent a number of cars per hour. Since $x = z - 50$, this means z must be at least 50. Note that z corresponds to the traffic past the proposed construction, which must be handled so that at least 50 cars per hour can pass the construction site. Indeed, taking $z = 50$ gives $(w, x, y, z) = (350, 0, 250, 50)$, which satisfies the conditions of the problem. Since $z \geq 50$ we can draw additional information from (2.13), namely that $w \geq 350$ (which was obvious from the start, since at least 350 cars per hour enter intersection A), and that $y \geq 250$ (which was not obvious before our analysis). ∎

Practice Problem 9 Figure 2.7 shows the number of vehicles per hour using certain portions of four one-way streets during rush hour. The city would like to repave Low Street between Vine and Elm. Determine the minimum number of vehicles per hour that must use this portion of Low Street in order to avoid congestion.

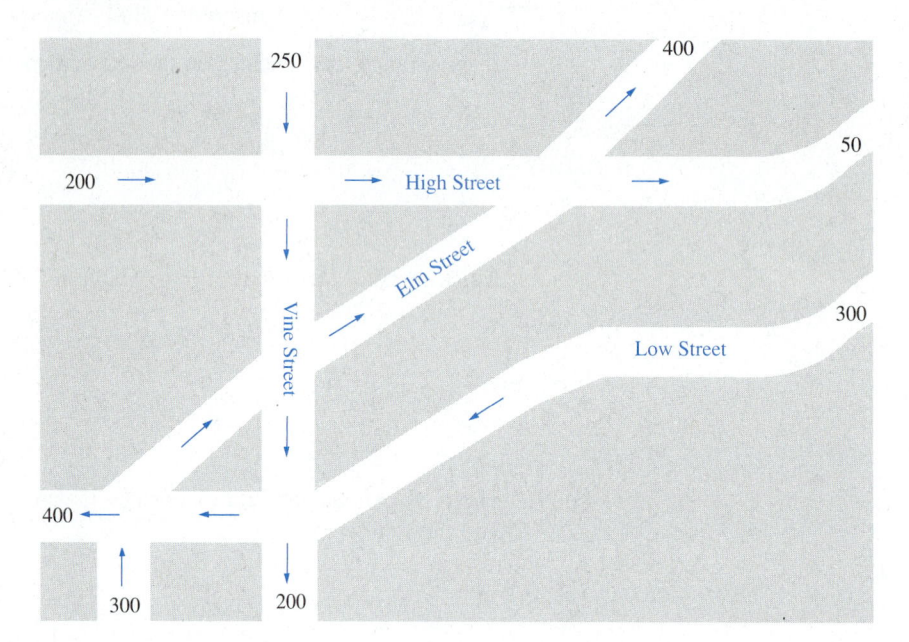

FIGURE 2.7

EXERCISES 2.3

1. The Allegheny Mining Company operates three mines which produce high-, medium-, and low-grade ore. The daily yield of each mine is given in the table below.

	Ore yield per day		
Mine	1	2	3
High-grade	2 tons	1 ton	0 tons
Medium-grade	2 tons	3 tons	1 ton
Low-grade	1 ton	2 tons	2 tons

The company must fill an order for a total of 40 tons of high-grade, 75 tons of medium-grade, and 50 tons of low-grade ore. How many days should each mine be operated to fill this order exactly?

2. The Consolidated Steel Corporation has been ordered to reduce the three main pollutants—particulates, sulfur oxide, and hydrocarbons—emitted at its Pittsburgh plant, by 5400, 6000, and 9600 units per month, respectively. Three filtering methods are avail-

able. The monthly reduction in emissions, per dollar spent, for each of the filtering methods is given below.

	Monthly reduction per dollar spent		
Method	A	B	C
Particulates	9 units	10 units	10 units
Sulfur oxide	15 units	10 units	5 units
Hydrocarbons	15 units	20 units	15 units

How much must the company allocate per month to each method, in order to meet the reduction order exactly?

3. An agricultural supply house makes three types of fertilizer. Type A contains 10% nitrogen and 2% phosphates by weight. Type B contains 5% nitrogen and 1% phosphates, and Type C contains 8% nitrogen and 4% phosphates. The company has received an order for 600 pounds of fertilizer which is to contain 9% nitrogen and 2% phosphates. Can the order be filled by mixing together the three available types? If so, how?

4. Each ounce of Food A contains 3 gm of carbohydrate, 2 gm of protein, and 4 gm of fat, while an ounce of Food B contains 3 gm of carbohydrate, 4 gm of protein, and 2 gm of fat. By combining these foods, is it possible to obtain exactly 75 gm of carbohydrate, 60 gm of protein, and 60 gm of fat?

5. Four grades of crude oil are to be blended in an oil storage tank. The cost and sulfur content of each grade are given in the table below.

Grade	A	B	C	D
Cost per barrel	$20	$25	$40	$50
Sulfur per barrel	125 gm	100 gm	75 gm	25 gm

The tank initially contains 12,000 barrels of crude which costs $15 per barrel and has a sulfur content of 225 gm per barrel. Suppose that x, y, z, and w barrels of the four respective grades are added to the tank to produce a blend which costs $25 per barrel and contains 150 gm per barrel of sulfur.

(a) What equations must x, y, z, and w satisfy?
(b) Solve these equations for x and y in terms of z and w.
(c) What values of z and w will yield a feasible solution?

6. A metallurgist has melted down three bronze alloys, whose composition by weight is given in the table below.

Alloy	A	B	C
Copper	95%	90%	80%
Tin	4%	2%	10%
Zinc	1%	8%	10%

A new alloy is to be formed by adding x, y, and z tons of the three respective alloys to a smelting kiln that already contains 100 tons of Alloy A.

(a) Is it possible to obtain a mixture whose composition is 90% copper, 5% tin, and 5% zinc?
(b) Is the solution unique?

7. A manufacturing firm that produces desk calculators and typewriters has three service departments—data processing, maintenance, and payroll. The table below gives the direct cost of each department and the proportion of services used by each department. Determine the total costs for each department.

User of Services	Direct cost	Provider of Service Data processing	Maintenance	Payroll
Data processing	$3000	0.3	0.1	0.3
Maintenance	$4000	0.2	0.4	0.2
Payroll	$1500	0.1	0.2	0.1
Calculators	$6000	0.2	0.2	0.2
Typewriters	$7000	0.2	0.1	0.2

8. A firm manufactures laboratory equipment and culture media, which it supplies to hospitals. Its three service departments (data processing, research and development (R&D), and electronics) provide services to the five departments as indicated in the table below. Using the direct costs in this table, determine the total costs for each department.

User of Services	Direct cost	Provider of Service Data processing	R&D	Electronics
Data processing	$4,000	0.4	0.0	0.2
R&D	$10,000	0.1	0.8	0.1
Electronics	$3,000	0.2	0.0	0.1
Equipment	$12,000	0.1	0.1	0.5
Media	$15,000	0.2	0.1	0.1

9. A men's clothing factory manufactures trousers, jackets, and suits. These three production departments use services from the accounts receivable, data processing, and maintenance departments as indicated in the table below. Determine the total costs for each department.

User of Services	Direct cost	Accounts receivable	Data processing	Maintenance
Accounts receivable	$22,000	0.1	0.2	0.1
Data processing	$40,000	0.2	0.1	0.3
Maintenance	$2,000	0.1	0.2	0.1
Trousers	$20,000	0.2	0.2	0.1
Jackets	$20,000	0.2	0.1	0.1
Suits	$30,000	0.2	0.2	0.3

10. The diagram at the top of the next page shows the number of vehicles per hour using certain portions of five one-way streets during rush hour. The city would liked to repave East Street between North Avenue and South Avenue. Determine the minimum number of vehicles per hour that must use this portion of East Street in order to avoid congestion.

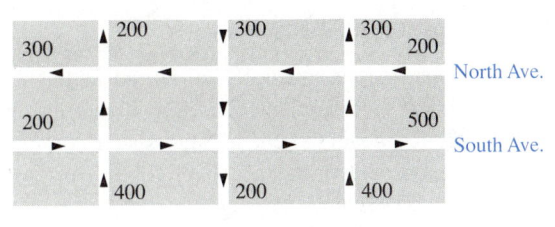

11. The diagram below shows the number of vehicles per hour using certain portions of five one-way streets during rush hour. Determine the minimum number of vehicles per hour that must use Third Avenue between Washington Street and Adams Street to avoid congestion.

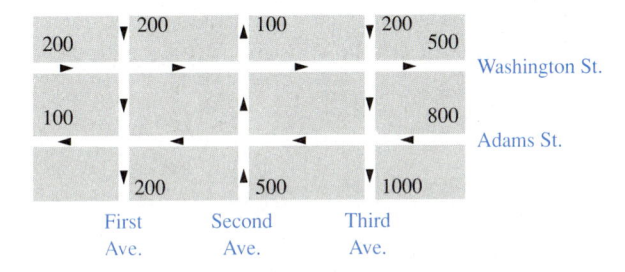

12. The diagram below shows the number of vehicles per hour using certain portions of five one-way streets during rush hour. Determine the minimum number of vehicles per hour that must use Pine Street between Jefferson Avenue and Monroe Avenue in order to avoid congestion.

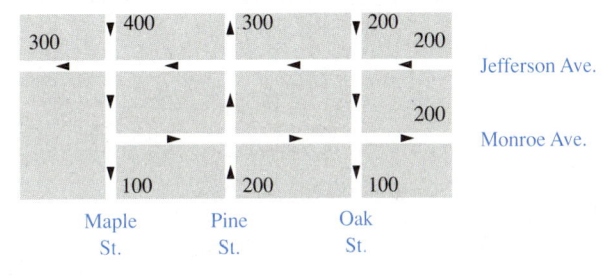

Answers to Practice Problems

7. (a) $70x + 80y = 1090$
$2x + 6y = 46$
(b) 11 slices of bread and 4 eggs

8. (a) $x_1 = 400 + .2x_1 + .1x_2$
$x_2 = 210 + .2x_1 + .2x_2$
(b) $770 to Direct Sales and $940 to Mail Order

9. 100

2.4 Matrices

The first use of matrices is generally attributed to the English mathematician Arthur Cayley in 1858. The widespread use of matrices is much more recent, however. In fact, the development of high-speed computers capable of doing matrix calculations is the primary reason for the present use of matrices in business and the social sciences.

In the remainder of this chapter we introduce matrices and the fundamental matrix operations and show their connection to the solution of linear systems.

MATRICES AND VECTORS

A **matrix** (plural: matrices) is simply a rectangular array of numbers, enclosed in brackets. (Some authors use parentheses.) The numbers are called the **entries** of the matrix. The following are three examples.

$$\begin{bmatrix} 3 & 5 & -2 \\ 4 & 0 & 0 \\ 0 & 22 & \frac{1}{2} \end{bmatrix} \quad \begin{bmatrix} 3.4 & 3.3 & 5.6 & 1.2 \\ 9.1 & 0.0 & -1.3 & 0.4 \end{bmatrix} \quad \begin{bmatrix} 76 \\ 44 \\ 9 \\ 0 \\ 3 \end{bmatrix}$$

The **size** of a matrix is determined by the number of rows and columns it has. A matrix with m rows and n columns is said to be an $m \times n$ **matrix** (read "m by n"). For example, the three matrices above are 3×3, 2×4, and 5×1 matrices. We have already dealt with matrices under a different name. Example 2.10 in the previous section involved the linear system

$$0.7x_1 - 0.3x_2 - 0.2x_3 = 1000$$
$$-0.3x_1 + 0.7x_2 - 0.2x_3 = 800$$
$$-0.1x_1 - 0.1x_2 + 0.6x_3 = 1200,$$

which we wrote in tableau form as follows.

$$\left[\begin{array}{rrr|r} 0.7 & -0.3 & -0.2 & 1000 \\ -0.3 & 0.7 & -0.2 & 800 \\ -0.1 & -0.1 & 0.6 & 1200 \end{array} \right]$$

This can just as well be regarded as a matrix, and in fact we call this the **augmented matrix** of the system of equations. Vertical or horizontal lines are often used to segregate certain entries of a matrix, as the vertical line in the last matrix separates the coefficients in our linear system from the constants. They have no effect on matrix definitions or operations.

Two matrices are considered to be **equal** when they have the same size and all their corresponding entries are equal. No two of the following matrices are equal.

$$\begin{bmatrix} 2 & 3 \\ 1 & 4 \\ 0 & 2 \end{bmatrix} \quad \begin{bmatrix} 2 & 1 & 0 \\ 3 & 4 & 2 \end{bmatrix} \quad \begin{bmatrix} 2 & 1 & 0 \\ 3 & 5 & 2 \end{bmatrix}$$

The first two are not equal because they have different sizes; the first is 3×2 and the second is 2×3. The second and third matrices are not equal because the entries in the second row and second column are not the same. In general the entry in row i and column j of a matrix is called the **i,j entry** of the matrix. For example the 3,2 entry of the first matrix above is 2, and its 1,2 entry is 3.

Often matrices are denoted by a single letter. For example we could define

$$A = \begin{bmatrix} 2 & 1 & 0 \\ 3 & 5 & 2 \end{bmatrix}.$$

We will usually use capital letters to denote matrices and small letters for their entries. Sometimes a system of double subscripts is used to name the entries in a matrix. We might write

$$A = \begin{bmatrix} 2 & 1 & 0 \\ 3 & 5 & 2 \end{bmatrix} = \begin{bmatrix} a_{11} & a_{12} & a_{13} \\ a_{21} & a_{22} & a_{23} \end{bmatrix},$$

where $a_{11} = 2$, $a_{12} = 1$, etc. Notice that in the notation a_{ij} the first subscript i is the row number, and the second subscript j is the column number. We may indicate this situation symbolically by writing $A = [a_{ij}]$; this means that A is a matrix whose i,j entry is denoted by a_{ij}.

EXAMPLE 2.12 Suppose we have

$$B = [b_{ij}] = \begin{bmatrix} 3 & 5 & 6 & -1 \\ 4 & -3 & 0 & 7 \end{bmatrix} = \begin{bmatrix} 5-x & 5 & 6 & -1 \\ 4 & -3 & 0 & 7 \end{bmatrix}.$$

(a) What is the size of B?
(b) What is the 2,3 entry of B? What is b_{14}?
(c) What is x?

Solution (a) Since B has 2 rows and 4 columns, B is a 2×4 matrix.
(b) The 2,3 entry of B is 0, and $b_{14} = -1$.
(c) Since for two matrices to be equal their corresponding entries must be the same, we have $3 = 5 - x$. Thus $x = 2$. ∎

Matrices with a single row or column are given special names. A matrix with a single row is called a **row vector,** and a matrix with a single column is called a **column vector.** For example, the 1×5 matrix C below is a row vector, and the 3×1 matrix D is a column vector.

$$C = \begin{bmatrix} 3 & 4 & 11 & -2 & 6 \end{bmatrix} \qquad D = \begin{bmatrix} 81 \\ 43 \\ 30 \end{bmatrix}.$$

MATRIX OPERATIONS

We compute the **sum** of two matrices of the same size by adding their corresponding entries. For example if

$$A = \begin{bmatrix} 1 & -1 & 3 \\ 4 & 5 & -2 \end{bmatrix} \quad \text{and} \quad B = \begin{bmatrix} 3 & 4 & -2 \\ 3 & 0 & -1 \end{bmatrix},$$

then

$$A + B = \begin{bmatrix} 1+3 & -1+4 & 3+(-2) \\ 4+3 & 5+0 & -2+(-1) \end{bmatrix} = \begin{bmatrix} 4 & 3 & 1 \\ 7 & 5 & -3 \end{bmatrix}.$$

We can indicate the definition of matrix addition symbolically by

$$[a_{ij}] + [b_{ij}] = [a_{ij} + b_{ij}].$$

Practice Problem 10 Compute

$$\begin{bmatrix} 3 & 0 \\ -1 & 5 \\ -2 & 1 \end{bmatrix} + \begin{bmatrix} 2 & 2 \\ 6 & 4 \\ 5 & -3 \end{bmatrix}.$$

EXAMPLE 2.13 The matrices P_0 and P_1 below give the number of millions of tons of various pollutants emitted into the air of the United States by various sources during the years 1970 and 1971, respectively.

	Transportation	Stationary fuel combustion	Industrial processes	Solid waste disposal	Miscellaneous	
Carbon monoxide	82.3	1.1	11.8	5.5	6.6	
Sulfur oxides	0.7	27.0	6.4	0.1	0.1	
Hydrocarbons	14.7	1.6	2.9	1.4	11.5	$= P_0$
Particulates	1.2	8.3	15.7	1.1	1.2	
Nitrogen oxides	9.3	10.1	0.6	0.3	0.1	

$$\begin{array}{c} \\ \text{Carbon monoxide} \\ \text{Sulfur oxides} \\ \text{Hydrocarbons} \\ \text{Particulates} \\ \text{Nitrogen oxides} \end{array} \begin{array}{c} \text{Transportation} \\ \begin{bmatrix} 77.5 \\ 1.0 \\ 14.7 \\ 1.0 \\ 11.2 \end{bmatrix} \end{array} \begin{array}{c} \text{Stationary} \\ \text{fuel} \\ \text{combustion} \\ 1.0 \\ 26.3 \\ 0.3 \\ 6.5 \\ 10.2 \end{array} \begin{array}{c} \text{Industrial} \\ \text{processes} \\ 11.4 \\ 5.1 \\ 5.6 \\ 13.5 \\ 0.2 \end{array} \begin{array}{c} \text{Solid} \\ \text{waste} \\ \text{disposal} \\ 3.8 \\ 0.1 \\ 1.0 \\ 0.7 \\ 0.2 \end{array} \begin{array}{c} \text{Miscel-} \\ \text{laneous} \\ 4.9 \\ 0.1 \\ 4.7 \\ 4.9 \\ 0.2 \end{array}\begin{array}{c}\\ \\ \\ \end{array} = P_1$$

Compute $P_0 + P_1$, and interpret the meaning of its entries.

Solution By adding the corresponding entries, we get

$$P_0 + P_1 =$$

$$\begin{array}{c} \\ \text{Carbon monoxide} \\ \text{Sulfur oxides} \\ \text{Hydrocarbons} \\ \text{Particulates} \\ \text{Nitrogen oxides} \end{array} \begin{array}{c} \text{Transportation} \\ \begin{bmatrix} 159.8 \\ 1.7 \\ 29.4 \\ 2.2 \\ 20.5 \end{bmatrix} \end{array} \begin{array}{c} \text{Stationary} \\ \text{fuel} \\ \text{combustion} \\ 2.1 \\ 53.3 \\ 1.9 \\ 14.8 \\ 20.3 \end{array} \begin{array}{c} \text{Industrial} \\ \text{processes} \\ 23.2 \\ 11.5 \\ 8.5 \\ 29.2 \\ 0.8 \end{array} \begin{array}{c} \text{Solid} \\ \text{waste} \\ \text{disposal} \\ 9.3 \\ 0.2 \\ 2.4 \\ 1.8 \\ 0.5 \end{array} \begin{array}{c} \text{Miscel-} \\ \text{laneous} \\ 11.5 \\ 0.2 \\ 16.2 \\ 6.1 \\ 0.3 \end{array}\begin{array}{c}\\ \\ \\ \end{array}.$$

The entries in this matrix represent the total number of millions of tons of pollutants emitted by different sources in the years 1970 and 1971 combined. ■

The **difference** between two matrices of the same size is formed by subtracting their corresponding entries. Thus

$$\begin{bmatrix} 2 & 3 & -1 \\ 5 & 3 & 9 \end{bmatrix} - \begin{bmatrix} 1 & 8 & -6 \\ 7 & -2 & 0 \end{bmatrix} = \begin{bmatrix} 2-1 & 3-8 & -1-(-6) \\ 5-7 & 3-(-2) & 9-0 \end{bmatrix}$$

$$= \begin{bmatrix} 1 & -5 & 5 \\ -2 & 5 & 9 \end{bmatrix}.$$

When we talk about matrices, the word **scalar** is used to denote a real number, as opposed to a matrix. If c is a scalar and A is a matrix, we define the **product** cA of c and A to be the matrix formed by multiplying each entry of A by c. For example

$$3 \begin{bmatrix} 2 & 3 & -1 \\ 5 & 3 & 9 \end{bmatrix} = \begin{bmatrix} 3(2) & 3(3) & 3(-1) \\ 3(5) & 3(3) & 3(9) \end{bmatrix} = \begin{bmatrix} 6 & 9 & -3 \\ 15 & 9 & 27 \end{bmatrix}.$$

Matrix subtraction and multiplication of a matrix by a scalar can be defined symbolically by

$$[a_{ij}] - [b_{ij}] = [a_{ij} - b_{ij}] \quad \text{and} \quad c[a_{ij}] = [ca_{ij}].$$

It is easily checked that

$$A - B = A + (-1)B.$$

Notice that matrices A and B can only be added or subtracted when they are of the same size.

Practice Problem 11 Compute $2A - 3B$, where

$$A = \begin{bmatrix} 1 & 3 & -1 \\ 0 & 2 & 1 \end{bmatrix} \quad \text{and} \quad B = \begin{bmatrix} -2 & 0 & 1 \\ 3 & -3 & 5 \end{bmatrix}.$$

EXAMPLE 2.14 (a) Use the matrices P_0 and P_1 of Example 2.13 to compute a matrix giving the decrease from 1970 to 1971 in the number of millions of tons of each pollutant from each source listed.

(b) Suppose the goal for 1990 is to have each source decrease the amount of each pollutant by 40% from its 1970 level. Compute a matrix giving the goal for each pollutant and source in millions of tons.

Solution (a) The decrease from 1970 to 1971 for each source and pollutant will be given by the entries in the matrix

$$P_0 - P_1 =$$

	Transportation	Stationary fuel combustion	Industrial processes	Solid waste disposal	Miscellaneous
Carbon monoxide	4.8	0.1	0.4	1.7	1.7
Sulfur oxides	−0.3	0.7	1.3	0.0	0.0
Hydrocarbons	0.0	1.3	−2.7	0.4	6.8
Particulates	0.2	1.8	2.2	0.4	−3.7
Nitrogen oxides	−1.9	−0.1	0.4	0.1	−0.1

Of course negative entries indicate an *increase* between 1970 and 1971.

(b) If the amount of each pollutant from each source is to decrease by 40%, then this amount will be 60%, or 0.6 times its 1970 level. Thus we compute the matrix

$$0.6 P_0 =$$

	Transportation	Stationary fuel combustion	Industrial processes	Solid waste disposal	Miscellaneous
Carbon monoxide	⎡ 49.38	0.66	7.08	3.30	3.96 ⎤
Sulfur oxides	⎢ 0.42	16.20	3.84	0.06	0.06 ⎥
Hydrocarbons	⎢ 8.82	0.96	1.74	0.84	6.90 ⎥.
Particulates	⎢ 0.72	4.98	9.42	0.66	0.72 ⎥
Nitrogen oxides	⎣ 5.58	6.06	0.36	0.18	0.06 ⎦

We could also have computed $P_0 - 0.4P_0$, which would have given the same result. ■

MATRIX ALGEBRA

In the last example we claimed that the matrices $0.6P_0$ and $P_0 - 0.4P_0$ were the same. We might try to justify this with the calculation

$$P_0 - 0.4P_0 = 1P_0 - 0.4P_0$$
$$= (1 - 0.4)P_0 \qquad (2.14)$$
$$= 0.6P_0.$$

If P_0 were a real number these steps would certainly be valid, but since P_0 is a matrix we need to be careful. Actually, matrices obey many algebraic laws similar to those for real numbers. Some of these laws involving matrix addition and multiplication by a scalar are given below. In these rules whenever matrices are added they are assumed to be of the same size. The symbol O stands for a **zero matrix** of the appropriate size, that is, a matrix in which all the entries are 0.

Laws for Matrix Addition and Multiplication by a Scalar

M1. $A + B = B + A$
M2. $A + (B + C) = (A + B) + C$
M3. $A + O = A$
M4. $k(A + B) = kA + kB$
M5. $(c + d)A = cA + dA$
M6. $c(dA) = (cd)A$
M7. $1A = A$

The reader should check that the first two equations of (2.14) are justified by M7 and M5. Other familiar-looking rules, such as $c(A - B) = cA - cB$ and $0A = O$, may be derived from M1 through M7. In the next section we will explain how to multiply matrices of appropriate sizes. We will find that matrix multiplication satisfies some, but not all, of the real number multiplication properties.

EXAMPLE 2.15 Find a 2 × 2 matrix X satisfying the equation

$$2(B + 3X) = 3B - 4(A - X),$$

where

$$A = \begin{bmatrix} 1 & -5 \\ 3 & 0 \end{bmatrix} \quad \text{and} \quad B = \begin{bmatrix} 8 & -6 \\ 4 & 2 \end{bmatrix}.$$

Solution First, we solve the equation for the unknown X.

$$2(B + 3X) = 3B - 4(A - X)$$
$$2B + 6X = 3B - 4A + 4X$$
$$2X = B - 4A$$
$$X = \frac{1}{2}(B - 4A) = \frac{1}{2}B - 2A$$
$$X = \frac{1}{2}\begin{bmatrix} 8 & -6 \\ 4 & 2 \end{bmatrix} - 2\begin{bmatrix} 1 & -5 \\ 3 & 0 \end{bmatrix}$$
$$= \begin{bmatrix} 4 & -3 \\ 2 & 1 \end{bmatrix} - \begin{bmatrix} 2 & -10 \\ 6 & 0 \end{bmatrix} = \begin{bmatrix} 2 & 7 \\ -4 & 1 \end{bmatrix}$$

We remark (without proof) that each of the steps used in solving the equation—such as multiplying out and transposing—can be justified on the basis of the laws M1–M7. ∎

EXERCISES 2.4

Exercises 1–16 refer to the matrices

$$A = \begin{bmatrix} 3 & -1 & 0 & 4 \\ -3 & 5 & 2 & 1 \end{bmatrix}, \quad B = \begin{bmatrix} 6 & 3 & 11 & 8 \\ 2 & -1 & 0 & 3 \end{bmatrix},$$

$$C = \begin{bmatrix} 2 & 4 \\ 6 & 7 \\ 3 & -2 \end{bmatrix}, \quad D = \begin{bmatrix} 4 & 9 \\ -1 & 0 \\ 2 & 3 \end{bmatrix}.$$

In Exercises 1–12 compute the expression, if it is defined.

1. $A + B$
2. $C + D$
3. $A + C$
4. $B + D$
5. $B - A$
6. $D - C$
7. $3B$
8. $-2C$
9. $2A + B$
10. $D + 3A$
11. $2B + 4A$
12. $5D - 2C$

In Exercises 13–16 solve for the matrix X.

13. $X + 2A = B$

14. $3B - X = A$

15. $3A + 2X = 4B$

16. $2(X - 2A) = 3B$

In Exercises 17–22 write the augmented matrix of the given linear system.

17. $x + 3y + z = 5$
$x + 7y - z = 0$

18. $x - y + 2z = 6$
$2x + 2y - 3z = 8$
$3x + 2y + z = 5$

19. $x + 3y = 9$
$x + 7z = 12$

20. $x + y = 1$
$y + z = 3$

21. $x_1 + 2x_3 = -3$
$x_2 - x_4 = 5$
$x_1 + 4x_3 - x_2 = 7$

22. $ a + b = c$
$2a - b + c = 4$
$4a + 4c = b$

23. Write a row vector with 5 entries, all 3's.

24. Write a column vector with 4 entries, all 2's.

25. Write a 3 × 2 matrix with all entries equal to 5.

26. Write a 2 × 4 matrix with all entries equal to 3.

The matrices E and F below give the average monthly cost in dollars of housing and food for a family of 4 in 1986 and 1987 in the cities of Elkville and Foxtown, respectively. All questions in Exercises 27–32 refer to monthly costs.

$$E = \begin{bmatrix} \text{housing} & \text{food} \\ 300 & 420 \\ 335 & 440 \end{bmatrix} \begin{matrix} \\ 1986 \\ 1987 \end{matrix} \qquad F = \begin{bmatrix} \text{housing} & \text{food} \\ 275 & 350 \\ 300 & 370 \end{bmatrix} \begin{matrix} \\ 1986 \\ 1987 \end{matrix}$$

27. What is the average housing cost for a family of four in Foxtown in 1987?

28. What is the average food cost for a family of four in Elkville in 1986?

29. Express in terms of E and F a matrix giving the average excess cost for housing and food for a family living in Elkville instead of Foxtown in 1986 and 1987.

30. Express in terms of E and F the average cost for three families for housing and food in Elkville in 1986 and 1987.

31. Express in terms of E and F the average housing and food cost per family for 2 families to live in Foxtown and 3 families in Elkville in 1986 and 1987.

32. Express in terms of E and F the housing and food cost for a family in Foxtown for 1986 and 1987 if the state welfare department pays 20% of their housing and food bills.

In Elkville, housing cost a family of four $300 per month in 1986 and $335 in 1987.

Answers to Practice Problems

10. $\begin{bmatrix} 5 & 2 \\ 5 & 9 \\ 3 & -2 \end{bmatrix}$

11. $\begin{bmatrix} 8 & 6 & -5 \\ -9 & 13 & -13 \end{bmatrix}$

2.5 Matrix Multiplication

In this section we will define the product of two matrices. The definition may seem unnatural and unnecessarily complicated at first, but it turns out to be very useful. We will start by defining the product AB when A is a $1 \times n$ matrix (row vector) and B is an $n \times 1$ matrix (column vector). The **product** AB is defined to be a 1×1 matrix with a single entry, formed by multiplying the corresponding entries of A and B and adding the resulting products. Symbolically,

$$[a_1 \ a_2 \ \ldots \ a_n] \begin{bmatrix} b_1 \\ b_2 \\ \cdot \\ \cdot \\ \cdot \\ b_n \end{bmatrix} = [a_1 b_1 + a_2 b_2 + \ldots + a_n b_n].$$

For example

$$[2 \ -3 \ 4 \ 1] \begin{bmatrix} 3 \\ 3 \\ 0 \\ 4 \end{bmatrix} = [2 \cdot 3 + (-3)3 + 4 \cdot 0 + 1 \cdot 4]$$

$$= [6 - 9 + 0 + 4] = [1].$$

Sums of products of this sort arise in many situations. For example, the Sporting Chance, a retail sporting goods store, finds that its downtown store has 20 metal, 12 graphite, and 48 wood tennis rackets in stock. We could summarize this as a row matrix, namely

$$\begin{array}{cccc} & \text{metal} & \text{graphite} & \text{wood} \\ R = [& 20 & 12 & 48 \ ]. \end{array}$$

The metal rackets retail for $60, the graphite for $110, and the wood rackets for $40. We put these prices in a column vector:

$$C = \begin{array}{c} \text{metal} \\ \text{graphite} \\ \text{wood} \end{array} \begin{bmatrix} 60 \\ 110 \\ 40 \end{bmatrix}.$$

Notice now that the total retail value of all the rackets is

$$20(\$60) + 12(\$110) + 48(\$40) = \$1200 + \$1320 + \$1920 = \$4440.$$

We can get the same answer by computing RC.

$$RC = [20 \ 12 \ 48] \begin{bmatrix} 60 \\ 110 \\ 40 \end{bmatrix}$$

$$= [20 \cdot 60 + 12 \cdot 110 + 48 \cdot 40] = [4440]$$

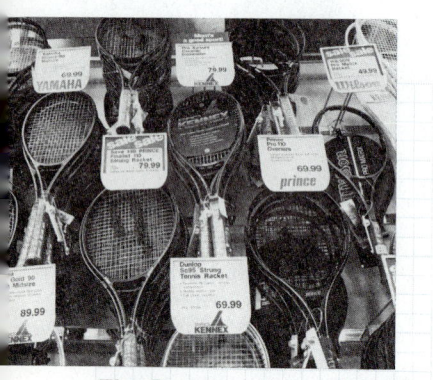

The Sporting Chance has 20 metal, 12 graphite, and 48 wood rackets.

EXAMPLE 2.16

Compute AB, where

$$A = \begin{bmatrix} -1 & 2 & 2 & 0 \end{bmatrix} \quad \text{and} \quad B = \begin{bmatrix} 4 \\ 3 \\ 2 \\ 5 \end{bmatrix}.$$

Solution The answer is the 1×1 matrix $[-1 \cdot 4 + 2 \cdot 3 + 2 \cdot 2 + 0 \cdot 5] = [6]$. ∎

We will now show how to multiply matrices that are not necessarily row or column vectors. We will give several examples before giving the formal definition of matrix multiplication.

The Sporting Chance actually has three stores in all. Besides the downtown store there is also a suburban store and a store in the mall. The numbers of metal, graphite, and wood tennis rackets at each store are listed in the following matrix.

$$\begin{array}{c} \\ \text{downtown} \\ \text{suburban} \\ \text{mall} \end{array} \begin{array}{ccc} \text{metal} & \text{graphite} & \text{wood} \end{array} \\ \begin{bmatrix} 20 & 12 & 48 \\ 16 & 9 & 30 \\ 25 & 18 & 36 \end{bmatrix}$$

To compute the total retail value of the rackets at each store, we need to do three calculations like the one above. We organize them as follows.

$$\begin{bmatrix} 20 & 12 & 48 \\ 16 & 9 & 30 \\ 25 & 18 & 36 \end{bmatrix} \begin{bmatrix} 60 \\ 110 \\ 40 \end{bmatrix} = \begin{bmatrix} 20 \cdot 60 + 12 \cdot 110 + 48 \cdot 40 \\ 16 \cdot 60 + 9 \cdot 110 + 30 \cdot 40 \\ 25 \cdot 60 + 18 \cdot 110 + 36 \cdot 40 \end{bmatrix} = \begin{bmatrix} 4440 \\ 3150 \\ 4920 \end{bmatrix}$$

We have introduced some horizontal lines in the matrix on the left to indicate how the rows are used one at a time, like row vectors. Each row on the left contributes to one entry in the 3×1 answer, which indicates that the rackets have a retail value of $4400 at the downtown store, $3150 at the suburban store, and $4920 at the mall store. This example shows how to compute a product AB, where B is a column vector with as many entries as the matrix A has columns.

EXAMPLE 2.17

Compute AB, where

$$A = \begin{bmatrix} 3 & 0 \\ 2 & 3 \\ -5 & 2 \\ 4 & -1 \end{bmatrix} \quad \text{and} \quad B = \begin{bmatrix} 5 \\ -2 \end{bmatrix}.$$

Solution Using the rows of A one at a time, we get

$$AB = \begin{bmatrix} 3 & 0 \\ 2 & 3 \\ -5 & 2 \\ 4 & -1 \end{bmatrix} \begin{bmatrix} 5 \\ -2 \end{bmatrix}$$

$$= \begin{bmatrix} 3 \cdot 5 + 0(-2) \\ 2 \cdot 5 + 3(-2) \\ (-5)5 + 2(-2) \\ 4 \cdot 5 + (-1)(-2) \end{bmatrix} = \begin{bmatrix} 15 \\ 4 \\ -29 \\ 22 \end{bmatrix}. \blacksquare$$

Practice Problem 12 Compute AB, where

$$A = \begin{bmatrix} 1 & 3 & 2 \\ 0 & -1 & 1 \end{bmatrix} \quad \text{and} \quad B = \begin{bmatrix} -1 \\ 3 \\ 0 \end{bmatrix}.$$

Now we return to our example of the sporting goods stores. For insurance purposes the Sporting Chance needs to know not only the retail, but also the total *wholesale* value of the tennis rackets at each store. The wholesale prices of metal, graphite, and wood rackets are \$35, \$60, and \$25, respectively. These and the retail prices can be summarized in a matrix as follows.

$$\begin{array}{c} \\ \text{metal} \\ \text{graphite} \\ \text{wood} \end{array} \begin{array}{cc} \text{retail} & \text{wholesale} \\ \begin{bmatrix} 60 & 35 \\ 110 & 60 \\ 40 & 25 \end{bmatrix} \end{array}$$

Now we calculate both the retail and wholesale values of the rackets at each store by using both columns of the last matrix. The calculation goes as follows.

$$\begin{array}{c} \\ \text{downtown} \\ \text{suburban} \\ \text{mall} \end{array} \begin{array}{ccc} \text{metal} & \text{graphite} & \text{wood} \\ \begin{bmatrix} 20 & 12 & 48 \\ 16 & 9 & 30 \\ 25 & 18 & 36 \end{bmatrix} \end{array} \begin{array}{cc} \text{retail} & \text{wholesale} \\ \begin{bmatrix} 60 & 35 \\ 110 & 60 \\ 40 & 25 \end{bmatrix} \end{array} \begin{array}{c} \\ \text{metal} \\ \text{graphite} \\ \text{wood} \end{array}$$

$$= \begin{bmatrix} 20 \cdot 60 + 12 \cdot 110 + 48 \cdot 40 & 20 \cdot 35 + 12 \cdot 60 + 48 \cdot 25 \\ 16 \cdot 60 + 9 \cdot 110 + 30 \cdot 40 & 16 \cdot 35 + 9 \cdot 60 + 30 \cdot 25 \\ 25 \cdot 60 + 18 \cdot 110 + 36 \cdot 40 & 25 \cdot 35 + 18 \cdot 60 + 36 \cdot 25 \end{bmatrix}$$

$$= \begin{array}{c} \\ \text{downtown} \\ \text{suburban} \\ \text{mall} \end{array} \begin{array}{cc} \text{retail} & \text{wholesale} \\ \begin{bmatrix} 4440 & 2620 \\ 3150 & 1850 \\ 4920 & 2855 \end{bmatrix} \end{array}$$

THE GENERAL DEFINITION OF MATRIX MULTIPLICATION

The entries of the **product** AB of the matrices A and B are computed by multiplying the rows of A, regarded as row vectors, by the columns of B, regarded as column vectors. More precisely, the i,j entry of AB is the product of the ith row of A and the jth column of B. This is indicated graphically in Figure 2.8.

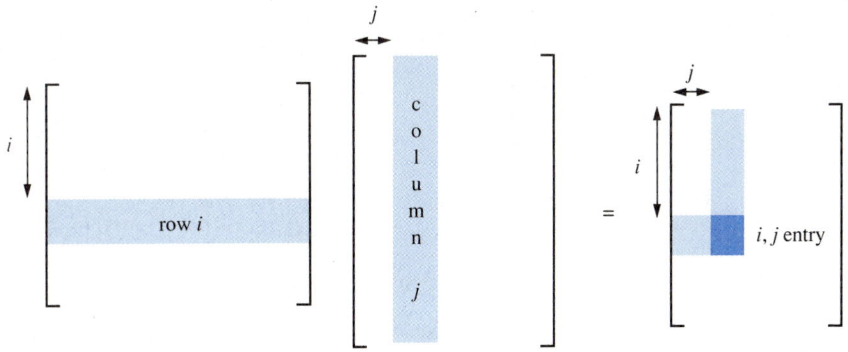

2.5 MATRIX MULTIPLICATION

FIGURE 2.8

Notice that in order to multiply a row by a column, they must have the same number of entries. This means that in order for the product AB to be defined, the number of columns of A must equal the number of rows of B.

Suppose A is an $m \times n$ matrix and B is an $r \times s$ matrix.

1. The product AB is defined if and only if $n = r$.
2. If AB is defined, then AB is an $m \times s$ matrix.

EXAMPLE 2.18 Compute any of the products AB, AC, BA, BC, CA, and CB that are defined, where

$$A = \begin{bmatrix} 3 & 0 \\ 4 & -2 \\ 1 & 6 \end{bmatrix}, \quad B = \begin{bmatrix} -3 & 8 \\ 5 & -1 \end{bmatrix}, \quad \text{and} \quad C = \begin{bmatrix} 0 & 2 & 9 \\ 4 & -5 & 7 \end{bmatrix}.$$

Solution Notice that A, B, and C are 3×2, 2×2, and 2×3 matrices, respectively. Since in order for an $m \times n$ matrix times an $r \times s$ matrix to be defined we need that $n = r$, only AB, AC, BC, and CA are defined. We will show the computation of AB and AC.

$$AB = \begin{bmatrix} 3 & 0 \\ 4 & -2 \\ 1 & 6 \end{bmatrix} \begin{bmatrix} -3 & 8 \\ 5 & -1 \end{bmatrix}$$

$$= \begin{bmatrix} 3(-3) + 0(5) & 3(8) + 0(-1) \\ 4(-3) + (-2)5 & 4(8) + (-2)(-1) \\ 1(-3) + 6(5) & 1(8) + 6(-1) \end{bmatrix}$$

$$= \begin{bmatrix} -9 & 24 \\ -22 & 34 \\ 27 & 2 \end{bmatrix}$$

$$AC = \begin{bmatrix} 3 & 0 \\ 4 & -2 \\ 1 & 6 \end{bmatrix} \begin{bmatrix} 0 & 2 & 9 \\ 4 & -5 & 7 \end{bmatrix}$$

$$= \begin{bmatrix} 3(0) + 0(4) & 3(2) + 0(-5) & 3(9) + 0(7) \\ 4(0) + (-2)4 & 4(2) + (-2)(-5) & 4(9) + (-2)7 \\ 1(0) + 6(4) & 1(2) + 6(-5) & 1(9) + 6(7) \end{bmatrix}$$

$$= \begin{bmatrix} 0 & 6 & 27 \\ -8 & 18 & 22 \\ 24 & -28 & 51 \end{bmatrix}$$

The reader should check that

$$BC = \begin{bmatrix} 32 & -46 & 29 \\ -4 & 15 & 38 \end{bmatrix} \quad \text{and} \quad CA = \begin{bmatrix} 17 & 50 \\ -1 & 52 \end{bmatrix}. \quad \blacksquare$$

Practice Problem 13 Compute AB, where

$$A = \begin{bmatrix} 1 & 2 \\ 3 & -1 \end{bmatrix} \quad \text{and} \quad B = \begin{bmatrix} 1 & 0 & -1 \\ 2 & 3 & 0 \end{bmatrix}.$$

PROPERTIES OF MATRIX MULTIPLICATION

We have seen that, given matrices A and B, the product AB may or may not be defined. And if AB is defined, BA may not be; in fact this was the case in Example 2.18. Even when two matrices can be multiplied in either order, the results are usually not the same. We saw in Example 2.18 that the products AC and CA were both defined, but not equal. In fact AC was a 3×3 matrix, while CA was a 2×2 matrix. According to law M1 of the previous section, whenever A and B have the same size, then $A + B = B + A$. As we have seen, however, the corresponding property for matrix multiplication need not be true, even when both products are defined. Nevertheless, matrix multiplication does obey many familiar rules.

Laws of Matrix Multiplication

M8. $A(BC) = (AB)C$
M9. $A(B + C) = AB + AC$
M10. $(A + B)C = AC + BC$
M11. $c(AB) = (cA)B = A(cB)$

It is assumed in the above laws that A, B, and C are matrices, c is a scalar, and the matrix sums and products indicated are all defined.

EXERCISES 2.5

Exercises 1–24 refer to the following matrices.

$$A = \begin{bmatrix} 3 & -2 & 2 \end{bmatrix} \quad B = \begin{bmatrix} -1 & 0 & 5 \end{bmatrix} \quad C = \begin{bmatrix} 4 \\ 3 \\ 0 \end{bmatrix} \quad D = \begin{bmatrix} 3 \\ -2 \\ 1 \end{bmatrix}$$

$$E = \begin{bmatrix} 2 & -1 & 5 \\ 3 & 1 & 0 \\ 4 & 0 & 3 \end{bmatrix} \quad F = \begin{bmatrix} 2 & 0 & -4 \\ 2 & 9 & 3 \\ 3 & 1 & 0 \end{bmatrix} \quad G = \begin{bmatrix} 4 & 0 \\ 0 & -2 \\ -3 & 6 \end{bmatrix} \quad H = \begin{bmatrix} 0 & 3 & -3 \\ 5 & 2 & 8 \end{bmatrix}$$

In Exercises 1–24 compute each indicated matrix product that is defined.

1. AB
2. BC
3. AD
4. BD
5. EC
6. ED
7. FC
8. FD
9. HC
10. HD
11. EF
12. FE
13. EG
14. GE
15. EH
16. HE
17. GH
18. HG
19. BE
20. BF
21. CA
22. CB
23. (EF)G
24. E(FG)

Exercises 25–27 refer to the following three matrices.

$$A = \begin{bmatrix} 2 & .5 \\ 2 & 3 \\ 1 & 2 \end{bmatrix} \begin{matrix} \text{cups sugar} \\ \text{cups flour} \\ \text{number eggs} \end{matrix} \quad B = \begin{bmatrix} 13 \\ 17 \end{bmatrix} \begin{matrix} \text{number cakes} \\ \text{number loaves} \end{matrix} \quad C = \begin{bmatrix} .10 & .08 & .06 \\ .09 & .08 & .07 \end{bmatrix} \begin{matrix} \text{Chicago cost} \\ \text{Detroit cost} \end{matrix}$$

(columns of A: cake, bread; columns of C: cup sugar, cup flour, one egg)

Matrix A gives the amount of sugar, flour and eggs needed to make a cake or a loaf of bread. For example a loaf of bread uses 2 eggs. Matrix B indicates that 13 cakes and 17 loaves of bread are needed. Matrix C gives the cost of a unit of each ingredient, both in Chicago and Detroit. In Exercises 25–27 indicate symbolically a matrix product presenting the information listed, then compute the product.

25. The amounts of sugar, flour, and eggs to make 13 cakes and 17 loaves of bread.
26. The cost of ingredients for making a cake and for making a loaf of bread in Chicago and Detroit.
27. The ingredient cost of making 13 cakes and 17 loaves of bread, both for Chicago and Detroit.

Exercises 28–31 refer to the following four matrices.

$$W = \begin{bmatrix} .7 & .4 \end{bmatrix} \begin{matrix} \text{germination} \end{matrix} \quad \begin{matrix} \text{bluegrass} & \text{rye} \end{matrix}$$

$$X = \begin{bmatrix} .75 & .60 \\ .25 & .40 \end{bmatrix} \begin{matrix} \text{bluegrass} \\ \text{rye} \end{matrix} \quad \begin{matrix} \text{mix A} & \text{mix B} \end{matrix}$$

$$Y = \begin{bmatrix} 30 \\ 50 \end{bmatrix} \begin{matrix} \text{lbs mix A} \\ \text{lbs mix B} \end{matrix} \quad Z = \begin{bmatrix} 3 & 2 \\ 4.5 & 3 \end{bmatrix} \begin{matrix} \text{wholesale} \\ \text{retail} \end{matrix} \quad \begin{matrix} \text{mix A} & \text{mix B} \end{matrix}$$

Matrix W gives the proportion of bluegrass and rye seed that will germinate. Matrix X indicates that mix A contains 75% bluegrass and 25% rye, while mix B contains 60% bluegrass and 40% rye. Matrix Y indicates the number of pounds of the two mixes to be used in a city park. Matrix Z gives the wholesale and retail cost in dollars of a pound of each mix. In Exercises 28–31 indicate symbolically a matrix product presenting the information listed, then compute the product.

28. The number of pounds of bluegrass and rye to be used in the park.
29. The wholesale and retail cost of the seed used in the park.
30. The proportion of mix A and mix B that will germinate.
31. The number of pounds of seed used in the park that will germinate.

Answers to Practice Problems 12. $\begin{bmatrix} 8 \\ -3 \end{bmatrix}$ 13. $\begin{bmatrix} 5 & 6 & -1 \\ 1 & -3 & -3 \end{bmatrix}$

2.6 Inverse Matrices

In this section we will explore more of the algebra of matrix multiplication and show its connection with the solution of linear systems.

SQUARE MATRICES

The product of two matrices of different sizes can have a different size than either of the original two. For example, the product of a 3×2 matrix and a 2×4 matrix is a 3×4 matrix. If we restrict ourselves to matrices with the same number of rows as columns, that is, to **square matrices,** then the situation is much simpler. The product of an $m \times m$ matrix and an $n \times n$ matrix is only defined when $m = n$, and then the result is another matrix of the same size. For square matrices of a given size, one matrix is special. The $n \times n$ **identity matrix,** I_n, is the matrix with 1's down the top-left to bottom-right diagonal and 0's everywhere else. For example,

$$I_2 = \begin{bmatrix} 1 & 0 \\ 0 & 1 \end{bmatrix} \quad \text{and} \quad I_4 = \begin{bmatrix} 1 & 0 & 0 & 0 \\ 0 & 1 & 0 & 0 \\ 0 & 0 & 1 & 0 \\ 0 & 0 & 0 & 1 \end{bmatrix}.$$

The special property of the identity matrix is the following law.

■ ■ ■ ■ ■

M12. If A is an $n \times n$ matrix, then $I_n A = A I_n = A$.

We illustrate this in the 2×2 case:

$$\begin{bmatrix} 1 & 0 \\ 0 & 1 \end{bmatrix} \begin{bmatrix} a & b \\ c & d \end{bmatrix} = \begin{bmatrix} 1 \cdot a + 0 \cdot c & 1 \cdot b + 0 \cdot d \\ 0 \cdot a + 1 \cdot c & 0 \cdot b + 1 \cdot d \end{bmatrix} = \begin{bmatrix} a & b \\ c & d \end{bmatrix}.$$

The reader should check that the result of this computation is the same if the matrix factors on the left are reversed. Sometimes we denote an identity matrix as simply I, instead of I_n.

LINEAR SYSTEMS IN MATRIX FORM

Linear systems can be expressed using matrices. We illustrate this with the following example involving a system of three equations in three unknowns.

EXAMPLE 2.19   Consider the 3 by 3 system

$$\begin{aligned} x - 2y + 3z &= 1 \\ 2x - y + 4z &= 25 \\ -x + 3y - 4z &= 5. \end{aligned} \quad (2.15)$$

If we let

$$A = \begin{bmatrix} 1 & -2 & 3 \\ 2 & -1 & 4 \\ -1 & 3 & -4 \end{bmatrix}, \quad X = \begin{bmatrix} x \\ y \\ z \end{bmatrix}, \quad \text{and} \quad B = \begin{bmatrix} 1 \\ 25 \\ 5 \end{bmatrix},$$

then the system (2.15) is equivalent to the single matrix equation

$$AX = B \quad (2.16)$$

because (2.16) says that

$$AX = \begin{bmatrix} 1 & -2 & 3 \\ 2 & -1 & 4 \\ -1 & 3 & -4 \end{bmatrix} \begin{bmatrix} x \\ y \\ z \end{bmatrix} = \begin{bmatrix} x - 2y + 3z \\ 2x - y + 4z \\ -x + 3y - 4z \end{bmatrix} = \begin{bmatrix} 1 \\ 25 \\ 5 \end{bmatrix} = B,$$

and these two 3×1 vectors can be equal only if all three corresponding entries are equal; that is, only if equations (2.15) hold. ∎

Equation (2.16) is called the **matrix form of the linear system.** We call A the **matrix of coefficients,** X the **vector of unknowns,** and B the **vector of right-hand constants.** Note that X and B are both written as *column* vectors.

Any m by n linear system can be expressed in matrix form as

$$AX = B,$$

where

A = matrix of coefficients ($m \times n$)
X = vector of unknowns ($n \times 1$)
B = vector of right-hand constants ($m \times 1$).

We saw in Section 2.1 that an $n \times n$ linear system normally has a unique solution. In other words, when A is square the matrix equation $AX = B$ usually has one and only one solution vector X. Before we discuss solving this equation by matrix methods, let us review briefly the Gauss-Jordan elimination procedure as applied to the linear system of Example 2.19.

The first step is to write the system in tableau form.

$$\begin{bmatrix} 1 & -2 & 3 & | & 1 \\ 2 & -1 & 4 & | & 25 \\ -1 & 3 & -4 & | & 5 \end{bmatrix}$$

This is just the matrix $[A|B]$, the augmented matrix of the system. It consists of the matrix A of coefficients, "augmented" by the vector B of right-hand constants.

When an n by n linear system has a unique solution, the Gauss-Jordan procedure amounts to applying elementary row operations to $[A|B]$, until finally the identity matrix I appears on the left side of the line. In our example the calculation goes as follows.

$$\begin{bmatrix} \textcircled{1} & -2 & 3 & | & 1 \\ 2 & -1 & 4 & | & 25 \\ -1 & 3 & -4 & | & 5 \end{bmatrix}$$

$$\begin{bmatrix} 1 & -2 & 3 & | & 1 \\ 0 & 3 & -2 & | & 23 \\ 0 & \textcircled{1} & -1 & | & 6 \end{bmatrix} \quad \begin{array}{l} (R_2 + (-2)R_1) \\ (R_3 + R_1) \end{array}$$

$$\begin{bmatrix} 1 & 0 & 1 & | & 13 \\ 0 & 0 & \textcircled{1} & | & 5 \\ 0 & 1 & -1 & | & 6 \end{bmatrix} \quad \begin{array}{l} (R_1 + 2R_3) \\ (R_2 + (-3)R_3) \end{array}$$

$$\begin{bmatrix} 1 & 0 & 0 & | & 8 \\ 0 & 0 & 1 & | & 5 \\ 0 & 1 & 0 & | & 11 \end{bmatrix} \quad \begin{array}{l} (R_1 + (-1)R_2) \\ \\ (R_3 + R_2) \end{array}$$

$$\begin{bmatrix} 1 & 0 & 0 & | & 8 \\ 0 & 1 & 0 & | & 11 \\ 0 & 0 & 1 & | & 5 \end{bmatrix} \quad \begin{array}{l} (R_3) \\ (R_2) \end{array}$$

The solution is therefore $x = 8$, $y = 11$, $z = 5$, or, in vector form,

$$X = \begin{bmatrix} x \\ y \\ z \end{bmatrix} = \begin{bmatrix} 8 \\ 11 \\ 5 \end{bmatrix}.$$

Note that the final tableau has the form $[I \mid C]$, where I is the identity matrix and $X = C$ is the unique solution. More generally, we have the following result.

2.6 INVERSE MATRICES

Theorem 2.1

Consider the equation

$$AX = B,$$

where A is a square matrix. If $[A \mid B]$ can be transformed into $[I \mid C]$ by elementary row operations, then $AX = B$ has the unique solution $X = C$.

In fact, it is not difficult to show that Theorem 2.1 holds not only when X and B are column vectors, but more generally whenever A is $n \times n$ and X and B are both $n \times k$ matrices. The next example illustrates this for $n = k = 2$.

EXAMPLE 2.20 Solve the matrix equation

$$\begin{bmatrix} 1 & 2 \\ 3 & 2 \end{bmatrix} \begin{bmatrix} x & y \\ z & w \end{bmatrix} = \begin{bmatrix} 7 & 4 \\ 9 & 8 \end{bmatrix}.$$

Solution This has the form $AX = B$, where X is an unknown 2×2 matrix. Accordingly, we form the augmented matrix $[A \mid B]$ and try to reach $[I \mid C]$ by elementary row operations.

$$[A \mid B]$$

$$\begin{bmatrix} ① & 2 & | & 7 & 4 \\ 3 & 2 & | & 9 & 8 \end{bmatrix}$$

$$\begin{bmatrix} 1 & 2 & | & 7 & 4 \\ 0 & ⊖4 & | & -12 & -4 \end{bmatrix} \quad (R_2 + (-3)R_1)$$

$$\begin{bmatrix} 1 & 2 & | & 7 & 4 \\ 0 & ① & | & 3 & 1 \end{bmatrix} \quad (-\tfrac{1}{4}R_2)$$

$$\begin{bmatrix} 1 & 0 & | & 1 & 2 \\ 0 & 1 & | & 3 & 1 \end{bmatrix} \quad (R_1 + (-2)R_2)$$

$$[I \mid C]$$

Since the identity matrix appears on the left, the solution appears on the right.

$$X = \begin{bmatrix} x & y \\ z & w \end{bmatrix} = \begin{bmatrix} 1 & 2 \\ 3 & 1 \end{bmatrix}$$

Check
$$AX = \begin{bmatrix} 1 & 2 \\ 3 & 2 \end{bmatrix} \begin{bmatrix} 1 & 2 \\ 3 & 1 \end{bmatrix} = \begin{bmatrix} 7 & 4 \\ 9 & 8 \end{bmatrix} = B. \blacksquare$$

MATRIX INVERSION

Let us now consider the matrix equation $AX = B$ from the point of view of matrix algebra. Since matrices behave somewhat like numbers, we should be able to "divide" both sides of the equation by the matrix A, to obtain the solution as $X = $ "B/A." We must proceed carefully, though, because matrix division is not defined in general, and even matrix multiplication does not obey all the rules of ordinary algebra, as we have seen.

The first step is to define the "reciprocal" of a square matrix A. Since the identity matrix I plays the role of the number 1 in matrix algebra, we define the **inverse** of A, written A^{-1}, to be a square matrix satisfying the conditions

$$AA^{-1} = I \quad \text{and} \quad A^{-1}A = I.$$

If such a matrix exists, we say that A is **invertible.** As we shall see, there are noninvertible square matrices. However, it is an important fact that *the inverse of a square matrix is unique when it exists.* That is, no matrix can have two different inverses.

Since the inverse of A is a solution of the matrix equation $AX = I$, we can find it by the method of Theorem 2.1. This procedure is called **matrix inversion.**

Matrix Inversion

To find the inverse of an $n \times n$ matrix A:

1. Form the augmented matrix $[A \mid I]$, where I is the $n \times n$ identity matrix.

2. Apply elementary row operations until I appears to the left of the line. The inverse A^{-1} then appears to the right of the line.

To illustrate the method, let us apply it to the coefficient matrix A in Example 2.19.

EXAMPLE 2.19 Revisited Invert the matrix

$$A = \begin{bmatrix} 1 & -2 & 3 \\ 2 & -1 & 4 \\ -1 & 3 & -4 \end{bmatrix}.$$

Solution We form the augmented matrix $[A \mid I]$ and try to reach $[I \mid A^{-1}]$ by elementary row operations.

$$[A \mid I]$$

$$\begin{bmatrix} \text{①} & -2 & 3 & | & 1 & 0 & 0 \\ 2 & -1 & 4 & | & 0 & 1 & 0 \\ -1 & 3 & -4 & | & 0 & 0 & 1 \end{bmatrix}$$

Note that the row operations are exactly the same as before.

$$\begin{bmatrix} 1 & -2 & 3 & | & 1 & 0 & 0 \\ 0 & 3 & -2 & | & -2 & 1 & 0 \\ 0 & \textcircled{1} & -1 & | & 1 & 0 & 1 \end{bmatrix} \quad \begin{array}{l} (R_2 + (-2)R_1) \\ (R_3 + R_1) \end{array}$$

$$\begin{bmatrix} 1 & 0 & 1 & | & 3 & 0 & 2 \\ 0 & 0 & \textcircled{1} & | & -5 & 1 & -3 \\ 0 & 1 & -1 & | & 1 & 0 & 1 \end{bmatrix} \quad \begin{array}{l} (R_1 + 2R_3) \\ (R_2 + (-3)R_3) \end{array}$$

$$\begin{bmatrix} 1 & 0 & 0 & | & 8 & -1 & 5 \\ 0 & 0 & 1 & | & -5 & 1 & -3 \\ 0 & 1 & 0 & | & -4 & 1 & -2 \end{bmatrix} \quad \begin{array}{l} (R_1 + (-1)R_2) \\ \\ (R_3 + R_2) \end{array}$$

$$\begin{bmatrix} 1 & 0 & 0 & | & 8 & -1 & 5 \\ 0 & 1 & 0 & | & -4 & 1 & -2 \\ 0 & 0 & 1 & | & -5 & 1 & -3 \end{bmatrix} \quad \begin{array}{l} (R_3) \\ (R_2) \end{array}$$

$$[I \mid A^{-1}]$$

The identity matrix now appears on the left side of the line, and so the inverse appears on the right.

$$A^{-1} = \begin{bmatrix} 8 & -1 & 5 \\ -4 & 1 & -2 \\ -5 & 1 & -3 \end{bmatrix}$$

Check

$$AA^{-1} = \begin{bmatrix} 1 & -2 & 3 \\ 2 & -1 & 4 \\ -1 & 3 & -4 \end{bmatrix} \begin{bmatrix} 8 & -1 & 5 \\ -4 & 1 & -2 \\ -5 & 1 & -3 \end{bmatrix} = \begin{bmatrix} 1 & 0 & 0 \\ 0 & 1 & 0 \\ 0 & 0 & 1 \end{bmatrix} = I \quad \blacksquare$$

It can be shown that for a square matrix X, the equation $AX = I$ implies $XA = I$, so that only the "one-sided" check of the inverse is necessary.

Practice Problem 14 Find A^{-1}, where

$$A = \begin{bmatrix} 0 & 1 & 3 \\ 0 & 1 & 2 \\ 1 & 0 & -1 \end{bmatrix}.$$

MATRIX SOLUTION OF $AX = B$

If A is invertible and we know A^{-1}, then the equation $AX = B$ can be solved very simply. A little matrix algebra confirms that a solution is $X = A^{-1}B$. For then

$$\begin{aligned} AX &= A(A^{-1}B) \\ &= (AA^{-1})B \quad \text{(by M8)} \\ &= IB \quad \text{(definition of } A^{-1}) \\ &= B. \quad \text{(by M12)} \end{aligned}$$

For the system $AX = B$ of Example 2.19, this gives

$$X = A^{-1}B = \begin{bmatrix} 8 & -1 & 5 \\ -4 & 1 & -2 \\ -5 & 1 & -3 \end{bmatrix} \begin{bmatrix} 1 \\ 25 \\ 5 \end{bmatrix} = \begin{bmatrix} 8 \\ 11 \\ 5 \end{bmatrix},$$

which agrees with the result we obtained by Gauss-Jordan elimination. Note that only a single matrix multiplication is required to solve $AX = B$, once the inverse of A has been calculated.

Matrix Solution of $AX = B$

If A is an invertible matrix, then the equation $AX = B$ has the unique solution $X = A^{-1}B$.

Practice Problem 15 Solve

$$\begin{bmatrix} 0 & 1 & 3 \\ 0 & 1 & 2 \\ 1 & 0 & -1 \end{bmatrix} X = \begin{bmatrix} 3 \\ -1 \\ 5 \end{bmatrix}.$$

(Note that the inverse of the matrix on the left was computed in Practice Problem 14.)

EXAMPLE 2.21 The Allegheny Mining Company produces two grades of ore. Grade 1 contains 25% copper and 50% iron by weight, while Grade 2 contains 30% copper and 20% iron. How many tons of each grade should be sent to the refinery to fill the following orders exactly, without wasting any copper or iron?

(a) 1000 tons of copper and 1000 tons of iron
(b) 1000 tons of copper and 1500 tons of iron
(c) 2000 tons of copper and 1000 tons of iron

Solution If the company sends x tons of Grade A ore and y tons of Grade B ore, the amount of copper and iron in the shipment is given by

$$\text{number of tons of copper} = .25x + .30y$$
$$\text{number of tons of iron} = .50x + .20y.$$

To fill an order for b_1 tons of copper and b_2 tons of iron, x and y must satisfy the equations

$$.25x + .30y = b_1$$
$$.50x + .20y = b_2,$$

or, in matrix form,

$$AX = \begin{bmatrix} .25 & .30 \\ .50 & .20 \end{bmatrix} \begin{bmatrix} x \\ y \end{bmatrix} = \begin{bmatrix} b_1 \\ b_2 \end{bmatrix} = B.$$

2.6 INVERSE MATRICES

Notice that *the matrix A stays the same for all three orders; only the vector B changes*. We will use our method to invert the matrix A.

$$[A \mid I]$$

$$\begin{bmatrix} \boxed{.25} & .30 & 1 & 0 \\ .50 & .20 & 0 & 1 \end{bmatrix}$$

$$\begin{bmatrix} \boxed{1} & 1.2 & 4 & 0 \\ .50 & .20 & 0 & 1 \end{bmatrix} \quad (4R_1)$$

$$\begin{bmatrix} 1 & 1.2 & 4 & 0 \\ 0 & \boxed{-.4} & -2 & 1 \end{bmatrix} \quad (R_2 + (-.5)R_1)$$

$$\begin{bmatrix} 1 & 1.2 & 4 & 0 \\ 0 & \boxed{1} & 5 & -2.5 \end{bmatrix} \quad (-2.5R_2)$$

$$\begin{bmatrix} 1 & 0 & -2 & 3 \\ 0 & 1 & 5 & -2.5 \end{bmatrix} \quad (R_1 + (-1.2)R_2)$$

$$[I \mid A^{-1}]$$

Since the identity matrix is on the left, the inverse is on the right.

$$A^{-1} = \begin{bmatrix} -2 & 3 \\ 5 & -2.5 \end{bmatrix}$$

Substituting the appropriate vector B in the equation $X = A^{-1}B$, we now obtain each of the three solutions.

(a) $\begin{bmatrix} x \\ y \end{bmatrix} = X = A^{-1}B = \begin{bmatrix} -2 & 3 \\ 5 & -2.5 \end{bmatrix} \begin{bmatrix} 1000 \\ 1000 \end{bmatrix} = \begin{bmatrix} 1000 \\ 2500 \end{bmatrix}$

(b) $\begin{bmatrix} x \\ y \end{bmatrix} = X = A^{-1}B = \begin{bmatrix} -2 & 3 \\ 5 & -2.5 \end{bmatrix} \begin{bmatrix} 1000 \\ 1500 \end{bmatrix} = \begin{bmatrix} 2500 \\ 1250 \end{bmatrix}$

(c) $\begin{bmatrix} x \\ y \end{bmatrix} = X = A^{-1}B = \begin{bmatrix} -2 & 3 \\ 5 & -2.5 \end{bmatrix} \begin{bmatrix} 2000 \\ 1000 \end{bmatrix} = \begin{bmatrix} -1000 \\ 7500 \end{bmatrix}$

Note that the solution in (c) calls for a negative amount of Grade 1 ore, which means that this order cannot be filled exactly. That is, there is no way to provide exactly 2000 tons of copper and 1000 tons of iron. ■

NONINVERTIBLE MATRICES

The method of matrix inversion can only be used to solve n by n linear systems, because only a square matrix can have an inverse. But we may even encounter problems in the n by n case if the coefficient matrix A is noninvertible. When we try to invert such a matrix, our procedure will eventually produce *a row in which all the entries to the left of the line are zero*. This tells us that the matrix has no inverse.

EXAMPLE 2.22 Invert the matrix

$$A = \begin{bmatrix} 1 & -1 & 0 \\ 1 & 0 & -1 \\ 0 & 1 & -1 \end{bmatrix}.$$

Solution We form $[A \mid I]$ as usual and try to obtain an identity matrix I on the left side.

$$\begin{bmatrix} \textcircled{1} & -1 & 0 & | & 1 & 0 & 0 \\ 1 & 0 & -1 & | & 0 & 1 & 0 \\ 0 & 1 & -1 & | & 0 & 0 & 1 \end{bmatrix}$$

$$\begin{bmatrix} 1 & -1 & 0 & | & 1 & 0 & 0 \\ 0 & \textcircled{1} & -1 & | & -1 & 1 & 0 \\ 0 & 1 & -1 & | & 0 & 0 & 1 \end{bmatrix} \quad (R_2 + (-1)R_1)$$

$$\begin{bmatrix} 1 & 0 & -1 & | & 0 & 1 & 0 \\ 0 & 1 & -1 & | & -1 & 1 & 0 \\ 0 & 0 & 0 & | & 1 & -1 & 1 \end{bmatrix} \quad \begin{array}{l}(R_1 + R_2) \\ \\ (R_3 + (-1)R_2)\end{array}$$

We cannot continue, because the third row of the left-hand matrix consists entirely of zeros. The matrix A is therefore noninvertible. ■

Practice Problem 16 Find the inverses of the following matrices, if possible.

(a) $\begin{bmatrix} 2 & 1 & 0 \\ 0 & 1 & -1 \\ 1 & 0 & 1 \end{bmatrix}$ (b) $\begin{bmatrix} 2 & 1 & 0 \\ 1 & 2 & 1 \\ 4 & 5 & 2 \end{bmatrix}$

EXERCISES 2.6

Exercises 1 and 2 refer to the following matrices.

$$A = \begin{bmatrix} 2 & 0 & -2 \\ 2 & 1 & 0 \\ 3 & 2 & 4 \end{bmatrix} \quad B = \begin{bmatrix} 4 & 1 & -3 & 0 & 5 \\ 0 & 3 & 2 & -1 & 0 \\ 3 & 0 & 0 & 1 & 2 \\ 2 & 0 & 1 & -1 & 3 \\ 3 & 3 & 4 & -3 & 0 \end{bmatrix}$$

2.6 INVERSE MATRICES

1. Write out I_3, and check that $AI_3 = A$ and $I_3A = A$.
2. Write out I_5, and check that $BI_5 = B$ and $I_5B = B$.

In Exercises 3–8 a linear system is given that can be expressed as $AX = B$ in matrix form. Tell what A, X, and B are.

3. $\begin{aligned} x + 2y + z &= 3 \\ 2x + 2z &= 5 \\ 3y - z &= 0 \end{aligned}$

4. $\begin{aligned} 3x + 2y + z &= 1 \\ x - z &= 4 \\ 5x &= 9 \end{aligned}$

5. $\begin{aligned} r + s + t &= 5 \\ s + t &= 8 \end{aligned}$

6. $\begin{aligned} u + v &= 6 \\ u - v &= -2 \\ 3u + v &= 10 \end{aligned}$

7. $\begin{aligned} x_1 + x_3 &= 5 \\ x_1 + x_2 &= x_3 \\ x_1 + x_3 &= 0 \end{aligned}$

8. $\begin{aligned} a + 2b + 3c + d &= 4 \\ a + c + 3 &= b \\ a + e &= 5 \end{aligned}$

In Exercises 9–12 rewrite $AX = B$ as a linear system of equations, where A, X, and B are as given.

9. $A = \begin{bmatrix} 2 & -2 \\ 3 & 5 \end{bmatrix}$, $X = \begin{bmatrix} x \\ y \end{bmatrix}$, $B = \begin{bmatrix} 0 \\ 2 \end{bmatrix}$

10. $A = \begin{bmatrix} 1 & 0 & 2 \\ 2 & 1 & 0 \end{bmatrix}$, $X = \begin{bmatrix} x \\ y \\ z \end{bmatrix}$, $B = \begin{bmatrix} 5 \\ -1 \end{bmatrix}$

11. $A = \begin{bmatrix} 1 & 2 \\ 2 & 1 \\ 0 & 3 \end{bmatrix}$, $X = \begin{bmatrix} u \\ v \end{bmatrix}$, $B = \begin{bmatrix} 0 \\ 2 \\ 5 \end{bmatrix}$

12. $A = \begin{bmatrix} 0 & 1 & 2 \\ 3 & 0 & -1 \\ -1 & 2 & 0 \end{bmatrix}$, $X = \begin{bmatrix} r \\ s \\ t \end{bmatrix}$, $B = \begin{bmatrix} 9 \\ 1 \\ 0 \end{bmatrix}$

In Exercises 13–22 find the inverse of the given matrix, if possible.

13. $\begin{bmatrix} 1 & 3 \\ 2 & 5 \end{bmatrix}$

14. $\begin{bmatrix} 2 & 7 \\ 1 & 4 \end{bmatrix}$

15. $\begin{bmatrix} 3 & 2 & 0 \\ 1 & 1 & 0 \\ 3 & 2 & 1 \end{bmatrix}$

16. $\begin{bmatrix} 1 & 1 & 1 \\ 0 & 1 & -2 \\ 1 & 1 & 2 \end{bmatrix}$

17. $\begin{bmatrix} 1 & -1 & 0 \\ 3 & -1 & 3 \\ 1 & -1 & 1 \end{bmatrix}$

18. $\begin{bmatrix} 1 & 2 & 0 \\ 2 & 1 & 1 \\ -1 & -5 & 1 \end{bmatrix}$

19. $\begin{bmatrix} 3 & 5 & 0 \\ 0 & 2 & -1 \\ 3 & 1 & 2 \end{bmatrix}$

20. $\begin{bmatrix} 3 & 2 & 1 \\ 2 & 2 & 1 \\ 2 & 1 & 1 \end{bmatrix}$

21. $\begin{bmatrix} 1 & 1 & 0 & 0 \\ 0 & 1 & 0 & 0 \\ 0 & 0 & 1 & 1 \\ 1 & 1 & 0 & 1 \end{bmatrix}$

22. $\begin{bmatrix} 1 & -1 & 0 & 0 \\ 0 & 1 & 0 & 0 \\ 3 & 0 & 1 & 0 \\ 2 & 0 & 0 & 1 \end{bmatrix}$

Use the given products to solve the linear systems in Exercises 23–30, using the matrix solution to $AX = B$.

$\begin{bmatrix} 1 & 0 & 1 \\ -1 & 1 & -1 \\ -2 & 0 & -1 \end{bmatrix} \begin{bmatrix} -1 & 0 & -1 \\ 1 & 1 & 0 \\ 2 & 0 & 1 \end{bmatrix} = I_3$

$\begin{bmatrix} 4 & 3 & -3 \\ -1 & 0 & 1 \\ -1 & -1 & 1 \end{bmatrix} \begin{bmatrix} 1 & 0 & 3 \\ 0 & 1 & -1 \\ 1 & 1 & 3 \end{bmatrix} = I_3$

23. $\begin{aligned} x + z &= 3 \\ -x + y - z &= 0 \\ -2x - z &= 5 \end{aligned}$

24. $\begin{aligned} x + z &= -1 \\ -x + y - z &= 2 \\ -2x - z &= -2 \end{aligned}$

25. $\begin{aligned} 4x + 3y - 3z &= 2 \\ -x + z &= 5 \\ -x - y + z &= -6 \end{aligned}$

26. $\begin{aligned} 4x + 3y - 3z &= 8 \\ -x + z &= 2 \\ -x - y + z &= 0 \end{aligned}$

27. $\begin{aligned} -u - w &= 3 \\ u + v &= 2 \\ 2u + w &= 1 \end{aligned}$

28. $\begin{aligned} a + 3c &= 0 \\ b - c &= 5 \\ a + b + 3c &= -2 \end{aligned}$

29. $\begin{aligned} x_1 + 3x_3 &= 1 \\ x_2 - x_3 &= 2 \\ x_1 + x_2 + 3x_3 &= -2 \end{aligned}$

30. $\begin{aligned} -r - t &= 2 \\ r + s &= 1 \\ 2r + t &= -3 \end{aligned}$

Answers to Practice Problems

14. $\begin{bmatrix} 1 & -1 & 1 \\ -2 & 3 & 0 \\ 1 & -1 & 0 \end{bmatrix}$

15. $X = \begin{bmatrix} 9 \\ -9 \\ 4 \end{bmatrix}$

16. (a) $\begin{bmatrix} 1 & -1 & -1 \\ -1 & 2 & 2 \\ -1 & 1 & 2 \end{bmatrix}$ (b) Doesn't exist

■ ■ ■ ■ ■ ■ ■ ■ **MATHEMATICS IN ACTION**

2.7 The Leontief Input-Output Model

The technique of matrix inversion gives us a powerful new tool for the solution of n by n linear systems. The method is particularly useful when we have a number of large systems to solve, all of which share the same coefficient matrix. In this section we encounter an important application of this type.

THE ANALYSIS OF NATIONAL ECONOMIES

The economies of modern industrial nations are extremely complex. When we break such systems down into their components, we find hundreds or even thousands of different industries, each supplying the others with goods and services needed in the production process. The coal industry, for example, supplies the steel industry with one of its main raw materials, but at the same time steel—in the form of machinery, vehicles, and so on—is needed to mine and transport coal. It is this constant flow of goods between industries that characterizes an interacting economy, and at the same time constitutes one of the main obstacles to effective economic planning. If we set arbitrary production goals for each industry we may wind up with shortages; and a price increase in one key industry may send a ripple through the entire economy.

Is it possible to describe mathematically the interaction of an economic system? This question occurred to Wassily Leontief, a Russian-born economist, while he was a student in Berlin in the 1920s. Using matrix algebra, Leontief developed a very elegant model for the working of an economy, known as *input-output theory*. After emigrating to the U.S. in 1931 he concentrated on the task of putting his theoretical ideas into practice. By the late 1940s, with the help of the newly developed computer, Leontief was able to distill the parameters he needed from a mountain of government statistics. His model proved to be remarkably accurate in predicting the needs of the postwar American economy, and today input-output theory is used to forecast the production needs of large corporations, as well as advanced and developing nations. The theory is also used to set prices for various commodities, to predict the effects of price changes, and to analyze the impact of production on the environment. In recognition of the widespread application of his ideas, Leontief was awarded the Nobel Prize for Economics in 1973.

INPUT-OUTPUT THEORY

We begin by dividing an economy into a certain number of **producing sectors** $P_1, P_2, \ldots, P_n$, each of which is thought of as producing a single good, commodity, or service. In his famous study of the 1947 U.S. economy, Leontief identified 500 industries, which he combined into 42 producing sectors such as "textile mill products," "petroleum and coal products," "electrical machinery," and so on.

Once the producing sectors are determined, we must decide on an appropriate unit of measurement for the output of each sector. These units may be either *physical* or *monetary*; for example, one unit of steel might represent 10,000 tons (physical) or $1 million (monetary). The advantage of monetary units is that the output of different sectors can be directly compared, but physical units are sometimes essential, as in consideration of pricing. For our present discussion it does not matter which type of unit we adopt, so we shall use the term *unit* without specifying whether it is physical or monetary.

Central to Leontief's theory is a certain set of ratios. If the sectors $P_1, \ldots, P_n$ produce respective goods $G_1, \ldots, G_n$, then the **input-output ratio** a_{ij} specifies the amount of G_i needed to produce one unit of G_j. The $n \times n$ matrix $A = [a_{ij}]$ whose entries are the input-output ratios is called the **input-output matrix** (or **technology matrix**) for the economy. The meaning of the input-output matrix is illustrated in the next example.

EXAMPLE 2.23 Consider a highly simplified economy with three producing sectors: agriculture, manufacturing, and energy. Suppose the input-output matrix is as follows.

$$A = \begin{bmatrix} .2 & .1 & .3 \\ .4 & .2 & .1 \\ .3 & .5 & .5 \end{bmatrix} \begin{matrix} \text{agr} \\ \text{mfg} \\ \text{engy} \end{matrix}$$

To output one unit of agr, mfg, engy — Input units needed

The entries in each *column* indicate how many input units are needed from each of the three sectors in order to produce one unit of output. For example, the second column (manufacturing) shows that we need .1, .2, and .5 units of input from the agriculture, manufacturing, and energy sectors, respectively, to produce one unit of manufactured goods. If the units are monetary, this means that for each $1 worth of manufactured goods produced, the manufacturing sector consumes 10¢ worth of agricultural products, 20¢ worth of (other) manufactured goods, and 50¢ worth of energy. ∎

In practice, these input-output ratios could be determined by looking at how much each sector buys from the others. For instance, if the manufacturing sector bought $45 billion worth of energy in a given year and produced $90 billion worth of goods, then each $1 worth of manufactured goods required 45/90 = .5 = 50¢ worth of energy. This is essentially how Leontief calculated his input-output ratios.

It is a fundamental assumption of the input-output model that *these ratios remain the same regardless of the levels of production*. Thus, if one unit of manufactured output requires .5 units of energy, then 1000 units of output should require 500 units of energy, and more generally x units of output should require $.5x$ units of energy. This assumption must be carefully investigated in real applications. For example, for large values of x there may be a more energy-efficient manufacturing process available. In practice, however, the assumption has been found to be reasonable, provided that production levels stay within certain limits. Thus, barring any sudden technological changes—such as the discovery of a new, cheap energy source—the input-output matrix will be a fairly stable feature of the economy.

GROSS PRODUCTION AND SURPLUS

Returning to Example 2.23, let us represent the total output of the three sectors by x_1, x_2, x_3.

x_1 = total units of G_1 (agriculture) produced

x_2 = total units of G_2 (manufactured goods) produced

x_3 = total units of G_3 (energy) produced

Then the flow of goods in the economy can be represented graphically by means of the diagram in Figure 2.9.

In addition to the three sectors and their gross outputs, the diagram shows the number of units of output sent from each sector to the others. For example, since we are producing x_1 units of agricultural products, the input-output matrix shows that the agriculture sector will need as input $.2x_1$ units from itself, $.4x_1$ units from manufacturing, and $.3x_1$ units from energy, as indicated in Figure 2.9. The other arcs are labeled similarly, corresponding to the input-output matrix A.

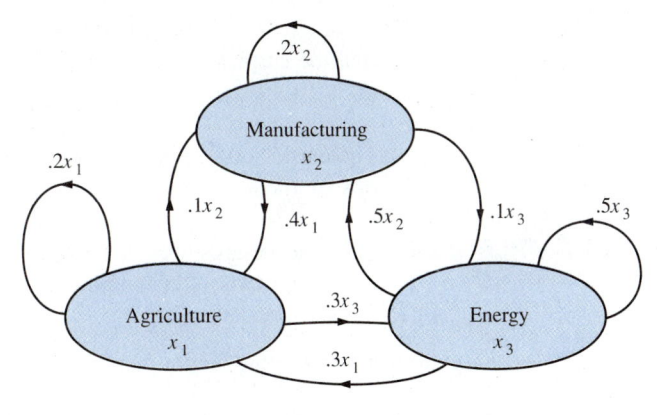

2.7 MATHEMATICS IN ACTION: THE LEONTIEF INPUT-OUTPUT MODEL

FIGURE 2.9

Looking at the three outgoing arcs from the agriculture sector, we see that the amount of agricultural output *used up internally by the three sectors* (the number of units needed for the production process itself) is given by the following equation:

$$u_1 = .2x_1 + .1x_2 + .3x_3.$$

There are similar equations for the other two sectors:

$$u_2 = .4x_1 + .2x_2 + .1x_3$$
$$u_3 = .3x_1 + .5x_2 + .5x_3.$$

In matrix terms these three equations say that

$$\begin{bmatrix} u_1 \\ u_2 \\ u_3 \end{bmatrix} = \begin{bmatrix} .2 & .1 & .3 \\ .4 & .2 & .1 \\ .3 & .5 & .5 \end{bmatrix} \begin{bmatrix} x_1 \\ x_2 \\ x_3 \end{bmatrix}$$

or

$$U = AX, \qquad (2.17)$$

where

$$U = \begin{bmatrix} u_1 \\ u_2 \\ u_3 \end{bmatrix} = \text{internal consumption vector,}$$

$$X = \begin{bmatrix} x_1 \\ x_2 \\ x_3 \end{bmatrix} = \text{gross production vector, \quad and}$$

$$A = \begin{bmatrix} .2 & .1 & .3 \\ .4 & .2 & .1 \\ .3 & .5 & .5 \end{bmatrix} = \text{input-output matrix.}$$

If we are given the gross production X, we can calculate $U = AX$ and then find the surplus, or net production, by subtracting.

$$S = X - U = \textbf{surplus or net production vector} \qquad (2.18)$$

The entries of S represent the amount of output available for consumer demand or export.

EXAMPLE 2.23
Continued

Suppose the gross production of the economy is 800 units of agricultural products, 900 units of manufactured goods, and 1500 units of energy. What will the net production be?

Solution The gross production vector is given by

$$X = \begin{bmatrix} x_1 \\ x_2 \\ x_3 \end{bmatrix} = \begin{bmatrix} 800 \\ 900 \\ 1500 \end{bmatrix}.$$

We substitute this into formula (2.17).

$$U = AX = \begin{bmatrix} .2 & .1 & .3 \\ .4 & .2 & .1 \\ .3 & .5 & .5 \end{bmatrix} \begin{bmatrix} 800 \\ 900 \\ 1500 \end{bmatrix} = \begin{bmatrix} 700 \\ 650 \\ 1440 \end{bmatrix}$$

Hence, by (2.18),

$$\text{net production} = S = X - U = \begin{bmatrix} 800 \\ 900 \\ 1500 \end{bmatrix} - \begin{bmatrix} 700 \\ 650 \\ 1440 \end{bmatrix} = \begin{bmatrix} 100 \\ 250 \\ 60 \end{bmatrix}.$$

In other words, at the given gross production the economy will be able to keep itself running and have 100 units of agriculture, 250 units of manufactured goods, and 60 units of energy left over. ∎

Practice Problem 17 Suppose in Example 2.23 the gross production is 1000 units of agricultural products, 1200 units of manufactured goods, and 2000 units of energy. What are the internal consumption vector U and the surplus production vector S?

MEETING A GIVEN DEMAND

We can also turn the problem around. Suppose we want the economy to produce just enough to meet a given **demand vector:**

$$D = \begin{bmatrix} d_1 \\ d_2 \\ d_3 \end{bmatrix} \begin{array}{l} \text{units of } G_1 \text{ (agriculture)} \\ \text{units of } G_2 \text{ (manufactured goods)} \\ \text{units of } G_3 \text{ (energy).} \end{array}$$

At what gross production X must the economy operate?

2.7 MATHEMATICS IN ACTION: THE LEONTIEF INPUT-OUTPUT MODEL

To answer this, we first observe that by (2.17) and (2.18)

$$\text{surplus} = X - U = X - AX.$$

Setting the surplus (net production) exactly equal to the demand, we obtain the matrix equation

$$\begin{aligned} D = S &= X - AX \\ &= IX - AX \quad \text{(by M12)} \\ &= (I - A)X \quad \text{(by M10).} \end{aligned}$$

If we multiply both sides on the left by $(I - A)^{-1}$, we arrive at the solution

$$X = (I - A)^{-1}D, \quad (2.19)$$

provided, of course, that the matrix $I - A$ is invertible.

EXAMPLE 2.23 Continued What must the gross production of the economy be to meet a demand for 100 units of agriculture, 200 units of manufactured goods, and 150 units of energy?

Solution The demand vector is given by

$$D = \begin{bmatrix} 100 \\ 200 \\ 150 \end{bmatrix}.$$

To find the gross production vector X using formula (2.19), we first compute the matrix $I - A$.

$$I - A = \begin{bmatrix} 1 & 0 & 0 \\ 0 & 1 & 0 \\ 0 & 0 & 1 \end{bmatrix} - \begin{bmatrix} .2 & .1 & .3 \\ .4 & .2 & .1 \\ .3 & .5 & .5 \end{bmatrix} = \begin{bmatrix} .8 & -.1 & -.3 \\ -.4 & .8 & -.1 \\ -.3 & -.5 & .5 \end{bmatrix}$$

Although the decimals make the computation tedious, this matrix can be inverted by the method of Section 2.6 to obtain

$$(I - A)^{-1} = \begin{bmatrix} 2.80 & 1.60 & 2.0 \\ 1.84 & 2.48 & 1.6 \\ 3.52 & 3.44 & 4.8 \end{bmatrix}.$$

Formula (2.19) now gives

$$X = (I - A)^{-1}D = \begin{bmatrix} 2.80 & 1.60 & 2.0 \\ 1.84 & 2.48 & 1.6 \\ 3.52 & 3.44 & 4.8 \end{bmatrix} \begin{bmatrix} 100 \\ 200 \\ 150 \end{bmatrix} = \begin{bmatrix} 900 \\ 920 \\ 1760 \end{bmatrix}.$$

That is, the three sectors must produce 900, 920, and 1760 units of output, respectively, in order to meet the specified demand. ■

CHAPTER 2 SYSTEMS OF LINEAR EQUATIONS

Practice Problem 18 In Example 2.23 what gross production is needed for a surplus of 200 units of agricultural products, 100 units of manufactured goods, and 300 units of energy?

The derivation of equation (2.19) can be generalized to any *n*-sector economy as follows.

Theorem 2.2

Suppose an economy has input-output matrix A such that $I - A$ is invertible. Then in order to meet a given demand vector D, the economy must operate at gross production X, where

$$X = (I - A)^{-1}D.$$

EXAMPLE 2.24

The economy of a Pacific island can be broken down into goods and services. To produce $1 worth of goods requires $.20 worth of goods and $.20 worth of services. To produce $1 worth of services requires $.70 worth of goods and $.20 worth of services. How many dollars worth of goods and services must be produced to have an export surplus of $40,000 worth of goods and $8000 worth of services?

Solution First we set up the input-output matrix A.

$$A = \begin{bmatrix} .2 & .7 \\ .2 & .2 \end{bmatrix} \begin{matrix} \text{goods} \\ \text{services} \end{matrix}$$

(To produce $1 worth of goods, services; Input value needed)

Thus we have

$$I - A = \begin{bmatrix} 1 & 0 \\ 0 & 1 \end{bmatrix} - \begin{bmatrix} .2 & .7 \\ .2 & .2 \end{bmatrix} = \begin{bmatrix} .8 & -.7 \\ -.2 & .8 \end{bmatrix}.$$

Using the method of the previous section we compute

$$(I - A)^{-1} = \begin{bmatrix} 1.6 & 1.4 \\ .4 & 1.6 \end{bmatrix}.$$

Now the demand vector is given by

$$D = \begin{bmatrix} 40{,}000 \\ 8{,}000 \end{bmatrix},$$

so by Theorem 2.2 we can compute the gross production vector

$$X = (I - A)^{-1}D = \begin{bmatrix} 1.6 & 1.4 \\ .4 & 1.6 \end{bmatrix} \begin{bmatrix} 40{,}000 \\ 8{,}000 \end{bmatrix} = \begin{bmatrix} 75{,}200 \\ 28{,}800 \end{bmatrix}.$$

2.7 MATHEMATICS IN ACTION: THE LEONTIEF INPUT-OUTPUT MODEL

Thus $75,200 worth of goods and $28,800 worth of services must be produced to supply the desired surplus. This can be checked by computing the internal consumption vector

$$U = AX = \begin{bmatrix} .2 & .7 \\ .2 & .2 \end{bmatrix} \begin{bmatrix} 75{,}200 \\ 28{,}800 \end{bmatrix} = \begin{bmatrix} 35{,}200 \\ 20{,}800 \end{bmatrix}.$$

Then the surplus is

$$S = X - U = \begin{bmatrix} 75{,}200 \\ 28{,}800 \end{bmatrix} - \begin{bmatrix} 35{,}200 \\ 20{,}800 \end{bmatrix} = \begin{bmatrix} 40{,}000 \\ 8{,}000 \end{bmatrix},$$

as was desired. ■

EXERCISES 2.7

Exercises 1–4 refer to an interacting economy with four producing sectors: metals, nonmetals, energy, and services, with the following input-output matrix, where the units are dollars.

To produce $1 worth of

metals	nonmetals	energy	services		
.32	.04	.09	.02	metals	
.05	.22	.41	.06	nonmetals	Input
.14	.03	.12	.02	energy	needed
.07	.11	.08	.05	services	

1. On what sector is the energy sector most dependent? Least dependent?
2. Which sector is most dependent on energy? Least dependent on energy?
3. If the nonmetals sector produces $7.5 million worth of output, how much input must it consume from each of the four sectors?
4. If the services sector produces $5.8 million worth of output, how much input must it consume from each of the four sectors?

Exercises 5–8 refer to an interacting economy with five producing sectors: agriculture, textiles, chemicals, machinery, and transportation, with the following input-output matrix, where the units are dollars.

To produce $1 worth of

agriculture	textiles	chemicals	machinery	transportation		
.25	.21	.12	0	0	agriculture	
0	.13	0	0	0	textiles	Input
.02	.09	.35	.03	0	chemicals	needed
.05	.03	.01	.08	.14	machinery	
.07	.10	.02	.04	.10	transportation	

5. On what sector is the textiles sector most dependent? Least dependent?
6. Which sector is most dependent on machinery? Least dependent on machinery?

7. If the agriculture sector produces $20 billion worth of output, how much input must it consume from each of the five sectors?

8. If the machinery sector produces $12 billion worth of output, how much input must it consume from each of the five sectors?

9. An economy has two interacting sectors, manufacturing and energy. It takes .3 units of manufacturing and .4 units of energy to produce one unit of manufactured goods, and it takes .2 units of manufacturing and .1 units of energy to produce one unit of energy. What is the input-output matrix for this economy?

10. An economy has two interacting sectors, agriculture and chemicals. In a typical year, the agriculture sector spends a total of $30 billion on its own goods and $6 billion on chemicals and produces $60 billion worth of output, while the chemicals sector spends $20 billion on its own products and $10 billion on agricultural goods and produces $50 billion worth of output. Using monetary units, what is the input-output matrix for this economy?

11. A two-sector economy has the following input-output matrix:

$$A = \begin{bmatrix} .02 & .08 \\ .12 & .03 \end{bmatrix}.$$

The gross production of the economy is 1250 units for the first sector and 1800 units for the second sector.
(a) Find the gross production vector X.
(b) Compute the internal consumption vector U.
(c) Compute the surplus, or net production, S.
(d) Make a diagram like Figure 2.9, showing the flow of goods in the economy.

12. A three-sector economy has the following input-output matrix:

$$A = \begin{bmatrix} 0 & .11 & .05 \\ .15 & .02 & .08 \\ .09 & .14 & .01 \end{bmatrix}.$$

The gross production of the economy is $x_1 = 1500$, $x_2 = 1400$, $x_3 = 1700$.
(a) Find the gross production vector X.
(b) Compute the internal consumption vector U.
(c) Compute the surplus, or net production, S.
(d) Make a diagram like Figure 2.9, showing the flow of goods in the economy.

13. In a two-sector economy each dollar of goods requires $.20 of goods and $.20 of services to produce, and each dollar of services requires $.30 of goods and $.30 of services to produce. What must the gross production of the economy be to meet an external demand for
(a) $5000 of goods and $6000 of services?
(b) $20,000 of goods and $10,000 of services?

14. In a two-sector economy each megawatt of electricity needs .5 barrels of oil to produce, and each barrel of oil needs .6 megawatts of electricity and .3 barrels of oil to produce. What must the gross production of the economy be to meet an external demand for
(a) 2000 megawatts of electricity and 2000 barrels of oil?
(b) 0 megawatts and 4000 barrels of oil?

15. A certain simple society produces housing, clothing, and food. Production of 1 unit of housing or 1 unit of clothing requires 0.2 units of housing, 0.2 units of clothing, and 0.1 units of food. Production of 1 unit of food requires 0.4 units of housing, 0.4 units of clothing, and 0.2 units of food. What gross production is necessary to meet an external demand for 10 units of housing, 10 units of clothing, and 20 units of food?

16. Suppose that a small country produces three forms of energy—electricity, natural gas, and oil. Production of 1 unit of electricity requires 0.4 units of electricity, 0.2 units of natural gas, and 0.1 units of oil. Production of 1 unit of natural gas requires 0.1 units of electricity, 0.1 units of natural gas, and 0.05 units of oil. Finally, production of 1 unit of oil requires 0.4 units of natural gas and 0.2 units of oil. What gross production is required to meet an external demand for 400 units of electricity, 200 units of natural gas, and 300 units of oil?

17. A small country produces three forms of energy—electricity, oil, and coal. Production of 1 unit of electricity requires 0.2 units of electricity, 0.2 units of oil, and 0.1 units of coal. Production of 1 unit of oil requires 0.4 units of electricity, 0.4 units of oil, and 0.2 units of coal. Finally, production of 1 unit of coal requires 0.1 units of electricity, 0.1 units of oil, and 0.3 units of coal. What gross production is necessary to meet an external demand for 100 units of electricity, 200 units of oil, and 100 units of coal?

18. Suppose that a simple economy produces transportation, food, and oil. Production of 1 unit of transportation or 1 unit of food requires 0.1 units of transportation, 0.25 units of food, and 0.1 units of oil. Production of 1 unit of oil requires 0.3 units of transportation, 0.25 units of food, and 0.3 units of oil. What gross production is necessary to meet an external demand for 40 units of transportation, 60 units of food, and 80 units of oil?

Answers to Practice Problems

17. $U = \begin{bmatrix} 920 \\ 840 \\ 1900 \end{bmatrix}$, $S = \begin{bmatrix} 80 \\ 360 \\ 100 \end{bmatrix}$

18. 1320 units of agricultural products, 1096 units of manufactured goods, and 2488 units of energy

CHAPTER 2 REVIEW

IMPORTANT TERMS

- elementary operations *(2.1)*
- elementary row operations
- Gauss-Jordan elimination
- linear equation
- m by n linear system
- pivot row and column
- pivot transformation
- tableau form
- basic and free variables *(2.2)*
- dependent system
- inconsistent system
- augmented matrix of a linear system *(2.4)*
- i, j entry
- product of a matrix by a scalar
- row and column vectors
- sum and difference of matrices
- zero matrix
- product of matrices *(2.5)*
- identity matrix *(2.6)*
- inverse matrix
- invertible matrix
- matrix of a linear system
- square matrix
- demand vector *(2.7)*
- input-output matrix
- input-output ratio
- internal consumption vector
- gross production vector
- surplus or net production vector

IMPORTANT FORMULAS

Laws of Matrix Algebra

- M1. $A + B = B + A$
- M2. $A + (B + C) = (A + B) + C$
- M3. $A + O = A$
- M4. $k(A + B) = kA + kB$
- M5. $(c + d)A = cA + dA$
- M6. $c(dA) = (cd)A$
- M7. $1A = A$
- M8. $A(BC) = (AB)C$
- M9. $A(B + C) = AB + AC$
- M10. $(A + B)C = AC + BC$
- M11. $c(AB) = (cA)B = A(cB)$
- M12. $IA = AI = A$

Matrix Solution to a Linear System

If A is an invertible matrix, then $AX = B$ has the unique solution $X = A^{-1}B$.

Internal Consumption and Surplus

If A is the input-output matrix, X is the gross production vector, U is the internal consumption vector, and S is the surplus vector, then $U = AX$ and $S = X - U$.

Gross Production

The gross production needed to satisfy a demand D is $X = (I - A)^{-1}D$, assuming that the matrix inverse exists.

REVIEW EXERCISES

In Exercises 1–3 solve the given linear system using Gauss-Jordan elimination.

1. $\quad 3x + y + z = 2$
   $\quadx - y + z = 0$
   $\quad -x + 4y - 2z = 3$

2. $\quad 2x + y - z + w = 3$
   $\quadx - 2y + w = 0$
   $\quad -x + 7y - z - 2w = 3$

3. $\quad a + 2b - 3c + d = 6$
   $\quad 2a - b + c - d = -1$
   $\quad 4a + 3b - 5c + d = 2$

4. The chemicals in a fertilizer are denoted by three numbers, giving the percentage by weight of nitrogen, phosphorus, and potassium in the fertilizer. For example, 5-10-5 fertilizer is 5% nitrogen, 10% phosphorus, and 5% potassium. A gardener has available three fertilizers: 5-15-10, 20-5-5, and 10-10-10. How many pounds of each type should she use to get a mixture containing 55 pounds of nitrogen, 70 pounds of phosphorus, and 55 pounds of potassium?

Exercises 5–12 refer to the following matrices.

$$A = \begin{bmatrix} 0 & 1 & 1 \\ 2 & 0 & 1 \\ -3 & 1 & 0 \end{bmatrix} \quad B = \begin{bmatrix} 2 & 1 & 0 \\ 1 & 1 & -1 \\ -1 & -2 & 3 \end{bmatrix} \quad C = \begin{bmatrix} 2 & 1 & 0 \\ 3 & -1 & 5 \end{bmatrix} \quad D = \begin{bmatrix} -2 \\ 4 \\ 3 \end{bmatrix}$$

In Exercises 5–12 compute the indicated matrix, if possible.

5. $A - 2B$ **6.** AB **7.** BD **8.** AC

9. $A + I_3$ **10.** CI_3 **11.** A^{-1} **12.** B^{-1}

13. Each of the linear systems in Exercises 1–3 can be expressed as $AX = B$ in matrix form. Tell what A, X, and B are in each case.

14. Find the inverse of the coefficient matrix and use it to solve the system

$$\begin{aligned} x + z &= b_1 \\ y + z &= b_2 \\ x + 2y + z &= b_3 \end{aligned}$$

for (a) $b_1 = 3$, $b_2 = 5$, $b_3 = -2$, and (b) $b_1 = 350$, $b_2 = 400$, $b_3 = 100$.

15. A simple economy is based on goods and services. To produce a unit of goods requires .7 units of goods and .1 units of services, while to produce a unit of services requires .4 units of goods and .7 units of services.

(a) What is the surplus production if the gross production is 1000 units of goods and 600 units of services?

(b) What gross production is needed to produce a surplus of 500 units of goods and 800 units of services?

3

LINEAR PROGRAMMING

3.1 Optimization with Linear Constraints
3.2 The Graphical Method
3.3 The Simplex Method
3.4 Maximization Problems
3.5 Minimization Problems and Duality
3.6 Artificial Variables
    Chapter Review

Probably the most widely applied mathematical theory in modern business and industry is one that did not even exist fifty years ago: the theory of *linear programming*. Today it is estimated that fully 25 percent of scientific computer use is devoted to linear programming, and the range of its applications seems almost unlimited.

In this chapter we study linear programming and present some basic applications of the theory. We start geometrically, then reexamine the subject from an algebraic viewpoint, explaining the *simplex algorithm,* a computational technique for solving linear programming problems efficiently and methodically.

3.1 Optimization with Linear Constraints

Many practical problems involve **optimization:** how can a certain quantity—profit or cost, for instance—be maximized or minimized, subject to given conditions? Questions of this type arise in every phase of industrial production. For example, materials must be obtained at minimum cost, subject to fixed production quotas; limited resources must be allocated in the most profitable way; finished goods must be transported as economically as possible, given fixed supplies and demands in different locations.

During World War II many optimization problems arose in connection with military logistics, and teams of research scientists were formed to investigate possible methods of solution. They noticed that many of their problems shared the same features, once they were formulated in mathematical terms. The name "linear program" was used to describe any abstract problem of maximization or minimization subject to linear constraints, and a new mathematical theory was born. By the 1950s, linear programming was being applied far beyond its original setting, to problems in business, industry, agriculture, and the natural and social sciences.

EXAMPLE 3.1 Albert's Smoke Shop has in stock 480 ounces of Virginia and 720 ounces of Latakia tobacco leaf. Albert sells two popular house blends in a standard 12-ounce can. Each can of Blend A (Albert's Mixture) contains 8 ounces of Virginia and 4 ounces of Latakia, and sells at $16. Each can of Blend B (Balkan Intrigue) contains 3 ounces of Virginia and 9 ounces of Latakia, and sells at $12. How many cans of each blend should Albert make in order to maximize his total sales revenue?

Formulation We assume that Albert can sell both blends at the stated prices, so that his problem is not limited demand, but rather his limited supply of tobacco leaf. The unknowns in the problem are

x = number of cans of Blend A to be produced

y = number of cans of Blend B to be produced.

We summarize the information we have in the following table.

	Blend A	Blend B	Available
Virginia per can	8 oz	3 oz	480 oz
Latakia per can	4 oz	9 oz	720 oz
Price per can	$16	$12	
Number of cans	x	y	

Albert's revenue R from x cans of Blend A at $16 and y cans of Blend B at $12 is

$$16x + 12y = R,$$

and this is the quantity he wants to maximize. The total number of ounces of Virginia needed is $8x + 3y$, and so we must have

$$8x + 3y \leq 480. \tag{3.1}$$

Likewise $4x + 9y$ ounces of Latakia will be used, so

$$4x + 9y \leq 720. \tag{3.2}$$

An implicit restriction on the variables is that neither can be negative, since x and y represent the number of cans of Blends A and B. Thus

$$x \geq 0, \tag{3.3}$$
$$y \geq 0. \tag{3.4}$$

We have reduced our problem to a purely mathematical formulation, namely:

Find x and y that maximize

$$16x + 12y = R,$$

subject to
$$8x + 3y \leq 480 \tag{3.1}$$
$$4x + 9y \leq 720 \tag{3.2}$$
$$x \geq 0 \tag{3.3}$$
$$y \geq 0. \tag{3.4}$$

We will give a geometric method of solving this problem in Section 3.2. ■

The mathematical problem above is known as a **linear program.** The quantity $16x + 12y = R$ that is to be optimized (in this case maximized) is called the **objective function,** and the inequalities (3.1) through (3.4) are called the **constraints** of the program. A point (x, y) is said to be a **feasible point** in case it satisfies all the constraints. For example, the point (40, 50) is feasible in the above program because

$$8(40) + 3(50) = 470 \leq 480$$
$$4(40) + 9(50) = 610 \leq 720$$
$$40 \geq 0,$$
$$50 \geq 0.$$

Thus Albert *could* make 40 cans of Blend A and 50 cans of Blend B, even though he would have tobacco left over, and so not maximize his revenue. On the other hand (50, 40) is *not* feasible, since for this point (3.1) becomes

$$8(50) + 3(40) = 520 \leq 480,$$

which is false. A point (x, y) that optimizes the objective function is called an **optimal point.**

Practice Problem 1 (a) Tell which of the following points are feasible in the program of Example 3.1: (50, 30), (30, 50), (45, 40). (b) Among the feasible points listed, which gives the best value of the objective function?

In Example 3.1 it could be argued that another implicit restriction on x and y is that they be whole numbers, since Albert only sells full, 12-ounce cans of Blends A and B. In general we will ignore such restrictions when setting up linear programming problems since they make solving the problems much more difficult. Usually reasonable answers can be found by treating the variables as real numbers, then rounding to integers after the best real solution is found.

EXAMPLE 3.2 Nutritionists indicate that an adequate daily diet should provide at least 75 gm of carbohydrate, 60 gm of protein, and 60 gm of fat. An ounce of Food A contains 3 gm of carbohydrate, 2 gm of protein, and 4 gm of fat, and costs 10¢, while an ounce of Food B contains 3 gm of carbohydrate, 4 gm of protein, and 2 gm of fat, and costs 15¢. How many ounces of each food should be combined per day to meet the nutritional requirements at lowest cost?

Formulation The unknowns are

$$x = \text{number of ounces of Food A per day},$$
$$y = \text{number of ounces of Food B per day}.$$

The information is summarized in the table below.

	Food A	Food B	Needed
Carbohydrate/oz	3 gm	3 gm	75 gm
Protein/oz	2 gm	4 gm	60 gm
Fat/oz	4 gm	2 gm	60 gm
Cost/oz	10¢	15¢	
Number of ounces	x	y	

The objective function is the total cost

$$10x + 15y = C \text{ (in cents)},$$

which we want to *minimize* in this example. The number of grams of carbohydrates in x ounces of Food A and y ounces of Food B is $3x + 3y$, which must be at least 75 to meet the daily requirement. Thus we have the constraint

$$3x + 3y \geq 75.$$

Similarly, the protein and fat requirements lead to the following constraints:

$$2x + 4y \geq 60,$$
$$4x + 2y \geq 60.$$

Adding the obvious requirement that neither x nor y can be negative, we arrive at the following linear program:

Minimize $\qquad 10x + 15y = C$
subject to $\qquad 3x + 3y \geq 75$
$\qquad\qquad\quad\; 2x + 4y \geq 60$
$\qquad\qquad\quad\; 4x + 2y \geq 60$
$\qquad\qquad\quad\quad\quad\;\; x \geq 0$
$\qquad\qquad\quad\quad\quad\;\; y \geq 0.$ ■

Practice Problem 2   A truck can carry at most 4000 pounds and 2000 cubic feet of cargo. A TV set weighs 40 pounds and occupies 3 cubic feet. A VCR weighs 25 pounds and occupies 2 cubic feet. The trucker gets $5 for each TV set and $4 for each VCR carried, and wants to load the truck so as to maximize revenue. Set up the corresponding linear program.

Remark   Notice that the first order of business when setting up a linear program is to *state clearly what the variables represent*. Without this step anything that follows is meaningless. In Example 3.1 statements such as

$$x = \text{Blend A}, \qquad y = \text{Blend B}$$

would be inadequate. Do x and y count cans, ounces, or what? *Failure to identify completely and explicitly what the variables represent leads to confusion for both the person setting up a linear program and anyone trying to understand his or her work.*

Making a table as in the previous examples can be helpful in determining the constraints and objective function of a linear program.

LINEAR INEQUALITIES

Although a linear program may involve any number of variables, if there are precisely two, as in our examples so far, the constraints of the program can be given a geometric interpretation. They have the form of **linear inequalities,** that is, inequalities of the form

$$ax + by \geq c$$

or

$$ax + by \leq c,$$

where a and b are not both 0. Of course the graph of the corresponding *equality,*

$$ax + by = c,$$

is a straight line, as we saw in Chapter 1. It turns out that

---

The graph of a linear inequality

$$ax + by \geq c \quad \text{or} \quad ax + by \leq c,$$

where a and b are not both 0, consists of all points in the plane on one side of the line with equation $ax + by = c$, along with the points on the line itself.

Consider, for example, the linear inequality

$$8x + 3y \leq 480, \tag{3.1}$$

which is one of the constraints in the linear program of Example 3.1. The line that is the graph of $8x + 3y = 480$ is graphed by finding its x- and y-intercepts, as shown in Figure 3.1(a).

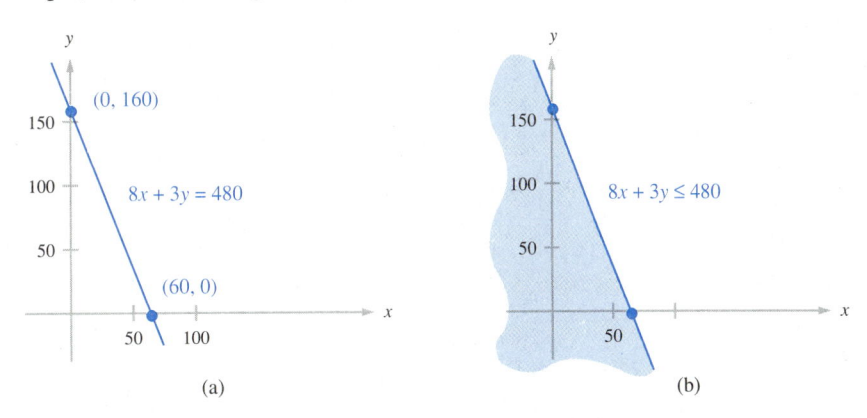

FIGURE 3.1

Figure 3.1(b) indicates with shading the set of points (x, y) satisfying the inequality

$$8x + 3y \leq 480. \tag{3.1}$$

Once one knows that the graph of a linear inequality consists of all points on one side of a line, graphing the inequality entails only graphing the line and deciding which side to shade. The latter decision can be made by testing in the inequality a single point P, not lying on the line. If the point's x- and y-coordinates satisfy the inequality, then the side of the line containing P is shaded; but if they do not, then the side of the line not containing P is shaded. For example, the origin $(0, 0)$ may be tested in the inequality (3.1). Since

$$8(0) + 3(0) \leq 480$$

is a true statement, the side of the line containing $(0, 0)$, that is, below and to its left, is the graph of the inequality. Such a set of points on one side of a line is called a **half-plane.**

EXAMPLE 3.3 Graph the inequalities (a) $3x - 4y \geq 12$ and (b) $4x + 5y \geq 0$.

Solution (a) We start graphing the line with equation $3x - 4y = 12$ by finding its intercepts, $(4, 0)$ and $(0, -3)$. Testing $(0, 0)$ gives us

$$3(0) - 4(0) \geq 12,$$

which is false. Thus we shade the side of the line *not* containing the origin, as shown in Figure 3.2(a).

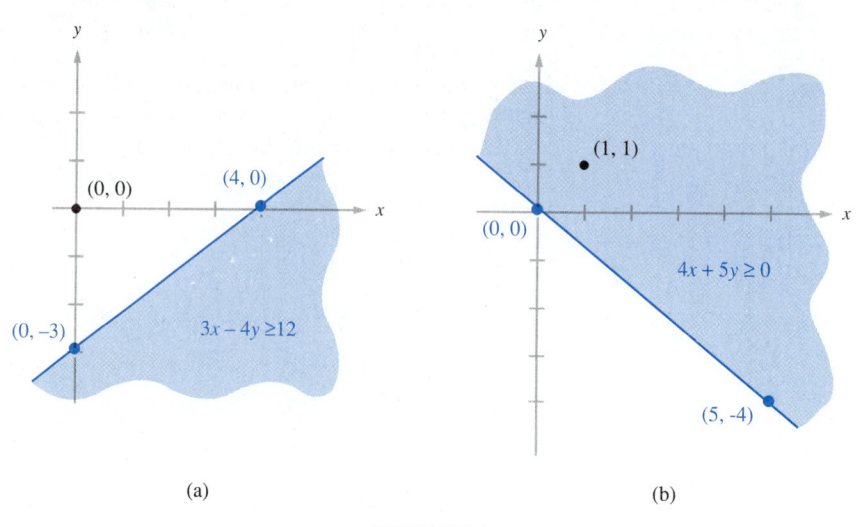

FIGURE 3.2

(b) Again we start by graphing the corresponding line, in this case with equation $4x + 5y = 0$. In this case, however, the point $(0, 0)$ is *on* this line. Thus we choose another point to test, say $(1, 1)$. Since

$$4(1) + 5(1) \geq 0$$

is true, we shade the side of the line containing $(1, 1)$ as shown in Figure 3.2(b). ∎

EXAMPLE 3.3 Continued

Graph the set of points satisfying *both* $3x - 4y \geq 12$ and $4x + 5y \geq 0$.

Solution Using Figure 3.2, we now graph *both* inequalities on the same set of axes in Figure 3.3(a). To save time in graphing, instead of shading each inequality, we use arrows to point from each line to the half-plane satisfying each inequality as in Figure 3.3(b). The set of points where the two half-planes overlap is shaded in Figure 3.3(c). Notice that only the portions of the lines $3x - 4y = 12$ and $4x + 5y = 0$ adjacent to the hatched region belong to the graph. ∎

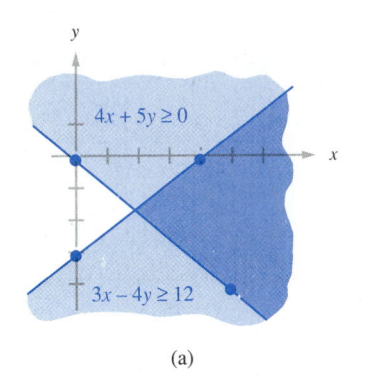

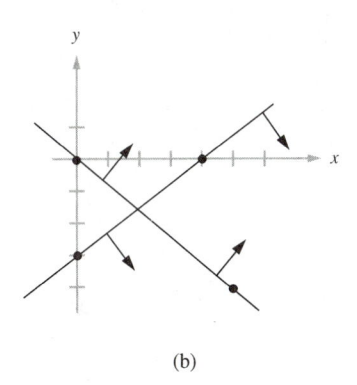

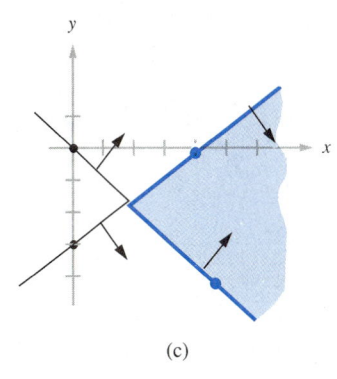

FIGURE 3.3

We see that to graph a system of linear inequalities we may use the following technique.

Graphing a System of Linear Inequalities

1. Graph the equation corresponding to each inequality.

2. Use a test point to decide which side of each line contains points satisfying the corresponding inequality, and indicate this side with an arrow.

3. Shade the region on the correct side of all the lines.

122　CHAPTER 3　LINEAR PROGRAMMING

Practice Problem 3　Graph the set of points satisfying the inequalities $2x + 3y \leq 12$ and $x \geq 2$.

Practice Problem 4　Graph the region of feasible points for the linear program of Example 3.2; that is, graph the system of linear inequalities forming its constraints. These constraints are

(1) $3x + 3y \geq 75$
(2) $2x + 4y \geq 60$
(3) $4x + 2y \geq 60$
(4) $\quad\quad x \geq 0$
(5) $\quad\quad y \geq 0.$

Regions formed by overlapping half-planes may have certain **corner points,** that is, points in all the half-planes and on more than one of the lines defining them. In Example 3.3 there is just one corner point, namely the point where the lines $3x - 4y = 12$ and $4x + 5y = 0$ intersect. (See Figure 3.4.) This point can be found by solving the system

$$3x - 4y = 12$$
$$4x + 5y = 0$$

as in Chapters 1 and 2. We find the corner point to be $(60/31, -48/31)$.

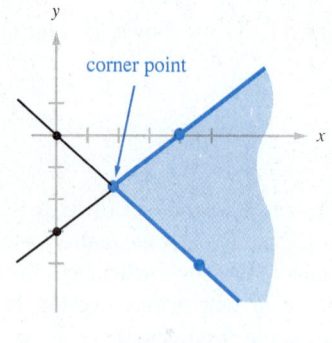

FIGURE 3.4

EXAMPLE 3.4　Graph the set of points satisfying

(1) $3x + 3y \leq 15$
(2) $\quad x + 2y \geq 2$
(3) $\quad\quad\quad y \geq x,$

and find all corner points.

Solution　The region is graphed in Figure 3.5.

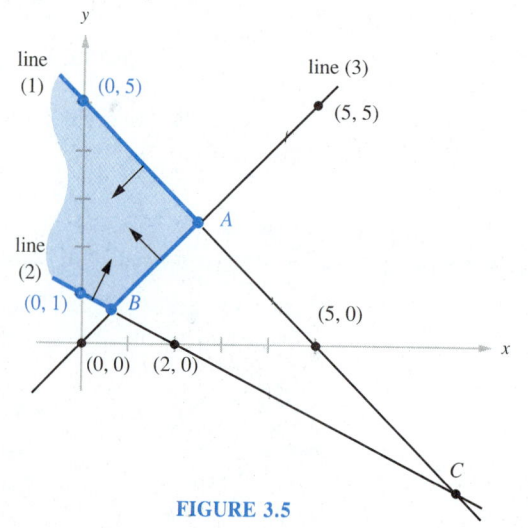

FIGURE 3.5

Corner point A is found to be (2.5, 2.5) by solving the system

$$(1) \quad 3x + 3y = 15,$$
$$(3) \quad y = x.$$

Likewise corner point B is found to be (2/3, 2/3) by solving the system

$$(2) \quad x + 2y = 2,$$
$$(3) \quad y = x.$$

Notice that the point labeled C is *not* a corner point, since it does not satisfy all the inequalities. ■

Practice Problem 5 How many corner points does the region graphed in Practice Problem 4 have?

EXERCISES 3.1

In Exercises 1 and 2 set up a linear program corresponding to the problem.

1. Karl's Koffeehaus sells two coffee blends. Each pound of the Haus blend contains 8 ounces of Brazilian and 8 ounces of Colombian coffee, while each pound of the Special blend contains 4 ounces of Brazilian and 12 ounces of Colombian. Karl has 1000 ounces of Brazilian and 1500 ounces of Colombian coffee in stock. He makes a profit of $1 on each pound of the Haus blend and $1.25 on each pound of the Special blend. He would like to maximize his profit by making up x pounds of the Haus blend and y pounds of the Special blend. This information is summarized in the following table.

	Haus	Special	Available
Brazilian per pound	8 oz	4 oz	1000 oz
Colombian per pound	8 oz	12 oz	1500 oz
Profit per pound	$1	$1.25	
Number of pounds	x	y	

2. A farmer grows beets for both their roots and green tops. Each acre of Big Red beet costs $400 to plant and yields 2000 pounds of roots and 1500 pounds of tops. Each acre of Cannonball beet costs $600 to plant and yields 2500 pounds of roots and 1200 pounds of tops. The farmer has a contract to supply 100,000 pounds of roots and 75,000 pounds of tops. He plants x acres of Big Red and y acres of Cannonball, and would like to minimize his total cost. This information is summarized in the following table.

	Big Red	Cannonball	Needed
Roots per acre	2000 lbs	2500 lbs	100,000 lbs
Tops per acre	1500 lbs	1200 lbs	75,000 lbs
Cost per acre	$400	$600	
Number of acres	x	y	

In Exercises 3 and 4 make a table summarizing the information given. Then set up the corresponding linear program.

3. A teacher is buying fried chicken for a class picnic. A Big Barrel of chicken contains 10 pieces of white meat and 10 pieces of dark meat, and costs $9. A Supersack contains 7 pieces of white meat and 4 pieces of dark meat, and costs $6. The class demands at least 100 pieces of white meat and 80 pieces of dark meat. The teacher would like to provide this at the least possible cost, buying x Big Barrels and y Supersacks.

4. Patty's Party Service gives birthday parties for $40 and Halloween parties for $50. A birthday party requires 20 balloons and 15 noisemakers, while a Halloween party requires 25 balloons and 20 noisemakers. Only 200 balloons and 150 noisemakers are available. Patty would like to schedule x birthday parties and y Halloween parties so as to take in as much money as possible.

In Exercises 5–12 set up a linear program corresponding to each problem. Explain what the variables represent in Exercises 9–12.

5. A steel mill makes two grades of stainless steel, both of which are sold in 100-pound bars. The standard-grade bar contains 90% steel and 10% chromium by weight, and sells for $110. The high-grade bar contains 80% steel and 20% chromium by weight, and sells for $120. If the company has a supply of 72,000 pounds of steel and 10,000 pounds of chromium, how many bars of each grade should it produce to maximize total revenue? Suppose it produces x standard-grade bars and y high-grade bars.

6. A farmer has a 2000-acre plot of land on which she intends to plant some combination of alfalfa and beets. Alfalfa requires $20 capital and 25 inches of water for each acre planted, while beets require $30 capital and 75 inches of water for each acre planted. The net profit per acre is $30 for alfalfa and $40 for beets. If the farmer has a total of $45,000 in capital and a total water allotment of 90,000 acre-inches, how many acres of each crop should she plant to maximize total profit? (*Note:* An acre-inch is the amount of water needed to irrigate an acre of land to a depth of one inch.) Suppose she grows x acres of alfalfa and y acres of beets.

7. The House of Coffee sells two different mixtures of Mocha, Java, and Santos beans. Each pound of Mixture A contains $\frac{1}{4}$ pound Mocha, $\frac{1}{2}$ pound Java, and $\frac{1}{4}$ pound Santos and sells for $3.50. Each pound of Mixture B contains $\frac{3}{5}$ pound Mocha, $\frac{1}{5}$ pound Java, and $\frac{1}{5}$ pound Santos and sells for $4.00. If the store has a supply of 300 pounds of Mocha, 250 pounds of Java, and 150 pounds of Santos beans, how many pounds of each mixture should be made in order to maximize total revenue? Suppose x pounds of Mixture A and y pounds of Mixture B are made.

8. World Travel Service has agreed to transport a group of 1800 football fans from Los Angeles to Chicago for one of the season's big games. The agency can charter two different types of aircraft for the trip. Type A has 20 first-class seats and 80 economy seats and costs $3000 to operate for one trip. Type B has 40 first-class seats and 60 economy seats and costs $4000 to operate. If 600 of the passengers want to go first-class while the remaining 1200 have ordered economy seats, how many planes of each type should be assigned to the charter to minimize total operating cost? Suppose x Type A and y Type B planes are used.

9. A gem dealer buys rough-cut amethysts and beryls of uniform size, which she finishes for sale to retail jewelers. Each amethyst requires one hour on a grinder, one hour on a

sander, and two hours on a polisher and yields a net profit of $75. Each beryl requires two hours on a grinder, five hours on a sander, and two hours on a polisher and yields a profit of $100. If the grinder, sander, and polisher are each available for 40 hours a week, how many amethysts and beryls should be finished each week to maximize total profit?

10. A cannery has decided to replace its old machines with newer equipment. The manager has been allocated $300,000 of capital for the new machines. He has 36 available operators and space for, at most, 28 new units. Of the two models under consideration, Unit A costs $10,000, requires one operator, and produces 600 cans per hour, while Unit B costs $15,000, requires two operators, and produces 1000 cans per hour. How many units of each type should be purchased in order to maximize can production?

11. A printing house orders newsprint in bulk rolls. The rolls supplied by Company A cost $100; each roll gives a two-day supply of paper and occupies 35 square feet of storage space. The rolls supplied by Company B cost $300; each roll gives a five-day supply of paper and occupies 25 square feet of storage space. The printing house has a total of 10,500 square feet of storage space and needs enough paper for 1000 days. How can this be achieved at lowest cost?

12. The Heine-Borel Corporation is trying to decide between two comprehensive insurance policies covering fire, theft, and liability. The Acme Insurance Company offers a policy that provides $10,000, $5000, and $20,000 of fire, theft, and liability coverage, respectively, for each unit bought, at a premium of $500 per unit. The Baltimore Insurance Company offers a policy that provides $10,000 each of fire, theft, and liability coverage for each unit bought, at a premium of $450 per unit. The corporation needs coverage of at least $400,000 each for fire and theft and $1,000,000 for liability. How many units of each policy should be purchased to minimize cost?

In Exercises 13–18 tell which of the points gives the best value of the objective function, among those points given that are feasible.

13. Points: (3, 5), (4, 4), (5, 3), (4, 5), (3, 6)

   Program: Maximize $5x + 6y$
   subject to $3x + 4y \leq 30$
   $5x + y \leq 35$
   $x \geq 0, y \geq 0.$

14. Points: (10, 20), (15, 12), (13, 17), (20, 0), (4, 25)

   Program: Maximize $7x + 3y$
   subject to $3x + 4y \leq 100$
   $x + 2y \geq 30$
   $x - y \geq -5$
   $x \geq 0, y \geq 0.$

15. Points: (2, −5), (5, 1), (7, 0), (3, 3), (1, 4)

   Program: Minimize $3x + 5y$
   subject to $x + y \geq 5$
   $3x + 2y \geq 12$
   $x \geq 0, y \geq 0.$

16. Points: (1.1, 3.2), (2.0, 4.3), (1.5, −1.5), (0.9, 3.3), (4.2, 1.0)

   Program: Minimize $10x + 12y$
   subject to $2x + 4y \geq 12$
   $3x - 2y \geq 0$
   $5x + y \leq 5$
   $x \geq 0, y \geq 0.$

17. Points: (1, 2, 3), (2, 3, 1), (2, 2, 2), (3, 0, 2), (1, 4, 0)

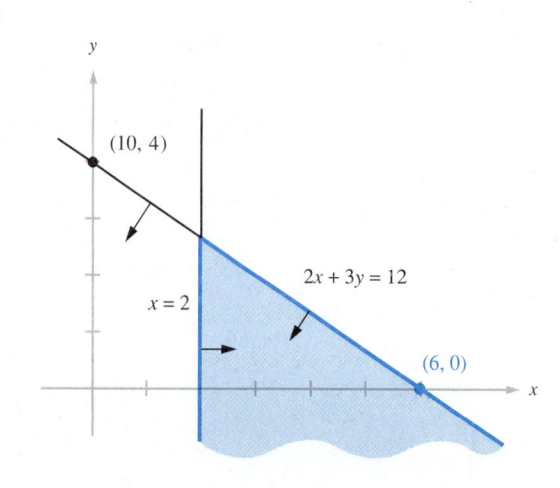

Program: Maximize $2x + 4y + 3z$
subject to $x + y + 2z \leq 6$
$2x + z \leq 7$
$x \geq 0, y \geq 0, z \geq 0.$

18. Points: (11, 0, 20), (30, 0, 0), (12, 10, 10), (0, 13, 18), (5, 15, 9)

Program: Minimize $x + 3y + 2z$
subject to $2x + y + 2z \geq 45$
$x + 2y + 2z \geq 40$
$x \geq 0, y \geq 0, z \geq 0.$

In Exercises 19–28 graph the given inequality.

19. $5x + 7y \geq 35$
20. $8x + 5y \geq 20$
21. $3x + 8y \leq 12$
22. $10x + 7y \leq 35$
23. $4x - 6y \leq 15$
24. $6x - 5y \geq -15$
25. $2x - 5y \geq 30$
26. $-8x + 6y \leq 36$
27. $3x + 5y \leq 0$
28. $12x - 7y \geq 0$

In Exercises 29–36 graph the system of inequalities and identify the coordinates of all corner points of the region graphed.

29. $4x + 5y \leq 10, x \geq 0$
30. $2x + 3y \geq 6, y \geq 0$
31. $3x + 2y \leq 12, x + y \geq 2, x \geq 0, y \geq 0$
32. $x + 2y \geq 1, x + 3y \leq 9$
33. $4x + 7y \leq 14, x \geq 0, y \geq 0$
34. $7x + 3y \geq 21, x \geq 0, y \geq 0$
35. $3x + 4y \leq 12, x - y \geq 3, x \geq 0, y \geq 0$
36. $x + y \geq 2, 2x + y \leq 10, x \geq 0, y \geq 0$

Answers to Practice Problems

1. (a) The points (30, 50) and (45, 40) are feasible. (b) The point (45, 40) is better.

2. If $x = $ the number of TV sets carried and
$y = $ the number of VCRs carried,

then the linear program is:

Maximize $5x + 4y = R$
subject to $40x + 25y \leq 4000$
$3x + 2y \leq 2000$
$x \geq 0$
$y \geq 0.$

3.

4.

(1) $3x + 3y = 75$
(2) $2x + 4y = 60$
(3) $4x + 2y = 60$

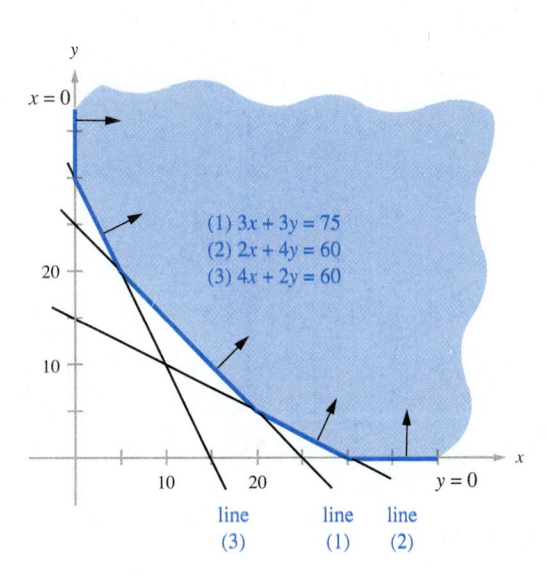

5. 4

3.2 The Graphical Method

In this section we will actually solve some linear programs. Our method involves graphing the set of points satisfying the constraints of the linear program. This set is called the **feasible region** of the program. Because our method is graphical we are limited to problems where there are two variables.

EXAMPLE 3.1 Revisited

(a) Graph the feasible region for Example 3.1, which involved Albert's Smoke Shop.
(b) Can Albert choose x and y so as to have a revenue of $1000? Of $1500?

Solution (a) Recall that the constraints in Example 3.1 are

$$8x + 3y \leq 480$$
$$4x + 9y \leq 720$$
$$x \geq 0$$
$$y \geq 0.$$

Using the methods of the last section we graph the feasible region as in Figure 3.6 on the next page.

(b) To determine whether Albert can achieve revenues of $1000 and $1500, we recall that his revenue is given by

$$16x + 12y = R.$$

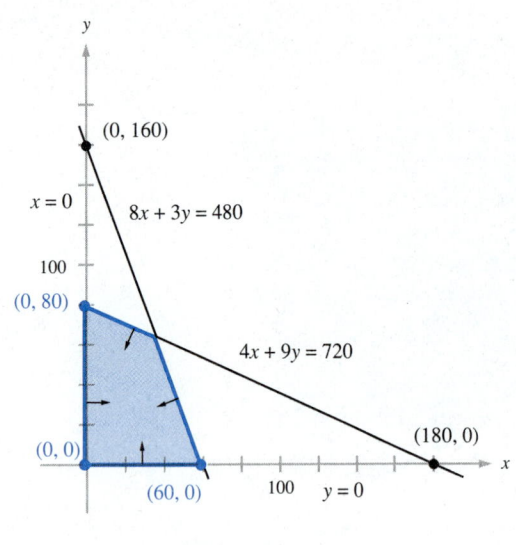

FIGURE 3.6

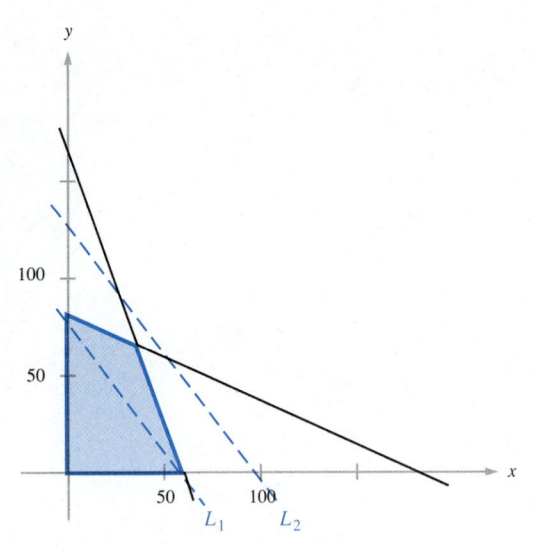

FIGURE 3.7

Thus the set of points giving a revenue of $1000 is exactly the set of points on the line L_1: $16x + 12y = 1000$. Likewise those points giving a revenue of $1500 lie on the line L_2: $16x + 12y = 1500$. Including these lines in our diagram yields Figure 3.7. Since there are feasible points on the line L_1, a revenue of $1000 is possible for Albert. The line L_2 does not touch the feasible region, however, so it is not possible for him to take in $1500. ■

OPTIMIZING THE OBJECTIVE FUNCTION

An examination of Figure 3.7 shows us how Albert may maximize his revenue. Clearly he can do better than $1000, since there are points (x, y) in the feasible region having both x and y bigger than points on L_1, and increasing both x and y must increase $16x + 12y = R$. The lines

$$L_1: 16x + 12y = 1000$$

$$L_2: 16x + 12y = 1500$$

appear to be parallel. Indeed, it is not hard to show that all lines with equations of the form

$$16x + 12y = R,$$

where R is some constant, are parallel to each other. These are called the **isolines** of the objective function. Increasing R simply moves the line upward and to the right, as Figure 3.8 shows.

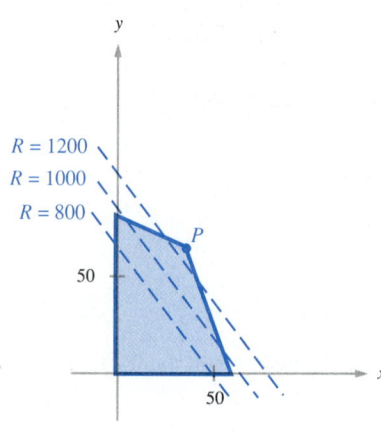

FIGURE 3.8

To achieve maximal revenue we need to find a line parallel to those shown that is as far to the top and right of the feasible region as possible while still touching it. Clearly in this case such a line will just touch the corner point P, and so the x- and y-coordinates of P will produce the most revenue for Albert. The point P is easily found, since it is on the lines with equations

$$8x + 3y = 480$$
$$4x + 9y = 720.$$

Solving these equations simultaneously gives the unique solution $x = 36$, $y = 64$. Thus Albert should make up 36 cans of Blend A and 64 cans of Blend B to maximize his revenue. This will bring in

$$R = 16(36) + 12(64) = \$1344.$$

THE GRAPHICAL METHOD

We summarize the graphical method of linear programming as follows.

1. Identify what variables x and y are sought.

2. Write down a linear program consisting of an objective function to be optimized and a set of linear constraints.

3. Graph the feasible region.

4. Set the objective function equal to two different arbitrarily chosen constants, and graph the corresponding isolines.

5. Imagine one of the isolines in step 4 moved parallel to itself so as to optimize the objective function at a point in the feasible region. *If this optimal point exists, it can always be taken to be a corner point of the feasible region.*

6. If necessary, solve simultaneously the equations of the lines intersecting at the optimal point to find the values of x and y there.

EXAMPLE 3.5 A university alumni association is planning a trip to a bowl game. Two types of buses are available. Type A can carry 40 people and 1000 pounds of baggage, and costs $2000. Type B can carry 50 people and 750 pounds of baggage, and costs $2400. A total of 800 people are going on the trip, and they have 18,000 pounds of baggage. How many of each type bus should be chartered to minimize the cost of the trip?

130 CHAPTER 3 LINEAR PROGRAMMING

Solution Let

$$x = \text{the number of Type A buses}$$
$$y = \text{the number of Type B buses.}$$

We summarize the given information in a table.

	Type A	Type B	Needed
People carried	40	50	800
Baggage carried	1000	750	18,000
Cost	$2000	$2400	
Number	x	y	

Our linear program is:

Minimize $\quad 2000x + 2400y = C$

subject to $\quad 40x + 50y \geq 800$

$\quad\quad\quad\quad\quad 1000x + 750y \geq 18{,}000$

$\quad\quad\quad\quad\quad\quad\quad\quad\quad x \geq 0$

$\quad\quad\quad\quad\quad\quad\quad\quad\quad y \geq 0.$

The feasible region is graphed in Figure 3.9.

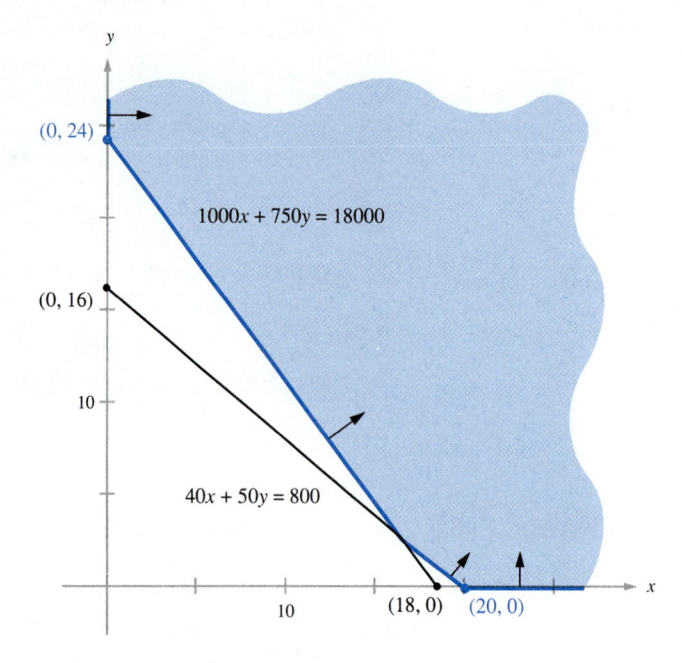

FIGURE 3.9

3.2 THE GRAPHICAL METHOD

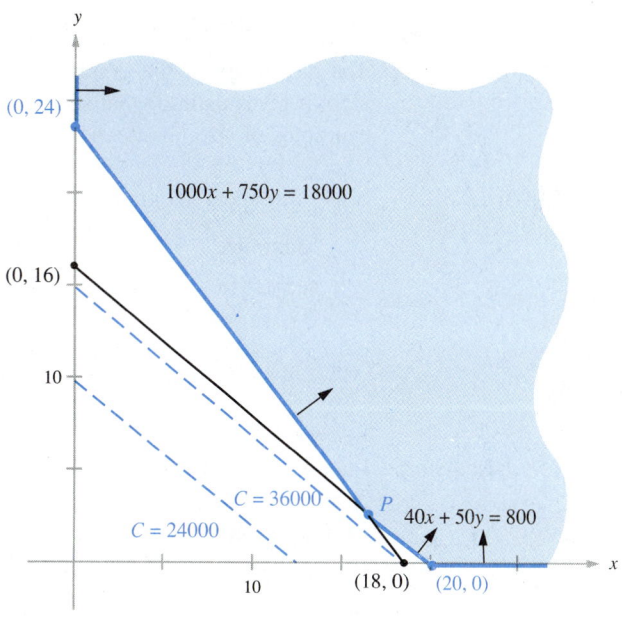

FIGURE 3.10

Now we graph two isolines $2000x + 2400y = C$, where we take C to be 24,000 and 36,000. (These values were chosen to make the intercepts of the corresponding isolines easy to compute.) The result is shown in Figure 3.10. Since we wish to minimize the cost C, we imagine these lines moved parallel to themselves as far to the lower left as possible while still touching the feasible region. The last point touched will be the corner point P where the two lines $40x + 50y = 800$ and $1000x + 750y = 18,000$ intersect. This is easily determined to be $(x, y) = (15, 4)$ by the methods of the previous chapter. Thus the alumni association should charter 15 Type A buses and 4 Type B buses. The cost will then be

$$C = 2000(15) + 2400(4) = \$39{,}600.\quad\blacksquare$$

Practice Problem 6 Use the graphical method to solve the following linear program:

Maximize $\quad 2x + y = R$
subject to $\quad 4x + 3y \leq 12$
$\quad\quad\quad\quad\ \ x + 3y \leq 6$
$\quad\quad\quad\quad\quad\quad\ \ x \geq 0$
$\quad\quad\quad\quad\quad\quad\ \ y \geq 0.$

SPECIAL CASES

The advantage of the graphical method of solving a linear program is the insight it gives us into what such problems are really about. We now give several examples of special situations that can arise.

EXAMPLE 3.6 Solve the linear program:

$$\begin{aligned} \text{Maximize} \quad & 6x + 3y = R \\ \text{subject to} \quad & x + 2y \geq 8 \\ & 2x + y \geq 10 \\ & x \geq 0 \\ & y \geq 0. \end{aligned}$$

Discussion Figure 3.11 shows the feasible region, along with isolines corresponding to $R = 12$ and $R = 24$. Since we are trying to maximize R, we want to find an isoline as far to the top and right as possible that still touches the feasible region. But this region is unbounded in this direction, and we can find lines

$$6x + 3y = R$$

with R as big as we please containing feasible points. Thus, there is no maximum value of R, and the linear program has no solution. ∎

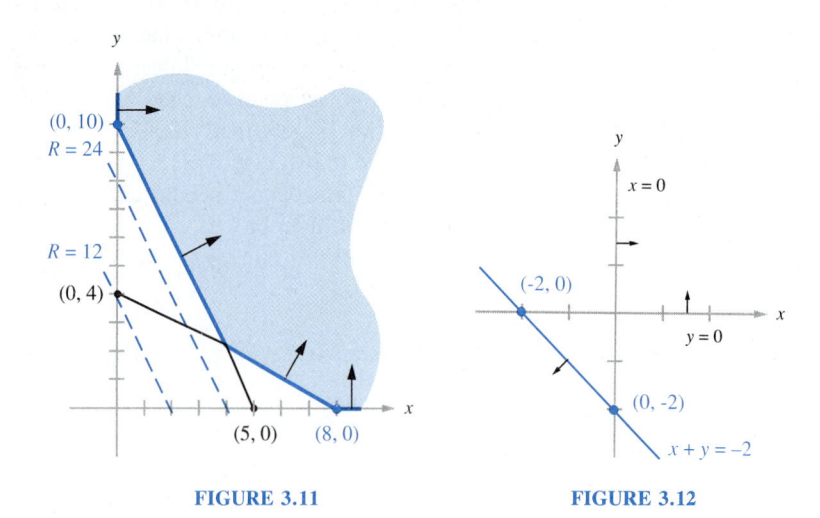

FIGURE 3.11

FIGURE 3.12

3.2 THE GRAPHICAL METHOD

EXAMPLE 3.7 Solve the linear program:

$$\text{Maximize} \quad 3x + 7y = R$$
$$\text{subject to} \quad x + y \leq -2$$
$$x \geq 0$$
$$y \geq 0.$$

Discussion Figure 3.12 shows the graphs of the lines corresponding to our constraints, with arrows indicating the half-planes defined. Clearly no points satisfy all the inequalities. The feasible region is empty, and there is no solution to the linear program. ∎

EXAMPLE 3.8 Solve the linear program:

$$\text{Maximize} \quad 7x + 9y = R$$
$$\text{subject to} \quad 10x + 13y \leq 130$$
$$x + 2y \leq 18$$
$$x \geq 0$$
$$y \geq 0.$$

Solution In Figure 3.13 we have graphed the feasible region, along with the isolines corresponding to $R = 45$ and $R = 63$.

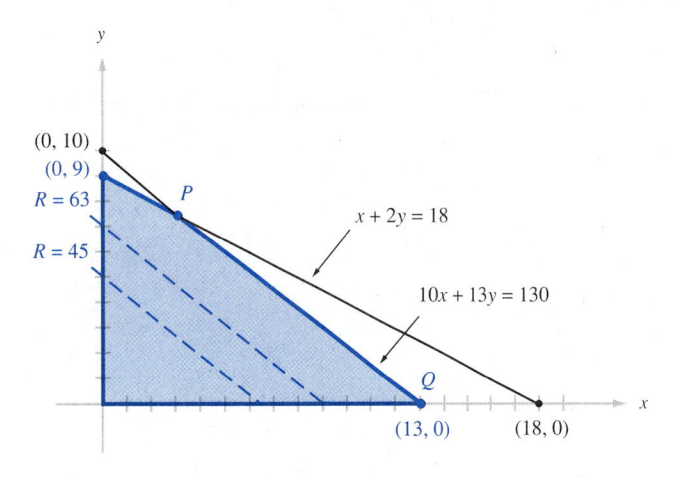

FIGURE 3.13

When we move an isoline to the top and right it is not clear whether the point P or the point Q will be the last feasible point touched. Our graph is not sufficiently precise to tell whether P or Q is optimal. Thus we resort to computing the objective function at both P and Q. We find P by solving the system

$$10x + 13y = 130$$
$$x + 2y = 18$$

getting $P = (26/7, 50/7)$. Likewise Q is the solution of the system

$$10x + 13y = 130$$
$$y = 0,$$

which is easily seen to be $(13, 0)$. Now we test these points in the objective function.

(x, y)	$7x + 9y = R$
$P = (26/7, 50/7)$	$7(26/7) + 9(50/7) = 90\frac{2}{7}$
$Q = (13, 0)$	$7(13) + 9(0) = 91$ ⟵ maximum

Since we want to maximize R, the optimal point is $(13, 0)$. ∎

EXAMPLE 3.9 Solve the linear program:

Maximize $\quad 6x + 3y = R$
subject to $\quad x + 2y \leq 8$
$\quad\quad\quad\quad\quad 2x + y \leq 10$
$\quad\quad\quad\quad\quad x \geq 0$
$\quad\quad\quad\quad\quad y \geq 0.$

Solution The feasible region and the lines $R = 12$ and $R = 24$ are graphed in Figure 3.14.

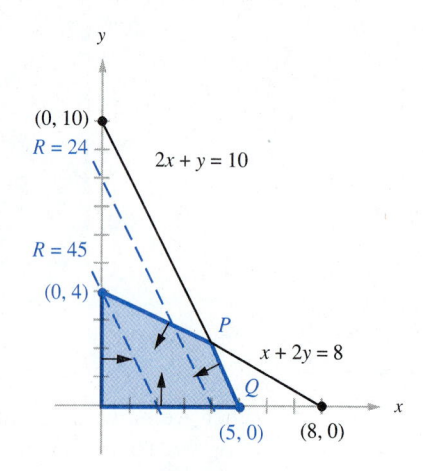

FIGURE 3.14

It is not clear from our graph whether point P or point Q will be the last point in the feasible region touched by an isoline moving to the top and right. We will decide by testing both points in the objective function $6x + 3y = R$. We can find P by solving the system

$$x + 2y = 8$$
$$2x + y = 10,$$

getting $P = (4, 2)$. Likewise Q satisfies the system

$$2x + y = 10$$
$$y = 0,$$

and so $Q = (5, 0)$. We compute R at both these points.

(x, y)	$6x + 3y = R$
$P = (4, 2)$	$6(4) + 3(2) = 30$
$Q = (5, 0)$	$6(5) + 3(0) = 30$

Since we get $R = 30$ at both P and Q, the isolines must be parallel to the line through P and Q. Any of the infinitely many points on the segment from P to Q gives the same maximal value of R. Note that two of the optimal points are corner points of the feasible region, namely P and Q. Even though there are many optimal points, there is only one maximal value of R, namely 30. ∎

LIMITATIONS OF THE GRAPHICAL METHOD

The main limitation of the graphical method is that it can only be used when there are two variables to be determined. (Although a geometric interpretation can be given to three-variable problems in which the feasible region is a solid bounded by planes, graphing in three dimensions is too difficult for this to be a practical method for solving three-variable problems.) Even in two-variable problems graphing may not be a reasonable method. For example, when there are many constraints the feasible region may be quite complicated, or if three or more lines intersect at points close to each other, it may be hard to tell which intersections are corners of the feasible region. Furthermore imagining an isoline moved parallel to itself involves an "eyeball" estimate that may not be accurate enough to produce the correct answer, as we saw in our last two examples. In the next section we will introduce an algebraic method of solving linear programming problems that applies no matter how many variables there are.

EXERCISES 3.2

Use the graphical method to solve Exercises 1–14. Give the optimal point and corresponding value of the objective function.

1. Maximize $3x + 2y = R$
   subject to $4x + y \leq 100$
   $x + 4y \leq 100$
   $x \geq 0, y \geq 0.$

2. Maximize $4x + 5y = R$
   subject to $5x + 3y \leq 28$
   $3x + 4y \leq 30$
   $x \geq 0, y \geq 0.$

3. Minimize $6x + 4y = C$
   subject to $5x + 3y \geq 15$
   $2x + 5y \geq 10$
   $x \geq 0, y \geq 0.$

4. Minimize $x + 2y = C$
   subject to $x + y \geq 30$
   $2x + y \geq 40$
   $x \geq 0, y \geq 0.$

5. Maximize $x + 2y = R$
   subject to $x + y \leq 50$
   $x - y \geq -30$
   $x \leq 40$
   $x \geq 0, y \geq 0.$

6. Minimize $x - y = C$
   subject to $4x + y \geq 4$
   $x + 2y \leq 4$
   $x \geq 0, y \geq 0.$

7. Exercise 5, Section 3.1
8. Exercise 6, Section 3.1
9. Exercise 7, Section 3.1
10. Exercise 8, Section 3.1
11. Exercise 9, Section 3.1
12. Exercise 10, Section 3.1
13. Exercise 11, Section 3.1
14. Exercise 12, Section 3.1

15. For an experiment on animal behavior a psychologist must train rats and mice to run two mazes. The amount of time each rat and each mouse is given in the mazes is shown below.

	Time per rat	Time per mouse	Available
Maze A	12 min	8 min	240 min
Maze B	10 min	15 min	300 min

Under these conditions what is the maximum number of animals that can be used in this experiment, and what combination of rats and mice will produce this maximum?

16. A manufacturer makes two sizes of glass vases. The taller ones, being more sturdy than the shorter ones, are less expensive to ship but more expensive to produce. If the production and shipping costs for each type of vase are as given in the table below, and if the manufacturer earns a $4 profit on the taller vases and a $3 profit on the shorter vases, what combination of vases will yield the maximum profit? What is this profit?

	Taller vase	Shorter vase	Available
Production cost	$10	$4	$6000
Shipping cost	$ 1	$2	$ 800

17. Using two foods, a hospital dietician must prepare a special meal that contains required amounts of vitamin A and calcium. The amount of vitamin A and calcium supplied by one ounce of each food and the required amounts are shown in the table below. (*Note:* i.u. = international units)

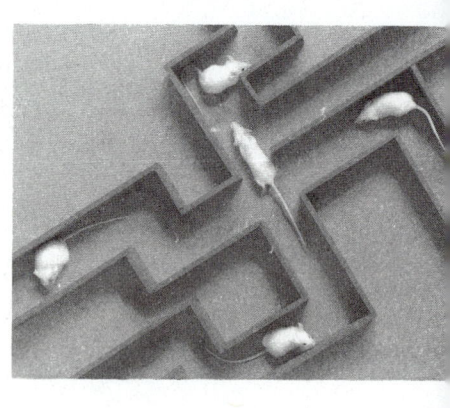

	Units per ounce		Needed
	Food 1	Food 2	
Vitamin A	800 i.u.	300 i.u.	6000 i.u.
Calcium	50 mg	150 mg	900 mg

If each ounce of Food 1 contains 200 calories and each ounce of Food 2 contains 150 calories, what combination of the foods will supply the required amounts of vitamin A and calcium with the fewest calories? What is the minimum number of calories?

18. A homeowner wants to mix two lawn fertilizers, Nourish and Greenup, to obtain a mixture of nitrogen and phosphoric acid for use on her lawn. The amounts of each chemical provided by the two fertilizers are given in the following table.

	Pounds per bag		Needed
	Nourish	Greenup	
Nitrogen	2.25	1.5	36 lbs
Phosphoric acid	0.25	0.5	7 lbs

If each bag of Nourish costs $5 and each bag of Greenup costs $6, what combination of the two fertilizers will provide the necessary amounts of nitrogen and phosphoric acid at the least cost?

19. An oil company operates two eastern refineries. The Albany facility produces 100 barrels of high-grade oil and 300 barrels of medium-grade oil per day. The Boston facility produces 200 barrels of high-grade and 100 barrels of medium-grade per day. The company has determined that the combined total output of high-grade oil should fall somewhere between 12,000 and 18,000 barrels (inclusive), and the combined output of medium-grade should fall between 24,000 and 30,000 barrels. If the Albany and Boston refineries cost $2000 and $4200 per day to operate, respectively, how many days should each be run to meet the requirement at lowest cost?

20. A chemical firm has been trying to reduce sulfur oxide emissions from its two Chicago plants. Plant I produces 10 tons of sulfuric acid, 6 tons of ammonia, and 7.5 tons of ammonium sulfate per day, with 100 pounds of sulfur oxide emitted. Plant II produces no sulfuric acid, 12 tons of ammonia, and 5 tons of ammonium sulfate per day, with 72 pounds of sulfur oxide emitted. How many days should each plant operate in order to produce at least 600 tons of each of the three chemicals with a minimum total emission of sulfur oxide?

Answer to Practice Problem 6. The maximum is 6 when $x = 3$ and $y = 0$.

3.3 The Simplex Method

In this section we will introduce an algebraic method, called the **simplex algorithm,** for solving linear programs. This method was developed by the American mathematician George B. Dantzig in the 1940s. It relies heavily on the theory of linear systems explained in Chapter 2, and in particular on the pivot transformation.

We first apply the simplex method to a linear program with two variables, so that our work will have a geometric interpretation. Let us consider the problem:

Maximize $\qquad 20x + 30y = R$

subject to $\qquad x + 2y \le 14$

$\qquad\qquad 3x + 2y \le 18$

$\qquad\qquad x \ge 0, \quad y \ge 0.$

The feasible region for this program is shown in Figure 3.15, along with the isolines corresponding to $R = 60$ and $R = 90$.

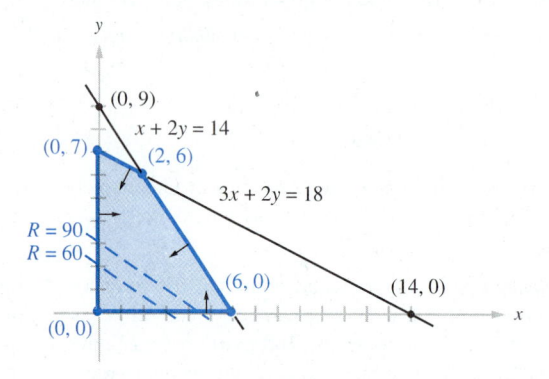

FIGURE 3.15

The feasible region is bounded by the lines $x + 2y = 14$, $3x + 2y = 18$, $x = 0$, and $y = 0$, corresponding to the constraints of the program. We know that the solution to the program will be one of the corners of this region, and each such corner is the intersection of a pair of these lines. We have marked the coordinates of all intersections of two of these lines in Figure 3.15. Four of these intersections are corners of the feasible region, but two, namely (0, 9) and (14, 0), are not. Since R is to be maximized, it appears that (2, 6) is the optimal point, although it is somewhat of a close call between (2, 6) and (0, 7).

Now we will show how the simplex method arrives at this solution algebraically. First we will replace the constraints

$$x + 2y \le 14$$
$$3x + 2y \le 18$$

with equations by introducing what are called **slack variables.** If $x + 2y$ is less than or equal to 14, then it falls short of 14 by some nonnegative quantity s; that is

$$x + 2y + s = 14, \qquad s \ge 0.$$

Likewise $3x + 2y \leq 18$ means that

$$3x + 2y + t = 18, \quad t \geq 0.$$

Thus our constraints can be replaced by a set of equations and inequalities involving two new slack variables s and t:

$$x + 2y + s = 14$$
$$3x + 2y + t = 18$$
$$x \geq 0, y \geq 0, s \geq 0, t \geq 0.$$

If we neglect the requirement that x, y, s, and t all be nonnegative, we have a system of two equations in four unknowns,

$$x + 2y + s = 14 \qquad (3.5)$$
$$3x + 2y + t = 18,$$

which will have infinitely many solutions. Notice that setting any one of the unknowns equal to zero restricts the point (x, y) to lie on one of the lines bounding the feasible region. For setting x or y equal to zero means that (x, y) is on the line $x = 0$ or $y = 0$, while setting s equal to zero makes the first equation $x + 2y = 14$, and setting t equal to zero makes the second equation $3x + 2y = 18$. Of course the feasible region is bounded by the lines $x = 0$, $y = 0$, $x + 2y = 14$, and $3x + 2y = 18$.

Setting *two* of the variables equal to zero thus means that (x, y) must be on two of the bounding lines; that is, (x, y) must be one of the six intersection points labeled in Figure 3.15. These include the corner points of the feasible region, at least one of which optimizes our objective function. Thus we need a method for finding which variables must be set equal to zero in order to specify an optimal point.

In Chapter 2 we saw how in the system of two equations in four unknowns (3.5) we could choose any two of the variables as *free* variables and solve for the other two variables (called *basic* variables) in terms of them. As the system stands, s and t are isolated, so we could easily solve for them:

$$s = 14 - x - 2y$$
$$t = 18 - 3x - 2y.$$

Here we are thinking of x and y as free and s and t as basic variables. Setting the free variables x and y both equal to 0 gives the point $(x, y) = (0, 0)$ in Figure 3.15. (Of course our graph does not show s and t.)

By performing a pivot transformation we could make a different variable basic. Recall from Section 2.1 that the tableau form of system (3.5) is

$$\begin{array}{c} \\ s \\ t \end{array} \begin{array}{cccc} x & y & s & t \\ \left[\begin{array}{cccc|c} \boxed{1} & 2 & \boxed{1} & 0 & 14 \\ 3 & 2 & 0 & \boxed{1} & 18 \end{array}\right] \end{array}$$

The columns corresponding to the basic variables s and t consist of a single 1 (outlined with a square), with the other entries 0. Setting both the free variables x and y equal to 0 makes $s = 14$ and $t = 18$. These values can be read from the rightmost column, with 14 in the row where there is a 1 in the s column, and 18 in the row where there is a 1 in the t column. We indicate this by marking each row on the left with the corresponding basic variable. These **row labels** indicate which basic variables take on the values in the rightmost column when all free variables equal 0.

We will pivot on the circled entry in row 1 and column 1. Since column 1 corresponds to the variable x, after pivoting x will become the basic variable corresponding to row 1.

$$\begin{array}{c} x \\ t \end{array} \begin{bmatrix} \begin{array}{cccc|c} x & y & s & t & \\ \boxed{1} & 2 & 1 & 0 & 14 \\ 0 & -4 & -3 & \boxed{1} & -24 \end{array} \end{bmatrix} \quad (R_2 + (-3)R_1)$$

Now x and t are the basic variables, and we can solve for them in terms of y and s:

$$x = 14 - 2y - s$$
$$t = -24 + 4y + 3s.$$

Setting the new free variables y and s equal to zero gives $x = 14$ and $t = -24$, as can be read from the rightmost column. This corresponds to the point $(x, y) = (14, 0)$ in Figure 3.15, again the intersection of two lines defining the feasible region.

Of course the point $(14, 0)$ is not even a corner of the feasible region, so our pivot does not seem to have done us much good. *The idea of the simplex method is to perform pivot operations in such an order that when all the pivots have been done, the solution to the linear program is obtained by setting the free variables equal to 0.*

THE SIMPLEX METHOD DEMONSTRATED

Now we will show how we can find the solution to our linear program by a sequence of pivots. To keep things simple *we will delay explaining how to choose the pivot entries until Section 3.4*. Before starting we will add another row to our tableau, a row corresponding to the equation

$$20x + 30y = R$$

defining the objective function. Thus our linear system is

$$\begin{aligned} x + 2y + s \quad\quad &= 14 \\ 3x + 2y \quad\quad + t &= 18 \\ 20x + 30y \quad\quad\quad &= R, \end{aligned}$$

and our tableau is

$$\begin{array}{c} s \\ t \end{array} \left[\begin{array}{cccc|c} x & y & s & t & \\ 1 & ② & \boxed{1} & 0 & 14 \\ 3 & 2 & 0 & \boxed{1} & 18 \\ \hline 20 & 30 & 0 & 0 & R \end{array} \right].$$

Here the horizontal line sets off the last row because the corresponding equation involving the objective function is special. *We will never pivot on an element in the bottom row.* We start by pivoting on the circled 2 in row 1 and column 2. Row 1 had corresponded to the basic variable s, but after the pivot s will be replaced by y (the variable of column 2) as a basic variable.

$$\left[\begin{array}{cccc|c} .5 & ① & .5 & 0 & 7 \\ 3 & 2 & 0 & 1 & 18 \\ \hline 20 & 30 & 0 & 0 & R \end{array} \right] \quad (.5R_1)$$

$$\begin{array}{c} y \\ t \end{array} \left[\begin{array}{cccc|c} x & y & s & t & \\ .5 & \boxed{1} & .5 & 0 & 7 \\ ② & 0 & -1 & \boxed{1} & 4 \\ \hline 5 & 0 & -15 & 0 & R - 210 \end{array} \right] \quad \begin{array}{l} \\ (R_2 + (-2)R_1) \\ (R_3 + (-30)R_1) \end{array}$$

Now y and t are basic variables and x and s are free. Setting x and s equal to 0 gives $x = 0$ (of course), and $y = 7$, since 7 is the element in the rightmost column opposite the 1 in the y-column. Thus we have moved to the point $(x, y) = (0, 7)$ in Figure 3.16, another corner of the feasible region.

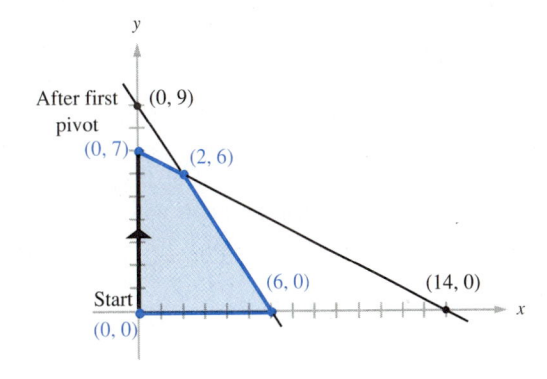

FIGURE 3.16

We continue by pivoting on the entry in row 2 and column 1. Note that column 1 corresponds to x, which will become the new basic variable for row 2.

$$\left[\begin{array}{cccc|c} .5 & 1 & .5 & 0 & 7 \\ \boxed{1} & 0 & -.5 & .5 & 2 \\ \hline 5 & 0 & -15 & 0 & R-210 \end{array}\right] \quad (.5R_2)$$

$$\begin{array}{c} \\ y \\ x \\ \end{array} \left[\begin{array}{cccc|c} x & y & s & t & \\ 0 & \boxed{1} & .75 & -.25 & 6 \\ \boxed{1} & 0 & -.5 & .5 & 2 \\ \hline 0 & 0 & -12.5 & -2.5 & R-220 \end{array}\right] \quad \begin{array}{c} \\ (R_1+(-.5)R_2) \\ \\ (R_3+(-5)R_2) \end{array}$$

Now x and y are the basic variables and s and t are free. Setting $s = t = 0$ gives $y = 6$ and $x = 2$, as we can read from the rightmost column. Thus we have reached the corner $(x, y) = (2, 6)$. Our progress from corner to corner of the feasible region is indicated in Figure 3.17.

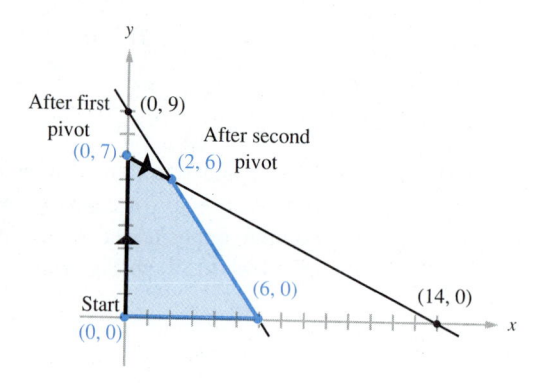

FIGURE 3.17

It turns out that now we can determine that $(2, 6)$ is the solution to our linear program without reference to the graph. Consider the bottom row of our last tableau. It translates into the equation

$$-12.5s - 2.5t = R - 220,$$

or

$$220 - 12.5s - 2.5t = R. \tag{3.6}$$

Recall that the point $(x, y) = (2, 6)$ corresponds to setting $s = t = 0$. Also recall that we must have $s \geq 0$ and $t \geq 0$. The only way we could change s and t would be to make one or both of them larger. *But that would decrease R, and R is the quantity we are trying to maximize.*

We see that the solution to our linear program must occur when $s = t = 0$, and so $(x, y) = (2, 6)$. Moreover we see from (3.6) that the maximum value of R is 220. Notice that this is precisely the number subtracted from R in the bottom right corner of our final tableau. In fact the rightmost column of this tableau, when properly interpreted, gives not only the maximal value of R but also the x and y coordinates achieving this value.

EXAMPLE 3.10 Solve the linear program:

$$\begin{aligned} \text{Maximize} \quad & 7x + 4y = R \\ \text{subject to} \quad & x + y \leq 160 \\ & 3x + 2y \leq 360 \\ & x \geq 0, y \geq 0, \end{aligned}$$

by introducing slack variables and pivoting on the entry in row 2 and column 1 of the resulting tableau.

Solution Introducing slack variables s and t leads to the system

$$\begin{aligned} x + y + s & = 160 \\ 3x + 2y + t & = 360 \\ 7x + 4y & = R \\ x \geq 0, y \geq 0, s \geq 0, t \geq 0, \end{aligned}$$

with tableau

$$\begin{array}{c} \\ s \\ t \\ \end{array} \begin{array}{c} \begin{array}{cccc} x & y & s & t \end{array} \\ \left[\begin{array}{cccc|c} 1 & 1 & \boxed{1} & 0 & 160 \\ \textcircled{3} & 2 & 0 & \boxed{1} & 360 \\ \hline 7 & 4 & 0 & 0 & R \end{array} \right]. \end{array}$$

We proceed to pivot on the circled entry. This pivot is in the column headed by x, which will replace t as the basic variable for the row of the pivot.

$$\left[\begin{array}{cccc|c} 1 & 1 & 1 & 0 & 160 \\ \textcircled{1} & \frac{2}{3} & 0 & \frac{1}{3} & 120 \\ \hline 7 & 4 & 0 & 0 & R \end{array} \right] \quad (\tfrac{1}{3}R_2)$$

$$\begin{array}{c} \\ s \\ x \\ \end{array} \begin{array}{c} \begin{array}{cccc} x & y & s & t \end{array} \\ \left[\begin{array}{cccc|c} 0 & \frac{1}{3} & \boxed{1} & -\frac{1}{3} & 40 \\ \boxed{1} & \frac{2}{3} & 0 & \frac{1}{3} & 120 \\ \hline 0 & -\frac{2}{3} & 0 & -\frac{7}{3} & R - 840 \end{array} \right] \end{array} \quad \begin{array}{l} (R_1 + (-1)R_2) \\ \\ (R_3 + (-7)R_2) \end{array}$$

Here the last row corresponds to the equation

$$840 - \frac{2}{3}y - \frac{7}{3}t = R.$$

Since neither y nor t can be negative, we see that R is maximal when $y = t = 0$. From the second row of our tableau, we see that $x = 120$ here. Thus the maximal value of R is 840 at the point $(x, y) = (120, 0)$. ∎

Practice Problem 7 Solve the linear program:

$$\begin{aligned} \text{Maximize} \quad & 4x + 3y = R \\ \text{subject to} \quad & x + 2y \leq 320 \\ & 5x + 4y \leq 1000 \\ & x \geq 0, y \geq 0 \end{aligned}$$

by introducing slack variables and pivoting on the entry in row 2 and column 1 of the corresponding tableau.

EXAMPLE 3.11 Solve the linear program:

$$\begin{aligned} \text{Maximize} \quad & 6x + 4y = R \\ \text{subject to} \quad & x + y \leq 100 \\ & 2x + y \leq 180 \\ & x + 3y \leq 240 \\ & x \geq 0, y \geq 0, \end{aligned}$$

by pivoting first on the entry in row 2 and column 1, and then on the entry in row 1 and column 2.

Solution We change the first three constraints to equations by introducing three slack variables s, t, and u. This gives the system

$$\begin{aligned} x + y + s \quad\quad\quad\quad &= 100 \\ 2x + y \quad\quad + t \quad\quad &= 180 \\ x + 3y \quad\quad\quad\quad + u &= 240 \\ 6x + 4y \quad\quad\quad\quad\quad\quad &= R \\ x \geq 0, y \geq 0, s \geq 0, t \geq 0, u \geq 0, \end{aligned}$$

and the tableau

$$\begin{array}{c} \quad\quad\; x \;\; y \;\; s \;\; t \;\; u \\ \begin{array}{c} s \\ t \\ u \end{array} \left[\begin{array}{ccccc|c} 1 & 1 & 1 & 0 & 0 & 100 \\ ② & 1 & 0 & 1 & 0 & 180 \\ 1 & 3 & 0 & 0 & 1 & 240 \\ \hline 6 & 4 & 0 & 0 & 0 & R \end{array} \right], \end{array}$$

where the indicated first pivot has been circled. It is in the column of x, which will replace t as the basic variable of row 2. The computation proceeds as follows.

$$\begin{bmatrix} 1 & 1 & 1 & 0 & 0 & | & 100 \\ \boxed{1} & .5 & 0 & .5 & 0 & | & 90 \\ 1 & 3 & 0 & 0 & 1 & | & 240 \\ \hline 6 & 4 & 0 & 0 & 0 & | & R \end{bmatrix} \quad (.5R_2)$$

$$\begin{array}{c} \\ s\\ x\\ u\\ \end{array} \begin{array}{cccccc} x & y & s & t & u & \\ \end{array}$$
$$\begin{array}{c} s\\ x\\ u\\ \end{array}\begin{bmatrix} 0 & \boxed{.5} & 1 & -.5 & 0 & | & 10 \\ 1 & .5 & 0 & .5 & 0 & | & 90 \\ 0 & 2.5 & 0 & -.5 & 1 & | & 150 \\ \hline 0 & 1 & 0 & -3 & 0 & | & R-540 \end{bmatrix} \begin{array}{l} (R_1+(-1)R_2) \\ \\ (R_3+(-1)R_2) \\ (R_4+(-6)R_2) \end{array}$$

Now we pivot on the .5 in row 1 and column 2, which will cause y to replace s as the basic variable for row 1.

$$\begin{bmatrix} 0 & \boxed{1} & 2 & -1 & 0 & | & 20 \\ 1 & .5 & 0 & .5 & 0 & | & 90 \\ 0 & 2.5 & 0 & -.5 & 1 & | & 150 \\ \hline 0 & 1 & 0 & -3 & 0 & | & R-540 \end{bmatrix} \quad (2R_1)$$

$$\begin{array}{c} \\ y\\ x\\ u\\ \end{array}\begin{array}{cccccc} x & y & s & t & u & \\ \end{array}$$
$$\begin{array}{c} y\\ x\\ u\\ \end{array}\begin{bmatrix} 0 & \boxed{1} & 2 & -1 & 0 & | & 20 \\ \boxed{1} & 0 & -1 & 1 & 0 & | & 80 \\ 0 & 0 & -5 & 2 & \boxed{1} & | & 100 \\ \hline 0 & 0 & -2 & -2 & 0 & | & R-560 \end{bmatrix} \begin{array}{l} \\ (R_2+(-.5)R_1) \\ (R_3+(-2.5)R_1) \\ (R_4+(-1)R_1) \end{array}$$

The last row corresponds to the equation
$$560 - 2s - 2t = R,$$
and so R takes on the maximum value 560 when $s = t = 0$. Also from the first two rows of the tableau we have $(x, y) = (80, 20)$ as the optimal point. ■

EXAMPLE 3.12 Solve the linear program:

Maximize $\quad x + y + z = R$
subject to $\quad x + 2y + z \leq 4$
$ 2x + y + 3z \leq 6$
$ x \geq 0, y \geq 0, z \geq 0,$

by pivoting first on the entry in row 2, column 1, and then on the entry in row 1, column 2.

Solution Notice that this is a three-variable problem, so a graphical solution is not practical. We introduce nonnegative slack variables s and t such that

$$x + 2y + z + s = 4 \quad \text{and} \quad 2x + y + 3z + t = 6,$$

leading to the tableau

$$\begin{array}{c} \\ s \\ t \end{array} \left[\begin{array}{ccccc|c} x & y & z & s & t & \\ 1 & 2 & 1 & \boxed{1} & 0 & 4 \\ \textcircled{2} & 1 & 3 & 0 & \boxed{1} & 6 \\ \hline 1 & 1 & 1 & 0 & 0 & R \end{array} \right].$$

When we pivot on the circled entry, x will replace t as the basic variable for row 2.

$$\left[\begin{array}{ccccc|c} 1 & 2 & 1 & 1 & 0 & 4 \\ \textcircled{1} & \frac{1}{2} & \frac{3}{2} & 0 & \frac{1}{2} & 3 \\ \hline 1 & 1 & 1 & 0 & 0 & R \end{array} \right] \quad (\tfrac{1}{2}R_2)$$

$$\begin{array}{c} \\ s \\ x \end{array} \left[\begin{array}{ccccc|c} x & y & z & s & t & \\ 0 & \textcircled{$\tfrac{3}{2}$} & -\tfrac{1}{2} & \boxed{1} & -\tfrac{1}{2} & 1 \\ \boxed{1} & \tfrac{1}{2} & \tfrac{3}{2} & 0 & \tfrac{1}{2} & 3 \\ \hline 0 & \tfrac{1}{2} & -\tfrac{1}{2} & 0 & -\tfrac{1}{2} & R-3 \end{array} \right] \begin{array}{l} (R_1 + (-1)R_2) \\ \\ (R_3 + (-1)R_2) \end{array}$$

Now we pivot on the 3/2 in row 1 and column 2, causing y to replace s as the basic variable for row 1.

$$\left[\begin{array}{ccccc|c} 0 & \textcircled{1} & -\tfrac{1}{3} & \tfrac{2}{3} & -\tfrac{1}{3} & \tfrac{2}{3} \\ 1 & \tfrac{1}{2} & \tfrac{3}{2} & 0 & \tfrac{1}{2} & 3 \\ \hline 0 & \tfrac{1}{2} & -\tfrac{1}{2} & 0 & -\tfrac{1}{2} & R-3 \end{array} \right] \quad (\tfrac{2}{3}R_1)$$

$$\begin{array}{c} \\ y \\ x \end{array} \left[\begin{array}{ccccc|c} x & y & z & s & t & \\ 0 & \boxed{1} & -\tfrac{1}{3} & \tfrac{2}{3} & -\tfrac{1}{3} & \tfrac{2}{3} \\ \boxed{1} & 0 & \tfrac{5}{3} & -\tfrac{1}{3} & \tfrac{2}{3} & \tfrac{8}{3} \\ \hline 0 & 0 & -\tfrac{1}{3} & -\tfrac{1}{3} & -\tfrac{1}{3} & R-\tfrac{10}{3} \end{array} \right] \begin{array}{l} \\ (R_2 + (-\tfrac{1}{2})R_1) \\ (R_3 + (-\tfrac{1}{2})R_1) \end{array}$$

The last row of this tableau implies the equation

$$\frac{10}{3} - \frac{1}{3}z - \frac{1}{3}s - \frac{1}{3}t = R,$$

from which we see that R is maximal when $z = s = t = 0$. The first two rows of the tableau then give $y = 2/3$ and $x = 8/3$. Thus R achieves the maximum value 10/3 when $(x, y, z) = (8/3, 2/3, 0)$. ■

EXERCISES 3.3

In Exercises 1–8 set up the simplex tableaus for the given linear program.

1. Maximize $6x + 4y = R$
   subject to $x + 3y \le 200$
   $2x + 5y \le 500$
   $x \ge 0, y \ge 0.$

2. Maximize $3x + 5y = R$
   subject to $4x + 3y \le 200$
   $x + y \le 60$
   $x \ge 0, y \ge 0.$

3. Maximize $3x + 8y = R$
   subject to $2x + 4y \le 1200$
   $x + 3y \le 1000$
   $x \ge 0, y \ge 0.$

4. Maximize $4x + 3y = R$
   subject to $2x + 3y \le 200$
   $2x + y \le 300$
   $x \ge 0, y \ge 0.$

5. Maximize $5x + 12y = R$
   subject to $3x + 4y \le 300$
   $x + 2y \le 120$
   $x \le 30$
   $x \ge 0, y \ge 0.$

6. Maximize $6x + 5y = R$
   subject to $x + 4y \le 240$
   $2x + 3y \le 200$
   $3x + 5y \le 360$
   $x \ge 0, y \ge 0.$

7. Maximize $5x + 4y + 3z = R$
   subject to $2x + y + z \le 30$
   $y + 3z \le 40$
   $x \ge 0, y \ge 0, z \ge 0.$

8. Maximize $4x + 2y + 3z = R$
   subject to $0.1x + 0.25y \le 40$
   $0.2x + 0.3y + 0.4z \le 100$
   $x \ge 0, y \ge 0, z \ge 0.$

In Exercises 9–12 pivot on the indicated entry of the tableau below, changing the appropriate row label. Perform each pivot on the original tableau shown. If the free variables are set equal to zero after the pivot, what is the point (x, y)?

$$\begin{array}{c} \\ s \\ t \\ \end{array} \begin{array}{c} \begin{array}{cccc} x & y & s & t \end{array} \\ \left[\begin{array}{cccc|c} 3 & 4 & 1 & 0 & 7 \\ 1 & 2 & 0 & 1 & 10 \\ \hline 3 & 2 & 0 & 0 & R \end{array} \right] \end{array}$$

9. The entry 3 in row 1 and column 1.

10. The entry 4 in row 1 and column 2.

11. The entry 1 in row 2 and column 1.

12. The entry 2 in row 2 and column 2.

In Exercises 13–16 pivot on the indicated entry of the tableau below, changing the appropriate row label. Perform each pivot on the original tableau shown. If the free variables are set equal to zero after the pivot, what is the point (x, y)?

$$\begin{array}{c} \\ x \\ t \\ \end{array} \left[\begin{array}{cccc|c} x & y & s & t & \\ 1 & \frac{2}{3} & \frac{1}{3} & 0 & \frac{8}{3} \\ 0 & \frac{8}{3} & -\frac{2}{3} & 1 & \frac{20}{3} \\ \hline 1 & 0 & -1 & 0 & R - 8 \end{array} \right]$$

13. The entry 2/3 in row 1 and column 2.

14. The entry 1/3 in row 1 and column 3.

15. The entry 8/3 in row 2 and column 2.

16. The entry $-2/3$ in row 2 and column 3.

In Exercises 17–24 solve the given linear program by pivoting on the indicated entries. Give the maximum value of R and the optimal point that produces it.

17. The entry in row 1 and column 1 of the tableau for Exercise 1.
18. The entry in row 2 and column 2 of the tableau for Exercise 2.
19. The entry in row 1 and column 2 of the tableau for Exercise 3.
20. The entry in row 1 and column 1 of the tableau for Exercise 4.
21. The entry in row 2 and column 2 of the tableau for Exercise 5.
22. The entry in row 2 and column 1 of the tableau for Exercise 6.
23. The entry in row 1 and column 1, then the entry in row 1 and column 2, of the tableau for Exercise 7.
24. The entry in row 1 and column 1, then the entry in row 2 and column 3, of the tableau for Exercise 8.

In Exercises 25 and 26 set up the simplex tableau and solve the linear program by pivoting on a single entry. Find the entry to pivot on by trial and error.

25. Maximize $4.5x + 3.5y = R$
    subject to $\quad 6x + 5y \leq 600$
    $\quad\quad\quad\quad\; 3x + 4y \leq 240$
    $\quad\quad\quad\quad\; x \geq 0, y \geq 0.$

26. Maximize $2.5x + 0.9y = R$
    subject to $\quad 5x + 2y \leq 3000$
    $\quad\quad\quad\quad\; 0.2x + 0.4y \leq 160$
    $\quad\quad\quad\quad\; x \geq 0, y \geq 0.$

Answer to Practice Problem   7. The final tableau is

$$\begin{array}{c} \\ s \\ x \\ \\ \end{array} \left[\begin{array}{cccc|c} x & y & s & t & \\ 0 & 1.2 & \boxed{1} & -0.2 & 120 \\ \boxed{1} & 0.8 & 0 & 0.2 & 200 \\ \hline 0 & -0.2 & 0 & -0.8 & R - 800 \end{array} \right],$$

and we conclude from it that R has the maximum value of 800 at $(x, y) = (200, 0)$.

3.4 Maximization Problems

In this section we will answer two questions that arise in applying the simplex method, namely,

(1) How do we choose the pivot entry of the simplex tableau?
(2) How do we know when we are finished, so that we can find the optimal point by setting the free variables equal to zero?

To illustrate the answers to these questions, we will return to the example of Albert's Smoke Shop (Example 3.1 from Sections 3.1 and 3.2), which led to the linear program

3.4 MAXIMIZATION PROBLEMS

Maximize $16x + 12y = R$

subject to
$$8x + 3y \leq 480 \tag{3.1}$$
$$4x + 9y \leq 720 \tag{3.2}$$
$$x \geq 0 \tag{3.3}$$
$$y \geq 0. \tag{3.4}$$

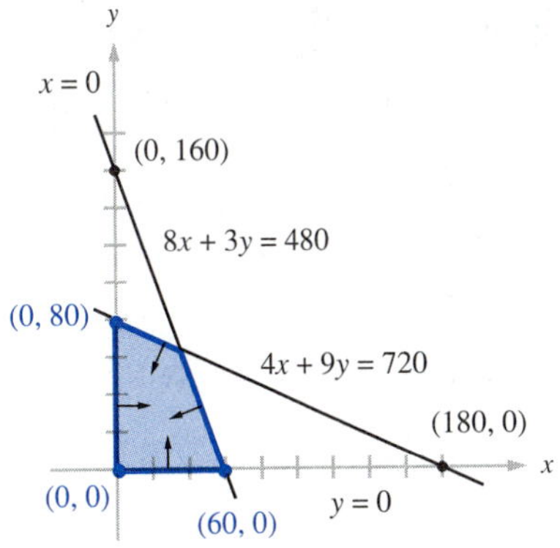

The feasible region is shown in Figure 3.18.

We will start by treating only **standard maximum programs**, that is, programs where

(a) The objective function is to be maximized.
(b) The variables are constrained to be nonnegative (as in (3.3) and (3.4)).
(c) All other constraints are of the $\leq$ type (as (3.1) and (3.2) are in this problem).
(d) The right-hand constants of the constraints are all nonnegative (480 and 720 in this problem).

We introduce slack variables s and t to transform the program to:

Maximize $16x + 12y = R$

subject to
$$8x + 3y + s = 480$$
$$4x + 9y + t = 720$$
$$x \geq 0, y \geq 0, s \geq 0, t \geq 0,$$

with the tableau

Tableau 1
$$\begin{array}{c} \\ s \\ t \\ \end{array} \left[\begin{array}{cccc|c} x & y & s & t & \\ \boxed{8} & 3 & \boxed{1} & 0 & 480 \\ 4 & 9 & 0 & \boxed{1} & 720 \\ \hline 16 & 12 & 0 & 0 & R \end{array} \right].$$

Note that the original row labels are simply the slack variables. It turns out that we can solve the program by pivoting first on the entry in row 1 and column 1, and then on the entry in row 2 and column 2. We will exhibit only the tableaus after each of these pivots, leaving it for the reader to check the arithmetic details. We pivot on the 8 in row 1 and column 1 (causing x to replace s as the label of row 1) to get

Tableau 2
$$\begin{array}{c} \\ x \\ t \\ \end{array} \left[\begin{array}{cccc|c} x & y & s & t & \\ 1 & \frac{3}{8} & \frac{1}{8} & 0 & 60 \\ 0 & \boxed{\frac{15}{2}} & -\frac{1}{2} & \boxed{1} & 480 \\ \hline 0 & 6 & -2 & 0 & R - 960 \end{array} \right].$$

Notice that setting the free variables y and s equal to 0 gives $x = 60$ and $t = 480$, so that we are now at the point $(x, y) = (60, 0)$ shown in Figure 3.19.

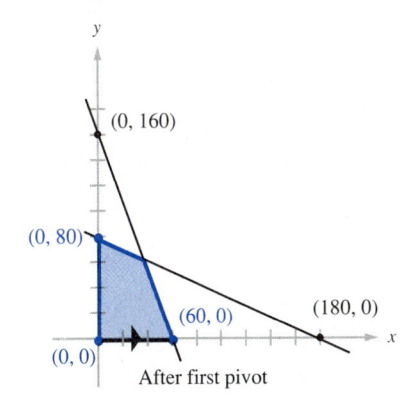

FIGURE 3.18

FIGURE 3.19 After first pivot

Now we pivot on the entry 15/2 in row 2 and column 2 (causing y to replace t as the label of row 2), to get

Tableau 3
$$\begin{array}{c} \\ x \\ y \\ \\ \end{array} \begin{array}{c} x & y & s & t \\ \left[\begin{array}{cccc|c} \boxed{1} & 0 & \frac{3}{20} & -\frac{1}{20} & 36 \\ 0 & \boxed{1} & -\frac{1}{15} & \frac{2}{15} & 64 \\ \hline 0 & 0 & -\frac{8}{5} & -\frac{4}{5} & R - 1344 \end{array}\right] \end{array}.$$

We know we are done at this point, because the last row of Tableau 3 is equivalent to the equation

$$R = 1344 - \frac{8}{5}s - \frac{4}{5}t,$$

so setting s and t both equal to 0 is the best we can do. This gives the maximal value of 1344 when $(x, y) = (36, 64)$. (See Figure 3.20.)

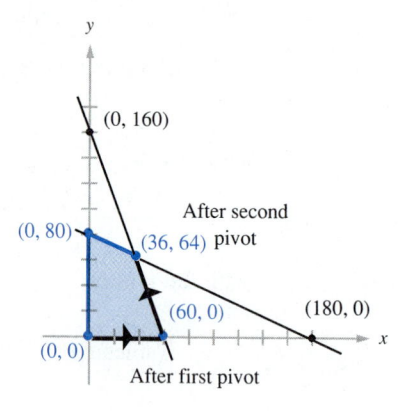

FIGURE 3.20

Why not stop at Tableau 2? Translating its last row into an equation provides the answer. An equivalent equation is

$$R = 960 + 6y - 2s. \tag{3.7}$$

It appears that making the free variable y positive instead of 0 would increase R. The positive coefficient 6 in (3.7) corresponds to the 6 in the last row of the Tableau 2. The entries in the bottom row of the simplex tableau to the left of the vertical line are called the **indicators** of the tableau. This brings us to our first rule.

■ ■ ■ ■ ■

The simplex algorithm is completed when no indicator is positive.

3.4 MAXIMIZATION PROBLEMS

Since we want to get rid of all positive indicators, it seems reasonable to pivot on an entry in a column with a positive indicator, so as to change it to 0. In Tableau 1 there are two positive indicators, 16 and 12. Which of their columns should we pivot in? Notice that the last row of Tableau 1 corresponds to the equation

$$16x + 12y = R.$$

Increasing x by 1 increases R by 16, but increasing y by 1 only increases R by 12. Since we are trying to maximize R, we adopt the following rule to choose the column to pivot in.

> Always pivot on a column with a positive indicator. If more than one column has a positive indicator, choose the column with the indicator that is largest. If there is a tie among columns, choose any one of them.

Applying this rule to Tableau 1, we choose our first pivot in column 1, and we mark this column with a star.

Now we must decide whether to pivot on the 8 or the 4. The effect of each of these pivots is easier to see if we multiply the first row by $\frac{1}{8}$ and the second row by $\frac{1}{4}$.

Tableau 1a
$$\left[\begin{array}{cccc|c} x & y & s & t & \\ \textcircled{1} & \frac{3}{8} & \frac{1}{8} & 0 & \frac{480}{8} = 60 \\ \textcircled{1} & \frac{9}{4} & 0 & \frac{1}{4} & \frac{720}{4} = 180 \\ \hline 16* & 12 & 0 & 0 & R \end{array} \right]$$

If the pivot entry is in the first row, we would complete the pivot transformation by subtracting this row from the second. This would make the entries to the right of the vertical line in these rows 60 and 120. Likewise to pivot using the 1 in the second row, we would subtract the second row from the first, making the entries to the right of the vertical line -120 and 180. But *we do not want negative entries to the right of the vertical line,* for then setting the free variables equal to 0 would give a negative basic variable, contradicting the original conditions of the program. This is what happened at our first attempt at a pivot in Section 3.3, when we moved to a point that was not in the feasible region.

Therefore in Tableau 1 we pivot on the 8 in the first row instead of the 4 because 480/8 is less than 720/4. This leads us to our third rule.

> Once the pivot column has been decided, divide each positive entry in that column above the horizontal line into the corresponding entry in the rightmost column. Pivot on the entry that gives the smallest positive quotient.

For example, knowing that in Tableau 1 we want to pivot on an entry in column 1, we divide the entries in that column into the corresponding rightmost entries, writing the results to the right of the display.

$$\text{Tableau 1} \quad \begin{array}{c} \\ \\ \end{array} \begin{bmatrix} x & y & s & t & \\ 8 & 3 & 1 & 0 & 480 \\ 4 & 9 & 0 & 1 & 720 \\ \hline 16^* & 12 & 0 & 0 & R \end{bmatrix} \quad \begin{array}{l} 480/8 = 60^* \\ 720/4 = 180 \end{array}$$

Since $60 < 180$ our pivot will be chosen from row 1. Thus we pivot on the 8 in the first row and column, producing, as we have already seen, the following tableau.

$$\text{Tableau 2} \quad \begin{bmatrix} x & y & s & t & \\ 1 & \frac{3}{8} & \frac{1}{8} & 0 & 60 \\ 0 & \frac{15}{2} & -\frac{1}{2} & 1 & 480 \\ \hline 0 & 6^* & -2 & 0 & R - 960 \end{bmatrix} \quad \begin{array}{l} 60/(3/8) = 160 \\ 480/(15/2) = 64^* \end{array}$$

Since the only positive indicator is the 6 in column 2, our next pivot will be chosen from this column. We divide the entries in the second column into the entries 60 and 480 in the rightmost column, getting the quotients 160 and 64. Since $64 < 160$ we mark the second row with a star, and pivot on the 15/2 in row 2 and column 2. This leads to Tableau 3 and the solution of the linear program, as we have already seen.

EXAMPLE 3.11
Revisited
Show how one would choose the pivots suggested in Example 3.11 of the previous section.

Solution   The initial tableau in Example 3.11 was

$$\begin{array}{c} \\ s \\ t \\ u \\ \text{Indicators} \end{array} \begin{bmatrix} \overbrace{x \quad y}^{\text{Free}} & \overbrace{s \quad t \quad u}^{\text{Basic}} & \\ 1 & 1 & 1 & 0 & 0 & 100 \\ 2 & 1 & 0 & 1 & 0 & 180 \\ 1 & 3 & 0 & 0 & 1 & 240 \\ \hline 6^* & 4 & 0 & 0 & 0 & R \end{bmatrix} \quad \begin{array}{l} 100/1 = 100 \\ 180/2 = 90^* \\ 240/1 = 240 \end{array}.$$

Since 6 is the largest indicator, we will choose a pivot in the column 1. Dividing the other entries of column 1 into the corresponding entries in the rightmost column gives the quotients 100, 90, and 240. Since 90 is the smallest of these, we pivot on the 2 in row 2 and column 1. As we saw in Section 3.3, this yields the tableau

3.4 MAXIMIZATION PROBLEMS

$$\begin{array}{c} \\ s \\ x \\ u \\ \\ \end{array} \begin{array}{c} x \quad\; y \quad\; s \quad\;\; t \quad u \\ \left[\begin{array}{ccccc|c} 0 & .5 & 1 & -.5 & 0 & 10 \\ 1 & .5 & 0 & .5 & 0 & 90 \\ 0 & 2.5 & 0 & -.5 & 1 & 150 \\ \hline 0 & 1* & 0 & -3 & 0 & R - 540 \end{array}\right] \end{array} \begin{array}{l} 10/.5 = 20* \\ 90/.5 = 180 \\ 150/2.5 = 60 \end{array}.$$

Now 1 is the only positive indicator, so we will pivot on an entry of column 2. Dividing the other entries of this column into 10, 90, and 150 gives 20, 180, and 60, and the smallest of these is 20. Thus we pivot on the entry .5 in row 1 and column 2. As we saw in Section 3.3, this produces the tableau

$$\begin{array}{c} \\ y \\ x \\ u \\ \\ \end{array} \begin{array}{c} x \;\; y \quad\; s \quad\;\; t \quad u \\ \left[\begin{array}{ccccc|c} 0 & 1 & 2 & -1 & 0 & 20 \\ 1 & 0 & -1 & 1 & 0 & 80 \\ 0 & 0 & -5 & 2 & 1 & 100 \\ \hline 0 & 0 & -2 & -2 & 0 & R - 560 \end{array}\right] \end{array}.$$

Since this has no positive indicators, we are done. We read from the tableau the solution $(x, y) = (80, 20)$, which yields the maximum $R = 560$. ∎

Practice Problem 8   What pivot should be chosen for the following tableau?

$$\left[\begin{array}{ccccc|c} 2 & 6 & 5 & 1 & 0 & 34 \\ 3 & 8 & 2 & 0 & 1 & 16 \\ 4 & 12 & 8 & 0 & 0 & 70 \\ \hline 3 & 5 & 7 & 0 & 0 & R \end{array}\right]$$

Practice Problem 9   Which are the basic variables and which are the free variables in the following tableau? What value of R results from setting all free variables equal to 0? What value of (x, y, z) produces this R?

$$\begin{array}{c} \\ z \\ s \\ x \\ \\ \end{array} \begin{array}{c} x \quad\; y \;\; z \;\; s \quad\;\; t \\ \left[\begin{array}{ccccc|c} 0 & 3 & 1 & 0 & 6 & 4 \\ 0 & -2 & 0 & 1 & 0 & 6 \\ 1 & 5 & 0 & 0 & -2 & 10 \\ \hline 0 & -3 & 0 & 0 & -5 & R - 22 \end{array}\right] \end{array}$$

We now summarize the method we have developed.

The Simplex Method for Standard Maximum Programs

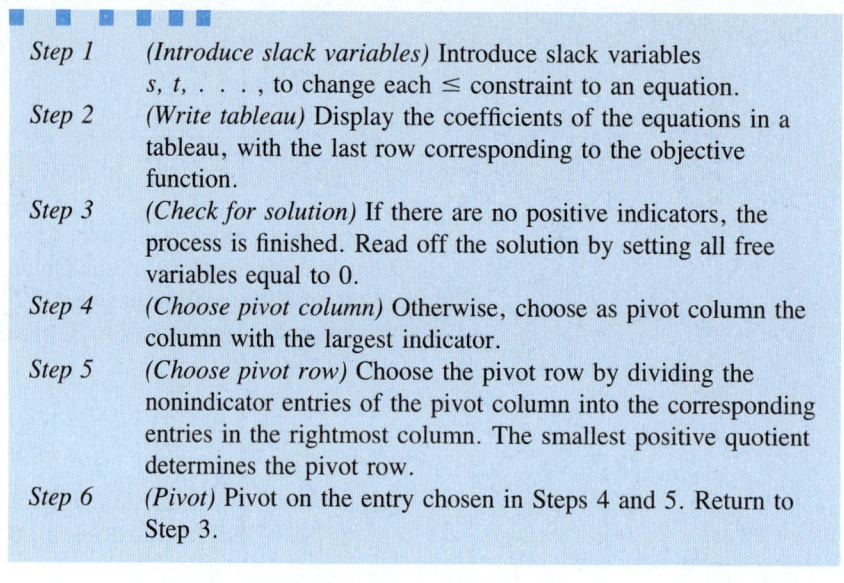

Step 1 *(Introduce slack variables)* Introduce slack variables $s, t, \ldots$, to change each $\leq$ constraint to an equation.

Step 2 *(Write tableau)* Display the coefficients of the equations in a tableau, with the last row corresponding to the objective function.

Step 3 *(Check for solution)* If there are no positive indicators, the process is finished. Read off the solution by setting all free variables equal to 0.

Step 4 *(Choose pivot column)* Otherwise, choose as pivot column the column with the largest indicator.

Step 5 *(Choose pivot row)* Choose the pivot row by dividing the nonindicator entries of the pivot column into the corresponding entries in the rightmost column. The smallest positive quotient determines the pivot row.

Step 6 *(Pivot)* Pivot on the entry chosen in Steps 4 and 5. Return to Step 3.

It should be noted that if in Step 5 there is a tie for the smallest positive quotient, complications may occur. We will not treat this possibility. The next example illustrates the simplex method from beginning to end.

EXAMPLE 3.13 A small furniture finishing factory produces desks, buffets, and cocktail tables. Each desk requires 0.3 hours of sanding, 0.1 hours of staining, and 0.2 hours of varnishing. Each buffet requires 0.2 hours of sanding, 0.4 hours of staining, and 0.4 hours of varnishing. In addition, each cocktail table requires 0.1 hours of sanding, 0.1 hours of staining, and 0.2 hours of varnishing. During a particular week the factory will have at most 44 hours of time available for sanding, at most 40 hours of time available for staining, and at most 48 hours available for varnishing. If the factory earns profits of \$10 on each desk, \$14 on each buffet, and \$7 on each cocktail table, how should production be scheduled in order to obtain a maximum profit?

Solution Let

$x =$ the number of desks to be made

$y =$ the number of buffets to be made

$z =$ the number of cocktail tables to be made.

Then we have the following standard maximum linear program:

Maximize $\quad 10x + 14y + 7z = P$

subject to
$$0.3x + 0.2y + 0.1z \leq 44$$
$$0.1x + 0.4y + 0.1z \leq 40$$
$$0.2x + 0.4y + 0.2z \leq 48$$
$$x \geq 0, y \geq 0, z \geq 0.$$

3.4 MAXIMIZATION PROBLEMS

Introducing nonnegative slack variables s, t, and u, we can rewrite the problem in the form

$$\text{Maximize} \quad 10x + 14y + 7z = P$$
$$\text{subject to} \quad 0.3x + 0.2y + 0.1z + s = 44$$
$$0.1x + 0.4y + 0.1z + t = 40$$
$$0.2x + 0.4y + 0.2z + u = 48$$
$$x \geq 0, y \geq 0, z \geq 0, s \geq 0, t \geq 0, u \geq 0.$$

Thus our simplex tableau is

$$\begin{array}{c} \\ s \\ t \\ u \\ \\ \end{array} \begin{array}{c} x \quad y \quad z \quad s \quad t \quad u \\ \left[\begin{array}{cccccc|c} 0.3 & 0.2 & 0.1 & 1 & 0 & 0 & 44 \\ 0.1 & \textcircled{0.4} & 0.1 & 0 & 1 & 0 & 40 \\ 0.2 & 0.4 & 0.2 & 0 & 0 & 1 & 48 \\ \hline 10 & 14^* & 7 & 0 & 0 & 0 & P \end{array}\right] \end{array} \quad \begin{array}{l} 44/0.2 = 220 \\ 40/0.4 = 100^* \\ 48/0.4 = 120 \end{array}$$

Since 14 is the largest indicator, we will pivot in column 2. The smallest of the corresponding quotients is the 100 in row 2, so we will pivot at the circled entry 0.4. This will change the label of row 2 from t to y.

$$\left[\begin{array}{cccccc|c} 0.3 & 0.2 & 0.1 & 1 & 0 & 0 & 44 \\ 0.25 & \textcircled{1} & 0.25 & 0 & 2.5 & 0 & 100 \\ 0.2 & 0.4 & 0.2 & 0 & 0 & 1 & 48 \\ \hline 10 & 14 & 7 & 0 & 0 & 0 & P \end{array}\right] \quad ((1/0.4)R_2)$$

$$\begin{array}{c} \\ s \\ y \\ u \\ \\ \end{array} \begin{array}{c} x \quad y \quad z \quad s \quad t \quad u \\ \left[\begin{array}{cccccc|c} 0.25 & 0 & 0.05 & 1 & -0.5 & 0 & 24 \\ 0.25 & 1 & 0.25 & 0 & 2.5 & 0 & 100 \\ 0.1 & 0 & 0.1 & 0 & -1 & 1 & 8 \\ \hline 6.5 & 0 & 3.5 & 0 & -35 & 0 & P - 1400 \end{array}\right] \end{array} \quad \begin{array}{l} (R_1 + (-0.2)R_2) \\ \\ (R_3 + (-0.4)R_2) \\ (R_4 + (-14)R_2) \end{array}$$

We repeat this tableau, indicating how we choose to pivot next on the entry in column 1 and row 3.

$$\left[\begin{array}{cccccc|c} 0.25 & 0 & 0.05 & 1 & -0.5 & 0 & 24 \\ 0.25 & 1 & 0.25 & 0 & 2.5 & 0 & 100 \\ 0.1 & 0 & 0.1 & 0 & -1 & 1 & 8 \\ \hline 6.5^* & 0 & 3.5 & 0 & -35 & 0 & P - 1400 \end{array}\right] \quad \begin{array}{l} 24/0.25 = 96 \\ 100/0.25 = 400 \\ 8/0.1 = 80^* \end{array}$$

The actual pivot transformation proceeds as follows.

$$\left[\begin{array}{cccccc|c} 0.25 & 0 & 0.05 & 1 & -0.5 & 0 & 24 \\ 0.25 & 1 & 0.25 & 0 & 2.5 & 0 & 100 \\ \textcircled{1} & 0 & 1 & 0 & -10 & 10 & 80 \\ \hline 6.5 & 0 & 3.5 & 0 & -35 & 0 & P - 1400 \end{array}\right] \quad (10R_3)$$

$$\begin{array}{c} \\ s \\ y \\ x \\ \\ \end{array} \begin{array}{c} x \quad y \quad\quad z \quad s \quad\quad t \quad\quad u \\ \left[\begin{array}{cccccc|c} 0 & 0 & -0.2 & 1 & 2 & -2.5 & 4 \\ 0 & 1 & 0 & 0 & 5 & -2.5 & 80 \\ 1 & 0 & 1 & 0 & -10 & 10 & 80 \\ \hline 0 & 0 & -3 & 0 & 30 & -65 & P - 1920 \end{array}\right] \end{array} \begin{array}{l} (R_1 + (-0.25)R_3) \\ (R_2 + (-0.25)R_3) \\ \\ (R_4 + (-6.5)R_3) \end{array}$$

We repeat this tableau, showing how we choose the next pivot to be the entry 2 in column 5 and row 1.

$$\left[\begin{array}{cccccc|c} 0 & 0 & -0.2 & 1 & ② & -2.5 & 4 \\ 0 & 1 & 0 & 0 & 5 & -2.5 & 80 \\ 1 & 0 & 1 & 0 & -10 & 10 & 80 \\ \hline 0 & 0 & -3 & 0 & 30^* & -65 & P - 1920 \end{array}\right] \begin{array}{l} 4/2 = 2^* \\ 80/5 = 16 \\ \text{(negative quotient)} \end{array}$$

The pivot transformation proceeds as follows.

$$\left[\begin{array}{cccccc|c} 0 & 0 & -0.1 & 0.5 & ① & -1.25 & 2 \\ 0 & 1 & 0 & 0 & 5 & -2.5 & 80 \\ 1 & 0 & 1 & 0 & -10 & 10 & 80 \\ \hline 0 & 0 & -3 & 0 & 30 & -65 & P - 1920 \end{array}\right] \quad (0.5R_1)$$

$$\begin{array}{c} \\ t \\ y \\ x \\ \\ \end{array} \begin{array}{c} x \quad y \quad\quad z \quad\quad s \quad\quad t \quad\quad u \\ \left[\begin{array}{cccccc|c} 0 & 0 & -0.1 & 0.5 & 1 & -1.25 & 2 \\ 0 & 1 & 0.5 & -2.5 & 0 & 3.75 & 70 \\ 1 & 0 & 0 & 5 & 0 & -2.5 & 100 \\ \hline 0 & 0 & 0 & -15 & 0 & -27.5 & P - 1980 \end{array}\right] \end{array} \begin{array}{l} \\ (R_2 + (-5)R_1) \\ (R_3 + 10R_1) \\ (R_4 + (-30)R_1) \end{array}$$

Since there are no positive indicators, we are done. From the last row we see that the maximum value of P is 1980, which occurs when the free variables z, s, and u are all 0. From the upper part of the tableau we see this makes $t = 2$, $y = 70$, and $x = 100$. The optimal point is $(x, y, z) = (100, 70, 0)$. The factory should make 100 desks, 70 buffets, and no cocktail tables and will then realize a profit of $1980. ∎

EXERCISES 3.4

In Exercises 1–10 state the row and column, if any, where a pivot should be made in the given tableau.

1. $\begin{bmatrix} 2 & 4 & 1 & 0 & | & 7 \\ 6 & 3 & 0 & 1 & | & 20 \\ \hline 5 & 3 & 0 & 0 & | & R \end{bmatrix}$

2. $\begin{bmatrix} 6 & 2 & 1 & 0 & | & 12 \\ 2 & 3 & 0 & 1 & | & 15 \\ \hline 6 & 9 & 0 & 0 & | & R \end{bmatrix}$

3. $\begin{bmatrix} 5 & 1 & 2 & 0 & | & 45 \\ 2 & 0 & 9 & 1 & | & 21 \\ \hline 1 & -2 & .5 & 0 & | & R-32 \end{bmatrix}$

4. $\begin{bmatrix} 12 & 0 & 7 & 1 & | & 33 \\ 6 & 1 & 4 & 0 & | & 18 \\ \hline 5 & 0 & 11 & -1 & | & R-22 \end{bmatrix}$

5. $\begin{bmatrix} 3.5 & 4 & 1 & 0 & 0 & | & 12 \\ 7 & 2 & 0 & 1 & 0 & | & 25 \\ 2 & 5 & 0 & 0 & 1 & | & 7 \\ \hline 3 & -6 & 0 & 0 & 0 & | & R \end{bmatrix}$

6. $\begin{bmatrix} 2 & 3 & 1 & 0 & 0 & | & 11 \\ 3 & -1 & 0 & 1 & 0 & | & 16 \\ 5 & 4 & 0 & 0 & 1 & | & 27 \\ \hline .3 & .2 & 0 & 0 & 0 & | & R \end{bmatrix}$

7. $\begin{bmatrix} \frac{2}{3} & \frac{1}{5} & 3 & 1 & 0 & | & 13 \\ 4 & \frac{2}{3} & -2 & 0 & 1 & | & 44 \\ \hline -3 & 1 & \frac{1}{2} & 0 & 0 & | & R \end{bmatrix}$

8. $\begin{bmatrix} 3 & \frac{3}{2} & 1 & 0 & | & 12 \\ -2 & 4 & 0 & 1 & | & 9 \\ \hline 4 & \frac{7}{2} & 0 & 0 & | & R \end{bmatrix}$

9. $\begin{bmatrix} 3 & 0 & 4 & 1 & | & 13 \\ 5 & 1 & 7 & 0 & | & 9 \\ \hline -2 & -1 & 0 & 0 & | & R-6 \end{bmatrix}$

10. $\begin{bmatrix} 3 & 5 & 0 & 1 & | & 33 \\ 0 & 4 & 1 & 0 & | & 21 \\ \hline 4 & 3 & 0 & 0 & | & R \end{bmatrix}$

In Exercises 11–14 final tableaus are given after the simplex method has been applied. Tell what the maximal value of R is, and what values of x and y or x, y, and z produce it.

11.
$\begin{array}{c} \\ t \\ y \end{array} \begin{array}{cccc} x & y & s & t \\ \end{array}$
$\begin{bmatrix} 4 & 0 & 3 & 1 & | & 5 \\ 5 & 1 & -2 & 0 & | & 7 \\ \hline -1 & 0 & -3 & 0 & | & R-13 \end{bmatrix}$

12.
$\begin{array}{c} \\ y \\ x \end{array} \begin{array}{cccc} x & y & s & t \\ \end{array}$
$\begin{bmatrix} 0 & 1 & 5 & -2 & | & 6 \\ 1 & 0 & 0 & 4 & | & 3 \\ \hline 0 & 0 & -3 & -2 & | & R-7 \end{bmatrix}$

13.
$\begin{array}{c} \\ z \\ x \end{array} \begin{array}{ccccc} x & y & z & s & t \\ \end{array}$
$\begin{bmatrix} 0 & 3 & 1 & 2 & 1 & | & 5 \\ 1 & -2 & 0 & 4 & 6 & | & 7 \\ \hline 0 & -1 & 0 & -3 & -3 & | & R-22 \end{bmatrix}$

14.
$\begin{array}{c} \\ t \\ y \end{array} \begin{array}{ccccc} x & y & z & s & t \\ \end{array}$
$\begin{bmatrix} 0 & 0 & 3 & 1 & 1 & | & 11 \\ 6 & 1 & -2 & 4 & 0 & | & 3 \\ \hline -1 & 0 & -4 & -1 & 0 & | & R-2 \end{bmatrix}$

In Exercises 15–18 solve the linear program by the simplex method and also by the graphical method. Indicate on your graph the edge-path described by the simplex pivots.

15. Maximize $5x + 2y = R$
subject to $2x + y \le 30$
$x + 2y \le 30$
$x \ge 0, y \ge 0.$

16. Maximize $10x + 15y = R$
subject to $x + 2y \le 4$
$x + y \le 3$
$x \ge 0, y \ge 0.$

17. Maximize $10x + 5y = R$
subject to $x + y \le 5$
$x \le 3$
$y \le 4$
$x \ge 0, y \ge 0.$

18. Maximize $2x - y = R$
subject to $x + y \le 10$
$x - y \le 4$
$x \le 6$
$x \ge 0, y \ge 0.$

In Exercises 19–26 solve the standard maximum program by the simplex method.

19. Maximize $\quad 3x - 2y = R$
    subject to $\quad x \leq 20$
    $\quad\quad\quad\quad\quad y \leq 20$
    $\quad\quad\quad\quad x - y \leq 10$
    $\quad\quad\quad -x + y \leq 10$
    $\quad\quad\quad x \geq 0,\, y \geq 0.$

20. Maximize $\quad 3x + 4y = R$
    subject to $\quad 2x + 3y \leq 180$
    $\quad\quad\quad\quad x - y \leq 15$
    $\quad\quad\quad\quad\quad y \leq 50$
    $\quad\quad\quad\quad x \leq 30$
    $\quad\quad\quad x \geq 0,\, y \geq 0.$

21. Maximize $\quad 2x + y + z = R$
    subject to $\quad x + 2y + 3z \leq 80$
    $\quad\quad\quad\quad x + y + z \leq 60$
    $\quad\quad\quad x \geq 0,\, y \geq 0,\, z \geq 0.$

22. Maximize $\quad 2x + 3y + 4z = R$
    subject to $\quad 2x + y + z \leq 40$
    $\quad\quad\quad\quad x + y + 2z \leq 50$
    $\quad\quad\quad x \geq 0,\, y \geq 0,\, z \geq 0.$

23. Maximize $\quad 2x + y - 3z = R$
    subject to $\quad x + y + z \leq 4$
    $\quad\quad\quad\quad x - y + z \leq 2$
    $\quad\quad\quad -x + y + z \leq 1$
    $\quad\quad\quad x \geq 0,\, y \geq 0,\, z \geq 0.$

24. Maximize $\quad 5x - 4y + z = R$
    subject to $\quad x - y + z \leq 20$
    $\quad\quad\quad\quad x + y \leq 10$
    $\quad\quad\quad\quad y + z \leq 15$
    $\quad\quad\quad x \geq 0,\, y \geq 0,\, z \geq 0.$

25. Maximize $\quad x + 2y - z + w = R$
    subject to $\quad 2x + y + 3z + w \leq 10$
    $\quad\quad\quad\quad x + z + 2w \leq 2$
    $\quad\quad\quad\quad y + 2z + w \leq 4$
    $\quad\quad\quad x \geq 0,\, y \geq 0,\, z \geq 0,\, w \geq 0.$

26. Maximize $\quad x + 2y + 4z + w = R$
    subject to $\quad x - y + z - w \leq 1$
    $\quad\quad\quad\quad x - z \leq 1$
    $\quad\quad\quad\quad y + w \leq 1$
    $\quad\quad\quad x \geq 0,\, y \geq 0,\, z \geq 0,\, w \geq 0.$

27. A farmer has 1000 acres of land on which he intends to plant asparagus, brussels sprouts, and cauliflower. These three crops require, respectively, 15, 25, and 20 inches of water per acre to grow to maturity, and the farmer has been allotted a total of 20,000 acre-inches of water for the season. (An acre-inch is the amount of water needed to irrigate an acre of land to a depth of one inch.) The respective profits on the three crops are $200, $250, and $220 an acre. How much of each crop should be planted to maximize total profit?

28. The Zeno Corporation makes three models of stereo receivers. There are three stages in the manufacture of each unit, as shown in the following table:

Model	Work-hours per unit		
	A	B	C
Assembly	0.2 wh	0.3 wh	0.1 wh
Wiring	0.2 wh	0.2 wh	0.1 wh
Testing	0.1 wh	0.2 wh	0.0 wh

The respective profits on the three models are $50, $75, and $30 per unit. The company has available a total of 40 work-hours of labor per day for assembly, 28 for wiring, and 16 for testing. How many receivers of each type should be made to maximize total profit?

By looking at the terminal tableau one can determine whether or not there is more than one optimal solution as follows: If every column headed by a free variable has a negative indicator, then the optimal solution is unique. But if some such column has indicator zero, there is more than one optimal solution. In fact another solution can be found by pivoting in this column. Exercises 29 and 30 illustrate this fact.

29. Show that the solution $(x, y, z) = (100, 70, 0)$ found in Example 3.13 is not unique by pivoting in column 3 of the final tableau. What is the new solution and value of the objective function?

30. Solve Example 3.9 of Section 3.2 by the simplex method. Then find another optimal point by pivoting in a column of the final tableau having indicator 0.

Answers to Practice Problems

8. The 5 in row 1 and column 3.

9. The variables x, z, and s are basic and y and t are free. Setting $y = t = 0$ gives $R = 22$. This corresponds to $(x, y, z) = (10, 0, 4)$.

3.5 Minimization Problems and Duality

Many linear programs that arise in practice involve minimizing the objective function, with constraints of the $\geq$ variety. In Section 3.2 we solved such a problem graphically in Example 3.5. The problem involved hiring buses for an alumni association excursion, and led to the linear program

$$\begin{aligned}
\text{Minimize} \quad & 2000x + 2400y = C \\
\text{subject to} \quad & 40x + 50y \geq 800 \\
& 1000x + 750y \geq 18{,}000 \\
& x \geq 0, y \geq 0.
\end{aligned}$$

We could summarize the information in this problem in a **compact tableau** as follows.

$$\begin{bmatrix} x & y & & \\ 40 & 50 & (\geq) & 800 \\ 1000 & 750 & (\geq) & 18{,}000 \\ \hline 2000 & 2400 & (\text{minimize}) & C \end{bmatrix} \quad (3.8)$$

Here the conditions $x \geq 0$ and $y \geq 0$ are understood. It turns out that corresponding to each minimization program with $\geq$ constraints, there is a maximization program with $\leq$ constraints, called the **dual program.** The numbers involved in the dual program are found by interchanging rows and columns in the compact tableau of the original program. For example the tableau just presented translates into

$$\begin{bmatrix} X & Y & & \\ 40 & 1000 & (\leq) & 2000 \\ 50 & 750 & (\leq) & 2400 \\ \hline 800 & 18{,}000 & (\text{maximize}) & R \end{bmatrix} \quad (3.9)$$

Notice how the first row of this tableau comes from the first column of (3.8), the second row comes from the second column of (3.8), etc. Tableau (3.9) corresponds to the linear program

Maximize $\quad 800X + 18{,}000Y = R$

subject to $\quad 40X + 1000Y \le 2000$

$\quad\quad\quad\quad\ 50X + 750Y \le 2400$

$\quad\quad\quad\quad X \ge 0,\ Y \ge 0.$

Of course this is a standard maximum program, solvable with the simplex method. It is a surprising fact that *applying the simplex method to this dual program leads also to a solution of the original minimization problem.*

Let us demonstrate the last statement. To solve the dual problem we introduce nonnegative slack variables s and t satisfying

$$40X + 1000Y + s \quad\quad\ = 2000$$
$$50X + 750Y \quad\quad + t = 2400,$$

which leads to the tableau

$$\begin{array}{c}\ \\ s \\ t \\ \ \end{array}\left[\begin{array}{cccc|c} X & Y & s & t & \\ 40 & \boxed{1000} & 1 & 0 & 2000 \\ 50 & 750 & 0 & 1 & 2400 \\ \hline 800 & 18000^* & 0 & 0 & R \end{array}\right] \begin{array}{l} 2000/1000 = 2^* \\ 2400/750 = 3.2 \end{array}$$

In the usual way we choose to pivot on the entry 1000 in row 1 and column 2 as follows.

$$\left[\begin{array}{cccc|c} X & Y & s & t & \\ .04 & \boxed{1} & .001 & 0 & 2 \\ 50 & 750 & 0 & 1 & 2400 \\ \hline 800 & 18000 & 0 & 0 & R \end{array}\right] \quad (.001 R_1)$$

$$\begin{array}{c} \ \\ Y \\ t \\ \ \end{array}\left[\begin{array}{cccc|c} X & Y & s & t & \\ .04 & 1 & .001 & 0 & 2 \\ \boxed{20} & 0 & -.75 & 1 & 900 \\ \hline 80^* & 0 & -18 & 0 & R - 36000 \end{array}\right] \begin{array}{l} 2/.04 = 50 \\ (R_2 + (-750)R_1)\ 900/20 = 45^* \\ (R_3 + (-18000)R_1) \end{array}$$

Our next pivot will be on the entry 20 in row 2 and column 1, producing the following tableaus.

$$\left[\begin{array}{cccc|c} X & Y & s & t & \\ .04 & 1 & .001 & 0 & 2 \\ \boxed{1} & 0 & -.0375 & .05 & 45 \\ \hline 80 & 0 & -18 & 0 & R - 36000 \end{array}\right] \quad (.05 R_2)$$

3.5 MINIMIZATION PROBLEMS AND DUALITY

$$\begin{array}{c} & \begin{array}{cccc} X & Y & s & t \end{array} & \\ \begin{array}{c} Y \\ X \end{array} & \left[\begin{array}{cccc|c} 0 & 1 & .0025 & -.002 & .2 \\ 1 & 0 & -.0375 & .05 & 45 \\ \hline 0 & 0 & -15 & -4 & R - 39600 \end{array} \right] & \begin{array}{c} (R_1 + (-.04)R_2) \\ \\ R_3 + (-80)R_2 \end{array} \end{array}$$

Since there are no positive indicators, we are done. From the last row of our tableau the maximum value of R is 39600. But *39600 is also the minimal value of C in our original problem.* The optimal point (x, y) from the original minimization problem can also be read from the last tableau. It is $(x, y) = (15, 4)$. The values of x and y can be found by simply taking the negatives of the indicators (in color) corresponding to the slack variables s and t in our final tableau. (In general the optimal values of the original variables will be the negatives of the indicators corresponding to the slack variables of the dual problem after the simplex method has been applied.) The reader should check that in Example 3.5 we did find graphically that the minimal value of C was 39600 at $(x, y) = (15, 4)$. Notice that row labels are not necessary when solving the dual program because the solution is read from the bottom row of the final tableau.

THE DUALITY THEOREM

The method just illustrated is justified by a theorem first proved by the famous Hungarian-born American mathematician John von Neumann.

The Duality Theorem

> Consider a linear program with $\geq$ constraints where the objective function C is to be minimized. Such a program has an optimal solution if and only if the dual program, which has $\leq$ constraints and an objective function R to be maximized, has an optimal solution. In this case the maximal value of R is the minimal value of C. Furthermore, if the dual program is solved with the simplex method, then the negatives of the indicators corresponding to the slack variables in the final tableau give the optimal values of the variables in the original problem.

EXAMPLE 3.2 Revisited In Example 3.2 in Section 3.1 we considered a problem of combining two foods. There we translated the problem into the linear program:

$$\begin{aligned} \text{Minimize} \quad & 10x + 15y = C \\ \text{subject to} \quad & 3x + 3y \geq 75 \\ & 2x + 4y \geq 60 \\ & 4x + 2y \geq 60 \\ & x \geq 0, \, y \geq 0. \end{aligned}$$

Let us solve this program by solving its dual.

Our minimizing program has the compact tableau

$$\begin{array}{c} x y \\ \left[\begin{array}{cc|l} 3 & 3 & (\geq)\ 75 \\ 2 & 4 & (\geq)\ 60 \\ 4 & 2 & (\geq)\ 60 \\ \hline 10 & 15 & (\text{minimize})\ C \end{array} \right] . \end{array}$$

Interchanging rows and columns gives us

$$\begin{array}{c} X Y Z \\ \left[\begin{array}{ccc|l} 3 & 2 & 4 & (\leq)\ 10 \\ 3 & 4 & 2 & (\leq)\ 15 \\ \hline 75 & 60 & 60 & (\text{maximize})\ R \end{array} \right] , \end{array}$$

the first two rows of which yield the system

$$3X + 2Y + 3Z + s = 10$$
$$3X + 4Y + 2Z + t = 15$$

when nonnegative slack variables s and t are introduced. Thus we have the simplex tableau

$$\begin{array}{c} X Y Z s t \\ \left[\begin{array}{ccccc|c} \circled{3} & 2 & 4 & 1 & 0 & 10 \\ 3 & 4 & 2 & 0 & 1 & 15 \\ \hline 75^* & 60 & 60 & 0 & 0 & R \end{array} \right] \end{array} \begin{array}{l} 10/3^* \\ 15/3 \end{array}.$$

By the methods of the previous section we see that we should pivot on the entry 3 in the first row and column.

$$\left[\begin{array}{ccccc|c} \circled{1} & \tfrac{2}{3} & \tfrac{4}{3} & \tfrac{1}{3} & 0 & \tfrac{10}{3} \\ 3 & 4 & 2 & 0 & 1 & 15 \\ \hline 75 & 60 & 60 & 0 & 0 & R \end{array} \right] \quad (\tfrac{1}{3} R_1)$$

$$\left[\begin{array}{ccccc|c} 1 & \tfrac{2}{3} & \tfrac{4}{3} & \tfrac{1}{3} & 0 & \tfrac{10}{3} \\ 0 & \circled{2} & -2 & -1 & 1 & 5 \\ \hline 0 & 10^* & -40 & -25 & 0 & R-250 \end{array} \right] \quad \begin{array}{l} (10/3)/(2/3) = 5 \\ (R_2 + (-3)R_1) \quad 5/2^* \\ (R_3 + (-75)R_1) \end{array}$$

The next pivot should be on the entry 2 in row 2 and column 2.

$$\left[\begin{array}{ccccc|c} 1 & \tfrac{2}{3} & \tfrac{4}{3} & \tfrac{1}{3} & 0 & \tfrac{10}{3} \\ 0 & \circled{1} & -1 & -\tfrac{1}{2} & \tfrac{1}{2} & \tfrac{5}{2} \\ \hline 0 & 10^* & -40 & -25 & 0 & R-250 \end{array} \right] \quad (\tfrac{1}{2} R_2)$$

$$\begin{array}{c} X Y Z s t \\ \left[\begin{array}{ccccc|c} 1 & 0 & 2 & \tfrac{2}{3} & -\tfrac{1}{3} & \tfrac{5}{3} \\ 0 & 1 & -1 & -\tfrac{1}{2} & \tfrac{1}{2} & \tfrac{5}{2} \\ \hline 0 & 0 & -30 & -20 & -5 & R-275 \end{array} \right] \end{array} \quad \begin{array}{l} (R_1 + (-\tfrac{2}{3})R_2) \\ (R_3 + (-10)R_2) \end{array}$$

Since there are no positive indicators, we are done. The maximum value of R, and thus the minimum value of C in the original problem, is 275. Furthermore, from the indicators corresponding to the slack variables s and t we see that C achieves its minimum at $(x, y) = (20, 5)$. In terms of the original problem this means that 20 ounces of Food A and 5 ounces of Food B should be used. See Figure 3.21 for a graphical interpretation of this problem. ■

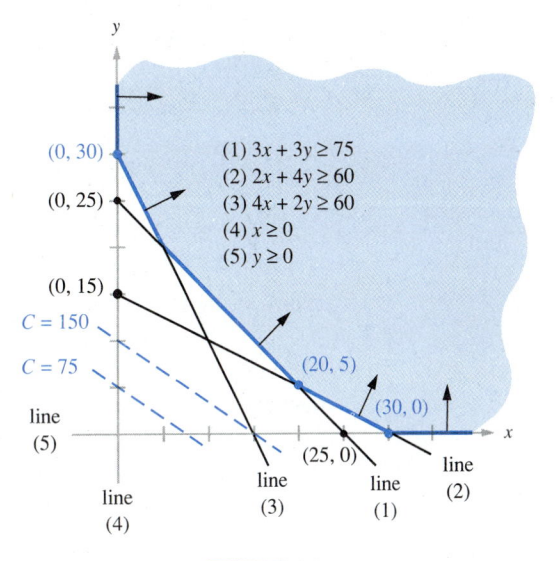

FIGURE 3.21

Practice Problem 10 What is the dual of the following linear program?

Minimize $\quad 2x + 5y + 3z = C$

subject to $\quad x + 2y + 4z \geq 12$

$\qquad\qquad\quad 3x + y + 2z \geq 15$

$\qquad\qquad\quad x \geq 0,\ y \geq 0,\ z \geq 0.$

EXAMPLE 3.14 During the ski season, World Travel Service offers a one-month vacation in Aspen, including round-trip fare and accommodations at the Aspen Lodge. The numbers of rooms needed at the lodge for each of the months January through March are as follows:

Month	Number of rooms needed
January	25
February	12
March	20

CHAPTER 3 LINEAR PROGRAMMING

The Lodge leases its rooms to World Travel at a discount rate of $800 for two consecutive months or $1000 for three consecutive months. What leasing plan will be least expensive?

Solution The Travel Service must decide how many rooms to lease for two months and how many for three months, and also which periods the various leases should cover. The three unknowns are illustrated in Figure 3.22.

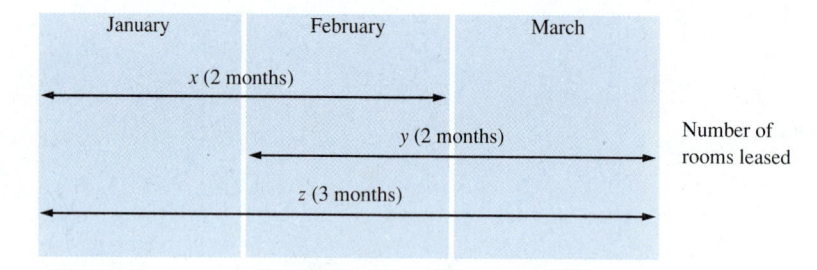

FIGURE 3.22

For example, x represents the number of rooms leased for two months beginning January 1. The total cost in dollars is given by

$$C = 800x + 800y + 1000z.$$

From Figure 3.22 we see that the total number of rooms available for January will be $x + z$. Thus, the condition that 25 rooms are needed for this month translates into the constraint

$$x + z \geq 25.$$

The requirements for February and March lead in the same way to the conditions

$$x + y + z \geq 12$$
$$y + z \geq 20.$$

The problem can thus be formulated as the three-variable linear program:

Minimize $\quad 800x + 800y + 1000z = C$
subject to $\quad x + z \geq 25$
$\quad x + y + z \geq 12$
$\quad y + z \geq 20$
$\quad x \geq 0, y \geq 0, z \geq 0,$

with compact tableau

$$\begin{bmatrix} x & y & z & \\ 1 & 0 & 1 & (\geq) \; 25 \\ 1 & 1 & 1 & (\geq) \; 12 \\ 0 & 1 & 1 & (\geq) \; 20 \\ \hline 800 & 800 & 1000 & \text{(minimize)} \; C \end{bmatrix}.$$

3.5 MINIMIZATION PROBLEMS AND DUALITY

The corresponding tableau for the dual program is

$$\left[\begin{array}{ccc|c} X & Y & Z & \\ 1 & 1 & 0 & (\leq) \ 800 \\ 0 & 1 & 1 & (\leq) \ 800 \\ 1 & 1 & 1 & (\leq) \ 1000 \\ \hline 25 & 12 & 20 & (\text{maximize}) \ R \end{array}\right].$$

We introduce nonnegative slack variables s, t, and u such that

$$\begin{aligned} X + Y + s &= 800 \\ Y + Z + t &= 800 \\ X + Y + Z + u &= 1000, \end{aligned}$$

giving the simplex tableau

$$\left[\begin{array}{cccccc|c} X & Y & Z & s & t & u & \\ ① & 1 & 0 & 1 & 0 & 0 & 800 \\ 0 & 1 & 1 & 0 & 1 & 0 & 800 \\ 1 & 1 & 1 & 0 & 0 & 1 & 1000 \\ \hline 25^* & 12 & 20 & 0 & 0 & 0 & R \end{array}\right] \quad \begin{array}{l} 800/1 = 800^* \\ \\ 1000/1 = 1000 \end{array}.$$

Pivoting on the entry 1 in row 1 and column 1 produces

$$\left[\begin{array}{cccccc|c} 1 & 1 & 0 & 1 & 0 & 0 & 800 \\ 0 & 1 & 1 & 0 & 1 & 0 & 800 \\ 0 & 0 & ① & -1 & 0 & 1 & 200 \\ \hline 0 & -13 & 20 & -25 & 0 & 0 & R - 20{,}000 \end{array}\right]. \quad \begin{array}{l} \\ (R_3 + (-1)R_1) \\ (R_4 + (-25)R_1) \end{array}$$

It is easily seen that the next pivot should be on the entry in row 3 and column 3, producing the tableau

$$\left[\begin{array}{cccccc|c} X & Y & Z & s & t & u & \\ 1 & 1 & 0 & 1 & 0 & 0 & 800 \\ 0 & 1 & 0 & 1 & 1 & -1 & 600 \\ 0 & 0 & 1 & -1 & 0 & 1 & 200 \\ \hline 0 & -13 & 0 & -5 & 0 & -20 & R - 24{,}000 \end{array}\right] \quad \begin{array}{l} \\ (R_2 + (-1)R_3) \\ \\ (R_4 + (-20)R_3). \end{array}$$

Since there are no more positive indicators, we are done. In the dual problem the maximum of R is 24,000, and so this is also the minimum value of C in the original problem. Furthermore, by taking the negatives of the indicators corresponding to the slack variables s, t, and u we see that C achieves its minimum when $(x, y, z) = (5, 0, 20)$. The Travel Service should rent 5 rooms for January and February, and 20 rooms for the 3-month period, for a total cost of $24,000. ∎

INTERPRETING THE DUAL VARIABLES

The variables in the dual of a linear program can be given a meaning. We consider again Example 3.5 of Section 3.2, which leads to the program:

$$\text{Minimize} \quad 2000x + 2400y = C$$
$$\text{subject to} \quad 40x + 50y \geq 800$$
$$1000x + 750y \geq 18{,}000$$
$$x \geq 0,\ y \geq 0.$$

Here x and y represent the number of Type A and Type B buses hired by an alumni association. These cost $2000 and $2400 per day, respectively, and the cost $2000x + 2400y$ is to be minimized. The first constraint comes about because the buses carry 40 and 50 people, respectively, and 800 people are going on the trip. Likewise the buses carry 1000 and 750 pounds of baggage, respectively, and 18,000 pounds must be carried.

The dual program is:

$$\text{Maximize} \quad 800X + 18{,}000Y = R$$
$$\text{subject to} \quad 40X + 1000Y \leq 2000$$
$$50X + 750Y \leq 2400$$
$$X \geq 0,\ Y \geq 0,$$

as we saw at the beginning of this section. The variables X and Y can be interpreted as follows. Suppose a rival transportation company is being formed to compete with the company that charters buses. The new company plans to charge a certain amount X for each person transported, and an amount Y for each pound of baggage carried. Recall that the bus company rents a Type A bus, carrying 40 people and 1000 pounds of baggage, for $2000. The new company will charge $40X + 1000Y$ to carry the same 40 people and 1000 pounds of baggage. Since it wants to be competitive with the first bus company, X and Y should be chosen such that

$$40X + 1000Y \leq 2000, \qquad (3.10)$$

for if we had $40X + 1000Y > 2000$, then customers would hire the Type A bus whenever it could be used to capacity. A similar analysis with respect to the Type B bus, which carries 50 people and 750 pounds of baggage for $2400, leads to the constraint

$$50X + 750Y \leq 2400. \qquad (3.11)$$

Of course (3.10) and (3.11) are exactly the constraints of the dual program.

Finally, the new company would like to maximize its revenue. Of course this will depend on how many passengers and pounds of baggage it carries. It is decided that the ratio of 800 people to 18,000 pounds of baggage of the alumni trip is about the usual proportion. Thus it is decided to maximize the revenue that would be generated for carrying 800 people and 18,000 pounds of baggage, which is

$$800X + 18{,}000Y = R.$$

This is exactly the objective function for the dual program.

At the beginning of this section we found the following final tableau in the simplex solution to this dual program.

$$\begin{array}{c} \\ Y \\ X \\ \end{array} \left[\begin{array}{cccc|c} X & Y & s & t & \\ 0 & 1 & .0025 & .002 & .2 \\ 1 & 0 & -0.375 & .05 & 45 \\ \hline 0 & 0 & -15 & -4 & R - 39600 \end{array} \right]$$

Thus the revenue is maximized when $(X, Y) = (45, .2)$. This means the new company should charge \$45 per passenger and 20¢ per pound of baggage. By doing so for the alumni trip it would take in \$39,600, the same as the minimal cost of renting buses found for the original problem. This is perhaps not surprising, since the new company's prices are set to be as high as possible without exceeding those of the first bus company.

LIMITATIONS OF THE DUAL PROGRAM METHOD

In order for the method of passing to the dual to be successful, the dual program must be a standard maximum program, since this was the only type treated in Section 3.4. In particular, the right-hand constants of the new constraints must all be nonnegative, since this is one of the conditions for a standard maximum program. This means that the coefficients of the objective function of the original minimization problem must all be nonnegative. In the next section we treat programs that cannot be transformed into standard maximum programs.

EXERCISES 3.5

In Exercises 1–6 write the dual of the given program.

1. Minimize $5x + 4y = C$
   subject to $x - y \geq -5$
   $x + 2y \geq 20$
   $x + y \geq 15$
   $x \geq 0, y \geq 0.$

2. Minimize $x + y = C$
   subject to $2x + y \geq 24$
   $x + 3y \geq 27$
   $x \geq 0, y \geq 0.$

3. Minimize $6x + 2y = C$
   subject to $x + 4y \geq 40$
   $4x + y \geq 40$
   $x + y \geq 25$
   $x \geq 0, y \geq 0.$

4. Minimize $5x + 4y = C$
   subject to $x + y \geq 20$
   $-x + y \geq -10$
   $x - y \geq -10$
   $x \geq 0, y \geq 0.$

5. Minimize $10x + 20y = C$
subject to $y \geq 5$
$-x + y \geq 0$
$5x + 2y \geq 20$
$x \geq 0, y \geq 0.$

6. Minimize $2x + 4y + z = C$
subject to $x + 2y + 3z \geq 80$
$x + y + z \geq 60$
$x \geq 0, y \geq 0, z \geq 0.$

In Exercises 7–12 solve the program in the listed exercise by applying the simplex algorithm to the dual program.

7. Exercise 1
8. Exercise 2
9. Exercise 3
10. Exercise 4
11. Exercise 5
12. Exercise 6

In Exercises 13–16 solve the program by applying the simplex algorithm to the dual program.

13. Minimize $5x + 4y + 3z = C$
subject to $2x + y + z \geq 40$
$x + y + 2z \geq 50$
$x \geq 0, y \geq 0, z \geq 0.$

14. Minimize $x + 4y + 3z = C$
subject to $x + y + z \geq 20$
$y + z \geq 15$
$z \geq 10, x \geq 0, y \geq 0, z \geq 0.$

15. Minimize $2x + y + z + 2w = C$
subject to $z + w \geq -50$
$x + z \geq 60$
$y + w \geq 40$
$x \geq 0, y \geq 0, z \geq 0, w \geq 0.$

16. Minimize $3x + y + 2z = C$
subject to $x + y \geq -10$
$y + z \geq 15$
$-x - z \geq -10$
$x \geq 0, y \geq 0, z \geq 0.$

In Exercises 17–22 formulate the given problem as a linear program and solve by applying the simplex algorithm to the dual program.

17. The heating plant for a large government building must produce at least 300 million Btu's of heat per hour. The plant is equipped to burn any combination of fuel oil and coal, but it must observe government pollution limits of at most 300 pounds of sulfur dioxide and 125 pounds of nitrogen dioxide emitted per hour. The cost, heat yield, and pollution content per ton for each of the fuels are summarized in the following table:

Fuel	Oil	Coal
Heat (million Btu/ton)	40	20
SO_2 (lb/ton)	20	30
NO_2 (lb/ton)	5	20
Cost ($/ton)	$290	$65

How many tons of each fuel should the plant burn per hour in order to meet all the requirements at minimum cost?

18. Nutritionists indicate that an adequate daily diet should provide at least 75 grams of carbohydrate, 60 grams of protein, and 60 grams of fat. Two foods are available; their nutritional yields and costs, per ounce, are as follows:

Food	A	B
Carb/oz	3 gm	3 gm
Protein/oz	2 gm	4 gm
Fat/oz	4 gm	2 gm
Cost/oz	10¢	15¢

How many ounces of each food should be combined per day to meet the nutritional requirements at lowest cost?

19. World Travel Service has agreed to transport a group of 1800 football fans from Los Angeles to Chicago for one of the season's big games. The agency can charter three different types of aircraft. Their operating costs for the trip, and their seating capacities, are as follows.

	Number of Seats per Plane		
Type	A	B	C
First-class	20	40	50
Economy	80	60	50
Cost	$3000	$4000	$5000

If 600 of the passengers want to go first-class while the remaining 1200 have ordered economy seats, how many planes of each type should be assigned to the charter to minimize the total operating cost?

20. The Allegheny Mining Company operates three mines which produce both high-grade and low-grade ore. The daily yield of each mine, and the daily operating cost, are given in the table below.

	Ore Yield per Day		
Mine	1	2	3
High-grade	2 tons	1 ton	0 tons
Low-grade	1 ton	2 tons	2 tons
Cost per day	$300	$200	$100

The company must fill an order for a total of 40 tons of high-grade ore and 50 tons of low-grade ore. How many days should each mine be operated in order to fill the order at lowest total cost?

21. A hospital dietician must prepare a special diet from two foods. The first food provides 1 unit of iron, 1 unit of magnesium, and 20 calories at a cost of 3¢ per gram. The second food provides 1 unit of iron, 4 units of magnesium, and 40 calories at a cost of 4¢ per gram. If the diet must provide at least 100 units of iron, 160 units of magnesium, and 2600 calories, what combination of the two foods will satisfy these conditions at the least cost? What is the least cost?

22. A jeweler makes rings, stickpins, and pendants from semiprecious stones. For each ring, 1 hour is required to cut the stones and 3 hours are needed to polish them; for each stickpin, 2 hours are required to cut the stones and 1 hour is needed to polish them; and for each pendant, 2 hours are required to cut the stones and 5 hours are required to polish them. During a particular week the jeweler wants to devote at least 9 hours to cutting and at least 30 hours to polishing. If it costs the jeweler $34 to cut and polish the stones for each ring, $28 to cut and polish the stones for each stickpin, and $60 to cut and polish the stones for each pendant, how many rings, stickpins, and pendants should be made in order to minimize the jeweler's costs? What is the minimum cost?

Answer to Practice Problem

10. Maximize $12X + 15Y = R$
    subject to $X + 3Y \leq 2$
    $2X + Y \leq 5$
    $4X + 2Y \leq 3$
    $X \geq 0, Y \geq 0$

3.6 Artificial Variables

So far we have only applied the simplex method to standard maximum linear programs, that is, programs with the four properties below.

(a) The objective function is to be maximized.
(b) All variables are to be nonnegative.
(c) The other constraints are of the $\leq$ variety.
(d) The righthand constants in the $\leq$ constraints are all nonnegative.

Restriction (a) is not hard to get around, since minimizing an objective function C is the same as maximizing the function $-C$. Thus any linear program can be written so as to involve maximizing the objective function. Likewise restriction (b) is automatic in almost all practical applications of linear programming; and we will not consider any programs where it does not hold.

Restrictions (c) and (d) are harder to deal with; and there are problems for which one or the other of them fails no matter how we set up the program. The algebraic fact that

$$A \geq B \quad \text{if and only if} \quad -A \leq -B \tag{3.12}$$

enables us to switch between $\geq$ and $\leq$ constraints, but the constant on the right may become negative in the process. For example, the $\geq$ constraint

$$3x + 5y \geq 7$$

is equivalent to the $\leq$ constraint

$$-3x - 5y \leq -7,$$

but a program with such a constraint would not be a standard maximum program because of restriction (d).

The reader may wonder why restriction (d) is necessary. The answer is that when the simplex method is applied to a standard maximum program, we start at the point where all the original variables are 0 and move by a sequence of pivots to an optimal corner of the feasible region. That is, the slack variables start out as basic and the original variables free. Setting the (free) original variables equal to 0 makes each slack variable equal to the corresponding constant in the rightmost column of the simplex tableau. But if one of these constants is negative, then so is the corresponding slack variable, contradicting the assumption made about the slack variables when they are introduced. What this means is that we are starting at a point that is not in the feasible region. We are in "left field", so to speak, and though pivoting may take us from vertex to vertex, there is no guarantee we will ever get to the feasible region.

In this section we will describe a method that can be used to solve linear programming problems that cannot be put into standard maximum form. The method will involve two phases. In **phase one** we will introduce not only slack variables, but also new "artificial variables." These artificial variables and a new objective function will be used to move to a corner of the feasible region. Then the artificial variables will be discarded, and in **phase two** we will use the simplex method and our original objective function to complete the solution of the problem.

We will assume that we have a linear program satisfying (a), (b), and (d), but not necessarily (c). That is:

1. If the problem is to minimize C, change it to the problem of maximizing $-C$.
2. If the right-hand constant of any constraint is negative, change it to a positive constant by multiplying both sides of the constraint by -1 and reversing the inequality sign.

Practice Problem 11 Change the following linear program to one satisfying (a), (b), and (d) at the beginning of this section.

$$
\begin{aligned}
\text{Minimize} \quad & -3x + y = C \\
\text{subject to} \quad & -x + y \geq -2 \\
& -2x - 3y \geq -6 \\
& x + y \geq 1 \\
& x \geq 0, y \geq 0.
\end{aligned}
$$

HOW PHASE ONE WORKS

We will demonstrate phase one with the program.

$$
\begin{aligned}
\text{Maximize} \quad & 3x - y = R \\
\text{Subject to} \quad & x - y \leq 2 \\
& 2x + 3y \leq 6 \\
& x + y \geq 1 \\
& x \geq 0, y \geq 0.
\end{aligned}
$$

The first two constraints, which are $\leq$ inequalities, we will treat as usual, introducing nonnegative slack variables s and t such that

$$x - y + s = 2$$
$$2x + 3y + t = 6.$$

Since according to the third constraint $x + y$ may exceed 1, it would be natural to introduce a nonnegative slack variable u such that

$$x + y - u = 1.$$

The problem with this is that setting the free variables x and y both equal to 0 gives $-u = 1$, or $u = -1$, contrary to the condition that u should be nonnegative. Thus at the same time we introduce u we will also introduce an **artificial variable** $a \geq 0$, such that

$$x + y - u + a = 1.$$

Thus we can have $x = y = u = 0$, so long as $a = 1$.

Now we have the system

$$x - y + s = 2$$
$$2x + 3y + t = 6$$
$$x + y - u + a = 1,$$

with all these variables nonnegative, and we wish to maximize

$$3x - y = R.$$

This gives the tableau

$$\begin{array}{c} \\ s \\ t \\ a \\ \\ \end{array} \begin{array}{c} x \quad y \quad s \quad t \quad u \quad a \\ \left[\begin{array}{cccccc|c} 1 & -1 & 1 & 0 & 0 & 0 & 2 \\ 2 & 3 & 0 & 1 & 0 & 0 & 6 \\ 1 & 1 & 0 & 0 & -1 & 1 & 1 \\ \hline 3 & -1 & 0 & 0 & 0 & 0 & R \end{array} \right]. \end{array}$$

Notice that row 3 has the label a rather than u, since the column headed by a contains a 1 in row 3 and zeros elsewhere. Our first order of business will be to get rid of the artificial variable a. As we have seen, setting x, y, and u equal to 0 makes $a = 1$. We would like to make $a = 0$, so that we could drop this variable. Thus we want to minimize a, or, since the simplex method is set up for maximization problems,

$$\text{maximize} \quad -a = Q. \tag{3.13}$$

Thus we have a new objective function that we will deal with before we try to maximize R. Maximizing Q will have the effect of moving to a corner of the original feasible region; that is, it will complete phase one of our method.

In order to maximize Q we will add a row corresponding to (3.13) to our tableau.

$$\begin{array}{c} \\ s \\ t \\ a \\ \\ \\ \end{array} \begin{array}{c} x \quad\; y \quad s \quad t \quad\; u \quad\; a \\ \left[\begin{array}{cccccc|c} 1 & -1 & 1 & 0 & 0 & 0 & 2 \\ 2 & 3 & 0 & 1 & 0 & 0 & 6 \\ 1 & 1 & 0 & 0 & -1 & 1 & 1 \\ \hline 3 & -1 & 0 & 0 & 0 & 0 & R \\ 0 & 0 & 0 & 0 & 0 & -1 & Q \end{array}\right] \end{array}$$

We have come far enough that we will summarize the steps necessary in phase one of our method. We have already illustrated steps 1 and 2 listed below.

Phase One of the Method of Artificial Variables

It is assumed that we have a linear program that is a maximization problem, and that the right-hand constants of all constraints are nonnegative.

Step 1 Make each constraint into an equation as follows. On the left side of each $\leq$ constraint introduce a slack variable with coefficient $+1$. On the left side of each $\geq$ constraint introduce a slack variable with coefficient -1, and also an artificial variable with coefficient $+1$.

Step 2 Set up the simplex tableau for the equations from step 1, along with the objective function. Add a row at the bottom with a -1 under each artificial variable, a 0 under every other variable, and the letter Q in the bottom right corner.

Step 3 To the last row of the tableau add each row above the horizontal line that corresponds to an equation with an artificial variable.

Step 4 Proceed with the simplex method as usual to get rid of all positive indicators in the bottom row of the tableau.

After phase one is completed we will throw away the last row of our tableau, along with all the columns corresponding to artificial variables, and finish the simplex algorithm in the usual way. This will be phase two.

Continuing with our example, we now apply step 3 to the tableau by adding row 3 to the last row.

$$\begin{array}{c} \\ s \\ t \\ a \\ \\ \\ \end{array} \begin{array}{c} x \quad\; y \quad s \quad t \quad\; u \quad a \\ \left[\begin{array}{cccccc|c} 1 & -1 & 1 & 0 & 0 & 0 & 2 \\ 2 & 3 & 0 & 1 & 0 & 0 & 6 \\ ① & 1 & 0 & 0 & -1 & 1 & 1 \\ \hline 3 & -1 & 0 & 0 & 0 & 0 & R \\ 1^* & 1 & 0 & 0 & -1 & 0 & Q+1 \end{array}\right] \end{array} \quad\begin{array}{l} 2/1 = 2 \\ 6/2 = 3 \\ 1/1 = 1^* \\ \\ (\mathbf{R_5 + R_3}) \end{array}$$

Notice that at this stage x, y, and u are free variables. Setting them all equal to 0 gives $s = 2$, $t = 6$, and $a = 1$. Although

$$(x, y, s, t, u, a) = (0, 0, 2, 6, 0, 1)$$

is feasible with respect to our new system, $(x, y) = (0, 0)$ is not a feasible point with respect to our original maximization problem. The feasible region of this problem is illustrated in Figure 3.23.

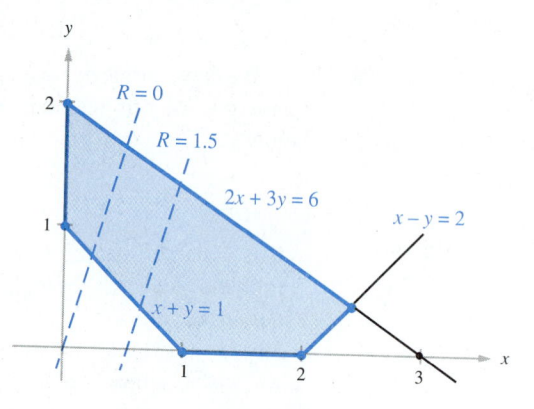

FIGURE 3.23

We proceed with step 4 to get rid of all positive indicators in the bottom row of our tableau. We arbitrarily choose to pivot in the first column instead of the second, and pick as the pivot the entry in row 3 and column 1 in the usual way. Pivoting produces the following tableau.

$$\begin{array}{c} \\ s \\ t \\ x \\ \\ \\ \end{array} \begin{bmatrix} x & y & s & t & u & a & \\ 0 & -2 & 1 & 0 & 1 & -1 & 1 \\ 0 & 1 & 0 & 1 & 2 & -2 & 4 \\ 1 & 1 & 0 & 0 & -1 & 1 & 1 \\ \hline 0 & -4 & 0 & 0 & 3 & -3 & R-3 \\ 0 & 0 & 0 & 0 & 0 & -1 & Q \end{bmatrix} \begin{array}{l} (R_1 + (-1)R_3) \\ (R_2 + (-2)R_3) \\ \\ \\ (R_4 + (-3)R_3) \\ (R_5 + (-1)R_3) \end{array}$$

At this point there are no positive indicators in the bottom row, so we have finished phase one. To start phase two, we eliminate the last row and the column corresponding to the artificial variable a.

$$\begin{array}{c} \\ s \\ t \\ x \\ \\ \end{array} \begin{bmatrix} x & y & s & t & u & \\ 0 & -2 & 1 & 0 & ① & 1 \\ 0 & 1 & 0 & 1 & 2 & 4 \\ 1 & 1 & 0 & 0 & -1 & 1 \\ \hline 0 & -4 & 0 & 0 & 3^* & R-3 \end{bmatrix} \begin{array}{l} 1/1 = 1^* \\ 4/2 = 2 \\ \text{(negative)} \\ \\ \end{array}$$

3.6 ARTIFICIAL VARIABLES

Notice that now y and u are free variables and x, s, and t are basic. Setting y and u equal to 0 gives $x = 1$, $s = 1$, and $t = 4$. Furthermore, the point $(x, y) = (1, 0)$ is a corner of our original feasible region. (See Figure 3.23.)

Pivoting on the next indicated pivot in row 1 and column 5 produces the following tableau.

$$\begin{array}{c} \\ u \\ t \\ x \\ \\ \end{array} \begin{array}{c} \begin{array}{ccccc} x & y & s & t & u \end{array} \\ \left[\begin{array}{ccccc|c} 0 & -2 & 1 & 0 & 1 & 1 \\ 0 & ⑤ & -2 & 1 & 0 & 2 \\ 1 & -1 & 1 & 0 & 0 & 2 \\ \hline 0 & 2* & -3 & 0 & 0 & R - 6 \end{array} \right] \end{array} \quad \begin{array}{l} \text{(negative)} \\ (R_2 + (-2)R_1) \quad 2/5* \\ (R_3 + R_1) \quad \text{(negative)} \\ (R_4 + (-3)R_1) \end{array}$$

Notice that now setting the free variables y and s equal to 0 gives $x = 2$. Thus we are at the point $(x, y) = (2, 0)$ in the feasible region.

Our last pivot is on the entry 5 in row 2 and column 2, and results in the following tableaus.

$$\left[\begin{array}{ccccc|c} 0 & -2 & 1 & 0 & 1 & 1 \\ 0 & ① & -.4 & .2 & 0 & .4 \\ 1 & -1 & 1 & 0 & 0 & 2 \\ \hline 0 & 2 & -3 & 0 & 0 & R - 6 \end{array} \right] \quad (.2R_2)$$

$$\begin{array}{c} \\ u \\ y \\ x \\ \\ \end{array} \begin{array}{c} \begin{array}{ccccc} x & y & s & t & u \end{array} \\ \left[\begin{array}{ccccc|c} 0 & 0 & .2 & .4 & 1 & 1.8 \\ 0 & 1 & -.4 & .2 & 0 & 0.4 \\ 1 & 0 & .6 & .2 & 0 & 2.4 \\ \hline 0 & 0 & -2.2 & -.4 & 0 & R - 6.8 \end{array} \right] \end{array} \quad \begin{array}{l} (R_1 + 2R_2) \\ (R_3 + R_2) \\ (R_4 + (-2)R_2) \end{array}$$

There are no more positive indicators, and we are done. The maximum value of R is 6.8 and occurs when $(x, y) = (2.4, 0.4)$. The geometrical progress of our two-phase method is shown in Figure 3.24.

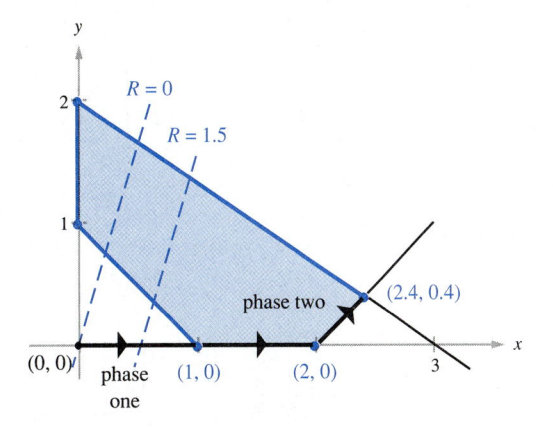

FIGURE 3.24

EXAMPLE 3.15 A shoe manufacturer makes two types of tennis shoes, the Court Ace and Great Get. The manufacturer has a contract with a large retailer to provide at least 5000 pairs of shoes, of which at least 2000 must be the Court Ace model. Only 18,000 minutes of machine time are available on the machines that make the tennis shoes, and a pair of the Court Ace model takes 2 minutes of machine time, while a pair of the Great Get model takes 3 minutes of machine time. The profit is $3 for each pair of Court Ace shoes, and $4 for each pair of Great Get shoes. Use the simplex method to determine how many pairs of each model should be made in order to maximize the manufacturer's profit.

Solution Let us suppose

$$x = \text{the number of pairs of Court Ace shoes made}$$
$$y = \text{the number of pairs of Great Get shoes made.}$$

Then the conditions of the problem lead us to the linear program:

Maximize $\quad 3x + 4y = P$

subject to
$$x + y \geq 5000$$
$$x \geq 2000$$
$$2x + 3y \leq 18000$$
$$x \geq 0, y \geq 0.$$

Since the constraints are of both the $\geq$ and $\leq$ variety, we will use the method of artificial variables. We introduce nonnegative slack variables s, t, and u and artificial variables a and b such that

$$x + y - s \qquad\qquad + a = 5000$$
$$x \qquad\qquad - t \qquad + b = 2000$$
$$2x + 3y \qquad\qquad + u \qquad = 18000.$$

Performing step 2 of phase one gives us the tableau

$$\begin{array}{c} a \\ b \\ u \\ \\ \\ \end{array} \left[\begin{array}{cccccc|c} x & y & s & t & u & a & b & \\ 1 & 1 & -1 & 0 & 0 & 1 & 0 & 5000 \\ 1 & 0 & 0 & -1 & 0 & 0 & 1 & 2000 \\ 2 & 3 & 0 & 0 & 1 & 0 & 0 & 18000 \\ \hline 3 & 4 & 0 & 0 & 0 & 0 & 0 & P \\ 0 & 0 & 0 & 0 & 0 & -1 & -1 & Q \end{array} \right].$$

Now we perform step 3 by adding rows 1 and 2 to the last row.

$$\begin{array}{c} a \\ b \\ u \\ \\ \\ \end{array} \left[\begin{array}{cccccc|c} x & y & s & t & u & a & b & \\ 1 & 1 & -1 & 0 & 0 & 1 & 0 & 5000 \\ ① & 0 & 0 & -1 & 0 & 0 & 1 & 2000 \\ 2 & 3 & 0 & 0 & 1 & 0 & 0 & 18000 \\ \hline 3 & 4 & 0 & 0 & 0 & 0 & 0 & P \\ 2^* & 1 & -1 & -1 & 0 & 0 & 0 & Q + 7000 \end{array} \right] \begin{array}{l} 5000/1 = 5000 \\ 2000/1 = 2000^* \\ 18000/2 = 9000 \end{array}$$

3.6 ARTIFICIAL VARIABLES

Here x and y are free variables so we are at the point $(x, y) = (0, 0)$. Our first pivot will be on the entry in row 2 and column 1. We will show the result of each pivot, but not the details of the calculations. Pivoting produces the following tableau.

$$
\begin{array}{c}
\\ a\\ x\\ u\\ \\ \\ \\
\end{array}
\begin{array}{c}
\begin{array}{cccccc} x & y & s & t & u & a & b \end{array} \\
\left[\begin{array}{cccccc|c}
0 & \textcircled{1} & -1 & 1 & 0 & 1 & -1 & 3000 \\
1 & 0 & 0 & -1 & 0 & 0 & 1 & 2000 \\
0 & 3 & 0 & 2 & 1 & 0 & -2 & 14000 \\
\hline
0 & 4 & 0 & 3 & 0 & 0 & -3 & P - 6000 \\
0 & 1^* & -1 & 1 & 0 & 0 & -2 & Q + 3000
\end{array}\right]
\end{array}
\quad
\begin{array}{l}
3000/1 = 3000^* \\
\text{(undefined)} \\
14000/3 \approx 4667
\end{array}
$$

Notice that now setting the free variables equal to 0 gives $(x, y) = (2000, 0)$, which is not feasible because $x + y < 5000$. Thus we continue with phase one by pivoting on the entry in row 1 and column 2 to obtain the next tableau.

$$
\begin{array}{c}
\\ y\\ x\\ u\\ \\ \\
\end{array}
\begin{array}{c}
\begin{array}{ccccccc} x & y & s & t & u & a & b \end{array} \\
\left[\begin{array}{ccccccc|c}
0 & 1 & -1 & 1 & 0 & 1 & -1 & 3000 \\
1 & 0 & 0 & -1 & 0 & 0 & 1 & 2000 \\
0 & 0 & 3 & -1 & 1 & -3 & 1 & 5000 \\
\hline
0 & 0 & 4 & -1 & 0 & -4 & 1 & P - 18000 \\
0 & 0 & 0 & 0 & 0 & -1 & -1 & Q
\end{array}\right]
\end{array}
$$

Since there are no more positive entries in the last row, phase one is completed. We erase the last row and the columns corresponding to the artificial variables a and b.

$$
\begin{array}{c}
\\ y\\ x\\ u\\ \\
\end{array}
\begin{array}{c}
\begin{array}{ccccc} x & y & s & t & u \end{array} \\
\left[\begin{array}{ccccc|c}
0 & 1 & -1 & 1 & 0 & 3000 \\
1 & 0 & 0 & -1 & 0 & 2000 \\
0 & 0 & \textcircled{3} & -1 & 1 & 5000 \\
\hline
0 & 0 & 4^* & -1 & 0 & P - 18000
\end{array}\right]
\end{array}
\quad
\begin{array}{l}
\text{(negative)} \\
\text{(undefined)} \\
5000/3^*
\end{array}
$$

Notice that now we are at the feasible point $(x, y) = (2000, 3000)$. The next pivot will be on the entry in row 3 and column 3. The result is the following tableau.

$$
\begin{array}{c}
\\ y\\ x\\ s\\ \\
\end{array}
\begin{array}{c}
\begin{array}{ccccc} x & y & s & t & u \end{array} \\
\left[\begin{array}{ccccc|c}
0 & 1 & 0 & \textcircled{$\frac{2}{3}$} & \frac{1}{3} & 14000/3 \\
1 & 0 & 0 & -1 & 0 & 2000 \\
0 & 0 & 1 & -\frac{1}{3} & \frac{1}{3} & 5000/3 \\
\hline
0 & 0 & 0 & \frac{1}{3}^* & -\frac{4}{3} & P - 74000/3
\end{array}\right]
\end{array}
\quad
\begin{array}{l}
(14000/3)/(2/3) = 7000^* \\
\text{(negative)} \\
\text{(negative)}
\end{array}
$$

Now we are at the point $(x, y) = (2000, 14000/3)$, and we pivot on the entry in row 1 and column 4.

$$\begin{array}{c} \\ t \\ x \\ s \\ \\ \end{array} \left[\begin{array}{ccccc|c} x & y & s & t & u & \\ 0 & \frac{3}{2} & 0 & 1 & \frac{1}{2} & 7000 \\ 1 & \frac{3}{2} & 0 & 0 & \frac{1}{2} & 9000 \\ 0 & \frac{1}{2} & 1 & 0 & \frac{1}{2} & 4000 \\ \hline 0 & -\frac{1}{2} & 0 & 0 & -\frac{3}{2} & P - 27000 \end{array} \right]$$

Finally there are no positive indicators, and we are done. We see that the maximum profit P is $27,000$, achieved when $(x, y) = (9000, 0)$. That is, 9000 pairs of Court Ace shoes should be made, and no Great Get shoes. Figure 3.25 shows geometrically how our method reaches the solution, pivoting twice in phase one and twice in phase two. ■

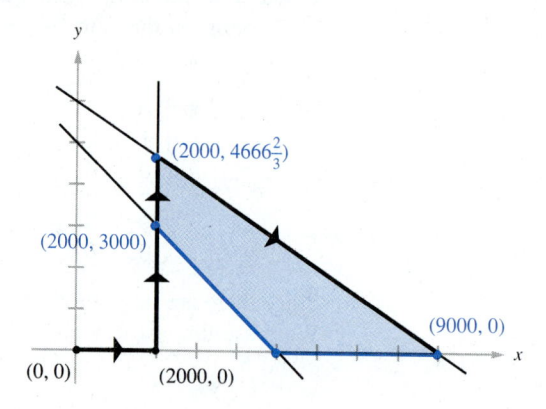

FIGURE 3.25

Practice Problem 12 Set up the initial phase one tableau for the following linear program:

$$\begin{aligned} \text{Maximize} \quad & x + 2y - 3z = R \\ \text{subject to} \quad & x \phantom{{}+{}} + z \leq 30 \\ & 2x + y + z \geq 10 \\ & x + 3y + 2z \geq 15 \\ & x \geq 0, \, y \geq 0, \, z \geq 0. \end{aligned}$$

What should be done in step 3?

LINEAR PROGRAMS WITH NO SOLUTION

The purpose of phase one of the method of artificial variables is to reach a point of the feasible region. If this region is empty, as in Example 3.7 of Section 3.2, then obviously phase one cannot be completed. We would rewrite the program of that example as

3.6 ARTIFICIAL VARIABLES

$$\text{Maximize} \quad 3x + 7y = R$$
$$\text{subject to} \quad -x - y \geq 2$$
$$x \geq 0, y \geq 0,$$

then introduce slack and artificial variables $s \geq 0$ and $a \geq 0$ such that

$$-x - y - s + a = 2.$$

The corresponding tableau is as follows.

$$a \begin{bmatrix} x & y & s & a & \\ -1 & -1 & -1 & 1 & 2 \\ \hline 3 & 7 & 0 & 0 & R \\ 0 & 0 & 0 & -1 & Q \end{bmatrix}$$

Step 3 of phase one gives the following tableau.

$$a \begin{bmatrix} x & y & s & a & \\ -1 & -1 & -1 & 1 & 2 \\ \hline 3 & 7 & 0 & 0 & R \\ -1 & -1 & -1 & 0 & Q+2 \end{bmatrix}$$

The last row corresponds to the equation

$$Q = -2 - x - y - s,$$

and since we are trying to maximize Q the best we can do is set $x = y = s = 0$. This makes $a = 2$. Since we cannot make the artificial variable a equal to 0, phase one cannot be completed.

In cases when the feasible region is nonempty, but the objective function cannot be optimized because the feasible region is unbounded, the simplex algorithm will not break down until phase two. Consider, for instance, Example 3.6 of Section 3.2, which is the program

$$\text{Maximize} \quad 6x + 3y = R$$
$$\text{subject to} \quad x + 2y \geq 8$$
$$2x + y \geq 10$$
$$x \geq 0, y \geq 0.$$

Introducing nonnegative variables $s, t, a,$ and b such that

$$x + 2y - s + a = 8$$
$$2x + y - t + b = 10,$$

leads to the following tableau.

$$\begin{array}{c} \\ a \\ b \\ \\ \\ \end{array} \begin{bmatrix} x & y & s & t & a & b & \\ 1 & 2 & -1 & 0 & 1 & 0 & 8 \\ 2 & 1 & 0 & -1 & 0 & 1 & 10 \\ \hline 6 & 3 & 0 & 0 & 0 & 0 & R \\ 0 & 0 & 0 & 0 & -1 & -1 & Q \end{bmatrix}$$

Then step 3 of phase one gives

$$\begin{array}{c} \\ a \\ b \\ \\ \\ \end{array} \begin{bmatrix} x & y & s & t & a & b & \\ 1 & 2 & -1 & 0 & 1 & 0 & 8 \\ ② & 1 & 0 & -1 & 0 & 1 & 10 \\ \hline 6 & 3 & 0 & 0 & 0 & 0 & R \\ 3^* & 3 & -1 & -1 & 0 & 0 & Q+18 \end{bmatrix} \quad \begin{array}{l} 8/1 = 8 \\ 10/2 = 5^* \end{array}$$

First we pivot on the entry in row 2 and column 1. The details are omitted.

$$\begin{array}{c} \\ a \\ x \\ \\ \\ \end{array} \begin{bmatrix} x & y & s & t & a & b & \\ 0 & \tfrac{3}{2} & -1 & \tfrac{1}{2} & 1 & -\tfrac{1}{2} & 3 \\ 1 & \tfrac{1}{2} & 0 & -\tfrac{1}{2} & 0 & \tfrac{1}{2} & 5 \\ \hline 0 & 0 & 0 & 3 & 0 & -3 & R-30 \\ 0 & \tfrac{3}{2}^* & -1 & \tfrac{1}{2} & 0 & -\tfrac{3}{2} & Q+3 \end{bmatrix} \quad \begin{array}{l} 3/(3/2) = 2^* \\ 5/(1/2) = 10 \end{array}$$

The next pivot entry is in row 1 and column 2.

$$\begin{array}{c} \\ y \\ x \\ \\ \\ \end{array} \begin{bmatrix} x & y & s & t & a & b & \\ 0 & 1 & -\tfrac{2}{3} & \tfrac{1}{3} & \tfrac{2}{3} & -\tfrac{1}{3} & 2 \\ 1 & 0 & \tfrac{1}{3} & -\tfrac{2}{3} & -\tfrac{1}{3} & \tfrac{2}{3} & 4 \\ \hline 0 & 0 & 0 & 3 & 0 & -3 & R-30 \\ 0 & 0 & 0 & 0 & -1 & -1 & Q \end{bmatrix}$$

Since there are no positive indicators in the bottom row, phase one is finished. Notice that we are at the point $(x, y) = (4, 2)$, a corner of the feasible region. (See Figure 3.11.)

Dropping the last row and the columns corresponding to artificial variables produces the tableau

$$\begin{array}{c} \\ y \\ x \\ \\ \end{array} \begin{bmatrix} x & y & s & t & \\ 0 & 1 & -\tfrac{2}{3} & ① & 2 \\ 1 & 0 & \tfrac{1}{3} & -\tfrac{2}{3} & 4 \\ \hline 0 & 0 & 0 & 3^* & R-30 \end{bmatrix} \quad \begin{array}{l} 2/(1/3) = 6^* \\ \text{(negative)} \end{array}$$

and we proceed to pivot on row 1 and column 4.

$$\begin{array}{c} \\ t \\ x \\ \end{array} \begin{array}{c} \\ \\ \\ \end{array} \left[\begin{array}{cccc|c} x & y & s & t & \\ 0 & 3 & -2 & 1 & 6 \\ 1 & 2 & -1 & 0 & 8 \\ \hline 0 & -9 & 6^* & 0 & R-48 \end{array} \right] \begin{array}{l} \text{(negative)} \\ \text{(negative)} \end{array}$$

Now a pivot in column 3 is indicated, but impossible because all the corresponding quotients are negative. Thus phase two breaks down, and the program has no solution.

Although we have illustrated the artificial variable algorithm with two-variable problems so that we could show pictorially how the method works, its main usefulness is with many-variable problems for which no graphical method is practical. The simplex algorithm proceeds by well-determined rules, and so may be programmed on a computer, where it is common to solve linear programming problems with hundreds of variables and constraints in seconds.

EQUALITY CONSTRAINTS

The artificial variable method may also be applied when some of the constraints are equations, rather than inequalities. All that need be done is to *introduce an artificial variable for each equality constraint*. (No slack variable is needed for such a constraint.) The following example illustrates the method.

EXAMPLE 3.16 Ralph plans to spend at most $1.40 for a 30-day supply of vitamin C pills. Healthcaps cost 4¢ each and provide 300 units of the vitamin. Likewise No-colds cost 3¢ and provide 200 units, and Orangos cost 5¢ and provide 450 units. How can he get the most vitamin C for his money?

Solution Suppose Ralph buys x Healthcaps, y No-colds, and z Orangos. Then we have the linear program:

Maximize $\quad 300x + 200y + 450z = R$
subject to $\quad 4x + 3y + 5z \leq 140$
$\quad\quad\quad\quad\; x + y + z = 30$
$\quad\quad\quad\quad\; x \geq 0, y \geq 0, z \geq 0.$

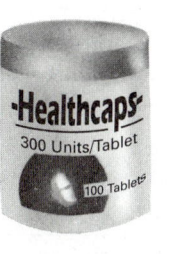

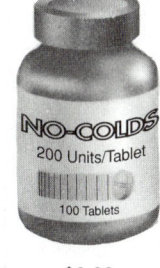

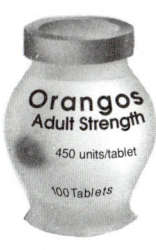

As usual we replace the inequality constraint with the equation

$$4x + 3y + 5z + s = 140, \qquad s \geq 0.$$

We introduce the artificial variable a to produce the equation

$$x + y + z + a = 30.$$

The reason for this is that the given equation $x + y + z = 30$ is not satisfied when we set the originally free variables x, y, and z equal to 0. The new equation is satisfied, however, if $x = y = z = 0$ and $a = 30$.

From this point we proceed as in the previous examples in this section. Our tableau is

$$\left[\begin{array}{ccccc|c} x & y & z & s & a & \\ 4 & 3 & 5 & 1 & 0 & 140 \\ 1 & 1 & 1 & 0 & 1 & 30 \\ \hline 300 & 200 & 450 & 0 & 0 & R \\ 0 & 0 & 0 & 0 & -1 & Q \end{array}\right].$$

Step 3 produces the tableau

$$\begin{array}{c} \\ s \\ a \end{array} \left[\begin{array}{ccccc|c} x & y & z & s & a & \\ 4 & 3 & 5 & 1 & 0 & 140 \\ ① & 1 & 1 & 0 & 1 & 30 \\ \hline 300 & 200 & 450 & 0 & 0 & R \\ 1* & 1 & 1 & 0 & 0 & Q + 30 \end{array}\right] \begin{array}{l} 140/4 = 35 \\ 30/1 = 30* \end{array}$$

We choose to pivot first in column 1, and thus in row 2. The result is as follows.

$$\begin{array}{c} \\ s \\ x \end{array} \left[\begin{array}{ccccc|c} x & y & z & s & a & \\ 0 & -1 & 1 & 1 & -4 & 20 \\ 1 & 1 & 1 & 0 & 1 & 30 \\ \hline 0 & -100 & 150 & 0 & -300 & R - 9000 \\ 0 & 0 & 0 & 0 & -1 & Q \end{array}\right]$$

Now phase one is finished, and we remove the last row and the column headed by a.

$$\begin{array}{c} \\ s \\ x \end{array} \left[\begin{array}{cccc|c} x & y & z & s & \\ 0 & -1 & ① & 1 & 20 \\ 1 & 1 & 1 & 0 & 30 \\ \hline 0 & -100 & 150* & 0 & R - 9000 \end{array}\right] \begin{array}{l} 20/1 = 20* \\ 30/1 = 30 \end{array}$$

Pivoting on the element in row 1 and column 3 produces the following tableau.

$$\begin{array}{c} \\ z \\ x \end{array} \left[\begin{array}{cccc|c} x & y & z & s & \\ 0 & -1 & 1 & 1 & 20 \\ 1 & ② & 0 & -1 & 10 \\ \hline 0 & 50* & 0 & -150 & R - 12000 \end{array}\right] \begin{array}{l} \text{negative} \\ 10/2 = 5* \end{array}$$

The last pivot is on the element in row 2 and column 2. We omit the intermediate step.

$$\begin{array}{c} \\ z \\ y \\ \end{array} \left[\begin{array}{ccc|c|c} x & y & z & s & \\ \frac{1}{2} & 0 & 1 & \frac{1}{2} & 25 \\ \frac{1}{2} & 1 & 0 & -\frac{1}{2} & 5 \\ \hline -25 & 0 & 0 & -125 & R - 12250 \\ \end{array} \right]$$

The maximum value of R is 12250, and occurs when $(x, y, z) = (0, 5, 25)$. Ralph should buy 5 No-colds and 25 Orangos. ∎

EXERCISES 3.6

In Exercises 1–8 introduce slack and artificial variables as appropriate for the two-phase method, and give the resulting tableau.

1. Maximize $3x + 5y = R$
   subject to $x + y \leq 20$
   $3x + y \geq 36$
   $x \geq 0, y \geq 0$.

2. Maximize $3x + 4y = R$
   subject to $x + y \leq 40$
   $2x + y \geq 50$
   $x \geq 0, y \geq 0$.

3. Maximize $3x + y = R$
   subject to $x + 2y \leq 200$
   $2x + y \geq 250$
   $x \geq 0, y \geq 0$.

4. Maximize $3x + 2y = R$
   subject to $2x + y \leq 75$
   $x + y \geq 60$
   $x \geq 0, y \geq 0$.

5. Minimize $3x + 6y = C$
   subject to $2x + y \leq 75$
   $x + y \geq 60$
   $x \geq 0, y \geq 0$.

6. Minimize $4x + y = C$
   subject to $x + 2y \leq 200$
   $2x + y \geq 250$
   $x \geq 0, y \geq 0$.

7. Minimize $6x + 4y = C$
   subject to $x + 2y \leq 150$
   $2x + 3y \geq 240$
   $x \geq 0, y \geq 0$.

8. Minimize $5x + 3y = C$
   subject to $4x + y \leq 400$
   $2x + y \geq 300$
   $x \geq 0, y \geq 0$.

In Exercises 9 and 10 a tableau for the two-phase method is presented. Apply the method, finding the maximum value of R and the optimal values of x and y.

9. $$\left[\begin{array}{ccccc|c} x & y & s & t & a & \\ 1 & 1 & 1 & 0 & 0 & 10 \\ 5 & 3 & 0 & -1 & 1 & 15 \\ \hline -2 & -1 & 0 & 0 & 0 & R \\ 0 & 0 & 0 & 0 & -1 & Q \\ \end{array} \right]$$

10. $$\left[\begin{array}{cccccc|c} x & y & s & t & a & b & \\ 1 & 2 & -1 & 0 & 1 & 0 & 2 \\ 2 & 1 & 0 & -1 & 0 & 1 & 2 \\ \hline -3 & -2 & 0 & 0 & 0 & 0 & R \\ 0 & 0 & 0 & 0 & -1 & -1 & Q \\ \end{array} \right]$$

184 CHAPTER 3 LINEAR PROGRAMMING

In Exercises 11–18 use the two-phase method to solve the program of the indicated exercise.

11. Exercise 1   **12.** Exercise 2   **13.** Exercise 3   **14.** Exercise 4

15. Exercise 5   **16.** Exercise 6   **17.** Exercise 7   **18.** Exercise 8

Solve the programs of Exercises 19–24 using the two-phase method.

19. Maximize $3x + 2y = R$
subject to $4x + y \le 400$
$2x + y \ge 300$
$x \ge 0, y \ge 0.$

20. Maximize $5x + 2y = R$
subject to $x + 2y \le 150$
$2x + 3y \ge 240$
$x \ge 0, y \ge 0.$

21. Minimize $2x + 5y = C$
subject to $x + 3y \le 180$
$3x + 5y \ge 480$
$x \ge 0, y \ge 0.$

22. Minimize $10x + 2y = C$
subject to $5x + 3y \le 600$
$3x + 2y \ge 390$
$x \ge 0, y \ge 0.$

23. Minimize $2x + 3y + 5z = C$
subject to $x - y - z \le 30$
$x - y + z \ge 40$
$x \ge 0, y \ge 0, z \ge 0.$

24. Minimize $2x + 6y + 3z = C$
subject to $x + y + z \ge 120$
$-2x + 2y + z \ge 108$
$x \ge 0, y \ge 0, z \ge 0.$

25. A manufacturer would like to produce a total of at least 30 ceramic tiles, which are available in three styles. The first style requires 5 minutes of shaping and no painting time; the second style requires 2 minutes of shaping and 3 minutes of painting; and the third style requires 2 minutes of shaping and 5 minutes of painting. Profits on the three styles are $3, $4, and $6 per tile, respectively. If at most 160 minutes are available for shaping and at most 125 minutes are available for painting, how many tiles of each style should be made for a maximum profit? What is the maximum profit?

26. A certain diet specifies a minimum daily requirement of 50 units of food supplement R and 60 units of supplement S. These supplements are obtained most economically by eating three foods. One gram of the first food provides 2 units of R and 5 units of S, one gram of the second food provides 1 unit of R and 3 units of S, and one gram of the third food provides 1 unit of R and 6 units of S. If a gram of each food costs 4¢, 5¢, and 3¢, respectively, what amounts of each food should be eaten in order to satisfy the minimum daily requirements at the least cost? What is the least daily cost?

27. The owner of a nut shop has 50 pounds of a mixture which is 40% cashews and 50% peanuts by weight. To this amount she adds certain quantities of the mixtures described below.

Mixture	A	B
Cashews	50%	80%
Peanuts	5%	10%
Cost per pound	$4	$6

The final blend is to contain at least 60% cashews and at most 20% peanuts. How can this be achieved at lowest cost?

28. A metallurgist wants to make a new bronze alloy by combining certain amounts of the three alloys described below (percentages by weight).

Alloy	A	B	C
Copper	95%	90%	75%
Zinc	1%	8%	5%
Cost per ton	$500	$450	$200

There must be exactly 10 tons of the new alloy, with at least 8 tons of copper and at most ½ ton of zinc.

In Exercises 29–32 solve the linear programs using the two-phase method.

29. Maximize $x + y + z = R$
subject to $2x + y + 3z \leq 6$
$x - y = 1$
$x \geq 0, y \geq 0, z \geq 0.$

30. Maximize $2x + y - z = R$
subject to $x + y + z = 12$
$-x + y \geq 1$
$x \geq 0, y \geq 0, z \geq 0.$

31. Minimize $x + y + 2z = C$
subject to $x + z \geq 3$
$x - y + z = 4$
$x \geq 0, y \geq 0, z \geq 0.$

32. Minimize $3x + z = C$
subject to $x + y = 10$
$x + y + z \geq 3$
$x \geq 0, y \geq 0, z \geq 0.$

Answers to Practice Problems

11. Maximize $3x - y = R$
subject to $x - y \leq 2$
$2x + 3y \leq 6$
$x + y \geq 1$
$x \geq 0, y \geq 0.$

12.

$$\begin{array}{c|cccccccc|c} & x & y & z & s & t & u & a & b & \\ \hline s & 1 & 0 & 1 & 1 & 0 & 0 & 0 & 0 & 30 \\ a & 2 & 1 & 1 & 0 & -1 & 0 & 1 & 0 & 10 \\ b & 1 & 3 & 2 & 0 & 0 & -1 & 0 & 1 & 15 \\ \hline & 1 & 2 & -3 & 0 & 0 & 0 & 0 & 0 & R \\ & 0 & 0 & 0 & 0 & 0 & 0 & -1 & -1 & Q \end{array}$$

Rows 2 and 3 should be added to the bottom row.

CHAPTER 3 REVIEW

IMPORTANT TERMS

- constraint *(3.1)*
- corner point
- feasible point
- half-plane
- linear inequality
- linear program
- objective function
- optimal point
- feasible region *(3.2)*
- isoline
- row label *(3.3)*
- simplex algorithm
- slack variable
- indicator *(3.4)*
- standard maximum program
- compact tableau *(3.5)*
- dual program
- artificial variable *(3.6)*
- two-phase method

REVIEW EXERCISES

In Exercises 1 and 2 introduce variables and set up the problem as a linear program.

1. To make a stuffed bear requires 15 minutes of sewing, 6 minutes of painting, and 12 minutes of finishing, while a stuffed dog requires 18 minutes of sewing, 5 minutes of painting, and 10 minutes of finishing. Only 300 minutes of sewing, 120 minutes of painting, and 240 minutes of finishing are available. If the profit is $3 on each bear and $3.20 on each dog, how many of each should be made to maximize the profit?

2. One cup of cooked snap beans contains 30 calories, 63 mg calcium, and 2 grams protein; one cup of canned red kidney beans contains 230 calories, 74 mg calcium, and 15 grams protein; and one cup of baked beans with pork and molasses contains 385 calories, 161 mg calcium, and 16 grams of protein. A mixture of the three beans is to be made containing at least 350 mg calcium, at least 50 grams protein, and as few calories as possible. How should this be done?

In Exercises 3 and 4 graph the given system of linear inequalities, and find the coordinates of all corner points of the region.

3. $2x + 3y \geq 6$
   $2x - 3y \leq 2$
   $y \leq 3$

4. $x + y \geq 10$
   $2x - y \geq 20$
   $x \geq 20$

In Exercises 5–8 solve the given linear program graphically, if possible. Otherwise tell why the program has no solution.

5. Minimize $4x + 5y = C$
   subject to $3x + 5y \geq 15$
   $12x + 5y \geq 24$
   $x \geq 0, y \geq 0.$

6. Minimize $-20x + 10y = C$
   subject to $2x + 5y \geq 10$
   $-x + y \geq 10$
   $x \geq 0, y \geq 0.$

7. Maximize $3x + 4y = R$
   subject to $x + 2y \leq 10$
   $3x + y \leq 15$
   $x + y \geq 8$
   $x \geq 0, y \geq 0$

8. Maximize $3x + 2y = R$
   subject to $7x + 8y \leq 28$
   $y \leq 2$
   $x - y \leq 2$
   $x \geq 0, y \geq 0.$

Solve the programs in Exercises 9 and 10 using the simplex algorithm.

9. Maximize $6x + 5y + z = R$
   subject to $x + 2z \leq 10$
   $2x + y + z \leq 12$
   $x \geq 0, y \geq 0, z \geq 0.$

10. Maximize $x + 2y + 2z = R$
    subject to $x + y + z \leq 6$
    $x + 2y - z \leq 8$
    $x \geq 0, y \geq 0, z \geq 0.$

Solve the programs in Exercises 11 and 12 by applying the simplex algorithm to the dual program.

11. Minimize $3x + 2y + z = C$
    subject to $x + 3y \geq 12$
    $x + 2y + 2z \geq 18$
    $x \geq 0, y \geq 0, z \geq 0.$

12. Minimize $x + 3y + z = C$
    subject to $x + y + z \geq 50$
    $2y + 4z \geq 100$
    $x \geq 0, y \geq 0, z \geq 0.$

Solve Exercises 13 and 14 by the two-phase method.

13. Maximize $3x + 2y = R$
    subject to $5x + 3y \leq 600$
    $3x + 2y \geq 390$
    $x \geq 0, y \geq 0.$

14. Maximize $4x + 6y = R$
    subject to $x + 3y \leq 180$
    $3x + 5y \geq 480$
    $x \geq 0, y \geq 0.$

15. A mining company operates three mines which produce titanium ore. The ore from Mine A is 5% titanium by weight and requires 4 work-hours per ton to mine. The ore from Mine B is 3% titanium and requires 3 work-hours per ton, while the ore from Mine C is 2% titanium and requires 1 work-hour per ton. The company has 800 work-hours of mining labor available per day, and its refinery can handle up to 500 tons of ore per day. How much ore should be taken daily from each mine in order to maximize the total amount of titanium produced?

16. An amusement park has "A" rides, which cost $1, and "B" rides, which cost $2. An economy book containing 10 A-ride tickets and 5 B-ride tickets can be purchased for $18 (a 10% discount), and a super economy book containing 25 A-ride tickets and 10 B-ride tickets can be purchased for $36 (a 20% discount). A teacher is bringing a class of 20 students to the park and each has been promised 5 A-rides and 3 B-rides. How many books and individual tickets of each type should be purchased in order to minimize the cost of the rides?

4

MATHEMATICS OF FINANCE

4.1 Simple Interest
4.2 Compound Interest
4.3 Annuities and Installment Loans
4.4 Sinking Funds and Amortization
Chapter Review

Anyone who has a savings account or makes installment purchases has contact with the world of finance. In recent years high interest rates have made knowledge of finance essential for every consumer. In this chapter we will study basic financial topics such as simple and compound interest, installment loans, and annuities. The mathematical foundation for this study involves the concept of a *difference equation*. As we will see, difference equations can be used to analyze not only financial matters but any periodic process.

Because of the nature of the calculations that are required in this chapter, we assume that the reader has access to a calculator with the capability of computing powers of any positive number. (The keys that perform these types of calculations are usually denoted y^x or x^y.)

4.1 Simple Interest

Interest is a fee paid for the use of someone else's money. Interest is paid not only by individuals, who borrow money by making installment purchases, but also by savings institutions and governmental units, which borrow money by offering savings accounts and issuing bonds. In this section we will develop the basic equations involving simple interest, which underlies almost all financial calculations. We begin by studying the special type of equation that occurs in all of the financial calculations that we will consider.

ARITHMETIC AND GEOMETRIC PROGRESSIONS

An infinite list of numbers

$$x_0, x_1, x_2, \ldots$$

is called a **sequence**. The individual numbers in the sequence are called its **terms**. The first term in the sequence, x_0, is called the **initial value** of the sequence. In this chapter we will be concerned with sequences that are defined by an equation of the form

$$x_{n+1} = ax_n + b, \qquad (4.1)$$

where a and b are constants. An equation of this type is called a **first-order linear difference equation**. Given the initial value x_0, we can use (4.1) to calculate all subsequent terms in the sequence as demonstrated in Example 4.1.

EXAMPLE 4.1 If a sequence defined by the difference equation $x_{n+1} = 2x_n - 3$ has an initial value of 7, find the next five terms.

Solution The first term in the sequence is the initial value x_0, which is 7. The next term x_1 is found by taking $n = 0$ in the defining equation $x_{n+1} = 2x_n - 3$; this yields

$$x_1 = 2x_0 - 3 = 2(7) - 3 = 11.$$

190   CHAPTER 4   MATHEMATICS OF FINANCE

Similarly, by taking n to be 1, 2, 3, and 4 in the defining equation, we obtain

$$x_2 = 2x_1 - 3 = 2(11) - 3 = 19,$$
$$x_3 = 2x_2 - 3 = 2(19) - 3 = 35,$$
$$x_4 = 2x_3 - 3 = 2(35) - 3 = 67,$$
$$x_5 = 2x_4 - 3 = 2(67) - 3 = 131.$$

Hence the first six terms of the sequence are 7, 11, 19, 35, 67, and 131. ∎

Practice Problem 1   If the sequence defined by the difference equation $x_{n+1} = -3x_n + 10$ has an initial value of 4, find the next five terms.

Although it is possible to find any term of (4.1) as in Example 4.1, the calculations will be quite tedious if n is large. Therefore it is desirable to have an explicit formula giving the value of x_n in terms of n. This will enable us to compute x_n directly without finding all the preceding terms in the sequence. For instance, we will see in Section 4.3 that the formula $x_n = 4 \cdot 2^n + 3$ gives the terms of the sequence in Example 4.1, as the reader can check for $n = 0, 1, 2, 3, 4, 5$. Thus we can compute $x_5 = 4 \cdot 2^5 + 3 = 4(32) + 3 = 131$ and $x_9 = 4 \cdot 2^9 + 3 = 4(512) + 3 = 2051$ directly.

In this section we will develop formulas for sequences defined by two special cases of (4.1). The first case that we will consider is the equation

$$x_{n+1} = x_n + b,$$

which occurs when $a = 1$ in (4.1). Each term of a sequence governed by this equation differs from its predecessor by a fixed amount b; such a sequence is known as an **arithmetic progression**. Given the initial value x_0, the equation $x_{n+1} = x_n + b$ generates the successive terms

$$x_1 = x_0 + b,$$
$$x_2 = x_1 + b = (x_0 + b) + b = x_0 + 2b,$$
$$x_3 = x_2 + b = (x_0 + 2b) + b = x_0 + 3b,$$
$$x_4 = x_3 + b = (x_0 + 3b) + b = x_0 + 4b,$$
$$\vdots$$

Thus for the equation $x_{n+1} = x_n + b$, the formula relating x_n to n is

$$x_n = x_0 + nb.$$

Theorem 4.1

The general term of the arithmetic progression defined by the difference equation $x_{n+1} = x_n + b$ and having initial value x_0 is $x_n = x_0 + nb$.

4.1 SIMPLE INTEREST

EXAMPLE 4.2 Find a formula for the general term of the sequence defined by the difference equation $x_{n+1} = x_n + 5$ and having initial term -8.

Solution The first few terms in this sequence are

$$-8, -3, 2, 7, 12, 17, \ldots.$$

This is an arithmetic progression because successive terms differ by 5. By Theorem 4.1 the general term of this progression is

$$x_n = -8 + 5n.$$

Therefore, for instance, we see that $x_{10} = 42$, $x_{50} = 242$, and $x_{100} = 492$. ■

Practice Problem 2 Find a formula for the general term of the sequence defined by the difference equation $x_{n+1} = x_n - 3$ and having initial term 36.

The second special case of (4.1) that we will consider is the equation

$$x_{n+1} = ax_n,$$

which occurs when $b = 0$ in (4.1). Each term of a sequence governed by this equation is obtained by multiplying its predecessor by the fixed amount a; such a sequence is called a **geometric progression.** Given the initial value x_0, the equation $x_{n+1} = ax_n$ generates the successive terms

$$x_1 = ax_0,$$
$$x_2 = ax_1 = a(ax_0) = a^2 x_0,$$
$$x_3 = ax_2 = a(a^2 x_0) = a^3 x_0,$$
$$x_4 = ax_3 = a(a^3 x_0) = a^4 x_0,$$
$$\vdots$$

Hence for the equation $x_{n+1} = ax_n$, the formula relating x_n to n is

$$x_n = a^n x_0.$$

Theorem 4.2 ■ ■ ■ ■ ■ ■
> The general term of the geometric progression defined by the difference equation $x_{n+1} = ax_n$ and having initial value x_0 is $x_n = a^n x_0$.

EXAMPLE 4.3 Find a formula for the general term of the sequence defined by the difference equation $x_{n+1} = 3x_n$ and having initial term 7.

Solution The first few terms in this sequence are

$$7, 21, 63, 189, 567, \ldots$$

This is a geometric progression, because each term is 3 times its predecessor.

CHAPTER 4 MATHEMATICS OF FINANCE

Theorem 4.2 shows that the general term of this sequence is

$$x_n = 3^n \cdot 7.$$

The terms in this sequence grow very quickly. For instance, x_{10} is almost half a million, x_{15} is more than 100 million, and x_{50} exceeds the number of grains of sand on all the beaches in the world! ■

Practice Problem 3   Find a formula for the general term of the sequence defined by the difference equation $x_{n+1} = 2x_n$ and having initial term 5.

Arithmetic and geometric progressions arise quite naturally in the mathematics of finance.

INTEREST

When money is invested in a savings account or loaned to a commercial borrower, the initial amount is called the **principal.** The fee charged for the use of the money is called **interest.** The amount of interest to be charged is usually stated as a percentage called the **interest rate.** Normally interest rates are stated as annual (yearly) rates, and we will follow common practice and assume that *all interest rates are annual rates unless it is explicitly stated otherwise.* The sum of the principal and the interest is called the **balance** of the account or loan.

Of the many ways that interest can be charged, the most basic is simple interest. **Simple interest** is charged as a percentage of the principal, per unit time. The amount of simple interest owed is given by the formula below.

Simple Interest Charge ■ ■ ■ ■ ■

$$I = Prt \qquad (4.2)$$

where I denotes the interest charge,
P denotes the principal,
r denotes the annual interest rate, and
t denotes the time in years for which the money is borrowed.

EXAMPLE 4.4   Suppose that an individual borrows $5000 at 10% simple interest. What amount of interest is due when the loan is repaid at the end of three years?

Solution   In this situation time is measured in years and the annual interest rate is 10%. So taking $P = \$5000$, $r = 0.10$, and $t = 3$ in the simple interest formula (4.2), we see that the amount of interest to be paid after three years is

$$I = Prt = \$5000(0.10)(3) = \$1500.$$

Thus the borrower must pay a total of $6500 at the end of three years, $5000 to repay the principal and an interest charge of $1500. ■

EXAMPLE 4.5 Suppose that an individual borrows $4000 at 12% simple interest. What amount of interest is due when the loan is repaid at the end of six months?

Solution In this case, time is measured in months and the annual simple interest rate is 12%. Since formula (4.2) requires that t be measured in years, we must convert 6 months to $6/12 = 1/2$ year. Thus we take $P = \$4000$, $r = 0.12$, and $t = 1/2$ in formula (4.2). Therefore

$$I = Prt = \$4000(0.12)\frac{1}{2} = \$240,$$

and hence the interest charge will be $240 at the end of six months. ■

Practice Problem 4 Suppose that $8000 is borrowed at 16% simple interest. What amount of interest is due when the loan is repaid at the end of 30 months?

More generally, suppose that a principal P earns simple interest at a rate of r per year, and let A_t denote the balance after t years. Now the balance after $t + 1$ years equals the balance after t years plus the interest charge for one year. Since the interest for one year is Pr by (4.2), we obtain the equation

$$A_{t+1} = A_t + Pr.$$

This is a difference equation of the form (4.1) with $a = 1$ and $b = Pr$. Since its initial value is $A_0 = P$, Theorem 4.1 gives the following formula:

$$A_t = A_0 + t(rP) = P + trP = P(1 + rt).$$

Simple Interest Amount

$$A = P(1 + rt) \qquad (4.3)$$

where A denotes the amount after t years,
P denotes the principal,
r denotes the annual interest rate, and
t denotes the time in years for which the money is borrowed.

Thus for the loan in Example 4.4 with $P = \$5000$, $r = 0.10$, and $n = 3$, formula (4.3) yields

$$A = \$5000[1 + 0.10(3)] = \$5000(1.30) = \$6500.$$

Consequently the amount to be repaid after three years (principal plus interest) is $6500, which agrees with our earlier calculation.

EXAMPLE 4.6 If a manufacturer borrows $250,000 for 9 months at 15% simple interest, how much will be owed when the loan is due?

Solution Since time is measured in months, we must convert it to 9/12 = 3/4 years as in Example 4.5. Thus we take $P = \$250,000$, $r = 0.15$, and $t = 3/4$ in (4.3).

$$A = P(1 + rt)$$
$$= \$250,000\left[1 + (0.15)\frac{3}{4}\right]$$
$$= \$250,000(1.1125) = \$278,125.$$

Hence the manufacturer must repay $278,125 after 9 months. Of this, $250,000 is the principal and the remaining $28,125 is interest. ■

Practice Problem 5 Suppose that $10,000 is borrowed at 15% simple interest. How much will be owed when the loan is due at the end of 18 months? How much of this amount is interest?

In some situations we must solve formula (4.3) for r, the simple interest rate.

EXAMPLE 4.7 A bond that is redeemable for $12,100 in 18 months was sold for $10,000. What annual simple interest rate does this bond pay?

Solution In this problem we must solve (4.3) for r knowing that $A = \$12,100$, $P = \$10,000$, and $t = 18/12 = 3/2$.

$$A = P(1 + rt)$$
$$\$12,100 = \$10,000\left[1 + r\left(\frac{3}{2}\right)\right]$$
$$1.21 = 1 + r\left(\frac{3}{2}\right)$$
$$0.21 = r\left(\frac{3}{2}\right)$$
$$0.14 = r$$

Thus the bond is paying 14% simple interest per year. ■

Practice Problem 6 A bond that is redeemable for $22,650 in 15 months was sold for $20,000. What annual simple interest rate does this bond pay?

EXERCISES 4.1

In Exercises 1–8 generate the first six terms of the sequence defined by each difference equation. Then determine whether the sequence is an arithmetic progression, a geometric progression, or neither.

1. $x_{n+1} = 5x_n - 1$, $x_0 = 3$
2. $x_{n+1} = \frac{1}{2}x_n$, $x_0 = 1$
3. $x_{n+1} = x_n + \frac{3}{2}$, $x_0 = 0$
4. $x_{n+1} = -0.2x_n + 40$, $x_0 = 5$

5. $x_{n+1} = -1.8x_n$, $x_0 = 625$
6. $x_{n+1} = 2x_n - 7$, $x_0 = 6$
7. $x_{n+1} = 1.1x_n + 100$, $x_0 = 5000$
8. $x_{n+1} = x_n - 3.2$, $x_0 = 24.8$

In Exercises 9–16 find a formula for the general term of the sequence satisfying each difference equation. Then use your formula to compute x_{10} and x_{20}.

9. $x_{n+1} = x_n + 6$, $x_0 = 2$
10. $x_{n+1} = 2x_n$, $x_0 = 3$
11. $x_{n+1} = x_n - \frac{5}{2}$, $x_0 = 35$
12. $x_{n+1} = -\frac{1}{3}x_n$, $x_0 = 1$
13. $x_{n+1} = 1.5x_n$, $x_0 = 1$
14. $x_{n+1} = x_n + 12$, $x_0 = -32$
15. $x_{n+1} = 0.9x_n$, $x_0 = 100$
16. $x_{n+1} = x_n - 0.18$, $x_0 = 5$

In Exercises 17–20 compute the amount of interest earned if a principal P is invested at simple interest for t years at an annual interest rate r.

17. $P = \$5000$, $t = 1\frac{1}{2}$, $r = 7\%$
18. $P = \$1,200$, $t = 3$, $r = 8\%$
19. $P = \$40,000$, $t = 6$, $r = 10\%$
20. $P = \$2000$, $t = \frac{1}{2}$, $r = 9\%$

In Exercises 21–24 compute the amount necessary to repay a loan of P dollars borrowed at simple interest for t years at an annual interest rate r.

21. $P = \$900$, $t = 4$, $r = 15\%$
22. $P = \$2400$, $t = 8$, $r = 16\%$
23. $P = \$3000$, $t = 5$, $r = 8\%$
24. $P = \$5000$, $t = 2$, $r = 9\%$

25. An individual borrows $6800 for 18 months at 14% simple interest. How much must be repaid when this loan is due? How much of this amount is interest?

26. A corporation borrows $2 million for 3½ years at 12% simple interest. How much must be repaid when this loan is due? How much of this amount is interest?

27. A lender offers $750 on the condition that the borrower repay $900 in 12 months. What simple interest rate is involved in this transaction?

28. A buyer pays $4000 for a bond that is redeemable at its face value of $5000 in another 15 months. What simple interest rate does this bond pay?

29. An item costing $50 was charged at a department store. Three months later the store issued a bill for $52. What annual rate of simple interest was the store charging?

30. If $3000 is borrowed at 12% simple interest, how much will be due in 1½ years?

31. A bank loaned $500 and collected $620 in payment 18 months later. What annual rate of simple interest did it charge?

32. If money is invested at 18% simple interest, how long will it take for the money to double?

Answers to Practice Problems
1. $-2, 16, -38, 124$, and -362
2. $x_n = 36 - 3n$
3. $x_n = 2^n \cdot 5$
4. $\$3200$
5. Of the $12,250 that is owed, $2250 is interest.
6. 10.6%

4.2 Compound Interest

Most transactions involving interest charges do not use simple interest, which is a percentage of the principal. Instead, the interest is computed as a percentage of the *balance* of the loan. In this situation, interest is charged not only on the original amount, but also on any previous interest charges. In this case the interest is said to have been **compounded,** and this type of interest is called **compound interest.**

EXAMPLE 4.8 Suppose that an individual borrows $5000 at 10% interest compounded annually. What will the total debt be at the end of three years?

Solution During the first year the interest charge is 10% of the principal. Using the formula for a simple interest amount, we see that at the end of the first year the borrower owes

$$P(1 + rt) = \$5000[1 + 0.10(1)] = \$5000(1.10) = \$5500.$$

The lender now treats this amount as the new principal during the second year. Thus the balance of the loan at the end of the second year is

$$P(1 + rt) = \$5500[1 + 0.10(1)] = \$5500(1.10) = \$6050.$$

Using $6050 as the principal for the third year, we see that the amount due after three years is

$$P(1 + rt) = \$6050[1 + 0.10(1)] = \$6050(1.10) = \$6655.$$

Note that this amount is considerably greater than the $6500 that was due in Example 4.4 when simple interest was used. In fact an interest rate of 10% compounded annually over a three-year period is equivalent to a simple interest rate of more than 11%. ■

In order to derive a general formula for the balance of a loan when interest is compounded, we will assume that a principal P earns interest at the rate of i per period. If A_n denotes the balance after n periods, then the interest charge for the next period will be iA_n. Hence the balance after $n + 1$ interest periods is

$$A_{n+1} = A_n + iA_n = (1 + i)A_n.$$

This is a difference equation of the form (4.1) with $b = 0$, and so the sequence of balances is a geometric progression with initial value $A_0 = P$. By Theorem 4.2 we see that $A_n = (1 + i)^n A_0 = P(1 + i)^n$. Thus we have the following general formula.

Compound Interest Amount

$$A = P(1 + i)^n \qquad (4.4)$$

where A denotes the amount after n periods,
P denotes the principal,
i denotes the interest rate per period, and
n denotes the number of interest periods for which the money is borrowed.

Applying this formula to the loan in Example 4.8 with $P = \$5000$, $i = 0.10$, and $n = 3$, we find that the amount due after three years is

$$A = P(1 + i)^n = \$5000(1 + 0.10)^3$$
$$= \$5000(1.10)^3 = \$5000(1.331) = \$6655.$$

Of course, this is the amount we obtained in Example 4.8 when we performed the calculation the long way.

The compound interest rate on savings accounts and installment loans is usually stated as an annual rate. If interest is compounded more than once a year, however, then the rate at which the balance grows is somewhat greater.

EXAMPLE 4.9 Compare the balance when $1000 is deposited for five years at 9% interest compounded (a) annually, (b) quarterly, and (c) monthly.

Solution (a) If interest is compounded annually, then the interest period is one year and the interest rate is 9% per period. Thus after five years the balance will be

$$A = P(1 + i)^5 = \$1000(1 + 0.09)^5$$
$$= \$1000(1.09)^5 \approx \$1000(1.53862)$$
$$= \$1538.62.$$

(b) With quarterly compounding the interest period is three months rather than one year. Thus one-fourth of the interest is paid every three months rather than paying the entire amount once a year. So because there are four quarters per year, the interest rate per period is $9\%/4 = 2.25\%$, and the principal is earning interest for $5(4) = 20$ quarters. Hence by (4.4) the balance after 20 quarters is

$$A = P(1 + i)^{20} = \$1000(1 + 0.0225)^{20}$$
$$= \$1000(1.0225)^{20} \approx \$1000(1.56051) = \$1560.51.$$

(c) If interest is compounded monthly, then the interest rate per month is $9\%/12 = 0.75\%$ and the principal will earn interest for $5(12) = 60$ months. Hence the balance after 60 months is

$$A = P(1 + i)^{60} = \$1000(1 + 0.0075)^{60}$$
$$= \$1000(1.0075)^{60} \approx \$1000(1.56568) = \$1565.68.$$

Practice Problem 7 If $6000 is invested in a certificate of deposit at 8% interest compounded quarterly and the interest is left to accumulate, how much will the certificate be worth after five years?

Notice that in Example 4.9 the balance grows more quickly when interest is compounded more frequently, but with a diminishing gain. That is, the largest balance occurs when interest is compounded most frequently. However, the difference between yearly compounding and quarterly compounding is about $22, whereas the difference between quarterly compounding and monthly compounding is only about $5. If interest were compounded daily (365 times per year), the balance would be $1568.23 after five years, which is less than $3 more than the balance obtained when interest is compounded monthly. Thus, unless very large sums of money are involved or the length of the loan is long, the difference between monthly compounding and daily compounding is insignificant.

EXAMPLE 4.10 At what annual rate must inflation continue so that in twelve years the purchasing power of a dollar will be half as much as it is now?

Solution To say that in twelve years the purchasing power of a dollar will be half as much as it is now means that goods costing $1 today will cost $2 in twelve years. Since the inflation rate i is in effect for each of the twelve years, the effects of inflation are "compounded" throughout the period, and we assume that the effects are compounded annually. Thus we must solve equation (4.4) for i if $A = \$2$, $P = \$1$, and $n = 12$.

$$A = P(1 + i)^n$$
$$\$2 = \$1(1 + i)^{12}$$
$$2 = (1 + i)^{12}$$
$$2^{1/12} = 1 + i$$
$$i = 2^{1/12} - 1 \approx 0.0595$$

Hence an inflation rate of approximately 5.95% will cut the purchasing power of a dollar by half in twelve years. ■

NOMINAL AND EFFECTIVE RATES

In Example 4.9 we saw that the same interest rate will yield different amounts depending on the frequency with which the interest is compounded. The stated rate of interest is called the **nominal rate.** Thus in Example 4.9 the nominal rate of interest is 9%, the stated annual rate. In order to compare the yields for different frequencies of compounding, it is useful to convert the nominal rate into an equivalent simple interest rate, called the **effective rate** (or **annual yield**). The effective rate can be computed by finding the percentage gain for an arbitrary principal, as in the example on the next page.

Fresh Anxiety About Inflation Deflates Rallies

MONDAY'S MARKETS

By DOUGLAS R. SEASE
Staff Reporter of THE WALL STREET JOURNAL

Renewed inflation worries pooped Wall Street's party.

After pushing stock and bond prices sharply higher Friday because a rip-roaring rise in wholesale prices has moderated, investors began the new week hunkered down for an expected dose of bad news about climbing consumer prices.

4.2 COMPOUND INTEREST

EXAMPLE 4.11 What is the effective rate of interest corresponding to a nominal rate of 8% compounded quarterly?

Solution To answer this question, we see how much of an increase an arbitrary principal experiences in one year. Suppose we choose a principal of $100. By (4.4) we see that the balance after one year is

$$A = P(1 + i)^4 = \$100(1.02)^4 \approx \$108.24.$$

Thus the $100 principal grew to $108.24 in one year, an increase of 8.24%. So the effective rate of interest for a nominal rate of 8% compounded quarterly is 8.24% per year. This means that each dollar invested at 8% compounded quarterly actually earns 8.24% in interest during a year. ∎

Practice Problem 8 Approximate the effective rate of interest corresponding to a nominal rate of 6% compounded monthly.

PRESENT VALUE

In our examples using simple and compound interest we have known the principal and been interested in finding the value of that principal at some future time. Other applications require knowing what principal P must be deposited now at a specified interest rate in order to accumulate a specific amount of money A at some future time. In this situation P is called the **present value** of A. To find present values, we use formulas (4.3) and (4.4) to solve for P.

Present Value of a Future Amount

$$\text{simple interest} \qquad P = \frac{A}{1 + rt} \qquad (4.5)$$

$$\text{compound interest} \qquad P = \frac{A}{(1 + i)^n} \qquad (4.6)$$

where P denotes the present value,
A denotes the future amount,
r denotes the annual interest rate,
i denotes the interest rate per period, and
n denotes the number of interest periods for which the money is borrowed.

EXAMPLE 4.12 How much money must be invested now in order to accumulate $12,000 in 15 years at the following interest rates:
(a) 18% simple interest?
(b) 10% interest compounded quarterly?

Solution (a) We must find the present value P of \$12,000 in 15 years at 18% simple interest. Using (4.5), we obtain

$$P = \frac{A}{1 + rt}$$

$$= \frac{\$12,000}{1 + 15(0.18)}$$

$$= \frac{\$12,000}{3.7} \approx \$3243.24.$$

(b) With quarterly compounding, the interest rate per period is $i = 0.10/4 = 0.025$ and the number of interest periods is $n = 15(4) = 60$. Thus by (4.6) we see that the present value of \$12,000 in 15 years at 10% interest compounded quarterly is

$$P = \frac{A}{(1 + i)^n}$$

$$= \frac{\$12,000}{(1.025)^{60}}$$

$$\approx \$2727.40.$$

Note that even though the interest rate in (b) is significantly lower than in (a), less money must be deposited in (b). This demonstrates again how much more quickly an amount grows when interest is compounded. ■

Practice Problem 9 Approximately how much money must be invested now in order to accumulate \$8000 in 6 years at the following interest rates:
(a) 15% simple interest? (b) 10% interest compounded quarterly?

EXERCISES 4.2

In Exercises 1–4 compute the amount in m years if a principal P is invested at a nominal annual interest rate of r compounded quarterly.

1. $P = \$6000$, $m = 5$, $r = 4\%$
2. $P = \$800$, $m = 2$, $r = 6\%$
3. $P = \$300$, $m = 1\frac{1}{2}$, $r = 5\%$
4. $P = \$400$, $m = 2\frac{1}{2}$, $r = 8\%$

In Exercises 5–8 compute the present value of an amount A invested for m years in an account paying interest compounded semiannually at a nominal annual rate r.

5. $A = \$1000$, $m = 3$, $r = 5\%$
6. $A = \$500$, $m = 8$, $r = 4\%$
7. $A = \$3000$, $m = 1$, $r = 6\%$
8. $A = \$2000$, $m = 2$, $r = 8\%$

In Exercises 9–12 compute the effective rate of interest for a nominal annual rate r under the given conditions.

9. $r = 6\%$ compounded monthly
10. $r = 8\%$ compounded quarterly
11. $r = 7\%$ compounded daily
12. $r = 9\%$ compounded monthly

13. A consumer makes a $250 purchase with a credit card that charges 18% interest compounded monthly. If no payments are made for three months, how much will be owed at that time?

14. A savings and loan pays 9% interest compounded quarterly on its 2½-year certificate of deposit. If $3200 is invested in such a certificate and interest is left to accumulate, how much will be available when the certificate matures?

15. Compare the balance when $600 is deposited for three years at 7.2% interest compounded (a) annually, (b) quarterly, (c) monthly, and (d) daily.

16. Which return is better for the investor: 15.6% compounded semiannually or 14.6% compounded daily?

17. In nine years a company expects to replace a piece of machinery that will cost $2.6 million. If money can be invested at 10% interest compounded quarterly, how much must be invested now so that the necessary amount will be available in nine years?

18. If money can be invested at 6% interest compounded quarterly, how much should be deposited in order to have $10,000 for college expenses in eight years?

19. Suppose that an executive deposits $20,000 in an account paying 8% interest compounded quarterly. How much money will be available at retirement in 25 years?

20. If you have $500 to invest for 2 years, which option is better: 6% interest compounded semiannually or 6¼% simple interest?

21. How much must be borrowed now at 18% simple interest so that the amount owed in six months is $100?

22. What annual simple interest rate is equivalent to a rate of 8.4% compounded monthly?

23. If money can be invested at 7% interest compounded quarterly, how much should be paid for a note that will be worth $5000 in six years?

24. An insurance company allows a policyholder the option of collecting $1000 now or $1125 in two years. If money can be invested at 6% interest compounded quarterly, which option is better for the policyholder?

25. A company with current sales of $3 billion is predicting sales of $4.1 billion in three years. What annual rate of increase in sales is this company expecting?

26. At what annual rate must inflation continue so that a dollar will lose half its value in ten years?

27. If food prices increase 4% per year, how much will food cost in 1995 that costs $100 in 1990?

28. A commodities speculator wants cash. He is willing to borrow money at the rate of 14.4% compounded monthly, and will repay the loan with the proceeds from the sale of property. If the property is expected to be sold in 15 months for $180,000, how much can the speculator afford to borrow?

29. A municipal bond selling for $8000 can be redeemed in 6 years for $10,000. If interest is compounded quarterly, what nominal annual interest rate is this bond paying?

30. In 1975 a town had a population of 20,000, and by 1985 it had grown to 30,000. Assuming that the town's annual percentage of growth remains constant throughout the period 1975–1995, estimate the town's population in 1995.

31. The consumer price index was 118.3 in 1970 (using 1965 as the base year for which the consumer price index is set at 100.0). Assuming that the annual rate of inflation was constant throughout this period, determine this annual inflation rate.

32. The consumer price index was 118.3 in 1970 and 159.1 in 1975 (using 1965 as the base year for which the consumer price index is set at 100.0). Assuming that the annual rate of inflation was constant throughout the period 1970–1975, determine this annual inflation rate.

33. Show that the value of an account in which a principal P is invested for n years at an annual rate of r compounded m times per year is

$$A = P\left(1 + \frac{r}{m}\right)^{mn}$$

34. Find the effective rate of interest corresponding to a nominal annual rate r compounded m times per year.

Answers to Practice Problems
7. $8915.68
8. 6.17%
9. (a) $4210.53 (b) $4423.00

4.3 Annuities and Installment Loans

Loans for the purchase of a home or an automobile are usually repaid in equal monthly installments. Likewise most life insurance companies offer whole life polices that are structured so that the policyholder makes fixed payments to the company for 20 or 30 years, after which time the sum of all the payments is available to the policyholder. Such a set of equal payments at regular intervals of time is called an **annuity.** In this section we will return to the general first-order linear difference equation (4.1) to analyze annuities.

THE GENERAL FIRST-ORDER LINEAR DIFFERENCE EQUATION

In Section 4.1 we saw how to solve the first-order linear difference equation

$$x_{n+1} = ax_n + b \tag{4.1}$$

in the special cases in which $a = 1$ or $b = 0$. If we are given such an equation with $a \neq 1$, we can reduce it to the case $b = 0$ by an appropriate substitution. Specifically, for $n = 0, 1, 2, \ldots$ we let

$$y_n = x_n - c,$$

where c is a constant that we will determine later. Since $y_n + c = x_n$ and $y_{n+1} + c = x_{n+1}$, substitution into (4.1) produces

$$y_{n+1} + c = a(y_n + c) + b,$$

or equivalently,
$$y_{n+1} + c = ay_n + (ac + b).$$

If we now choose c so that $c = ac + b$, then the preceding equation will have the same constant term on both sides. Now if $c = ac + b$, then:
$$c - ac = b$$
$$c(1 - a) = b$$
$$c = \frac{b}{1 - a}.$$

Thus if we choose
$$c = \frac{b}{1 - a},$$
the equation $y_{n+1} + c = ay_n + (ac + b)$ simplifies to
$$y_{n+1} = ay_n.$$

By Theorem 4.2 the general term of a sequence defined by this equation has the form
$$y_n = a^n y_0.$$

Substituting $y_n = x_n - c$ and $y_0 = x_0 - c$, we obtain
$$x_n - c = a^n(x_0 - c),$$
or
$$x_n = a^n(x_0 - c) + c.$$

Thus we have obtained the following result.

Theorem 4.3

If $a \neq 1$, then the general term of the sequence satisfying the difference equation
$$x_{n+1} = ax_n + b \tag{4.1}$$
and having initial value x_0 is
$$x_n = a^n(x_0 - c) + c, \tag{4.7}$$
where $c = \dfrac{b}{1 - a}.$

EXAMPLE 4.13

Find a formula for the general term of the sequence with initial value 7 that satisfies the difference equation $x_{n+1} = 2x_n - 3$.

Solution Since the given equation is of the form $x_{n+1} = ax_n + b$ with $a = 2$, we can apply Theorem 4.3. In this case we have

$$c = \frac{b}{1-a} = \frac{-3}{1-2} = 3.$$

Thus the desired formula is

$$\begin{aligned}x_n &= a^n(x_0 - c) + c \\ &= 2^n(7 - 3) + 3 \\ &= 4 \cdot 2^n + 3.\end{aligned}$$

We can see from this formula that the terms of the sequence x_n increase without bound as n increases. ∎

Practice Problem 10 Find a formula for the general term of the sequence with initial value 8 that satisfies the difference equation $x_{n+1} = 3x_n - 4$.

EXAMPLE 4.14

Suppose that a lumber company owns a stand of 8000 redwood trees. Each year the company plans to harvest 10% of these trees and plant 650 new ones.

(a) How many trees will there be in the stand after 10 years? After 20 years?
(b) How many trees will there be in the long run?

Solution (a) Let x_n denote the number of trees in the stand after n years. During year $n + 1$, 10% of the trees existing in year n will be harvested; this number is $0.10x_n$. Since 650 additional trees will be planted during year $n + 1$, the number of trees after $n + 1$ years will be described by the equation

$$x_{n+1} = x_n - 0.10x_n + 650$$

or

$$x_{n+1} = 0.90x_n + 650.$$

Since this is a difference equation of the form $x_{n+1} = ax_n + b$ with $a = 0.90$ and $b = 650$, we can use Theorem 4.3 to find a formula for x_n in terms of n. Now

$$c = \frac{b}{1-a} = \frac{650}{1-0.90} = 6500.$$

Hence by Theorem 4.3

$$\begin{aligned}x_n &= a^n(x_0 - c) + c \\ &= (0.9)^n(8000 - 6500) + 6500 \\ &= 1500(0.9)^n + 6500.\end{aligned}$$

Therefore after 10 years the number of trees will be
$$x_{10} = 1500(0.9)^{10} + 6500 \approx 7023,$$
and after 20 years the number of trees will be
$$x_{20} = 1500(0.9)^{20} + 6500 \approx 6682.$$

(b) As n increases, the quantity $(0.9)^n$ will decrease steadily to zero. Hence the formula
$$x_n = 1500(0.9)^n + 6500$$
implies that the number of trees in the stand will approach 6500. (Note that as the number of trees in the stand approaches 6500, the number of trees being harvested each year becomes equal to the number of new trees being planted.) ■

THE FUTURE VALUE OF AN ANNUITY

Suppose that equal payments are made at regular intervals into an annuity to accumulate money. It is possible that the interval between payments may be different from the interest period. For instance, we may deposit $100 *per month* into a savings account paying *quarterly* interest. However, for installment loans the two time periods are always the same. Thus for simplicity *we will consider only annuities in which the regular payments are made at the end of each interest period*. We would like to be able to calculate the amount of money in the annuity at any time. Since we are determining the total sum of money in the annuity at some future time, this sum is called the **future value of the annuity**.

To analyze this situation, we let p be the regular periodic payment, i be the interest rate per period, and F_n be the amount of money accumulated in the account after n payments. Now the accumulated amount after $n + 1$ payments equals the amount accumulated after n payments plus interest plus the regular periodic payment. Hence we have the following equation.
$$F_{n+1} = F_n + iF_n + p$$
$$= (1 + i)F_n + p$$

This is a difference equation of the form $x_{n+1} = ax_n + b$ with $a = 1 + i$ and $b = p$. Also,
$$c = \frac{b}{1 - a} = -\frac{p}{i}.$$

Thus by Theorem 4.3 we have
$$F_n = (1 + i)^n \left(F_0 + \frac{p}{i} \right) - \frac{p}{i}.$$

In the important special case in which the initial value of the fund is $F_0 = 0$, we obtain the following result.

Future Value of an Annuity

$$F = p\left[\frac{(1+i)^n - 1}{i}\right] \qquad (4.8)$$

where F denotes the future value after n payments,
p denotes the amount of the periodic payment,
i denotes the interest rate per period, and
n denotes the number of payments.

The expression in brackets in formula (4.8) occurs often in financial analysis and is denoted by the symbol $s_{\overline{n}|i}$, which is read "s angle n at i." Thus $s_{\overline{n}|i}$ denotes the future value of an annuity consisting of n payments of \$1 with an interest rate of i per period. Note that with this notation (4.8) can be written as

$$F = p s_{\overline{n}|i}.$$

EXAMPLE 4.15 An individual makes quarterly payments of \$100 into a retirement fund which earns 9% interest, compounded quarterly. How much will the fund contain after 30 years?

Solution Applying (4.8) with $p = \$100$, $i = 0.09/4 = 0.0225$, and $n = 30(4) = 120$, we obtain

$$\begin{aligned} F &= p\left[\frac{(1+i)^n - 1}{i}\right] \\ &= \$100\left[\frac{(1.0225)^{120} - 1}{0.0225}\right] \\ &\approx \$59{,}737.89. \end{aligned}$$

Thus the fund will contain approximately \$59,737.89 after 30 years. ■

Practice Problem 11 In order to obtain money for a down payment on their first house, a newly married couple has decided to deposit \$150 per month into a savings account paying 6% interest compounded monthly. How much money will this account contain after five years?

THE PRESENT VALUE OF AN ANNUITY

In Section 4.2 we learned how to find the value of the principal P that, if invested now, will accumulate to a particular amount A at some future time. (Recall that the amount P is called the *present value* of A.) Likewise it is often necessary to find the amount of money that must be deposited now in order to provide enough money to make a series of future periodic payments. For example, instant winners of the Illinois State Lottery receive $1 million in 20 yearly payments of $50,000. Lottery officials must determine how much money must be invested in order for these payments to be made. More specifically, the amount of money P necessary to make n periodic payments p from an account earning interest at the rate of i per period before exhausting the account is called the **present value of an annuity**. By (4.6) we have

$$P = \frac{A}{(1 + i)^n}.$$

So taking $A = F$ and substituting from (4.8) gives

$$P = \frac{p\left[\frac{(1 + i)^n - 1}{i}\right]}{(1 + i)^n} = p\left[\frac{1 - \frac{1}{(1 + i)^n}}{i}\right] = p\left[\frac{1 - (1 + i)^{-n}}{i}\right].$$

We have obtained the following formula.

Present Value of an Annuity

$$P = p\left[\frac{1 - (1 + i)^{-n}}{i}\right] \qquad (4.9)$$

where P denotes the present value of the annuity,
p denotes the amount of the periodic payment,
i denotes the interest rate per period, and
n denotes the number of payments.

Again the expression in brackets in formula (4.9) has a special symbol because of the frequency with which it occurs in financial analysis; it is denoted $a_{\overline{n}|i}$, which is read "a angle n at i." Thus $a_{\overline{n}|i}$ denotes the present value of an annuity consisting of n payments of $1 with an interest rate of i per period. Note that with this notation (4.9) can be written as

$$P = p a_{\overline{n}|i}.$$

EXAMPLE 4.16   A teenager's grandparents have decided to give her $5000 per year for the next four years to pay for her college education. How much must they deposit in an account paying 8% interest compounded annually in order to begin making these $5000 payments next year?

Solution This question asks for the present value of 4 yearly payments of $5000. Using (4.9) with $p = \$5000$, $i = 0.08$, and $n = 4$, we see that

$$P = p\left[\frac{1-(1+i)^{-n}}{i}\right]$$

$$= \$5000\left[\frac{1-(1.08)^{-4}}{0.08}\right]$$

$$\approx \$16{,}560.63.$$

Therefore, by depositing this amount, the grandparents will be able to withdraw $5000 each year for the next 4 years and will leave no money in the account after the last withdrawal. ∎

EXAMPLE 4.17 If money can be invested at the rate of 6% compounded quarterly, which is worth more: a gift of $32,000 now, or an installment gift of $1000 every three months for ten years?

Solution The installment gift is an annuity with $4(10) = 40$ payments of $1000. By (4.9) its present value is

$$P = p\left[\frac{1-(1+i)^{-n}}{i}\right]$$

$$= \$1000\left[\frac{1-(1.015)^{-40}}{0.015}\right]$$

$$\approx \$29{,}915.85.$$

Since the present value of the installment is approximately $29,915.85, the outright gift of $32,000 is worth more. ∎

EXAMPLE 4.18 The Gilmore family is considering the purchase of their first home. A conventional 30-year mortgage is available at a rate of 10.8% compounded monthly. After considering their financial status, the loan officer at their bank has advised the Gilmores that they can afford a down payment of $10,000 and mortgage payments of $1000 per month. What is the most expensive house they can afford?

Solution The most expensive house that the Gilmores can afford will cost $10,000 (their down payment) plus the present value of their mortgage payments. Using (4.9), we see that this present value is

$$P = p\left[\frac{1-(1+i)^{-n}}{i}\right]$$

$$= \$1000\left[\frac{1-(1.009)^{-360}}{0.009}\right]$$

$$\approx \$106{,}696.04.$$

Hence the most expensive house that the Gilmores can afford costs approximately $116,700. ∎

Practice Problem 12   Instant winners of the Illinois State Lottery receive $1 million in 20 yearly payments of $50,000. If money can be invested at 6% interest compounded annually, how much must be deposited in order for these payments to be made?

EXERCISES 4.3

In Exercises 1–8 find a formula for the general term of the given difference equation with initial value x_0. Then evaluate x_{10}.

1. $x_{n+1} = 3x_n + 4$, $x_0 = 5$
2. $x_{n+1} = \frac{1}{2}x_n + 1$, $x_0 = 34$
3. $x_{n+1} = -0.6x_n + 24$, $x_0 = 10$
4. $x_{n+1} = 2x_n - 3$, $x_0 = 8$
5. $x_{n+1} = x_n - 5$, $x_0 = 6$
6. $x_{n+1} = -\frac{2}{3}x_n + 5$, $x_0 = 4$
7. $x_{n+1} = 1.01x_n + 25$, $x_0 = 100$
8. $x_{n+1} = 1.2x_n - 2.4$, $x_0 = 12$

In Exercises 9–14 compute the future value of an annuity after n payments if p is the amount of the periodic payment and i is the interest rate per period.

9. $n = 12$, $p = \$200$, $i = 2\%$
10. $n = 18$, $p = \$500$, $i = 3\%$
11. $n = 36$, $p = \$100$, $i = 0.5\%$
12. $n = 25$, $p = \$1000$, $i = 7\%$
13. $n = 24$, $p = \$600$, $i = 1.2\%$
14. $n = 30$, $p = \$400$, $i = 9\%$

In Exercises 15–20 compute the present value of an annuity after n payments if p is the amount of the periodic payment and i is the interest rate per period.

15. $n = 48$, $p = \$200$, $i = 0.5\%$
16. $n = 10$, $p = \$12{,}000$, $i = 7\%$
17. $n = 20$, $p = \$3000$, $i = 2\%$
18. $n = 24$, $p = \$500$, $i = 1.1\%$
19. $n = 18$, $p = \$25$, $i = 1.5\%$
20. $n = 16$, $p = \$80$, $i = 0.8\%$

21. Suppose that $500 is deposited initially in an account that earns 8% interest compounded quarterly, and at the end of each quarter an additional $50 is deposited into the account. Let B_n denote the balance of the account after n quarters.
    (a) Write a difference equation that expresses B_{n+1} in terms of B_n.
    (b) Find a formula for the general term of the difference equation in part (a).
    (c) How much will this account be worth at the end of five years?

22. A university has raised $1.5 million for a scholarship endowment fund. This money is invested at 10% interest compounded annually, and at the end of each year $120,000 is withdrawn from the fund for scholarships. Let B_n denote the balance in the fund after n years.
    (a) Write a difference equation that expresses B_{n+1} in terms of B_n.
    (b) Find a formula for the general term of the difference equation in part (a).
    (c) How much will the fund contain at the end of ten years?

23. The Small Business Administration offers business loans at 9% interest compounded monthly for 7 years. If the owner of a restaurant can afford monthly payments of $800, what is the maximum amount the owner can borrow?

24. If monthly deposits of $75 are made into an account paying 9% interest compounded monthly, how much will the account contain after ten years?

25. A life insurance policy calls for quarterly payments of $110 to be made for 25 years. When the last payment is made, the insured may elect to receive as a lump sum payment an amount equal to all the payments plus 6% interest compounded quarterly. How much will this lump sum payment be?

26. A medical institute is attempting to raise funds to support a $25,000 research fellowship, to be awarded every year for the next twenty years. If the institute can invest its money at 11% interest compounded annually, how much money must be raised to support this project?

27. In his will, a rich uncle stipulated that his niece should receive $10,000 per year for 25 years, with the first payment made one year after his death. If money can be invested at 9% interest compounded annually, what is the equivalent cash value of this inheritance?

28. If monthly deposits of $25 are made into a Christmas Club account paying 6% interest compounded monthly, how much money will be available after the twelfth deposit?

29. Beginning on his thirtieth birthday, an accountant deposited $2000 per year into an individual retirement account (IRA). If this money earns 8.5% interest compounded annually, how much will the account be worth on his sixtieth birthday?

30. A publishing house sweepstakes winner will receive $1000 per month for the next ten years. How much must the company deposit at 9% interest compounded monthly in order to be able to meet these payments if the first payment will be made next month?

31. A newspaper ad states that a new car can be purchased for a down payment of $2000.00 and monthly payments of $189.23. If interest is charged at the rate of 3.9% compounded monthly for 60 months, how much does the car cost?

32. The owner of oriental rugs worth $7500 pays $85 per quarter for insurance. Show that if these premiums were invested instead in an account paying 10% interest compounded quarterly, then after twelve years the account would contain enough money to cover the loss of the rugs.

33. A company must decide whether to replace a piece of machinery or to keep it for another three years. The cost of leasing replacement machinery for three years will be $2800, and there will be no additional expenses during this period because the machinery will be serviced by the lessor. On the other hand, if the machinery is not replaced, the company can expect to make semiannual repairs costing $500 each. If money can be deposited at 6% interest compounded semiannually, which option is cheaper for the company?

34. The owner of an insurance policy has the option of collecting $25,000 now or receiving yearly payments of $5000 for the next six years. If money can be invested at 6% interest compounded annually, which option is worth more?

35. Karen has $3000 in a savings account paying 5% interest compounded quarterly. Beginning with the next interest period, she intends to make quarterly deposits of $600 into this account. How much money will be in this account after four years?

36. An insurance company would like to establish a permanent scholarship for an actuarial science major. If money can be invested at an annual rate of 10%, how much money must be deposited in order to provide $2000 to each year's recipient?

37. Theorem 4.3 gives the general term of a sequence satisfying the difference equation $x_{n+1} = ax_n + b$ when $a \neq 1$. What is the general term of such a sequence if $a = 1$?

38. Suppose the sequence $x_0, x_1, x_2, \ldots$ satisfies the difference equation $x_{n+1} = ax_n + b$, where $a \neq 1$. Let $c = b/(1 - a)$.
   (a) Show that if $x_0 = c$, then x_n is a constant.
   (b) Show that if $|a| < 1$, then x_n approaches c as n increases.

39. As a birthday present for her grandson's twelfth birthday, a wealthy widow has decided to establish a trust fund to pay for his college education. She would like the trust to provide $8000 payable on each of his 18th, 19th, 20th, and 21st birthdays. If the trust's money can be invested at the rate of 8% compounded annually, how much money must she place in the trust on his twelfth birthday in order for it to provide the four $8000 payments?

40. Evaluate $1/a_{\overline{n}|i} - 1/s_{\overline{n}|i}$.

Answers to Practice Problems
10. $x_n = 6 \cdot 3^n + 2$
11. $10,465.50
12. $573,496.06

4.4 Sinking Funds and Amortization

In Section 4.3 we learned how to compute the future and present value of an annuity using formulas (4.8) and (4.9). These formulas can also be used when we know the total amount of money for the annuity and must determine the size of the periodic payment.

SINKING FUNDS

Businesses and governments are frequently faced with the problem of accumulating capital to meet future expenses. Consider, for instance, a local government that issues bonds maturing in 20 years to fund a $10 million courthouse and jail. The revenue from the sale of the bonds is immediately available to pay for the construction project, and the government will have 20 years to save enough money to pay off the bonds when they mature. A fund of money that is being accumulated to meet some future expense (such as the redeeming of the bonds) is called a **sinking fund.** Often sinking funds are established in the form of annuities. For instance, the local government in our example may decide that it would like to make 20 equal deposits into a sinking fund, one per year, to accumulate enough capital to redeem its bonds. It needs to know the amount of the yearly deposit that will raise the needed $10 million after the last payment is made.

This type of calculation requires solving the future value of an annuity formula (4.8) for the periodic payment p. Performing this calculation produces the following formula.

Sinking Fund Payment

$$p = \frac{iF}{(1 + i)^n - 1} \quad (4.10)$$

where p denotes the amount of the periodic payment,
F denotes the future amount to be accumulated,
i denotes the interest rate per period, and
n denotes the number of payments.

This formula can also be written as

$$p = \frac{F}{s_{\overline{n}|i}}.$$

EXAMPLE 4.19 A local government would like to raise $10 million in 20 years by making equal yearly deposits into a sinking fund. If interest is earned at the rate of 8% compounded annually, how much must the yearly deposit be in order to have the needed $10 million after the last payment is made?

Solution The size of the yearly payment can be found by using the sinking fund payment formula with $F = \$10,000,000$, $i = 0.08$, and $n = 20$.

$$p = \frac{iF}{(1 + i)^n - 1}$$

$$= \frac{0.08(\$10,000,000)}{(1.08)^{20} - 1}$$

$$\approx \$218,522.09.$$

Hence yearly payments of $218,522.09 will grow to $10 million in 20 years. ∎

4.4 SINKING FUNDS AND AMORTIZATION

EXAMPLE 4.20 Under a payroll savings plan, an employee can set aside a part of her monthly salary. If these savings earn 8.4% interest compounded monthly, how much must be set aside per month in order to accumulate $12,000 in two years?

Solution Using the sinking fund payment formula again, we see that the monthly amount to be set aside is

$$p = \frac{iF}{(1+i)^n - 1}$$

$$= \frac{0.007(\$12{,}000)}{(1.007)^{24} - 1}$$

$$\approx \$460.92.$$

Practice Problem 13 An airline estimates that its present fleet of airplanes will have to be replaced in eight years at a cost of $120 million. In order to have this money available, the airline intends to make quarterly payments into an account earning 9% interest compounded quarterly. What should the quarterly payments be in order to pay for the replacement of the planes at the end of eight years?

$120 million needed in 8 years, 9% interest rate, compounded quarterly

AMORTIZATION

Home mortgages and many other loans are repaid in installments. That is, the borrower makes regular payments of a fixed amount, part of which repays the amount borrowed and the remainder of which is interest. This method of repaying a loan is called **amortization.** As an illustration of amortization, suppose that a consumer borrows $4500 for a home improvement loan at an interest rate of 12% compounded monthly. The loan is to be repaid by twelve monthly payments of $400. The following table, called an **amortization schedule,** shows the interest charge and balance of the loan for each month that the money is borrowed.

Month	Old Balance	Interest	Payment	New Balance
1	$4500.00	$45.00	$400	$4145.00
2	$4145.00	$41.45	$400	$3786.45
3	$3786.45	$37.86	$400	$3424.31
4	$3424.31	$34.24	$400	$3058.55
5	$3058.55	$30.59	$400	$2689.14
6	$2689.14	$26.89	$400	$2316.03
7	$2316.03	$23.16	$400	$1939.19
8	$1939.19	$19.39	$400	$1558.58
9	$1558.58	$15.59	$400	$1174.17
10	$1174.17	$11.74	$400	$785.91
11	$785.91	$7.86	$400	$393.77
12	$393.77	$3.94	$397.71	$0.00

Each new balance is computed by adding the current interest charge (1% of the old balance) to the old balance and subtracting the current payment. (The monthly payments are always $400 except for the final payment, which is slightly less.) Note that in later months the interest charge decreases, so that a larger portion of each monthly payment is used to repay the amount borrowed. The total interest paid on this loan is the sum of the twelve monthly interest charges, which is $297.71. An easier way to compute the total interest paid is by means of the relation

$$\text{total interest paid} = \text{sum of payments} - \text{amount borrowed}.$$

In this case the sum of the payments is

$$11(\$400.00) + \$397.71 = \$4797.71;$$

so the total interest paid is

$$\$4797.71 - \$4500.00 = \$297.71.$$

How was the size of the monthly payment determined so that it repays not only the amount borrowed but also the accumulated interest? Since the twelve monthly payments constitute an annuity which has a present value of $4500 (the amount borrowed), the size of the payment can be found by applying the present value of an annuity formula (4.9). Specifically, we must solve (4.9) for the amount of the periodic payment p. Thus we obtain the following formula.

Amortization

$$p = \frac{iA}{1 - (1 + i)^{-n}} \quad (4.11)$$

where p denotes the amount of the periodic payment,
A denotes the amount borrowed,
i denotes the interest rate per period, and
n denotes the number of payments.

This formula can also be written as

$$p = \frac{A}{a_{\overline{n}|i}}.$$

Often students experience difficulty in determining whether the sinking fund payment formula or the amortization formula should be used. To distinguish between these two, note whether the total amount of money in the problem is a future amount or a present amount. *If it is a future amount, then the sinking fund payment formula should be used, but if it is a present amount, then the amortization formula should be used.*

4.4 SINKING FUNDS AND AMORTIZATION

EXAMPLE 4.21 A home buyer is purchasing a $120,000 house. The down payment will be $20,000, and the remainder will be financed by a 30-year mortgage at a rate of 9.6% interest compounded monthly. What will the monthly payment be?

Solution The buyer will be borrowing $120,000 - $20,000 = $100,000$. Using the amortization formula with $A = \$100,000$, $i = 0.008$, and $n = 30(12) = 360$, we see that

$$p = \frac{iA}{1 - (1+i)^{-n}}$$

$$= \frac{0.008(\$100,000)}{1 - (1.008)^{-360}}$$

$$\approx \$848.16.$$

Hence the monthly payment will be approximately $848.16. ∎

EXAMPLE 4.22 Because of special incentives in 1987 to attract buyers, it was possible to finance a car at 3.9% interest compounded monthly. If the buyer of a car costing $13,300 made a $1000 down payment and financed the rest at this rate over 48 months, how much is the monthly payment? What is the total amount of interest that the buyer will pay?

Solution By using the amortization formula with $A = \$12,300$, $i = 0.00325$, and $n = 48$, we see that the monthly payment is

$$P = \frac{iA}{1 - (1+i)^{-n}}$$

$$= \frac{0.00325(\$12,300)}{1 - (1.00325)^{-48}}$$

$$\approx \$277.17.$$

Therefore

$$\text{total interest paid} = \text{sum of payments} - \text{amount borrowed}$$
$$= 48(\$277.17) - \$12,300$$
$$= \$1004.16. \quad \blacksquare$$

EXAMPLE 4.21 Revisited In recent years home buyers are increasingly choosing 15-year mortgages instead of the more traditional 25-year and 30-year mortgages. For the home buyer in Example 4.21, compare the monthly payments and the total amounts of interest paid if a 15-year mortgage is chosen instead of a 30-year mortgage.

Solution For a 15-year mortgage, the amortization formula gives

$$p = \frac{iA}{1 - (1 + i)^{-n}}$$

$$= \frac{0.008(\$100{,}000)}{1 - (1.008)^{-180}}$$

$$\approx \$1050.27.$$

Hence the monthly payment for a 15-year mortgage will be approximately $1050.27 instead of the $848.16 monthly payment for a 30-year mortgage. The total interest charge for the 30-year mortgage is

$$\text{total interest paid} = \text{sum of payments} - \text{amount borrowed}$$
$$= 360(\$848.16) - \$100{,}000$$
$$= \$205{,}337.60.$$

On the other hand, for the 15-year mortgage the total interest charge is

$$\text{total interest paid} = \text{sum of payments} - \text{amount borrowed}$$
$$= 180(\$1050.27) - \$100{,}000$$
$$= \$89{,}048.60.$$

Thus $116,289 in additional interest is paid with the 30-year mortgage. It is easy to see why the 15-year mortgages are becoming popular! ∎

Practice Problem 14 A Pontiac Fiero is advertised with a sale price of $9946. A prospective buyer anticipates that his present car will bring $1500 as a trade-in. If the Fiero can be financed for 60 months at 9.9% interest compounded monthly, how large will the prospective buyer's monthly payment be?

EXERCISES 4.4

In Exercises 1–6 compute the value of the sinking fund payment needed to accumulate the amount F in n payments if interest is paid at the rate of i per period.

1. $F = \$5000$, $n = 16$, $i = 1.5\%$
2. $F = \$10{,}000$, $n = 10$, $i = 3\%$
3. $F = \$2000$, $n = 8$, $i = 2\%$
4. $F = \$8000$, $n = 36$, $i = 0.5\%$
5. $F = \$180{,}000$, $n = 24$, $i = 0.8\%$
6. $F = \$16{,}000$, $n = 60$, $i = 0.7\%$

In Exercises 7–12 compute the value of the payment necessary to amortize the amount A in n payments if the interest rate is i per period.

7. $A = \$6000$, $n = 36$, $i = 1\%$
8. $A = \$2000$, $n = 8$, $i = 2\%$
9. $A = \$10{,}000$, $n = 60$, $i = 0.75\%$
10. $A = \$500$, $n = 12$, $i = 1.5\%$
11. $A = \$80{,}000$, $n = 180$, $i = 1.1\%$
12. $A = \$90{,}000$, $n = 300$, $i = 0.9\%$

13. An executive intends to retire in twenty years. At the time of her retirement she would like to have a retirement account of $1 million. In order to accumulate this amount, she intends to make yearly deposits into an account paying 8.25% interest compounded annually. How large should her yearly deposits be?

14. Work question 13 if the executive intends to retire in 25 years.

15. A home buyer borrows $90,000 for 25 years at 12% interest compounded monthly.
    (a) How large will the monthly payment be?
    (b) How much interest will be paid during the course of this mortgage?

16. Work question 15 if the mortgage lasts for 15 years instead of 25 years.

17. The Small Business Administration offers business loans at 14% interest compounded monthly. If the owner of a hardware store wants to borrow $400,000 for five years, how large will the monthly payments be to repay the loan?

18. What amount of money should be deposited every six months into an account paying 8% interest compounded semiannually in order to accumulate $5000 in three years?

19. A university has issued bonds worth $12.5 million in order to pay for the construction of a new arena. In order to have this money available, it will make annual deposits into an account paying 7.5% interest compounded annually. How large must its payments be in order to redeem the bonds at the end of 20 years?

20. A couple puts 20% down on furniture costing $1800. The balance is to be financed in equal payments over 24 months. If interest is charged at the rate of 18% compounded monthly, what will the monthly payment be?

21. In order to purchase a new car, Ms. Hawkins borrowed $9800 and agreed to repay the loan with 48 equal monthly payments. If interest is charged at the rate of 10.8% compounded monthly, how much will her payments be?

22. A vacationer traveling to Hawaii paid for an $800 airline ticket with a credit card charging monthly interest of 1.5%. If the vacationer decides to repay the loan in six equal monthly payments, how large should each payment be?

23. In ten years a municipality must retire an airport bond issue in the amount of $1,250,000. To meet this obligation it intends to make semiannual payments into an account paying 7% interest compounded semiannually. How large should the payments be in order to retire the debt?

24. A homeowner has decided to set aside a certain amount each month to meet his annual $1300 real estate tax bill. If money can be invested at 6% compounded monthly, how much should be set aside each month?

25. Mr. and Mrs. Washington are interested in building their dream house, which will cost $160,000. They intend to use the $40,000 equity in their present house as a down payment on the new one and will finance the rest with a 30-year mortgage. If the present interest rate is 11.4% compounded monthly, how large will their monthly payments be on the new house?

26. To prepare for their daughter's college education, a couple has decided to deposit a fixed amount each year on her birthday into an account earning interest at the rate of 8% compounded quarterly. If these deposits start on her first birthday and continue until their child turns eighteen, how much should be deposited in order to accumulate $20,000?

27. In three years a small company would like to purchase a minicomputer costing $80,000. To raise this money, the company has decided to make quarterly payments into an account paying 8% interest compounded quarterly. How much should each payment be in order to raise the amount needed?

28. Mr. Gonzalez intends to buy a Buick Skyhawk for $10,219. His present car is worth $2200 as a trade-in, and he intends to borrow the rest at 10.2% interest compounded monthly. If he repays this loan with sixty equal monthly payments, how much interest will he pay?

29. Alice charged stereo equipment costing $1000 to her credit card, which charges interest at the rate of 1.5% per month. Prepare an amortization schedule that will repay this loan in 6 months.

30. A $6000 automobile loan was taken at an interest rate of 3.9% compounded monthly. Create an amortization schedule to repay this loan in 12 months.

31. An executive intends to retire in twenty years. During his retirement, he would like to receive a monthly income of $5000 for fifteen years. In order to make this possible, he intends to deposit equal amounts each month until his retirement into an account paying 7.2% interest compounded monthly. How large should his monthly deposit be? (*Hint:* First determine the lump sum of money he must have at the time of his retirement in order to receive $5000 per month for fifteen years.)

32. A government-subsidized student loan program provides a college student $2000 per year for four years. One year after receiving the last $2000 check, the student must begin to repay the loan by making the first of five equal yearly payments. If interest is charged at the rate of 5% compounded annually, how large must the monthly payments be to repay this loan? What total interest will the student pay under this program?

33. On his twelfth birthday, a boy received $5000 from his grandmother to be used for his college education. The money will be paid in four equal installments on his 18th, 19th, 20th and 21st birthdays. If the money is invested at the rate of 8% compounded annually, how large will each installment be?

34. A professional athlete estimates that his playing career will last another five years. For each of these five years he has decided to put some of his salary into a retirement fund earning 10% interest compounded annually. When he retires, he wants to be able to draw $30,000 per year from this fund for 35 years. How much should each of his five payments be?

35. Suppose that a loan of L dollars is to be repaid by a sequence of n payments of p dollars each with an interest rate of i per period.
  (a) Write a difference equation describing the balance due after $k + 1$ payments.
  (b) Find a formula for the balance due after the kth payment.

Answers to Practice Problems   **13.** $2,600,897.92
                                    14. $179.04

CHAPTER 4 REVIEW

IMPORTANT TERMS

- arithmetic progression *(4.1)*
- balance of a loan
- difference equation
- geometric progression
- initial value of a sequence
- interest
- interest rate
- principal
- sequence
- simple interest
- terms of a sequence
- compound interest *(4.2)*
- effective rate of interest
- nominal interest rate
- present value of an amount
- annuity *(4.3)*
- future value of an annuity
- present value of an annuity
- amortization *(4.4)*
- amortization schedule
- sinking fund

IMPORTANT FORMULAS

Simple Interest Charge

$$I = Prt \qquad (4.2)$$

where *I* denotes the interest charge,
  P denotes the principal,
  r denotes the annual interest rate, and
  t denotes the time in years for which the money is borrowed.

Simple Interest Amount

$$A = P(1 + rt) \qquad (4.3)$$

where *A* denotes the amount after *t* years,
  P denotes the principal,
  r denotes the annual interest rate, and
  t denotes the time in years for which the money is borrowed.

Compound Interest Amount

$$A = P(1 + i)^n \qquad (4.4)$$

where *A* denotes the amount after *n* periods,
  P denotes the principal,
  i denotes the interest rate per period, and
  n denotes the number of interest periods for which the money is borrowed.

Present Value of a Future Amount

for simple interest $\qquad P = \dfrac{A}{1 + rt} \qquad (4.5)$

for compound interest $\qquad P = \dfrac{A}{1 + i^n} \qquad (4.6)$

where *P* denotes the present value,
  A denotes the future amount,
  r denotes the annual interest rate,
  i denotes the interest rate per period, and
  n denotes the number of interest periods for which the money is borrowed.

Future Value of an Annuity

$$F = p\left[\frac{(1 + i)^n - 1}{i}\right] \quad (4.8)$$

where F denotes the future value after n payments,
p denotes the amount of the periodic payment,
i denotes the interest rate per period, and
n denotes the number of payments.

Present Value of an Annuity

$$P = p\left[\frac{1 - (1 + i)^{-n}}{i}\right] \quad (4.9)$$

where P denotes the present value of the annuity,
p denotes the amount of the periodic payment,
i denotes the interest rate per period, and
n denotes the number of payments.

Sinking Fund Payment

$$p = \frac{iF}{(1 + i)^n - 1} \quad (4.10)$$

where p denotes the amount of the periodic payment,
F denotes the future amount to be accumulated,
i denotes the interest rate per period, and
n denotes the number of payments.

Amortization

$$p = \frac{iA}{1 - (1 + i)^{-n}} \quad (4.11)$$

where p denotes the amount of the periodic payment,
A denotes the amount borrowed,
i denotes the interest rate per period, and
n denotes the number of payments.

REVIEW EXERCISES

In Exercises 1–8 find a formula for the general term of the given difference equation with initial value x_0. Then evaluate x_8.

1. $x_{n+1} = x_n + 2$, $x_0 = -1$
2. $x_{n+1} = 0.8x_n$, $x_0 = 125$
3. $x_{n+1} = 4x_n + 9$, $x_0 = 0$
4. $x_{n+1} = \frac{1}{2}x_n + 1$, $x_0 = 256$
5. $x_{n+1} = -2x_n$, $x_0 = 3$
6. $x_{n+1} = 1.02x_n - 50$, $x_0 = 1000$
7. $x_{n+1} = -\frac{2}{3}x_n + 5/3$, $x_0 = 8$
8. $x_{n+1} = x_n + 0.75$, $x_0 = 0.5$

9. A company borrowed $75,000 for six months at 16% simple interest.
   (a) How much must be repaid when the loan is due?
   (b) How much of the amount in part (a) is interest?

10. Suppose that $6810 is required to repay a loan of $6000. If the money is borrowed for nine months, what simple interest rate is being charged?

11. Suppose that $2500 is deposited for two years in an account paying interest at the nominal annual rate of 9%. Compute the balance of the account if interest is compounded (a) annually, (b) quarterly, and (c) monthly.

12. What is the effective rate of interest corresponding to 15% interest compounded monthly?

13. How much invested now will grow to $30,000 in seven years at 12% interest compounded quarterly?

14. Suppose that you hold a note that promises to pay $15,000 in three years. If money can be invested at 7.2% interest compounded monthly, what is the present value of this note?

15. Valerie has just received an inheritance of $25,000 from her aunt. She intends to deposit the money in an account that earns 8.4% interest compounded monthly. At the end of each month she will withdraw $700 to cover her living expenses at college.
    (a) Write a difference equation involving B_n, the balance of the account after n months.
    (b) Find a formula for the general term of the difference equation in part (a).
    (c) How much money will the account contain when Valerie graduates in three years?

16. A buyer makes a $1500 down payment on a car costing $7200. The balance will be repaid in equal monthly installments over three years. If the interest rate is 10.2% compounded monthly, what is the monthly payment?

17. A savings and loan offers 30-year mortgages at 11.1% interest compounded monthly. If a home buyer can afford a down payment of $20,000 and monthly payments of $625, what is the most expensive house he can buy?

18. A manufacturer estimates that his plant's present machinery will need to be replaced in ten years at a cost of $750,000. In order to have this amount of money available, he has decided to make quarterly deposits into an account earning 8% interest compounded quarterly. How much must be deposited each quarter?

19. A power company spends $1.5 million per month for fuel. If it can invest money at 7.5% interest compounded monthly, how much money does the company need now in order to pay its fuel costs for the next twelve months?

20. Mr. and Mrs. Cohen intend to purchase a lot costing $33,800 in a new subdivision. They have arranged to borrow this amount at 10.5% interest compounded monthly. If they repay this loan with equal monthly payments over an eight-year period, how much interest will they pay on this loan?

21. Mr. Snedeker opened an individual retirement account on his thirtieth birthday and intends to continue depositing $2000 on every birthday until he becomes 65. If his IRA earns 8.5% interest compounded annually, how much will Mr. Snedeker have available after his last deposit?

22. After Mr. Snedeker in question 21 retires, what amount can he withdraw from his IRA each year if he wants to remove all the money from this account in twenty equal withdrawals?

23. Mr. Armstrong pays $180 per month on his car loan, which is amortized over 48 months at 9.6% interest compounded monthly. How much interest will Mr. Armstrong pay during the course of this loan?

24. In question 23, what will the unpaid balance of Mr. Armstrong's car loan be after three years?

SETS AND COUNTING TECHNIQUES

5.1 Sets and Venn Diagrams
5.2 Counting with Venn Diagrams and Tree Diagrams
5.3 The Multiplication and Addition Principles
5.4 Permutations and Combinations
5.5 Pascal's Triangle and the Binomial Theorem

Chapter Review

Counting is one of the most basic of all mathematical skills, one which most persons learn before beginning their formal schooling. This type of counting involves the enumeration of a specific collection of displayed objects. Other counting problems are more difficult because we must determine the number of objects of a certain type without having them exhibited before us. In this chapter we will develop some sophisticated methods for performing the latter type of counting. These techniques will be quite useful in our study of probability (Chapter 6). We will begin by introducing the concept of a set, which will also play an important role in the next chapter.

5.1 Sets and Venn Diagrams

We will regard a **set** as a collection of objects called its **elements** or **members**. If an object x is an element of a set S, we write $x \in S$; otherwise we write $x \notin S$. A set can be specified by listing its elements between braces. For example, the set S having as its elements the letters a, e, i, o, and u can be written as

$$S = \{a, e, i, o, u\}.$$

Thus $a \in S$ and $i \in S$ but $c \notin S$.

Another way to describe a set is by characterizing its elements in terms of some property that they possess. For example, the set

$$A = \{1, 3, 5, 7, 9\}$$

consists of all the odd positive integers less than 10. This set can be written as

$$A = \{x | x \text{ is an odd positive integer less than } 10\},$$

which is read "the set of all elements x such that x is an odd positive integer less than 10." Here the symbol x denotes a typical element of the set and can be replaced by any other letter without changing the set being described.

We call sets A and B **equal** when A and B contain precisely the same elements. If A and B are equal, we write $A = B$; otherwise we write $A \neq B$.

EXAMPLE 5.1 The set

$$B = \{n | n \text{ is an even positive integer less than } 9\}$$

consists of the four elements 2, 4, 6, and 8. Thus B can also be written as

$$B = \{2, 4, 6, 8\}. \blacksquare$$

EXAMPLE 5.2 The set

$$\{\text{Alaska, California, Hawaii, Oregon, Washington}\}$$

is equal to the set

$$\{s | s \text{ is a state bordering the Pacific Ocean}\}. \blacksquare$$

EXAMPLE 5.3 The set

$$\{y | y \text{ is a positive integer less than or equal to } 100\}$$

can be written

$$\{1, 2, 3, \ldots, 100\},$$

where the three dots indicate that the pattern established by the elements 1, 2, and 3 continues through the element 100. $\blacksquare$

A set which contains all the objects under consideration is called a **universal set** and will usually be denoted U. Since there can be many possible universal sets for a particular problem, we must always specify the particular one that we have in mind. The set that contains no elements is called the **empty set** and is denoted $\emptyset$.

EXAMPLE 5.4 If the universal set U is the set of all real numbers, determine the elements that belong to the following sets:

(a) $\{x | x^2 + x - 2 = 0\}$ (b) $\{x | x^2 - 2x + 1 \geq 0\}$ (c) $\{x | x^2 + 1 = 0\}$.

Solution (a) The given set consists of all real numbers x such that $x^2 + x - 2 = 0$. Since $x^2 + x - 2 = (x - 1)(x + 2)$, we see that this equation has solutions $x = 1$ and $x = -2$. Hence

$$\{x | x^2 + x - 2 = 0\} = \{1, -2\}.$$

(b) Because $x^2 - 2x + 1 = (x - 1)^2$ and the square of every real number is nonnegative, *every* real number satisfies the condition $x^2 - 2x + 1 \geq 0$. Thus

$$\{x | x^2 - 2x + 1 \geq 0\} = U.$$

(c) If $x^2 + 1 = 0$, then $x^2 = -1$. But since the square of every real number is nonnegative, no real number satisfies the equation $x^2 + 1 = 0$. Therefore

$$\{x | x^2 + 1 = 0\} = \emptyset. \blacksquare$$

Practice Problem 1 If the universal set U is the set of all real numbers, determine the elements that belong to the following sets:

(a) $\{x | x^2 - 2x + 1 < 0\}$ (b) $\{x | x^2 = 1\}$ (c) $\{x | (x - 2)^2 \geq 0\}$.

If every element of set A is contained in set B, then we say that A is a **subset** of B or that A is **included** in B. In this case we write $A \subseteq B$. For instance, if

$$A = \{2, 3, 5\} \quad \text{and} \quad B = \{1, 2, 3, 4, 5\},$$

then A is a subset of B because each element of A is also an element of B. But B is not a subset of A since there are elements of B that are not elements of A (namely, 1 and 4).

EXAMPLE 5.5 List all the subsets of the set $S = \{a, b, c\}$.

Solution Since S contains three elements, a subset of S must contain 0, 1, 2, or 3 elements. Thus there are eight subsets of S:

$$\emptyset, \{a\}, \{b\}, \{c\}, \{a, b\}, \{a, c\}, \{b, c\}, \{a, b, c\}. \quad \blacksquare$$

Notice that in Example 5.5 the empty set $\emptyset$ and S itself are two of the subsets of the given set S. In fact, for any set A we have $\emptyset \subseteq A$ and $A \subseteq A$.

SET OPERATIONS

Two sets A and B can be combined in various ways to form new sets. The **union** of A and B, denoted $A \cup B$, is the set consisting of all the elements that belong to either A or B or both. The **intersection** of A and B, denoted $A \cap B$, is the set consisting of all the elements that belong to both A and B. And the **complement** of A is the set denoted $\overline{A}$ that consists of all the elements in the universal set that do *not* belong to A. Symbolically, we have

union	$A \cup B = \{x \mid x \in A \text{ or } x \in B\}$,	
intersection	$A \cap B = \{x \mid x \in A \text{ and } x \in B\}$,	and
complement	$\overline{A} = \{x \mid x \in A\}$.	

If A and B have no elements in common, that is, if $A \cap B = \emptyset$, then A and B are called **disjoint.**

For three or more sets the union and intersection are defined analogously. Thus $A \cup B \cup C$ is the set consisting of the elements that are in at least one of the sets A, B, or C, and $A \cap B \cap C \cap D$ is the set consisting of the elements that are in all four of the sets A, B, C, and D.

EXAMPLE 5.6 Let $A = \{1, 2, 3, 4\}$, $B = \{2, 4, 6\}$, and $C = \{1, 3, 5, 7\}$. If the universal set is

$$U = \{1, 2, 3, 4, 5, 6, 7, 8, 9, 10\},$$

find $A \cup B$, $A \cap B$, $\overline{A}$, $A \cup C$, $B \cap C$, $\overline{(A \cap C)}$, and $A \cup B \cup C$.

Solution We see from the preceding definitions that

$$A \cup B = \{1, 2, 3, 4, 6\},$$
$$A \cap B = \{2, 4\},$$
$$\overline{A} = \{5, 6, 7, 8, 9, 10\},$$
$$A \cup C = \{1, 2, 3, 4, 5, 7\}$$
$$B \cap C = \emptyset,$$
$$\overline{(A \cap C)} = \overline{\{1, 3\}} = \{2, 4, 5, 6, 7, 8, 9, 10\}, \text{ and}$$
$$A \cup B \cup C = \{1, 2, 3, 4, 5, 6, 7\}.$$

Since $B \cap C = \emptyset$, the sets B and C are disjoint. ■

Practice Problem 2 Let $A = \{1, 2, 8\}$, $B = \{2, 3, 4, 6\}$, and $C = \{3, 4, 6, 7\}$. If the universal set is

$$U = \{1, 2, 3, 4, 5, 6, 7, 8\},$$

find $A \cup B$, $A \cap B$, $\overline{B}$, $A \cap C$, $B \cup C$, $\overline{(B \cup C)}$, and $A \cup B \cup C$.

The three set operations defined above can be illustrated by use of **Venn diagrams,** which are named for the English logician John Venn (1834–1923). In a Venn diagram the universal set is depicted by a rectangle, and the sets under consideration are represented by regions within this rectangle. Figure 5.1(a) shows two overlapping circles that represent subsets A and B of the

(a) Sets A and B

(b) $A \cup B$ is shaded

(c) $A \cap B$ is shaded

(d) $\overline{A}$ is shaded

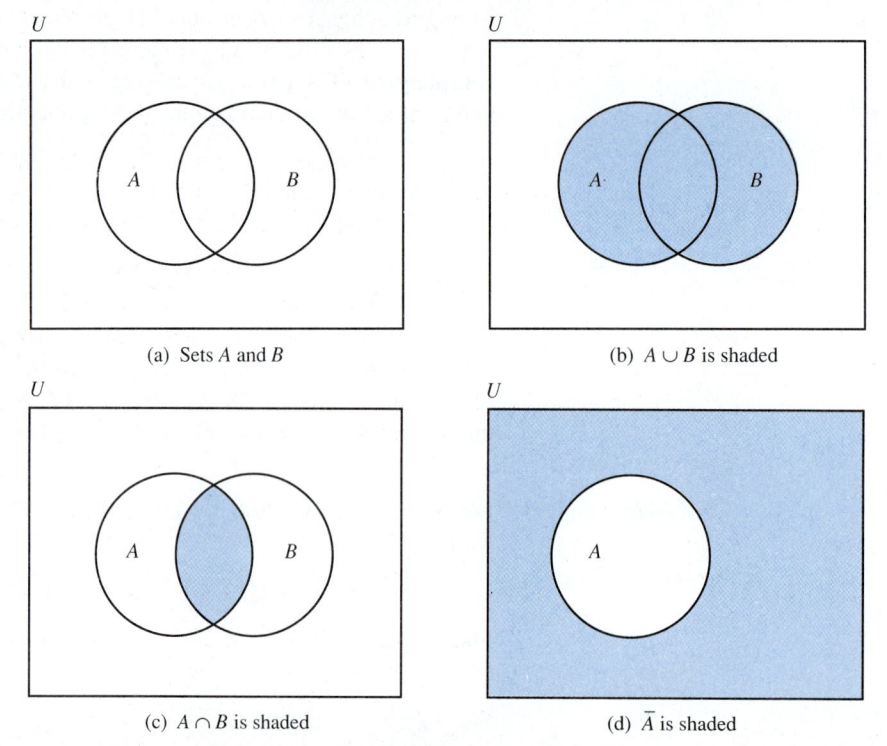

FIGURE 5.1

universal set U. The union, intersection, and complement operations can be illustrated by shading the appropriate regions as in Figures 5.1(b), (c), and (d).

Venn diagrams are helpful in visualizing the logical combinations of several properties. Consider, for instance, the sets A of all adults and M of all males if the universal set is the set of all living people. The sets A and M divide U into four disjoint regions, which are numbered I, II, III, and IV in Figure 5.2.

Each of the four numbered regions can be described as an intersection of A or $\overline{A}$ with M or $\overline{M}$:

Region I, which is $A \cap \overline{M}$, contains all women (that is, adult nonmales).

Region II, which is $A \cap M$, contains all men (that is, adult males).

Region III, which is $\overline{A} \cap M$, contains all boys (that is, nonadult males).

Region IV, which is $\overline{A} \cap \overline{M}$, contains all girls (that is, nonadult nonmales).

If we start with three sets instead of only two, the corresponding Venn diagram will divide the universal set into eight disjoint regions as in the example below.

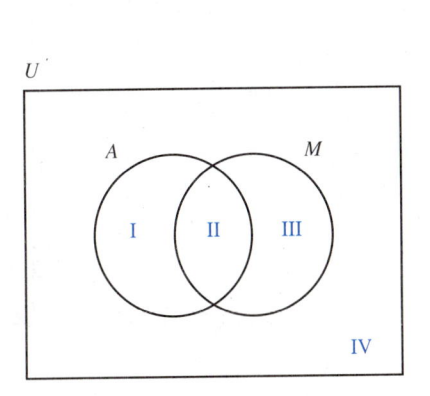

FIGURE 5.2

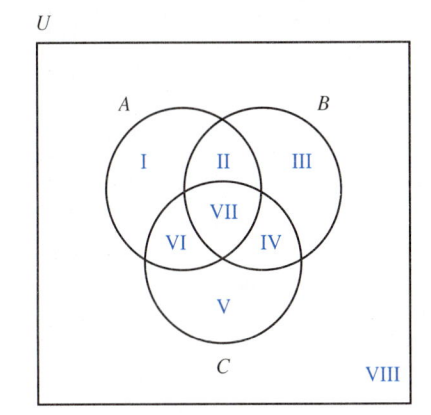

FIGURE 5.3

EXAMPLE 5.7 A nutritionist is conducting a survey of college athletes to determine which athletes regularly take vitamins A, B, and C. Let U denote the set of all athletes surveyed, and let A, B, and C denote the sets of athletes who regularly take vitamins A, B, and C, respectively. Use a Venn diagram to depict the sets of athletes who regularly take

(a) at least one of the three vitamins and
(b) exactly one of the three vitamins.

Solution The three sets A, B, and C are pictured as circular regions in Figure 5.3. Note that these three sets divide U into eight disjoint subsets, numbered I through VIII. Each of these subsets is an intersection of A or $\overline{A}$ with B or $\overline{B}$ and C or

$\overline{C}$. Region I, for instance, depicts the subset $A \cap \overline{B} \cap \overline{C}$ and represents the athletes who regularly take vitamin A but not B or C. Also region IV depicts the subset $\overline{A} \cap B \cap C$, which consists of the athletes who regularly take vitamins B and C but not A. Likewise region VIII depicts the subset $\overline{A} \cap \overline{B} \cap \overline{C}$, which consists of the athletes who do not take any of the three vitamins regularly.

(a) The athletes who regularly take at least one of vitamins A, B, or C are those who belong to one of the subsets numbered I through VII. These are just the elements of $A \cup B \cup C$, which is shaded in Figure 5.4.

(b) The athletes who regularly take exactly one of the vitamins are found in subsets I, III, and V. Thus the desired set is the union of these three subsets, which is

$$(A \cap \overline{B} \cap \overline{C}) \cup (\overline{A} \cap B \cap \overline{C}) \cup (\overline{A} \cap \overline{B} \cap C).$$

This set is shaded in Figure 5.5. ∎

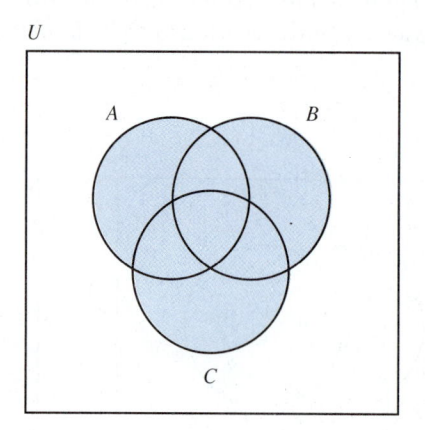

FIGURE 5.4

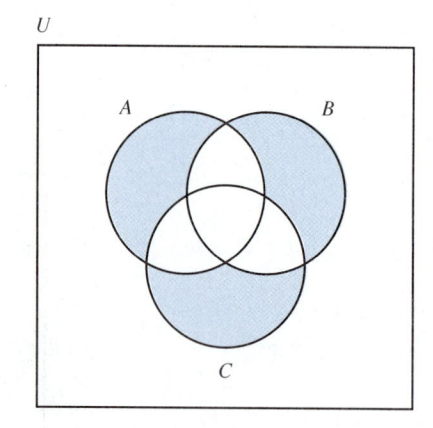

FIGURE 5.5

Practice Problem 3 (a) Describe regions III and VI in Figure 5.3 as intersections of A or $\overline{A}$ with B or $\overline{B}$ and C or $\overline{C}$. (b) Use a Venn diagram to depict the sets of athletes in Example 5.7 who regularly take exactly two of the three vitamins.

EXAMPLE 5.8 Human blood is classified according to the presence or absence of three antigens called A, B, and Rh. As we have seen, these three sets of antigens will divide human blood into eight types. These types are called AB+, AB−, A+, A−, B+, B−, O+, and O−. In this classification the letters A and B denote the presence of the corresponding antigens, and O denotes the absence of both the A and B antigens. The presence or absence of the Rh antigen is denoted by + or −, respectively. Thus blood of type AB− contains antigens

A and B but not Rh, blood of type A− contains only the A antigen, and blood of type O+ contains only the Rh antigen. The Venn diagram in Figure 5.6 displays the eight blood types. ∎

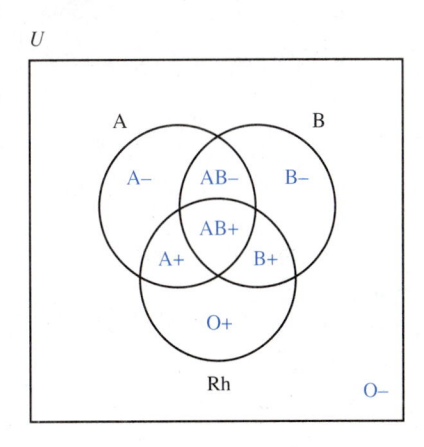

FIGURE 5.6

EXERCISES 5.1

In Exercises 1–10, list the elements in each of the following subsets of the universal set $U = \{1, 2, 3, \ldots\}$ *of all positive integers.*

1. $\{x \in U | 2 < x \leq 8\}$
2. $\{x \in U | x^2 - 5x + 6 = 0\}$
3. $\{x \in U | x^2 - 11x + 30 = 0\}$
4. $\{x \in U | (x - 2)(x - 9)(x - 17) = 0\}$
5. $\{x \in U | 2x + 8 > 5x - 1 \text{ or } 3x + 5 = 20\}$
6. $\{x \in U | \text{both } 24/x \text{ and } 60/x \text{ are integers}\}$
7. $\{x \in U | (x + 10)/x \text{ is an integer}\}$
8. $\{x \in U | x = 5n + 3 \text{ for some integer } n \leq 4\}$
9. $\{x \in U | x \text{ is odd and } x^2 \text{ is even}\}$
10. $\{x \in U | x \leq 5 \text{ or } 2x + 3 = 17\}$

Let $A = \{u, w, x, y\}$, $B = \{u, v, y\}$, and $C = \{v, x, z\}$ be subsets of the universal set $U = \{u, v, w, x, y, z\}$. *In Exercises 11–20, list the elements in each of the following sets.*

11. $A \cap B$
12. $A \cup C$
13. $A \cap B \cap C$
14. $\overline{B}$
15. $\overline{A} \cup \overline{C}$
16. $\overline{(A \cup C)}$
17. $\overline{(A \cup B)}$
18. $(A \cap \overline{B}) \cup (\overline{A} \cap B)$
19. $A \cup (B \cap C)$
20. $A \cap \overline{B} \cap C$

Let $A = \{1, 2, 3, 4\}$, $B = \{5, 6, 7, 8\}$, and $C = \{1, 2, 7, 8\}$ be subsets of the universal set $U = \{1, 2, 3, \ldots, 9, 10\}$. In Exercises 21–32, determine whether each statement is true or false.

21. $\emptyset \subseteq B$
22. $\overline{C} \subseteq \overline{C}$
23. $2 \in A \cap (B \cup C)$
24. $7 \notin (A \cup B) \cap \overline{C}$
25. $A \cap C \subseteq \overline{B}$
26. $\emptyset \in B \cup C$
27. $A \cup B \cup C = U$
28. $A \cap B \cap C = \emptyset$
29. $7 \subseteq B \cap C$
30. $C \neq \{2, 8, 7, 1\}$
31. A and B are disjoint.
32. $\overline{A}$ and $\overline{B}$ are disjoint.

33. List all the subsets of $S = \{a, b, c, d\}$ containing
   (a) exactly one element.
   (b) exactly two elements.
   (c) exactly three elements.

34. List all the subsets of $S = \{1, 2, 3, 4, 5\}$ containing
   (a) no elements.
   (b) exactly three elements.
   (c) exactly five elements.

35. In History 152, a midsemester examination and a final examination were given. Let M and F denote the sets of students who passed the midsemester and final examinations, respectively. Use set operations to describe the following sets of students, and illustrate each set with a Venn diagram.
   (a) Those who passed the midsemester examination but failed the final exam.
   (b) Those who passed both examinations.
   (c) Those who passed exactly one of the examinations.
   (d) Those who failed both examinations.

36. As part of its monthly unemployment figures, the census bureau compiles information about the employment of married couples. Let H denote the set of couples in which the husband is employed and W denote the set of couples in which the wife is employed. Use set operations to describe the following sets, and illustrate each with a Venn diagram.
   (a) Couples in which both spouses are employed.
   (b) Couples in which only the wife is employed.
   (c) Couples in which neither spouse is employed.
   (d) Couples in which exactly one spouse is employed.

37. A marketing consultant classifies persons by sex, marital status, and employment status. Let F, S, and E denote the sets of females, single persons, and employed persons, respectively. Use set operations to describe the following sets, and illustrate each one with a Venn diagram.
   (a) Working wives.
   (b) Unemployed bachelors.
   (c) Single women.
   (d) Persons who are single or employed.

38. Let U denote the set of students at Illinois State University, and let M, B, and E denote the subsets of U consisting of the students who are taking a course in mathematics, busi-

ness, and economics, respectively. Use set operations to describe the following sets, and illustrate each one with a Venn diagram.

(a) Students taking a business course or an economics course.
(b) Students taking a mathematics course and a business course.
(c) Students taking an economics course but not taking a mathematics course.
(d) Students not taking any courses in mathematics, business, or economics.
(e) Students taking a course in exactly one of these three subjects.
(f) Students taking a course in exactly two of these three subjects.

Answers to Practice Problems
1. (a) $\emptyset$ (b) $\{-1, 1\}$ (c) U
2. $A \cup B = \{1, 2, 3, 4, 6, 8\}$, $A \cap B = \{2\}$, $\overline{B} = \{1, 5, 7, 8\}$, $A \cap C = \emptyset$, $B \cup C = \{2, 3, 4, 6, 7\}$, $\overline{(B \cup C)} = \{1, 5, 8\}$, and $A \cup B \cup C = \{1, 2, 3, 4, 6, 7, 8\}$
3. (a) Regions III and VI denote the sets $\overline{A} \cap B \cap \overline{C}$ and $A \cap \overline{B} \cap C$, respectively.
   (b) Regions II, IV, and VI in Figure 5.3 contain the athletes who regularly take exactly two of the three vitamins. Thus the desired Venn diagram is shown below.

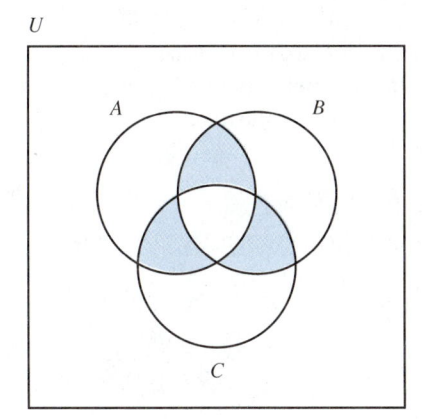

5.2 Counting with Venn Diagrams and Tree Diagrams

We saw in Section 5.1 that Venn diagrams are helpful in visualizing different combinations of properties in which we are interested. We begin this section by discussing the use of Venn diagrams to solve certain counting problems.

We will denote the number of elements in a finite set S by $|S|$. Thus if

$$A = \{1, 2, 3, 4\}, \quad B = \{3, 4, 6\},$$

and the universal set is

$$U = \{1, 2, 3, 4, 5, 6, 7, 8\},$$

then $|A| = 4$ and $|B| = 3$. Moreover, since

$$\overline{A} = \{5, 6, 7, 8\} \quad \text{and} \quad \overline{B} = \{1, 2, 5, 7, 8\},$$

we see that $|\overline{A}| = 4$ and $|\overline{B}| = 5$. Note that

$$|\overline{A}| = |U| - |A| \quad \text{and} \quad |\overline{B}| = |U| - |B|.$$

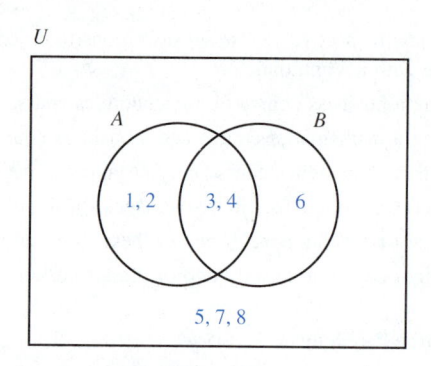

FIGURE 5.7

Now consider the problem of determining the number of elements in the union of two sets. In the example above, $A \cup B = \{1, 2, 3, 4, 6\}$. Note that $|A \cup B| \neq |A| + |B|$. The Venn diagram in Figure 5.7 shows why these numbers are different: the elements 3 and 4, which belong to both A and B, are counted only once by $|A \cup B|$ but twice by $|A| + |B|$. Hence $|A| + |B|$ counts each element of $A \cap B$ twice, and so $|A \cup B| = |A| + |B| - |A \cap B|$. We call this the **principle of inclusion-exclusion**.

This discussion illustrates the following general relationships.

Theorem 5.1

Let A and B be any subsets of a finite universal set U. Then

(a) $|\overline{A}| = |U| - |A|$

(b) $|A \cup B| = |A| + |B| - |A \cap B|$.   *(principle of inclusion-exclusion)*

EXAMPLE 5.9

In a certain group of 50 people, 16 have blond hair and 20 have blue eyes. If exactly 9 of these persons have both blond hair and blue eyes, determine

(a) the number of people without blue eyes,
(b) the number of people with blond hair or blue eyes, and
(c) the number of people with neither blond hair nor blue eyes.

Solution Let U denote the set of all 50 people, and let H and E denote the subsets of U consisting of the people having blond hair and blue eyes, respectively.

(a) The set of people without blue eyes is $\overline{E}$, and so the number of people without blue eyes is $|\overline{E}|$. By Theorem 5.1(a) this number is

$$|\overline{E}| = |U| - |E| = 50 - 20 = 30.$$

(b) The set of people with blond hair or blue eyes is $H \cup E$. By the principle of inclusion-exclusion, Theorem 5.1(b), this number is

$$|H \cup E| = |H| + |E| - |H \cap E| = 16 + 20 - 9 = 27.$$

(c) Since the set of people with neither blond hair nor blue eyes is $\overline{H \cup E}$, we can use part (b) and Theorem 5.1(a) to find the desired number. Thus there are

$$|\overline{H \cup E}| = |U| - |H \cup E| = 50 - 27 = 23$$

persons in this group with neither blond hair nor blue eyes.

The Venn diagram in Figure 5.8 shows the number of elements in the various subsets determined by H and E. ■

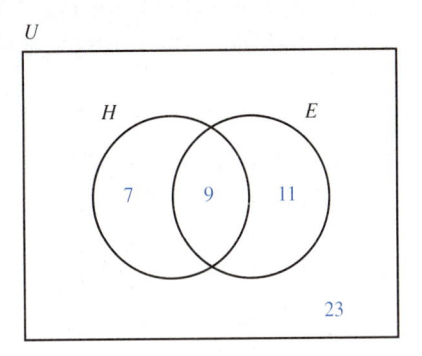

FIGURE 5.8

Practice Problem 4 In a group of 100 executives of small companies, 36 read *Forbes* and 28 read *Business Week*. If 7 of these executives read both *Forbes* and *Business Week*, how many of the executives

(a) do not read *Forbes*?
(b) read *Forbes* or *Business Week*?
(c) read neither *Forbes* nor *Business Week*?

The preceding ideas can be used to analyze data obtained from surveys. To illustrate the technique, let us consider the following data from a survey of 350 athletes.

90 regularly take vitamin A.
88 regularly take vitamin B.
97 regularly take vitamin C.
53 regularly take vitamins A and B.
55 regularly take vitamins A and C.
57 regularly take vitamins B and C.
32 regularly take all three vitamins.

CHAPTER 5 SETS AND COUNTING TECHNIQUES

How many of these athletes

(a) take only vitamin A regularly?
(b) take only one vitamin regularly?
(c) take none of the three vitamins regularly?

To answer these questions, we use the Venn diagram from Example 5.7, which is shown in Figure 5.9. We start at the center of the Venn diagram and work outward. Since 32 athletes take all three vitamins, we place this number in the region representing $A \cap B \cap C$, as shown in Figure 5.10(a). We are told that 53 athletes regularly take both vitamins A and B, and 32 of these also take vitamin C. Thus there must be $53 - 32 = 21$ who take A and B but not C. Similarly, $55 - 32 = 23$ take A and C but not B, and $57 - 32 = 25$ take B and C but not A. We place these numbers in the corresponding regions of the Venn diagram, as shown in Figure 5.10(b).

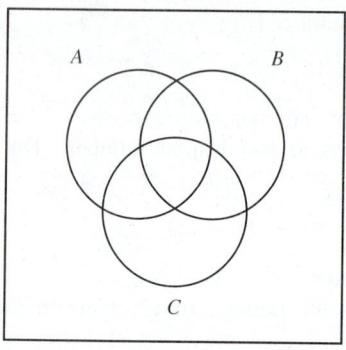

FIGURE 5.9

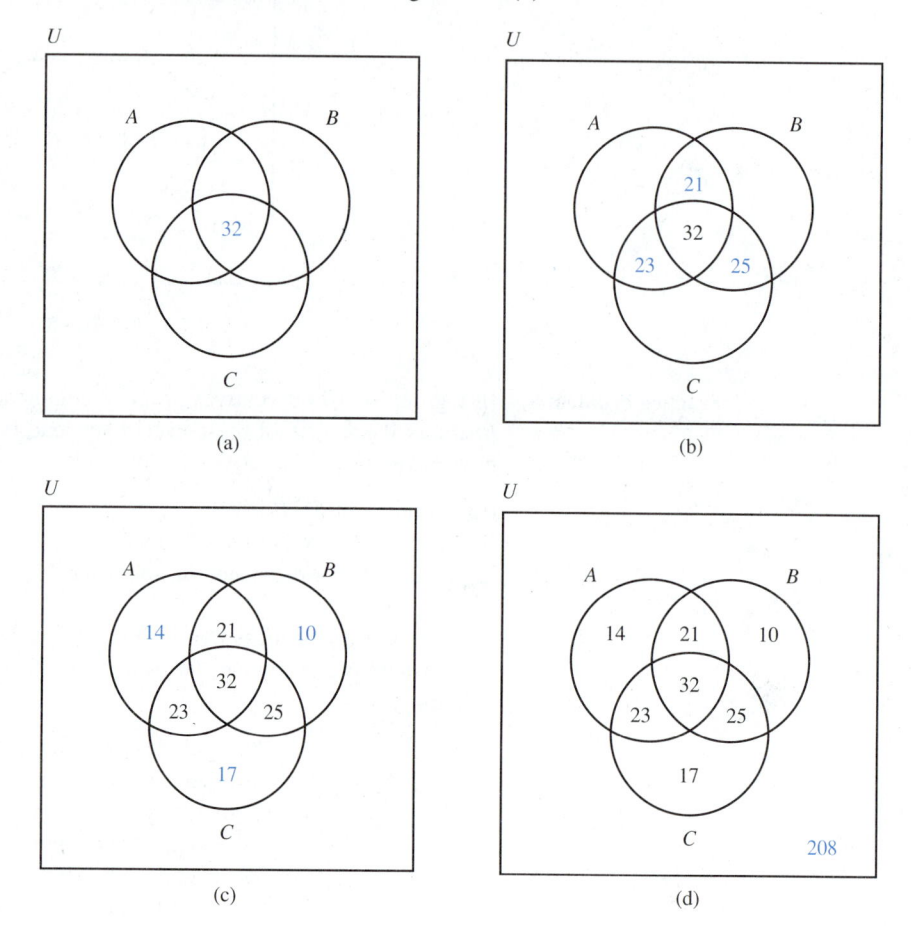

FIGURE 5.10

We see in Figure 5.10(b) that 21 + 32 + 23 = 76 of the athletes who take vitamin A regularly have been accounted for. Since there are 90 athletes in all who take vitamin A, there must be 90 − 76 = 14 who take vitamin A but not B or C. Similarly 88 − 78 = 10 athletes take only vitamin B, and 97 − 80 = 17 take only vitamin C. We now place these numbers in the corresponding regions of the Venn diagram, as shown in Figure 5.10(c).

In Figure 5.10(c) we see that $A \cup B \cup C$ contains a total of

$$14 + 21 + 10 + 23 + 32 + 25 + 17 = 142$$

elements, leaving 350 − 142 = 208 athletes who take none of the three vitamins regularly. The completed Venn diagram is shown in Figure 5.10(d). Thus we see that:

(a) 14 of these athletes take only vitamin A regularly.
(b) 14 + 10 + 17 = 41 of these athletes take only one vitamin regularly.
(c) 208 of these athletes take none of the vitamins regularly.

EXAMPLE 5.10 A firm purchased three mailing lists from a consultant dealing in such lists. The price is 10¢ per *distinct* name. The first list contains 1500 names, the second 3300, and the third 2800. A computer check shows that the first and second lists contain 382 names in common, the first and third contain 417 names in common, the second and third contain 741 names in common, and 219 names occur on all three lists. How much should the firm pay for these lists?

Solution Let U denote the set of all the names on the three lists. By proceeding as in Figure 5.10, we obtain the Venn diagram in Figure 5.11.

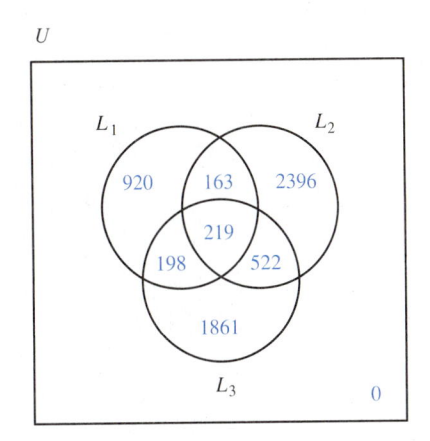

FIGURE 5.11

If we add all of the numbers in this diagram, we see that there are 6279 distinct names on the three lists. Thus the firm should pay $627.90 for the lists. ∎

Practice Problem 5 Consider the following data from a survey of students at a certain college.

> 244 were taking a business course.
> 208 were taking a mathematics course.
> 152 were taking an economics course.
> 72 were taking courses in business and mathematics.
> 46 were taking courses in business and economics.
> 60 were taking courses in mathematics and economics.
> 24 were taking courses in business, mathematics, and economics.
> 150 were not taking a course in business, mathematics, or economics.

Determine:

(a) the number of students taking a business course who were not taking a course in either mathematics or economics,
(b) the number of students taking a course in exactly two of the three disciplines,
(c) the number of students who were surveyed.

TREE DIAGRAMS

Many counting problems involve a sequence of choices. Consider, for instance, the problem of counting the number of teams consisting of one man and one woman that can be formed from among three men and two women. A pictorial device called a **tree diagram** is helpful in keeping track of the different ways in which a sequence of choices can be made. The tree diagram in Figure 5.12 shows the 6 possible teams that can be formed by choosing one man and then one woman.

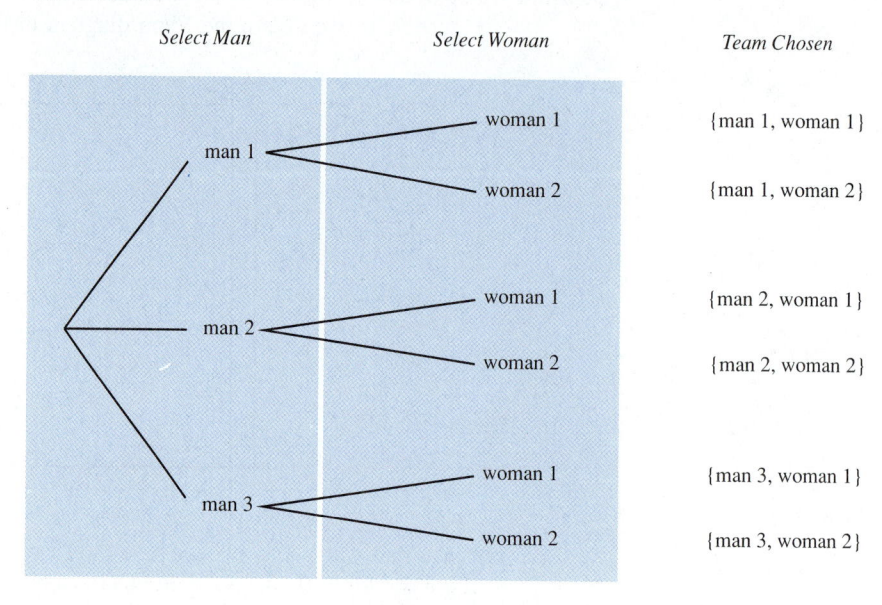

Select Man   *Select Woman*   *Team Chosen*

man 1 — woman 1   {man 1, woman 1}
man 1 — woman 2   {man 1, woman 2}

man 2 — woman 1   {man 2, woman 1}
man 2 — woman 2   {man 2, woman 2}

man 3 — woman 1   {man 3, woman 1}
man 3 — woman 2   {man 3, woman 2}

FIGURE 5.12

5.2 COUNTING WITH VENN DIAGRAMS AND TREE DIAGRAMS

EXAMPLE 5.11 A new subcompact car is available in red, blue, or gray. The buyer also has a choice of automatic or manual transmission and bench or bucket seats. In how many different ways can these three options be ordered?

Solution There are three choices of color. For each choice of color there are two choices of transmission followed by two choices of seats. Thus in ordering these three options a buyer follows one of the twelve paths through the tree diagram in Figure 5.13. For example, the buyer who selects a blue car with automatic transmission and bucket seats follows the path shown in color.

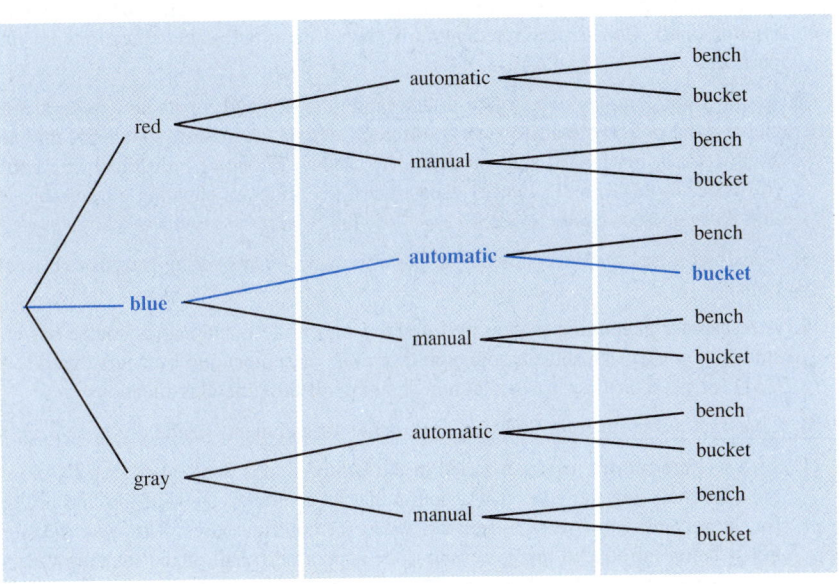

FIGURE 5.13

Practice Problem 6 In Example 5.5 we listed all the subsets of the set $S = \{a, b, c\}$. Draw a tree diagram to consider systematically all the possibilities.

EXERCISES 5.2

1. Construct a tree diagram showing the possible ways to order a sweater that is available in three colors (red, green, or blue) and three sizes (small, medium, or large).
2. A die is a cube with faces numbered 1 through 6. Construct a tree diagram showing the possible results when a single die is rolled and then a single coin is flipped.
3. Construct a tree diagram showing the possible results when a coin is flipped three times in succession.

4. Baron, in reference [1], hypothesizes that a subject cannot attend to more than one task at a time. An experiment is presented in which there are 3 possible states of attention (focused on the first task, the second task, or neither task), 2 possible responses to the first task (correct or incorrect), and 2 possible responses to the second task (correct or incorrect). Construct a tree diagram showing the possible cases for this experiment.

5. At a certain ice cream shop, a sundae can be ordered with one, two, or three scoops of ice cream, a choice of hot fudge or marshmallow, and a choice of whipped cream or nuts. Construct a tree diagram showing the possible ways of ordering a sundae.

6. Broadhurst, in reference [2], describes an experiment with male albino rats. The experiment involves 4 levels of motivation (air deprivation for 0, 2, 4 or 8 seconds), 3 levels of difficulty (easy, moderate, and difficult), and two levels of emotionality (emotional or unemotional). Construct a tree diagram showing the possible classifications for this experiment.

7. A soft drink company would like to determine the demand for its new mineral water. It can test the product in any one of 5 different regions (northeast, southeast, midwest, southwest, or northwest) and can bottle the product in one of 3 different size bottles (6-ounce, 8-ounce, or 12-ounce). Construct a tree diagram showing all possible ways that these choices can be made.

8. Construct a tree diagram showing all possible ways of answering four true-false questions.

9. A researcher examining student performance in a finite mathematics course has classified students by class (freshman, sophomore, junior, or senior) and by course grade (A, B, C, D, or F). Construct a tree diagram showing all possible classifications.

10. Construct a tree diagram showing all possible arrangements of the digits 1, 2, 3, and 4.

11. Two welfare-reform measures failed in the United States Senate last year by close margins. In analyzing the votes it was found that 45 senators voted for the first bill and 47 for the second. Moreover, 39 senators voted for both measures. This year a compromise bill is being introduced that is certain to be supported by all of the senators who voted for either of the two bills last year. If this bill needs 51 votes to pass, will it succeed?

12. A medical research team is comparing two different diagnostic tests for diabetes. When both tests were given to a group of known diabetics, test A indicated diabetes in 94% and test B indicated diabetes in 91%. Moreover, both tests indicated diabetes in 87% of the group. In what percentage did neither test indicate diabetes?

13. In a recent mayoral election the winner received 7431 votes. After analyzing the results, it was found that the winner received 4957 votes from liberals and 3893 votes from women. If the winner received 2315 votes from liberal women, how many votes did the winner get from men who are not liberal?

14. In a survey of 100 moviegoers it was found that 44 liked films directed by Bergman, 37 liked films directed by Fellini, and 23 liked both men's films. How many of these moviegoers liked neither man's films?

15. The following data were obtained from 25 fast-food restaurants in a certain city:

        12 served hamburgers.
        10 served roast beef sandwiches.
         9 served pizza.

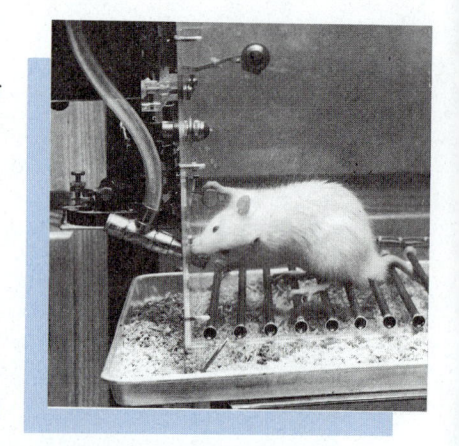

4 served hamburgers and roast beef sandwiches.
2 served hamburgers and pizza.
3 served roast beef sandwiches and pizza.
1 served all three of these foods.

(a) How many of these restaurants served hamburgers but neither roast beef sandwiches nor pizza?

(b) How many of these restaurants served pizza but neither hamburgers nor roast beef sandwiches?

16. The following information was obtained from persons at a shopping mall:

563 were adults.
414 were female.
310 had come alone.
296 were women (adult females).
213 adults had come alone.
124 females had come alone.
92 women had come alone.
54 were neither adult nor female and had not come alone.

(a) How many men (adult males) had come alone?

(b) How many children had come alone?

(c) How many people were questioned?

17. A congressman sent a questionnaire to his constituents, and his staff prepared the following summary of the replies for release to the local media.

3688 replies were received.
2471 wanted welfare reform.
2952 wanted a balanced budget.
1936 wanted gun control.
1997 wanted welfare reform and a balanced budget.
1713 wanted welfare reform and gun control.
1504 wanted a balanced budget and gun control.
1376 wanted welfare reform, a balanced budget, and gun control.

In preparing an article about the questionnaire, the local newspaper needs the additional information below. Supply this information.

(a) How many of the respondents favored none of the three issues?

(b) How many of the respondents favored only one of the three issues?

18. On May 7, 1970, Miami University in Oxford, Ohio was closed for ten days because of student protests following the invasion of Cambodia by the United States. An attitude questionnaire was prepared to sample student views regarding the closing of the university, and the results were reported in reference [4]. Among the statements to which the students responded were the three below.

(1) Miami University will have to change drastically its educational policies in the future.

(2) Trouble will never end at Miami until the whole university administrative structure is overturned.

(3) Violence seems to be the only way to get the administration to listen to us.

The following data were reported from the 434 responses.

193 persons thought the first statement was probably true.
82 persons thought the second statement was probably true.
112 persons thought the third statement was probably true.
62 persons thought the first two statements were probably true.
69 persons thought the first and third statements were probably true.
36 persons thought the last two statements were probably true.
30 persons thought all three statements were probably true.

How many respondents indicated that none of the three statements were probably true?

19. Among a random sample of 400 blacks, the following distribution of antigens can be expected:

140 will have antigen A.
100 will have antigen B.
340 will have antigen Rh.
20 will have both antigens A and B.
119 will have both antigens A and Rh.
85 will have both antigens B and Rh.
17 will have all three antigens.

Using the figure at the right, determine the number of persons of each blood type.

20. Among a random sample of 2000 whites, the following distribution of antigens can be expected:

860 will have antigen A.
340 will have antigen B.
1700 will have antigen Rh.
100 will have both antigens A and B.
731 will have both antigens A and Rh.
289 will have both antigens B and Rh.
85 will have all three antigens.

Using the figure above, determine the number of persons of each blood type.

21. An insurance company claimed to have 900 new policyholders, of which

796 bought automobile insurance.
402 bought life insurance.
667 bought home insurance.
347 bought automobile and life insurance.
580 bought automobile and home insurance.
291 bought life and home insurance.
263 bought automobile, life, and home insurance.

Explain why the state insurance regulatory office ordered an audit of the company's records.

22. A congresswoman hired an opinion research firm to survey voter attitudes in her district. Among the results in the report were these: 80% of the voters favored gun control, 53% favored nuclear power, and 69% favored tighter pollution limits. Both gun control and nuclear power were favored by 21% of those surveyed, both gun control and tighter

pollution limits were favored by 46%, both nuclear power and tighter pollution limits were favored by 34%, and 9% favored all three. After studying the report, the congresswoman accused the research firm of incompetence, claiming that these survey results are impossible. Why did she reach this conclusion?

23. Determine the set S if $|S| = 0$.

24. What can be said about sets A and B if $|A \cup B| = |A| + |B|$?

25. If $|U| = 100$, $|A| = 36$, $|B| = 61$, and $|A \cap B| = 23$, determine each of the following.
    (a) $|\overline{A}|$ (b) $|\overline{B}|$ (c) $|A \cup B|$

26. If $|U| = 160$, $|A| = 87$, $|B| = 52$, and $|A \cap B| = 39$, determine each of the following.
    (a) $|\overline{A}|$ (b) $|\overline{B}|$ (c) $|A \cup B|$

27. In a sociological study, 500 married couples were asked if they were satisfied with their marriages. In 210 cases both spouses were satisfied, while in 65 cases neither spouse was satisfied. Twice as many husbands as wives responded that they were satisfied with their marriages. How many husbands and how many wives were satisfied with their marriages? (*Hint:* Put unknowns into the regions of a Venn diagram determined by the sets of husbands and satisfied persons.)

28. During a broadcast of a baseball game, the announcer wanted to know the number of games last year in which the team's ace pitcher had lost a game that he failed to complete. The team's yearbook contained the following statistics about this pitcher.

    Games started: 36;   Wins: 17;   Losses: 8;
    Complete games: 14;   Complete games won: 9.

    Determine the answer to the announcer's question. (*Hint:* A complete game must result in either a win or a loss.)

Answers to Practice Problems

4. (a) 64   (b) 57   (c) 43
5. (a) 150   (b) 106   (c) 600
6. One possible tree diagram is shown here. The last column contains the eight possible subsets: $\emptyset$, $\{a\}$, $\{b\}$, $\{c\}$, $\{a, b\}$, $\{a, c\}$, $\{b, c\}$, $\{a, b, c\}$.

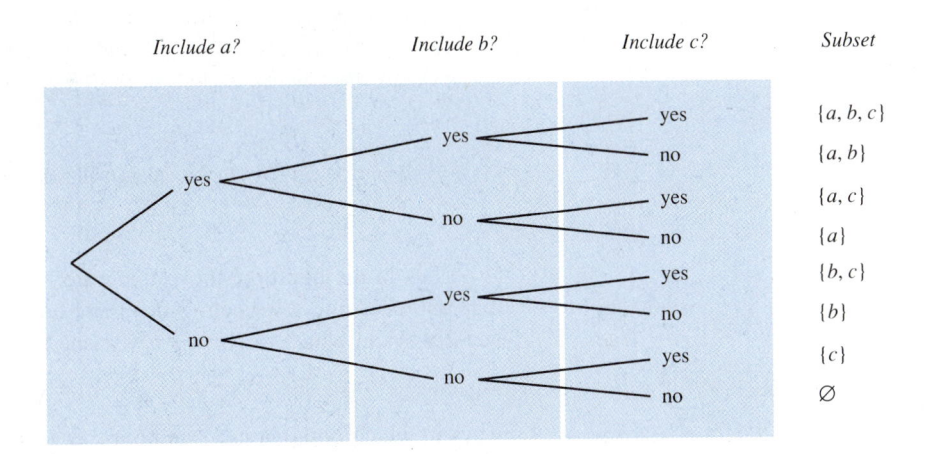

Include a?	Include b?	Include c?	Subset
yes	yes	yes	$\{a, b, c\}$
		no	$\{a, b\}$
	no	yes	$\{a, c\}$
		no	$\{a\}$
no	yes	yes	$\{b, c\}$
		no	$\{b\}$
	no	yes	$\{c\}$
		no	$\emptyset$

5.3 The Multiplication and Addition Principles

In this section we will discuss two basic principles of counting. These principles form the foundation for all of the counting problems that we will consider. Moreover, as we will see in Chapter 6, they lead to two important rules for computing probabilities.

THE MULTIPLICATION PRINCIPLE

We have seen that many counting problems involve a sequence of choices, each of which can be made in several ways. For example, the following question, which we considered in Section 5.2, is of this type.

> In order to resolve a labor dispute, a team of federal mediators is to be selected from 3 men and 2 women. If the team must consist of exactly one man and one woman, how many different teams can be chosen?

Since the number of men and women involved is small, it is easy to answer this question by listing all of the possibilities. As the tree diagram in Figure 5.12 shows, there are 6 possible teams that can be formed by selecting one man and one woman; these are

{man 1, woman 1},   {man 2, woman 1},   {man 3, woman 1},
{man 1, woman 2},   {man 2, woman 2},   {man 3, woman 2}.

However, there is a simple rule called the multiplication principle which can be applied in such cases to count the number of possibilities without actually enumerating them.

Theorem 5.2 (the multiplication principle)

Suppose that there are k operations to be performed in sequence, and the first operation can occur in n_1 ways, and for each of these the second operation can occur in n_2 ways, and for each of these the third operation can occur in n_3 ways, and so forth. Then there are

$$n_1 \cdot n_2 \cdots n_k$$

ways to perform all k of the operations in sequence.

Thus in the question above there are two choices to be made: choosing a man and choosing a woman. Since there are 3 ways in which to choose a man and 2 ways in which to choose a woman, there are

$$3 \cdot 2 = 6$$

ways in which both choices can be made. These correspond to the six possible teams of mediators.

EXAMPLE 5.12 A woman has 6 different pairs of slacks, 8 different blouses, 5 different pairs of shoes, and 3 different purses. How many outfits consisting of one pair of slacks, one blouse, one pair of shoes, and one purse can she create?

Solution In this situation there are four separate choices to be made: selecting a pair of slacks, selecting a blouse, selecting a pair of shoes, and selecting a purse. Since there are 6 ways to select slacks, 8 ways to select a blouse, 5 ways to select shoes, and 3 ways to select a purse, there are

$$6 \cdot 8 \cdot 5 \cdot 3 = 720$$

different ways in which all four choices can be made. Each of these choices corresponds to an outfit that can be created. ∎

6 pairs of slacks
8 blouses
5 pairs of shoes,
and 3 purses

EXAMPLE 5.13 An automobile license plate consists of three letters (A–Z) followed by three digits (0–9). How many different license plates are possible?

Solution In constructing a license plate, six separate choices must be made: select three letters and three digits. Since there are 26 ways in which to choose each letter and 10 ways to choose each digit, the multiplication principle shows that the number of possible ways of making all six choices is

$$26 \cdot 26 \cdot 26 \cdot 10 \cdot 10 \cdot 10 = 17{,}576{,}000.$$

Thus there are 17,576,000 possible license plates consisting of three letters followed by three digits. ∎

Practice Problem 7 An examiner wishes to create a test consisting of six questions. She has 8 variations of the first question, 7 variations of the second question, 10 variations of the third question, 5 variations of the fourth question, 6 variations of the fifth question, and 8 variations of the sixth question. How many different tests can be created by selecting one of the variations of each question?

EXAMPLE 5.14 At Wendy's Old Fashioned Hamburger Restaurants a hamburger can be ordered with any combination of 8 toppings—cheese, ketchup, lettuce, mayonnaise, mustard, onion, pickles, and tomatoes. How many different ways are there to order a hamburger and toppings?

Solution Since there are eight possible toppings, there are eight separate choices to be made. Moreover, there are two ways to order each topping, with or without. We will assume that the order in which the toppings are applied is irrelevant. Then the multiplication principle shows that the number of different ways to make all eight choices is

$$2 \cdot 2 \cdot 2 \cdot 2 \cdot 2 \cdot 2 \cdot 2 \cdot 2 = 2^8 = 256.$$

Consequently there are 256 different ways to order a hamburger and toppings at Wendy's. ∎

Practice Problem 8 A quiz is to be composed of ten questions to be answered true or false. How many different ways are there to answer all ten questions?

Notice that in order to use the multiplication principle, the number of ways to perform an operation must not depend on how the previous operations are performed. This restriction may complicate the solution of a problem, as in the example below.

EXAMPLE 5.15 Suppose that we are to use the digits 1–7 without repetition to make five-digit numbers.

(a) How many different five-digit integers can be made?
(b) How many of the numbers in part (a) begin with 6?
(c) How many of the numbers in part (a) contain both 1 and 2?

Solution (a) We will construct five-digit numbers by choosing a value for each of the five digits in the number. This amounts to filling each of the blanks below

$$\underset{1}{\underline{}} \; \underset{2}{\underline{}} \; \underset{3}{\underline{}} \; \underset{4}{\underline{}} \; \underset{5}{\underline{}}$$

with one of the digits 1–7. Now there are 7 ways in which the first digit can be selected because any of the digits 1–7 can be used. There are only 6 ways to choose the second digit, however, because the first digit cannot be repeated. Similar reasoning shows that there are 5 ways to choose the third digit, 4 ways to choose the fourth digit, and 3 ways to choose the fifth digit. Hence the multiplication principle shows that the number of possible ways of making all five choices is

$$7 \cdot 6 \cdot 5 \cdot 4 \cdot 3 = 2520.$$

So this is the number of five-digit numbers that can be formed from the digits 1–7 without repetition.

(b) To answer question (b), we proceed as above except that there is only one way to choose the first digit (since it must be 6). Therefore of the 2520 integers in part (a),

$$1 \cdot 6 \cdot 5 \cdot 4 \cdot 3 = 360$$

begin with 6.

(c) The method used in part (a) cannot be used to count the number of five-digit numbers containing both 1 and 2. The reason this method fails is that the number of choices for the fourth and fifth digits will depend on earlier choices. For example, if the first three digits are 231, then there are 4 ways in which the fourth digit can be chosen (namely, 4, 5, 6, or 7); but if the first three digits are 567, then there are only two ways in which the fourth digit can be chosen (namely, 1 or 2).

Consequently we must look for another approach. Since the digits 1 and 2 must be used, we begin by deciding where to put them. We can then fill the three remaining positions with any of the digits 3–7. Thus we will proceed as follows.

Choose a position for the 1 (in 5 possible ways).
Choose a position for the 2 (in 4 possible ways).
Fill the first unfilled position (using one of 5 possible digits).
Fill the second unfilled position (using one of 4 possible digits).
Fill the third unfilled position (using one of 3 possible digits).

Thus by the multiplication principle there are

$$5 \cdot 4 \cdot 5 \cdot 4 \cdot 3 = 1200$$

ways of making all five choices. Hence 1200 of the integers in part (a) contain both of the digits 1 and 2. ∎

As the last example shows, use of the multiplication principle requires careful attention. Although the principle is easy to understand, its application may require a bit of ingenuity. In order to use the multiplication principle correctly, you must clearly understand the objects being counted and create a systematic procedure for generating all of them. Be certain to specify precisely what is being chosen at each step, and only when you are convinced that your procedure is sound should you start multiplying numbers.

Practice Problem 9 How many ways are there to seat four married couples in a row of eight seats if

(a) each husband is to sit beside his wife? (*Hint:* When the first seat is filled, who can sit in the second seat?)
(b) each husband is to sit beside his wife and the first and last seats are to be filled by men?

THE ADDITION PRINCIPLE

Consider the following question.

> After much pleading by Alison, her parents have agreed to permit her to have a pet. However, the type of pet she chooses must be approved by her mother or father. Her mother, a cat fancier, will permit an Angora, Manx, or Siamese. Her father, a dog lover, will permit a cocker spaniel, collie, Dalmatian, Kerry blue terrier, or schnauzer. If Alison must choose her pet from one of these types of animals, how many different pets can she choose?

The answer is obviously 8, for Alison can choose among 3 different types of cats and 5 different types of dogs.

This question can be formulated as a question about sets by letting

$$C = \{\text{Angora, Manx, Siamese}\}$$

and

$$D = \{\text{cocker spaniel, collie, Dalmatian, Kerry blue terrier, schnauzer}\}$$

be the sets of pets acceptable to her mother and father, respectively. Then the number of different pets that Alison can choose is $|C \cup D|$, the number of elements in $C \cup D$. This number can be determined by using the principle of inclusion-exclusion [Theorem 5.1(b)]. For in this case $C \cap D = \emptyset$, and so

$$|C \cup D| = |C| + |D| - |C \cap D| = 3 + 5 - 0 = 8.$$

Notice that when the sets A and B are disjoint, the principle of inclusion-exclusion reduces to $|A \cup B| = |A| + |B|$. In fact, an analogous statement is true for a collection of k sets if all pairs of the sets are disjoint:

$$|A_1 \cup A_2 \cup \cdots \cup A_k| = |A_1| + |A_2| + \cdots + |A_k|.$$

This observation can be formulated in the following way.

Theorem 5.3 (the addition principle)

Suppose that there are k sets, each pair of which is disjoint. (That is, no two of the sets have an element in common.) If the first set contains n_1 elements, the second n_2 elements, and so forth, then the union of the sets contains

$$n_1 + n_2 + \cdots + n_k$$

elements.

5.3 THE MULTIPLICATION AND ADDITION PRINCIPLES

EXAMPLE 5.16 The snack bar at a movie theatre sells 5 different sizes of popcorn, 12 different candy bars, and 4 different beverages. In how many different ways can one snack be selected?

Solution There are three sets to consider: the set of popcorn snacks, the set of candy bars, and the set of beverages. Clearly each pair of these sets is disjoint. Thus since they contain 5, 12, and 4 elements, respectively, the number of elements in their union is

$$5 + 12 + 4 = 21.$$

This is the number of different ways exactly one snack can be chosen. ■

Practice Problem 10 A college student needs to choose one more course to complete next semester's schedule. She is considering 4 business courses, 7 physical education courses, and 3 economics courses. How many different courses can she select?

The addition principle is often used together with the multiplication principle. Recall that when using the multiplication principle, the number of ways to perform an operation must not depend on how previous operations are performed. If this condition is not met, it may be possible to subdivide the problem into cases where the multiplication principle can be used. If no two cases produce the same result, then they can be combined using the addition principle. The following examples illustrate this technique.

EXAMPLE 5.17 In some states a license plate may be of three different types: three letters followed by three digits, two letters followed by four digits, or no letters and six digits. How many such license plates are there?

Solution Although each license plate contains six characters, the number of choices for the first three characters depends on the type of license plate being constructed. For example, the first character on a plate containing three letters and three digits can be any of the 26 letters A–Z, but the first character on a plate containing no letters and six digits must be one of the ten digits 0–9. However, the number of ways to choose each character is the same for all plates of the same type.

Consequently we consider three cases: plates containing three letters followed by three digits, plates containing two letters followed by four digits, and plates containing no letters and six digits. By Example 5.13 there are

$$26 \cdot 26 \cdot 26 \cdot 10 \cdot 10 \cdot 10 = 17{,}576{,}000$$

plates containing three letters and three digits. Similar reasoning shows that there are

$$26 \cdot 26 \cdot 10 \cdot 10 \cdot 10 \cdot 10 = 6{,}760{,}000$$

plates containing two letters and four digits and

$$10 \cdot 10 \cdot 10 \cdot 10 \cdot 10 \cdot 10 = 1,000,000$$

plates containing no letters and six digits. Since no plate can be of two different types, the addition principle shows that there are

$$17,576,000 + 6,760,000 + 1,000,000 = 25,336,000$$

plates containing three letters followed by three digits, two letters followed by four digits, or no letters and six digits. ∎

EXAMPLE 5.18 How many integers between 1500 and 8000 (inclusive) contain no repeated digits?

Solution Notice that the number of ways in which the second (hundreds) digit can be chosen depends on the choice of the first (thousands) digit. For example, if the first digit is 6, then there are 9 possible choices for the second digit (any digit except 6). On the other hand, if the first digit is 1, then the second digit must be 5–9; so there are only 5 possible choices for the second digit. Thus we will consider two cases according to whether the first digit is 1 or not.

If the first digit is 1, then the second digit must be 5–9, and the third and fourth digits can be any unused value. Thus by the multiplication principle there are

$$1 \cdot 5 \cdot 8 \cdot 7 = 280$$

numbers between 1500 and 8000 that begin with 1 and have no repeated digits. On the other hand, if the first digit is not 1, then it must be 2, 3, 4, 5, 6, or 7. In this case the second, third, and fourth digits can be any unused value. So there are

$$6 \cdot 9 \cdot 8 \cdot 7 = 3024$$

numbers between 1500 and 8000 that do not begin with 1 and have no repeated digits. Clearly no integer is of both types because an integer cannot begin with 1 and also not begin with 1. Thus by the addition principle there are

$$280 + 3024 = 3304$$

integers between 1500 and 8000 that have no repeated digits. ∎

Practice Problem 11 In Bogart's restaurant the entrees include prime rib, filet mignon, ribeye steak, scallops, and a fish-of-the-day. Each dinner is served with salad and a vegetable. Customers may choose from 4 salad dressings and 5 vegetables, except that the seafood dishes are served with wild rice instead of the choice of vegetable. In how many different ways can a dinner be ordered?

EXERCISES 5.3

1. A men's clothing store has a sale on selected suits and blazers. If there are 30 different suits and 40 different blazers on sale, in how many different ways can a customer purchase exactly one item that is on sale?

2. A restaurant offers a choice of three green vegetables or a potato prepared in one of 5 ways. In how many different ways can one vegetable be ordered?

3. There are 16 nominees for the outstanding teacher award from the College of Arts and Sciences, 4 from the College of Business, and 3 from the College of Education. In how many different ways can the outstanding teacher award be given?

4. An actuarial science student has job offers from 4 insurance companies with headquarters in Chicago, 2 companies with headquarters in Bloomington, and 3 companies with headquarters in Boston. How many different choices of company are available to her?

5. A particular automobile manufacturer makes 4 subcompact models, 6 compact models, and 5 intermediate-sized models. In how many different ways can a car be ordered from this manufacturer?

6. Susan wants to buy one birthday present for her boyfriend. She is considering 5 different record albums, 3 different sweaters, and 6 different shirts. From these possibilities, how many different presents can she select?

7. A high school basketball player has been offered scholarships at 8 eastern universities, 5 southern universities, 6 midwestern universities, and 2 western universities. How many choices of scholarship are possible?

8. A student needing a physical education course can take 6 different sections of tennis, 3 sections of bowling, 4 sections of golf, and 8 sections of swimming. In how many different ways can a section of physical education be selected?

9. At a motel restaurant breakfast consists of a choice of cereal, scrambled eggs, fried eggs, or soft-boiled eggs; a choice of toast or muffins; and a choice of orange, tomato, or apple juice. In how many different ways can breakfast be ordered?

10. From among sixteen finalists in a contest the judges must select a winner and a runner-up. How many possible outcomes are there?

11. A tourist can take one of three different routes from her hotel to the museum and one of five routes from the museum to the park. In how many ways can she go from her hotel to the park with a stop at the museum?

12. An organization with 25 members is electing a president, secretary, and treasurer (all different). Assuming there are no ties, how many outcomes can the election have?

13. A consumer rating service is testing twelve new cars and will rate the three best in order of preference. How many different ratings are possible?

14. In how many different sequences can fifteen numbered billiard balls be hit into the pockets of a pool table?

15. In how many different ways can each of six court cases be assigned to one of four judges if a judge may hear more than one case?

16. Five rooms are to be painted one of twelve different colors. In how many different ways can the rooms be painted if
    (a) different rooms can be painted the same color?
    (b) each room must be a different color?

17. In a baseball league with six teams, each team is scheduled to play every other team four times, twice on each team's home field. How many games will be played in this league?

18. In a chess tournament with eight players, each player is scheduled to play two games against every other player, once as white and once as black. How many different games will be played in this tournament?

19. In how many different ways can a student answer each question on a twelve-question examination if each question is
    (a) to be answered true or false?
    (b) to be answered with one of five possible multiple-choice responses?

20. A die is a cube with faces numbered 1 through 6.
    (a) How many different sequences of numbers are possible if one die is rolled n times?
    (b) How many different results are possible if n differently colored dice are rolled at the same time?

21. At Avanti's a pizza can be ordered in three different sizes with any combination of seven toppings: green pepper, ham, hamburger, mushrooms, onions, pepperoni, and sausage. How many different pizzas can be ordered?

22. The following items can be ordered as optional equipment on a certain model of new car: air conditioning, automatic transmission, bucket seats, power steering, AM/FM radio, and rear window defroster. How many different ways are there to order the optional equipment on this car?

23. (a) A local telephone number is a sequence of seven digits (0–9) with the restriction that the first and second digits cannot be 0 or 1. How many local telephone numbers are possible?
    (b) An area code is a three-digit number that cannot begin with 0 or 1 and must have 0 or 1 as its middle digit. How many telephone numbers consisting of a local number and an area code are possible?

24. In the United States, radio station call letters consist of a sequence of three or four letters beginning with the letter K or W.
    (a) How many call letters are possible?
    (b) How many of the call letters contain no repeated letters?
    (c) How many of the call letters end in Z and contain no repeated letters?

25. How many five-digit numbers consisting of only even digits or only odd digits can be formed from the digits 2, 3, 4, 5, 6, 7, and 8?

26. How many four-digit numbers (integers between 999 and 10,000) either start with a 3 or do not contain a 3?

27. In how many different orders can three married couples be seated in a row of six seats under the following conditions?
    (a) Anyone may sit in any seat.
    (b) The first and last seats must be filled by men.
    (c) Each husband must sit beside his wife.
    (d) All members of the same sex are seated in adjacent seats.
    (e) Men and women are seated alternately.

28. Three women and two men are to be presented awards. In how many different orders can the awards be presented under the following conditions?
    (a) The awards can be presented in any order.
    (b) The awards are presented to the women before the men.
    (c) The first award is presented to a woman, and subsequent awards are given alternately to a man and then a woman.
    (d) The first and last awards are made to women.
    (e) The first and last awards are made to men.

29. Two different awards are to be given to business students. The 14 nominees include 5 accounting majors, 3 marketing majors, and 6 business administration majors.
    (a) In how many different ways can the awards be given if it is possible for the same student to win both awards?
    (b) In how many different ways can the awards be presented if they must be given to two different students?
    (c) In how many different ways can the awards be given to students with different majors?

30. A committee is to be formed from among 7 physicians, 10 engineers, and 5 lawyers. How many committees consisting of two members from different professions are possible?

31. A restaurant menu lists six appetizers, eight entrees, and four desserts.
    (a) How many complete meals consisting of one appetizer, one entree, and one dessert can be ordered?
    (b) How many meals consisting of one appetizer and one entree can be ordered?
    (c) How many meals consisting of one appetizer and one entree and either no dessert or one dessert can be ordered?

32. How many subsets are there of a set with n elements?

33. Consider all the five-digit numbers that do not begin with 0.
    (a) How many such numbers are there?
    (b) How many of these numbers contain only odd digits?
    (c) How many of them contain five different odd digits?
    (d) How many of these numbers begin and end with an even digit?
    (e) How many of them contain the digit 5 at least once? (*Hint:* Determine how many do not contain the digit 5.)

34. The letters in the word COMPUTERS are to be used no more than once each to form sequences of four letters.
    (a) How many different sequences can be formed?
    (b) How many of these sequences contain no vowels?
    (c) How many of these sequences begin with a consonant and end with a vowel?
    (d) How many of these sequences contain the letter M? (*Hint:* Determine the number of sequences that do not contain M.)

Answers to Practice Problems   **7.** 134,400   **8.** 1024
9. (a) 384   (b) 96   **10.** 14
11. 68

5.4 Permutations and Combinations

Two types of counting problems occur so frequently that they deserve special attention. These problems are:

(1) How many different arrangements (ordered lists) of r objects can be formed from a set of n distinct objects?
(2) How many different selections (unordered lists) of r objects can be made from a set of n distinct objects?

In this section we will develop formulas for solving these two types of counting problems.

PERMUTATIONS

To illustrate the ideas involved, let us consider a newly formed club having five members. At the first meeting it is decided to elect a slate of officers consisting of a president, vice-president, and treasurer (all different). How many slates are possible?

A slate of officers can be formed by making three successive choices: choosing a president, choosing a vice-president, and choosing a treasurer. The multiplication principle shows that the number of ways in which all three choices can be made is

$$5 \cdot 4 \cdot 3 = 60.$$

Thus there are 60 possible slates of officers. If we represent the five members of the club by the letters A, B, C, D, and E, then each slate of officers is represented by an arrangement of three of these letters. The 60 possible slates are listed below.

ABC	ABD	ABE	ACD	ACE	ADE	BCD	BCE	BDE	CDE
ACB	ADB	AEB	ADC	AEC	AED	BDC	BEC	BED	CED
BAC	BAD	BAE	CAD	CAE	DAE	CBD	CBE	DBE	DCE
BCA	BDA	BEA	CDA	CEA	DEA	CDB	CEB	DEB	DEC
CAB	DAB	EAB	DAC	EAC	EAD	DBC	EBC	EBD	ECD
CBA	DBA	EBA	DCA	ECA	EDA	DCB	ECB	EDB	EDC

Notice that the slate ABC (in which A is president, B is vice-president, and C is treasurer) is different from the slate BCA (in which B is president, C is vice-president, and A is treasurer). This is true even though the same three individuals are involved in each slate. Since the order of choice determines the office to be held, we are interested not only in *which* individuals are chosen but also in the *specific order* in which they are chosen.

Ordered arrangements of this type arise often in applications and are called **permutations.** We will denote the number of permutations of r objects selected from a set of n distinct objects by $P(n, r)$. Since the list above contains all

permutations of the letters A, B, C, D, and E taken three at a time, we see that
$$P(5, 3) = 5 \cdot 4 \cdot 3 = 60.$$

More generally, suppose that there are n distinct objects and we are given an integer r such that $1 \leq r \leq n$. To form a permutation of r of these n objects, we must fill in each of the blanks below with a different object.

$$\underline{}\ \underline{}\ \underline{}\ \cdots\ \underline{}\ \underline{}$$
$$\ \ 1\ \ \ \ 2\ \ \ \ 3\ \ \ \ \ \ \ \ r-1\ \ \ r$$

The first object can be chosen in n ways, the second in $n - 1$ ways, the third in $n - 2$ ways, and so forth. When we reach the last blank, we will have used $r - 1$ of the n objects, leaving $n - (r - 1) = n - r + 1$ objects from which to fill the last blank. Thus by the multiplication principle the number of permutations of n objects taken r at a time is

$$n(n - 1)(n - 2) \cdots (n - r + 1).$$

Notice that this "descending product" starts with n and contains exactly r factors.

Theorem 5.4

For any positive integers r and n such that $1 \leq r \leq n$,
$$P(n, r) = \underbrace{n(n - 1)(n - 2) \cdots (n - r + 1)}_{r \text{ factors}}.$$

Thus, for instance, we find by Theorem 5.4 that
$$P(10, 1) = 10,$$
$$P(10, 2) = 10 \cdot 9 = 90,$$
$$P(10, 3) = 10 \cdot 9 \cdot 8 = 720, \quad \text{and}$$
$$P(10, 4) = 10 \cdot 9 \cdot 8 \cdot 7 = 5040.$$

EXAMPLE 5.19 A pianist participating in a Chopin competition has decided to perform five of the fourteen Chopin waltzes. How many different programs are possible consisting of five waltzes played in a certain order?

Solution The pianist must select five different waltzes from fourteen and arrange them in a specific order. By Theorem 5.4 the number of such ordered arrangements (permutations) is

$$P(14, 5) = \underbrace{14 \cdot 13 \cdot 12 \cdot 11 \cdot 10}_{5 \text{ factors}} = 240{,}240. \ \blacksquare$$

EXAMPLE 5.20 In how many ways can six people be seated in a row of six chairs?

Solution A seating arrangement is merely a permutation of the people to be seated. Hence by Theorem 5.4 there are

$$P(6, 6) = 6 \cdot 5 \cdot 4 \cdot 3 \cdot 2 \cdot 1 = 720$$

possible seating arrangements. ∎

EXAMPLE 5.21 How many three-digit integers can be formed using the digits 1–7 without repetition?

Solution A three-digit integer using the digits 1–7 without repetition is simply an ordered arrangement of three of the digits 1–7. Thus by Theorem 5.4 there are

$$P(7, 3) = 7 \cdot 6 \cdot 5 = 210$$

such integers. ∎

Practice Problem 12 Nine athletes are entered in the conference high jump competition. In how many different ways can the gold, silver, and bronze medals be awarded?

FACTORIAL NOTATION

Computing the number of permutations of n objects taken r at a time involves products of consecutive integers. There is a convenient notation for representing such products.

If n is a positive integer, the symbol $n!$ (read **n factorial**) denotes the product of the positive integers less than or equal to n. That is,

$$n! = n(n - 1)(n - 2) \cdots 3 \cdot 2 \cdot 1.$$

In addition, we define

$$0! = 1.$$

The first few values of $n!$ are

$$0! = 1,$$
$$1! = 1,$$
$$2! = 2 \cdot 1 = 2,$$
$$3! = 3 \cdot 2 \cdot 1 = 6,$$
$$4! = 4 \cdot 3 \cdot 2 \cdot 1 = 24, \text{ and}$$
$$5! = 5 \cdot 4 \cdot 3 \cdot 2 \cdot 1 = 120.$$

For larger integers, $n!$ can be computed using the relation

$$n! = n \cdot (n - 1)!,$$

which follows immediately from the definition of the factorial symbol. Thus

$$6! = 6(5!) = 6(120) = 720,$$
$$7! = 7(6!) = 7(720) = 5040,$$
$$8! = 8(7!) = 8(5040) = 40{,}320,$$

and so forth. Notice that the value of $n!$ grows very quickly; in fact, 10! exceeds 3 million and 13! exceeds 6 billion.

The number of permutations of n objects taken r at a time can be expressed in a very compact form using factorial notation. For by Theorem 5.4

$$P(n, r) = n(n - 1)(n - 2) \cdots (n - r + 1)$$
$$= \frac{n(n - 1)(n - 2) \cdots (n - r + 1)(n - r)(n - r - 1) \cdots 2 \cdot 1}{(n - r)(n - r - 1) \cdots 2 \cdot 1}$$

Thus we have the following formula.

$$P(n, r) = \frac{n!}{(n - r)!}$$

Notice that $P(n, n) = n!$. In addition, if $r = 0$, the expression

$$\frac{n!}{(n - r)!}$$

has a value of 1. Accordingly we define

$$P(n, 0) = 1.$$

Although the formula

$$P(n, r) = \frac{n!}{(n - r)!}$$

is important theoretically and provides a convenient way of writing $P(n, r)$, it should not normally be used for computations because it requires more calculations than the expression in Theorem 5.4:

$$P(n, r) = n(n - 1)(n - 2) \cdots (n - r + 1).$$

COMBINATIONS

Now we will consider the second type of counting problem mentioned above. Suppose that the club with five members A, B, C, D, and E decides to appoint an executive committee consisting of three members having equal status instead of electing a president, a vice-president, and a treasurer. How many different executive committees can be formed?

If the three members of the executive committee are chosen in order, one after the other, the result will be one of the 60 permutations obtained before. (This list is reproduced below.)

ABC	ABD	ABE	ACD	ACE	ADE	BCD	BCE	BDE	CDE
ACB	ADB	AEB	ADC	AEC	AED	BDC	BEC	BED	CED
BAC	BAD	BAE	CAD	CAE	DAE	CBD	CBE	DBE	DCE
BCA	BDA	BEA	CDA	CEA	DEA	CDB	CEB	DEB	DEC
CAB	DAB	EAB	DAC	EAC	EAD	DBC	EBC	EBD	ECD
CBA	DBA	EBA	DCA	ECA	EDA	DCB	ECB	EDB	EDC

Now, however, since the members of the executive committee all have the same status, we are only interested in knowing which three individuals are chosen, not the specific order of their selection. Hence all of the permutations in any column above represent exactly the same executive committee. For instance, the permutations ACD, ADC, CAD, CDA, DAC, and DCA (shown in color above) all represent the committee with A, C, and D as its members.

We call such an unordered list of objects a **combination.** Thus a combination of n objects taken r at a time is simply a set containing r of the given n objects. The number of combinations of r objects selected from a set of n distinct objects will be denoted* by $C(n, r)$.

From the list above we see that each combination corresponds to

$$P(3, 3) = 3!$$

permutations, namely those in the same column of the list. Hence

$$C(5, 3) = \frac{P(5, 3)}{3!} = \frac{5 \cdot 4 \cdot 3}{3 \cdot 2 \cdot 1} = 10.$$

Thus there are 10 combinations of A, B, C, D, and E taken 3 at a time.

In general, each permutation of n objects taken r at a time can be obtained by first choosing r of the n objects and then arranging them in order. Since there are $C(n, r)$ ways to choose r objects from among n distinct objects and $P(r, r)$ ways to arrange these r objects in order, the multiplication principle shows that

$$P(n, r) = C(n, r) \cdot P(r, r).$$

Therefore

$$C(n, r) = \frac{P(n, r)}{P(r, r)} = \frac{P(n, r)}{r!}.$$

---

*Another common notation for $C(n, r)$ is $\binom{n}{r}$.

Theorem 5.5

For any positive integers r and n such that $0 \leq r \leq n$,

$$C(n, r) = \frac{P(n, r)}{r!}.$$

Thus, for instance, we find by Theorem 5.5 that

$$C(10, 0) = \frac{P(10, 0)}{0!} = \frac{1}{1} = 1,$$

$$C(10, 1) = \frac{P(10, 1)}{1!} = \frac{10}{1} = 10,$$

$$C(10, 2) = \frac{P(10, 2)}{2!} = \frac{10 \cdot 9}{2 \cdot 1} = 45,$$

$$C(10, 3) = \frac{P(10, 3)}{3!} = \frac{10 \cdot 9 \cdot 8}{3 \cdot 2 \cdot 1} = 120, \text{ and}$$

$$C(10, 4) = \frac{P(10, 4)}{4!} = \frac{10 \cdot 9 \cdot 8 \cdot 7}{4 \cdot 3 \cdot 2 \cdot 1} = 210.$$

Since

$$P(n, r) = \frac{n!}{(n-r)!},$$

we can also write the formula for $C(n, r)$ as shown below.

$$C(n, r) = \frac{n!}{r!(n-r)!}$$

But as was true for the corresponding permutation formula, this formula should not normally be used for computations.

EXAMPLE 5.22 How many ways are there to select a subcommittee of five members from among a committee of 12?

Solution The number of committees is just the number of combinations of 12 members taken 5 at a time. By Theorem 5.5 this number is

$$C(12, 5) = \frac{P(12, 5)}{5!} = \frac{12 \cdot 11 \cdot 10 \cdot 9 \cdot 8}{5 \cdot 4 \cdot 3 \cdot 2 \cdot 1} = 792. \blacksquare$$

EXAMPLE 5.23 How many ways are there to select 4 novels from a list of 16 novels to be read for a literature class?

Solution Since we are interested in knowing only which novels are selected (and not in their order of selection), this is a problem involving combinations. By Theorem 5.5 the answer is

$$C(16, 4) = \frac{P(16, 4)}{4!} = \frac{16 \cdot 15 \cdot 14 \cdot 13}{4 \cdot 3 \cdot 2 \cdot 1} = 1820.$$ ■

Practice Problem 13 Suppose that eight people raise their glasses in a toast. If every person clinks glasses exactly once with everyone else, how many clinks will be heard?

In practice one of the most difficult aspects of solving a counting problem is to determine whether to use permutations or combinations. Remember that although both involve choosing objects from a set of distinct objects, *permutations are used when the order of selection matters, and combinations are used when the order of selection does not matter*.

EXAMPLE 5.24 An investor is going to invest $16,000 in four stocks from a list of twelve prepared by his broker. How many different investments are possible if

(a) $4000 is to be invested in each stock?
(b) $6000 is to be invested in one stock, $5000 in another, $3000 in a third, and $2000 in the fourth?
(c) $5000 is to be invested in each of two stocks and $3000 is to be invested in each of two others?

Solution (a) Since the same amount of money is to be invested in each stock, the order in which the four stocks are selected is unimportant. Thus part (a) is a problem involving *combinations,* and so there are

$$C(12, 4) = \frac{P(12, 4)}{4!} = \frac{12 \cdot 11 \cdot 10 \cdot 9}{4 \cdot 3 \cdot 2 \cdot 1} = 495$$

different investments that can be made.

(b) In this case we will choose four stocks in sequence. In the first stock chosen we will invest $6000, in the second we will invest $5000, in the third we will invest $3000, and in the fourth we will invest $2000. Thus the order of selection matters, and so part (b) is a problem involving *permutations*. Hence there are

$$P(12, 4) = 12 \cdot 11 \cdot 10 \cdot 9 = 11{,}880$$

different investments that can be made.

(c) In this case we will divide the problem into two parts: first we will choose the two stocks in which $5000 will be invested, and then we will choose the two stocks in which $3000 will be invested. As in part (a) there are

$$C(12, 2) = \frac{P(12, 2)}{2!} = \frac{12 \cdot 11}{2 \cdot 1} = 66$$

ways to select the two stocks in which $5000 will be invested. Now we must select two of the remaining ten stocks in which to invest $3000 each. This selection can be made in

$$C(10, 2) = \frac{P(10, 2)}{2!} = \frac{10 \cdot 9}{2 \cdot 1} = 45$$

different ways. Thus by the multiplication principle there are

$$66 \cdot 45 = 2970$$

different investments that can be made. ■

Often, as in part (c) of Example 5.24, it is necessary to use permutations or combinations together with the addition or multiplication principles. The following examples are of this type.

EXAMPLE 5.25 A committee of four is to be chosen from among five women and six men. How many different committees are possible that contain at least three women?

Solution A committee containing at least three women must consist of either three women and one man or four women and no men. Thus we will consider two cases. The number of committees containing three women and one man can be computed using the multiplication principle because we can construct such a committee by first choosing the women and then choosing the man. Since there are $C(5, 3)$ ways to select three women from among five and $C(6, 1)$ ways to select one man from among six, there are

$$C(5, 3) \cdot C(6, 1) = 10 \cdot 6 = 60$$

committees containing three women and one man. Similar reasoning shows that there are

$$C(5, 4) \cdot C(6, 0) = 5 \cdot 1 = 5$$

committees containing four women and no men. Hence by the addition principle there are

$$60 + 5 = 65$$

possible committees containing at least three women. ■

EXAMPLE 5.26 An election is being held to fill two faculty seats and three student seats on the University Curriculum Committee. From among the five candidates for the faculty seat, the person receiving the most votes will serve a three-year term, and the runner-up will serve a two-year term. From among the seven student candidates, the three winners will each serve one-year terms. How many different election results are possible?

Solution The possible election results can be obtained by first choosing the faculty winners and then choosing the student winners. Because the order of finish is important for the faculty seats, the number of possible election results for the faculty seats involves *permutations*. Thus there are

$$P(5, 2) = 5 \cdot 4 = 20$$

possible results in the faculty election. On the other hand, the student winners all have the same length of term; so the selection of the student winners involves *combinations*. Hence there are

$$C(7, 3) = \frac{P(7, 3)}{3!} = \frac{7 \cdot 6 \cdot 5}{3 \cdot 2 \cdot 1} = 35$$

possible results in the student election. It now follows from the multiplication principle that there are

$$20 \cdot 35 = 700$$

possible election results. ■

Practice Problem 14 How many different lists of five letters contain three different consonants and two different vowels? (Regard "y" as a consonant.)

EXERCISES 5.4

Evaluate each of the numbers in Exercises 1–16.

1. $P(7, 5)$
2. $P(6, 6)$
3. $P(16, 3)$
4. $P(15, 5)$
5. $4!$
6. $7!$
7. $8!$
8. $10!$
9. $C(9, 6)$
10. $C(15, 5)$
11. $C(8, 4)$
12. $C(6, 3)$
13. $P(n, 0)$
14. $P(n, n)$
15. $C(n, n)$
16. $C(n, 0)$

17. (a) How many permutations are there of the letters $a, b, c, d,$ and e taken two at a time? List them all.

    (b) How many combinations are there of the letters $a, b, c, d,$ and e taken two at a time? List them all.

18. (a) How many permutations are there of the numbers 1, 2, 3, and 4 taken three at a time? List them all.

(b) How many combinations are there of the numbers 1, 2, 3, and 4 taken three at a time? List them all.

19. Twenty applicants for an executive position are to be interviewed to narrow the list of candidates to the top five. How many possible results are there if
    (a) the top five are ranked in order of preference?
    (b) the top five are unranked?

20. A newly formed consumer action group has thirty members. In how many ways can the group elect
    (a) a president, vice-president, secretary, and treasurer (all different)?
    (b) an executive committee consisting of five members?

21. Six speakers are scheduled to address a convention. In how many different orders can the speakers appear?

22. In how many different ways can the letters of the word CREAM be arranged?

23. A nautical signal consisting of five flags strung vertically on a rope is to be constructed. How many different signals can be made if there are nine flags of different colors?

24. Five chairs are placed in a row.
    (a) How many different ways are there of seating five persons in these chairs?
    (b) If there are eight people available, in how many different ways can the five seats be filled?

25. For marketing purposes a company has divided the United States into eight regions. It wishes to test a new product in three of these regions. How many different ways are there to select these three regions?

26. In how many different ways can the Supreme Court render a 5-to-4 decision?

27. In the state senate, thirteen of the twenty members come from urban districts, and the others come from rural districts. A committee of six is to be appointed to study educational reform.
    (a) How many committees can be formed?
    (b) If the committee is to be equally divided between urban and rural members, how many committees are possible?

28. A grievance committee consisting of five members is to be selected from among nine women and seven men.
    (a) How many different committees can be formed?
    (b) How many of these committees contain three women and two men?

29. Twelve of the houses on a particular street receive cable television and eight do not. A researcher studying the amount of time that people watch television intends to visit four of the houses that receive cable television and three that do not. How many such selections of houses are there?

30. A candy store classifies its candies as creams (ten types), nuts (seven types), and chews (eight types). A customer has ordered an assortment to consist of six types of creams, four types of nuts, and five types of chews. How many such assortments are possible?

31. A five-member committee is to be selected from among four representatives of management and five representatives of labor. In how many different ways can the committee be formed under the following circumstances?
    (a) Anyone is eligible to serve on the committee.
    (b) The committee must consist of three representatives of management and two representatives of labor.
    (c) The committee must consist of two representatives of management and three representatives of labor.
    (d) The committee must contain at least three representatives of management.

32. A woman must choose two pairs of shoes and either two skirts and three blouses or two dresses to pack for a business trip. If she has five pairs of shoes, four skirts, eight blouses, and six dresses from which to choose, in how many different ways can she pack her suitcase?

33. Of the 2170 stocks on the New York Stock Exchange, 1438 advanced, 522 declined, and 210 were unchanged during the last week of May, 1987. In how many ways can this happen? [Write your answer in terms of $C(n, r)$.]

34. In a literature class of twelve graduate students, the instructor will choose three students to analyze *Howards End*, four students to analyze *Room with a View*, and five students to analyze *A Passage to India*. In how many different ways can the students be chosen?

35. A Chinese restaurant offers a "family dinner" consisting of any four items on its menu. Its advertising claims that over 300 different family dinners are available. If this claim is true, what is the smallest number of items that can be listed on its menu?

36. The figure below shows a portion of the map of a city. How many different routes are there from City Hall to the Post Office without moving south or west?

37. Suppose that a coin is flipped n times and the resulting sequence of heads and tails is recorded.
    (a) How many sequences are possible?
    (b) If $0 \leq k \leq n$, how many of the sequences in part (a) contain exactly k heads?
    (c) A sequence is an ordered arrangement. Why is it not possible to use permutations to solve part (b)?

38. Use Theorem 5.5 to show that $C(n, 0) = 1$, $C(n, n) = 1$, and $C(n, r) = C(n, n - r)$. Justify these results in terms of the meaning of combinations.

39. Use Theorem 5.5 to show that $r \cdot C(n, r) = n \cdot C(n - 1, r - 1)$ for all integers r and n such that $1 \leq r \leq n$.

40. Show that $C(n, r) \leq C(n, r + 1)$ for all integers r and n such that $0 < r < n/2$.

41. Three Americans, four Russians, and two Chinese delegates are attending a conference on international trade.
    (a) In how many different ways can these nine delegates line up for a group photograph with all persons of the same nationality standing together?
    (b) How many of the ways in part (a) do not have a Russian standing beside a Chinese?

42. A book store wants to create a window display containing a row of two cookbooks, five novels, and three biographies. The books are to be chosen from among five cookbooks, eight novels, and seven biographies released in the last month.
    (a) How many such displays are possible?
    (b) How many such displays are possible if, in addition, all books of the same category must be side-by-side?

43. How many ways are there to seat n people around a *circular* table, taking into account only the position of each person with respect to the others?

44. (a) How many distinct numbers can be formed using the digits 1, 2, 3, 4, 5, 9, 9, 9, 9, 9, 9 if no two 9's can be adjacent?
    (b) How many distinct numbers can be formed using the digits 1, 2, 3, 4, 5, 6, 7, 8, 9, 9, 9, 9, 9 if no two 9's can be adjacent?

Answers to Practice Problems 12. 504
13. 28
14. 1,596,000

5.5 Pascal's Triangle and the Binomial Theorem

The numbers $C(n, r)$ defined in Section 5.4 arise naturally not only in counting problems but also in algebra, where they are referred to as **binomial coefficients**. The binomial coefficients occur when expressions of the form $(x + y)^n$ are multiplied out. In this section we will discuss the evaluation of such expressions.

Let us begin by multiplying out $(x + y)^n$ for several values of n to see if we can find a pattern.

$$(x + y)^0 = 1$$
$$(x + y)^1 = x + y$$
$$(x + y)^2 = x^2 + 2xy + y^2$$
$$(x + y)^3 = x^3 + 3x^2y + 3xy^2 + y^3$$
$$(x + y)^4 = x^4 + 4x^3y + 6x^2y^2 + 4xy^3 + y^4$$
$$(x + y)^5 = x^5 + 5x^4y + 10x^3y^2 + 10x^2y^3 + 5xy^4 + y^5$$

Note that in each case the expansion of $(x + y)^n$ contains $n + 1$ terms of the form

$$cx^{n-r}y^r$$

for $r = 0, 1, 2, \ldots, n$, where the coefficient c is some positive integer that depends on the values of n and r.

PASCAL'S TRIANGLE

We will begin by concentrating on the coefficients. The array of coefficients obtained by expanding the expressions $(x + y)^n$ for all nonnegative integers n is called **Pascal's triangle** after the French mathematician and philosopher Blaise Pascal (1623–62), who discovered many of the triangle's properties. Thus the beginning of Pascal's triangle is shown below.

```
(row 0)                    1
(row 1)                  1   1
(row 2)                1   2   1
(row 3)              1   3   3   1
(row 4)            1   4   6   4   1
(row 5)          1   5  10  10   5   1
```

Notice that each row begins and ends with 1. Moreover, every other entry in a row is the sum of the two closest entries in the row above. Thus, for example, the third entry in row 5 is the sum of the second and third entries in row 4:

$$10 = 4 + 6.$$

This property enables us to generate the rest of the triangle very easily. For instance, row 6 is obtained by adding consecutive entries in row 5 and then putting a 1 at both ends.

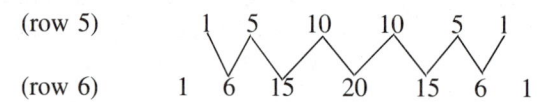

```
(row 5)        1   5  10  10   5   1
(row 6)      1   6  15  20  15   6   1
```

EXAMPLE 5.27 Find row 7 of Pascal's triangle.

Solution The row begins and ends with 1. The other entries are obtained by adding adjacent entries in row 6. The result is shown below.

$$1 \quad 7 \quad 21 \quad 35 \quad 35 \quad 21 \quad 7 \quad 1 \quad \blacksquare$$

Practice Problem 15 Find row 8 of Pascal's triangle.

5.4 PASCAL'S TRIANGLE AND THE BINOMIAL THEOREM

Surprisingly, the numbers in Pascal's triangle turn out to be the numbers $C(n, r)$ from Section 5.4. For example, the numbers in row 4 of Pascal's triangle are

$$C(4, 0) = 1 \quad C(4, 1) = 4 \quad C(4, 2) = 6 \quad C(4, 3) = 4 \quad C(4, 4) = 1.$$

It is not difficult to see why the coefficient of the term $x^{n-r}y^r$ in the expansion of $(x + y)^n$ is $C(n, r)$. Since

$$(x + y)^n = \underbrace{(x + y)(x + y) \cdots (x + y)}_{n \text{ factors}},$$

a term of the form $x^{n-r}y^r$ results whenever we choose the y term from exactly r of the factors $x + y$. And by Theorem 5.5 the number of ways to choose exactly r terms of y from among the n factors of $x + y$ is $C(n, r)$.

THE BINOMIAL THEOREM

Having determined the coefficients that occur in the expansion of $(x + y)^n$, it is now easy to evaluate this expression.

Theorem 5.6 (the binomial theorem)

The coefficient of $x^{n-r}y^r$ in the expansion of $(x + y)^n$ is $C(n, r)$. Thus

$$(x + y)^n = C(n, 0)x^n + C(n, 1)x^{n-1}y^1 + \cdots + C(n, n)y^n.$$

Observe that in the expansion of $(x + y)^n$ the powers of x decrease from n to 0 while the powers of y increase from 0 to n. Moreover, the sum of the x and y exponents in each term is n. The coefficients of the terms are just the numbers from row n of Pascal's triangle.

EXAMPLE 5.28 Expand $(x + y)^5$ in powers of x and y.

Solution Applying the binomial theorem with $n = 5$, we obtain

$$(x + y)^5 = C(5, 0)x^5 + C(5, 1)x^4y^1 + C(5, 2)x^3y^2 + C(5, 3)x^2y^3 \\ + C(5, 4)x^1y^4 + C(5, 5)y^5.$$

Thus, using row 5 of Pascal's triangle, we have

$$(x + y)^5 = x^5 + 5x^4y + 10x^3y^2 + 10x^2y^3 + 5xy^4 + y^5. \blacksquare$$

EXAMPLE 5.29 Expand $(2a - b)^4$ in powers of x and y.

Solution Taking $n = 4$ in the binomial theorem gives

$$(x + y)^4 = C(4, 0)x^4 + C(4, 1)x^3y^1 + C(4, 2)x^2y^2 + C(4, 3)x^1y^3 + C(4, 4)y^4$$
$$= x^4 + 4x^3y + 6x^2y^2 + 4xy^3 + y^4.$$

If we now substitute $x = 2a$ and $y = -b$ in the expansion above, we obtain

$$(2a - b)^4 = (2a)^4 + 4(2a)^3(-b) + 6(2a)^2(-b)^2 + 4(2a)(-b)^3 + (-b)^4$$
$$= 16a^4 - 32a^3b + 24a^2b^2 - 8ab^3 + b^4. \blacksquare$$

Practice Problem 16 Expand $(x - 2y)^5$ in powers of x and y.

In Chapter 6 we will use the binomial theorem to find a particular term in the expansion of $(x + y)^n$, as in the example below.

EXAMPLE 5.30 Find the coefficient of the term involving $x^{16}y^4$ in the expansion of $(x - 3y)^{20}$.

Solution The binomial theorem states that the term involving $x^{16}y^4$ in the expansion of $(x - 3y)^{20}$ is

$$C(20, 4)x^{16}(-3y)^4.$$

Since the value of only one binomial coefficient is needed, we will use Theorem 5.5 (rather than Pascal's triangle) to evaluate $C(20, 4)$. In this way we find that

$$C(20, 4) = \frac{P(20, 4)}{4!} = \frac{20 \cdot 19 \cdot 18 \cdot 17}{4 \cdot 3 \cdot 2 \cdot 1} = 4845.$$

Thus the term involving $x^{16}y^4$ in the expansion of $(x - 3y)^{20}$ is

$$C(20, 4)x^{16}(-3y)^4 = 4845x^{16}(81y^4) = 392{,}445x^{16}y^4,$$

and so the desired coefficient is 392,445. $\blacksquare$

Practice Problem 17 Find the coefficient of the term involving p^5q^{11} in the expansion of $(2p - q)^{16}$.

EXERCISES 5.5

1. Compute row 9 of Pascal's triangle.
2. Compute row 10 of Pascal's triangle.

In Exercises 3–14, use the binomial theorem to expand the given expression.

3. $(x + y)^6$
4. $(x + y)^7$
5. $(x - y)^7$
6. $(x - y)^6$
7. $(a + 4b)^4$
8. $(3p + q)^5$
9. $(3u - v)^5$
10. $(r - 2s)^6$
11. $(2x - \frac{1}{2})^6$
12. $(x - 1)^9$
13. $(1 - x)^8$
14. $(-2 + y)^8$

In Exercises 15–26, find the indicated coefficient.

15. the coefficient of x^{12} in $(x + 1)^{20}$
16. the coefficient of x^{18} in $(x + 1)^{28}$
17. the coefficient of x^{17} in $(x - 1)^{30}$
18. the coefficient of x^{15} in $(x - 1)^{24}$
19. the coefficient of $x^{14}y^8$ in $(x + y)^{22}$
20. the coefficient of x^6y^8 in $(x + y)^{14}$
21. the coefficient of $x^{19}y^{13}$ in $(x - y)^{32}$
22. the coefficient of $x^{10}y^{11}$ in $(x - y)^{21}$
23. the coefficient of $x^{12}y^6$ in $(x + 2y)^{18}$
24. the coefficient of x^4y^{15} in $(2x + y)^{19}$
25. the coefficient of $x^{10}y^5$ in $(x - \frac{1}{2}y)^{15}$
26. the coefficient of x^3y^{17} in $(\frac{1}{3}x - y)^{20}$

27. From among a group of eight disgruntled workers, a grievance committee of at least one and at most three persons is to be formed. How many different committees can be formed?

28. The university's alumni service award can be given to at most four people per year. This year there are seven nominees. In how many different ways can the recipients of the award be chosen?

29. Show that $C(n, 0) + C(n, 1) + C(n, 2) + \cdots + C(n, n) = 2^n$ for all positive integers n by using the binomial theorem.

30. Show that $C(n, 0) - C(n, 1) + C(n, 2) - \cdots + (-1)^n C(n, n) = 0$ for all positive integers n by using the binomial theorem.

31. How does the equation in Exercise 29 relate to the number of subsets of a set with n elements?

32. Let S be a set containing n elements. Show that the number of subsets of S containing an even number of elements equals the number of subsets of S containing an odd number of elements. (*Hint:* Use Exercise 30.)

33. Use Theorem 5.5 to show that $C(n + 1, r) = C(n, r - 1) + C(n, r)$.

Answers to Practice Problems   15. 1 8 28 56 70 56 28 8 1
16. $(x - 2y)^5 = x^5 - 10x^4y + 40x^3y^2 - 80x^2y^3 + 80xy^4 - 32y^5$
17. $-139{,}776$

CHAPTER 5 REVIEW

IMPORTANT TERMS

- complement of a set ($\overline{A}$) (5.1)
- disjoint sets
- element of a set
- empty set ($\emptyset$)
- equality of sets ($A = B$)
- intersection of sets ($A \cap B$)
- set
- subset ($A \subseteq B$)
- union of sets ($A \cup B$)
- universal set
- Venn diagram
- number of elements in a set ($|S|$) (5.2)
- tree diagram
- combination (5.4)
- n factorial ($n!$)
- permutation
- binomial coefficients (5.5)
- Pascal's triangle

IMPORTANT FORMULAS

Theorem 5.1

Let A and B be any subsets of a finite universal set U. Then
(a) $|\overline{A}| = |U| - |A|$
(b) $|A \cup B| = |A| + |B| - |A \cap B|$.   *(principle of inclusion-exclusion)*

Theorem 5.2 (the multiplication principle)

Suppose that there are k operations to be performed in sequence, and the first operation can be performed in n_1 ways, and for each of these the second operation can occur in n_2 ways, and for each of these the third operation can occur in n_3 ways, and so forth. Then there are

$$n_1 \cdot n_2 \cdot \cdots \cdot n_k$$

ways to perform all k of the operations in sequence.

Theorem 5.3 (the addition principle)

Suppose that there are k sets, each pair of which is disjoint. (That is, no two of the sets have an element in common.) If the first set contains n_1 elements, the second n_2 elements, and so forth, then the union of the sets contains

$$n_1 + n_2 + \cdots + n_k$$

elements.

Factorial notation

For any positive integer n,
$$0! = 1$$
$$n! = n(n-1)(n-2) \cdots 3 \cdot 2 \cdot 1.$$

Theorem 5.4 (number of permutations of n objects taken r at a time)

For any positive integers r and n such that $0 \leq r \leq n$,

$$P(n, r) = n(n-1)(n-2) \cdots (n-r+1)$$
$$= \frac{n!}{(n-r)!}.$$

Theorem 5.5 (number of combinations of n objects taken r at a time)

For any positive integers r and n such that $0 \leq r \leq n$,

$$C(n, r) = \frac{P(n, r)}{P(r, r)} = \frac{P(n, r)}{r!}.$$

Theorem 5.6 (the binomial theorem)

The coefficient of $x^{n-r}y^r$ in the expansion of $(x + y)^n$ is $C(n, r)$. Thus
$$(x + y)^n = C(n, 0)x^n + C(n, 1)x^{n-1}y^1 + C(n, 2)x^{n-2}y^2 + \cdots + C(n, n)y^n.$$

REVIEW EXERCISES

1. Let $A = \{2, 5, 6, 8\}$ and $B = \{1, 2, 3, 4, 5, 6\}$ be subsets of the universal set $U = \{1, 2, 3, 4, 5, 6, 7, 8, 9, 10\}$. Compute each of the following sets.
   (a) $A \cup B$ (b) $A \cap B$ (c) $\overline{A}$ (d) $A \cup \overline{B}$

2. Let $A = \{1, 2, 3\}$, $B = \{2, 3, 4, 5\}$, and $C = \{4, 5, 6, 7, 8\}$ be subsets of the universal set $U = \{1, 2, 3, 4, 5, 6, 7, 8, 9, 10\}$. Determine if the following statements are true or false.
   (a) $8 \in \overline{A} \cup B$
   (b) $5 \notin A \cup \overline{(B \cap C)}$
   (c) $A \cap B = \emptyset$
   (d) $C \subseteq \overline{A}$
   (e) $A \cup B \cup C = U$
   (f) $A \subseteq \{x \in U | x < 5\}$
   (g) $\{3\} \not\subseteq B$
   (h) $|C| = |\overline{C}|$

3. When a radio station surveyed 250 listeners of its evening programming, it found that 83 were female, 96 were teenagers, and 67 were both (that is, teenage females). How many of those surveyed were nonteenage males?

4. The following information was obtained about the houses in a particular subdivision.

   > 195 have central air conditioning.
   > 92 have a finished basement.
   > 201 have a two-car garage.
   > 72 have central air conditioning and a finished basement.
   > 151 have central air conditioning and a two-car garage.
   > 66 have a finished basement and a two-car garage.
   > 54 have central air conditioning, a finished basement, and a two-car garage.
   > 7 did not have central air conditioning, a finished basement, or a two-car garage.

   (a) How many houses in this subdivision have a two-car garage but neither central air conditioning nor a finished basement?
   (b) How many houses in this subdivision have a finished basement and a two-car garage but not central air conditioning?
   (c) How many houses are in this subdivision?

5. A hospital classifies its patients according to sex and blood type. Draw a tree diagram showing all of the possible classifications. (See Example 5.8 for the possible blood types.)

6. Evaluate each of the following numbers.
   (a) $5!$ (b) $9!$ (c) $P(11, 0)$ (d) $C(9, 8)$
   (e) $P(8, 4)$ (f) $C(7, 3)$ (g) $P(7, 7)$ (h) $C(12, 0)$

7. A police artist makes a composite sketch of a crime suspect's face by asking witnesses to select one each of ten noses, twelve pairs of eyes, eight mouths, and seven hairstyles. How many different faces are possible?

8. In reference [3] Kotler discusses the division of markets into segments based on geographic, demographic, and psychographic variables. He lists 5 geographic variables, 11 demographic variables, and 8 psychographic variables. How many different divisions are possible using exactly one variable of each type?

9. How many different rearrangements of the letters in SPRING are there?

10. An investor intends to buy 100 shares of each of four stocks selected from a list of twenty. How many different selections are possible?

11. First, second, and third prizes are to be awarded in a pie-baking contest. If there are 12 entries, in how many different ways can the prizes be given?

12. The mathematics department must assign one course to each of nine part-time instructors. The courses to be assigned consist of four sections of college algebra, three sections of finite math, and two sections of calculus. How many different teaching assignments are possible? (Consider two teaching assignments the same if each teacher teaches the same course.)

13. Six liberal candidates and five conservative candidates are running for four seats on a city council. Assume that no two candidates receive the same number of votes.
    (a) How many different orders of finish are possible?
    (b) How many different sets of winners are possible?
    (c) In how many of the sets in part (b) are four liberals elected?
    (d) In how many of the sets in part (b) are two liberals and two conservatives elected?
    (e) In how many of the sets in part (b) are at least two liberals elected?

14. Write rows 0, 1, 2, 3, 4, 5, and 6 of Pascal's triangle.

15. Expand $(3r + s)^6$ in powers of r and s.

16. Determine the coefficient of the term $x^{10}y^5$ in the expansion of $(x - 2y)^{15}$.

REFERENCES

1. Baron, Jonathan, "Division of Attention in Successiveness Discrimination," in *Attention and Performance IV,* Sylvan Kornblum, ed. New York: Academic Press, 1973, pp. 703–711.
2. Broadhurst, P. L., "Emotionality and the Yerkes-Dodson Law," *Journal of Experimental Psychology,* vol. 54(1957), pp. 345–352.
3. Kotler, Philip, *Marketing Management,* 3rd ed. Englewood Cliffs, N.J.: Prentice-Hall, 1976.
4. Rudestam, Kjell Erik and Bruce John Morrison, "Student Attitudes Regarding the Temporary Closing of a Major University," *American Psychologist,* vol. 26(1971), pp. 519–525.

6

PROBABILITY

6.1 Computing Probabilities by Counting
6.2 Assigning Probabilities
6.3 Basic Laws of Probability
6.4 Conditional Probability, the Intersection Rule, and Independent Events
6.5 Bayes's Formula
6.6 Bernoulli Trials
Chapter Review

Many historians regard an exchange of letters in 1654 between the great French mathematicians Blaise Pascal and Pierre de Fermat as the origin of probability theory. Their correspondence was prompted by two gambling problems posed to Pascal by the Chevalier de Méré, a noted dilettante in the French court. From this origin around the gambling tables of seventeenth century Europe, the theory of probability has gradually progressed to a position of prominence in modern science. Many of the greatest mathematicians in history, including Pascal, Bernoulli, Euler, Laplace, and Gauss, contributed to its early development. Yet the theory was rarely applied outside of gambling until the nineteenth century, when it began to be recognized as a powerful tool. Today probabilistic methods permeate the physical, biological, behavioral, and management sciences. In this chapter we will examine the basic principles of this theory and explore some of its applications.

6.1 Computing Probabilities by Counting

Uncertainties are part of life. When we flip a coin or take an airplane flight or start a new business, to some extent we must simply hope for the best. These actions are examples of *random processes*, in the sense that the outcome in each case cannot be predicted with absolute certainty. Probability can be defined as the mathematical theory of random processes, in other words, as the "science of uncertainty." It is paradoxical that mathematics, the most exact of the sciences, has something to say about the very nature of uncertainty. In this section we will begin to investigate this curious relationship.

THE CLASSICAL DEFINITION OF PROBABILITY

Among the many ways to think about probability, the oldest and most widely used is the *relative frequency* interpretation: The probability of an event represents the proportion of the time that the event tends to occur over the long run. Hence the probability of an event will be a number between 0 and 1 inclusive that measures the event's likelihood of happening. Thus, for instance, when a symmetric coin is flipped, we say that the probability of heads appearing is 1/2. This means that in a large number of flips the coin will tend to land with heads showing half of the time.

In many situations it is intuitively plausible that all possible outcomes are equally likely. In this case it is possible to calculate probabilities theoretically.

EXAMPLE 6.1 What is the probability that a 5 will appear in one roll of a uniform die?

Solution When a uniform die is rolled, each of its six faces is equally likely to appear. Thus the probability that a 5 appears is 1/6. (See Figure 6.1.) ∎

6.1 COMPUTING PROBABILITIES BY COUNTING

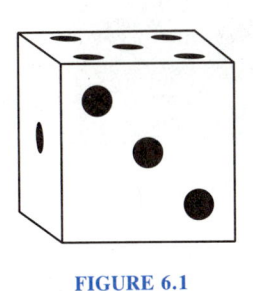

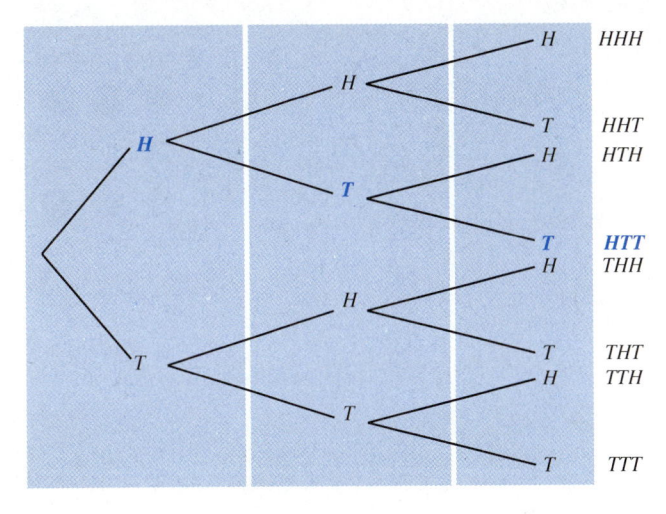

FIGURE 6.1

FIGURE 6.2

EXAMPLE 6.2 If a symmetric coin is flipped three times, what is the probability that it will land heads on the first flip and tails on the second and third flips?

Solution The possible sequences of heads and tails can be found by a tree diagram (see Figure 6.2, where H denotes heads and T denotes tails). Since the coin is symmetric, it is just as likely to land heads as tails on any flip. Thus each of the paths in this tree diagram is equally likely. Since there are eight sequences of heads and tails, the probability that the sequence H, T, T occurs is 1/8. ∎

Practice Problem 1 The spinner in Figure 6.3 is to be spun twice.

(a) Construct a tree diagram showing the possible outcomes.
(b) Determine the probability that both spins are 2's.

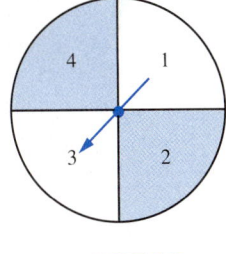

FIGURE 6.3

EXAMPLE 6.3 A pair of uniform dice, one white and one black, are rolled. What is the probability that 1 appears on each die?

Solution There are 6 values possible on the white die, and for each of these there are 6 possible values for the black die. (See Figure 6.4.) Hence the multiplication principle shows that there are $6 \cdot 6 = 36$ possible pairs of numbers that can result when these dice are rolled.

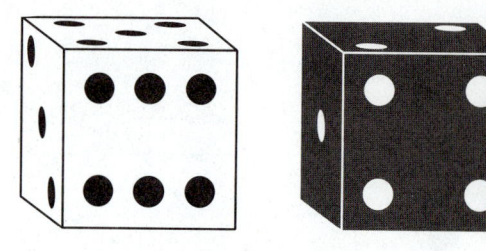

FIGURE 6.4

These 36 possible pairs can be found using a tree diagram similar to Figure 6.2. The results are recorded in the table below.

		\			Black			
			1	2	3	4	5	6
	1		(1, 1)	(1, 2)	(1, 3)	(1, 4)	(1, 5)	(1, 6)
	2		(2, 1)	(2, 2)	(2, 3)	(2, 4)	(2, 5)	(2, 6)
	3		(3, 1)	(3, 2)	(3, 3)	(3, 4)	(3, 5)	(3, 6)
White	4		(4, 1)	(4, 2)	(4, 3)	(4, 4)	(4, 5)	(4, 6)
	5		(5, 1)	(5, 2)	(5, 3)	(5, 4)	(5, 5)	(5, 6)
	6		(6, 1)	(6, 2)	(6, 3)	(6, 4)	(6, 5)	(6, 6)

Since the dice are uniform, we assume that each of these 36 pairs of numbers is equally likely to occur. Thus the probability is 1/36 that 1 appears on each die. ∎

SAMPLE SPACES AND EVENTS

We see from the table in Example 6.3 that, when rolling a pair of uniform dice, the only way to obtain a sum of 2 is for a 1 to appear on each die. Hence the probability of obtaining a sum of 2 is 1/36. Notice that there are only 11 possible sums that can occur if a pair of dice are rolled (namely, 2, 3, 4, . . ., 12). Thus if we had considered only the *sums* that can occur (instead of considering the actual pairs of numbers), we might have incorrectly concluded that the probability of obtaining a sum of 2 is 1/11. To avoid this type of mistake, we must be certain to identify all of the possible results in the situation of interest, and we must be sure that these results are all equally likely to occur.

We call a situation having various possible results an **experiment**. Thus we consider all of the following to be experiments: flipping a coin, rolling a pair of dice, observing the frequency of automobile accidents in a particular age group of drivers, or counting the number of customers in line at a supermarket.

By a **sample space** for an experiment we mean a set containing possible results of the experiment, which are called **outcomes.** *The outcomes must be* *chosen to have the property that when the experiment is performed, one and only one of the outcomes occurs.*

What we consider as outcomes of an experiment will depend on the type of situations that are of interest. For example, a particular marketing experiment may require observing the customers in a particular store. But, depending on the situation, we may wish to observe the number of customers, the interval between arrival of customers, or the yearly income of the customers.

It is important to understand that there may be different ways of choosing the outcomes (and hence the sample space) of an experiment. For example, in the experiment of rolling a single die, one way of choosing the outcomes is to select the possible numbers that can result. For this choice the sample space is

$$\{1, 2, 3, 4, 5, 6\}.$$

Notice that when a die is rolled, *exactly one* of these outcomes must occur. On the other hand, we may be interested in knowing only if an even or odd number is rolled; in this case the sample space is

$$\{\text{even, odd}\}.$$

Again notice that *one and only one* of these outcomes must occur when a die is rolled. However, neither

$$A = \{1, \text{even}, 5\} \quad \text{nor} \quad B = \{1, 2, 4, 6, \text{odd}\}$$

is a sample space for this experiment: The outcome 3 is not contained in set A, and two outcomes in set B (namely, 1 and odd) occur when a 1 is rolled.

EXAMPLE 6.4 Construct a sample space for the experiment of flipping a coin, assuming that the coin will not land on edge.

Solution In this case there are two outcomes, namely heads and tails. Thus a sample space for this experiment is

$$\{\text{heads, tails}\}. \blacksquare$$

EXAMPLE 6.5 Construct a sample space for the experiment of flipping a coin three times and observing the sequence of heads and tails that appears.

Solution In this experiment there are eight possible outcomes, shown in the tree diagram of Figure 6.2. The corresponding sample space is

$$\{HHH, HHT, HTH, HTT, THH, THT, TTH, TTT\}. \blacksquare$$

EXAMPLE 6.6 An insurance company is going to promote two of its executives to manage regional offices. Those under consideration are Mr. Adams, Mrs. Bethea, Ms. Choi, Mr. Dunn, and Mr. Eovaldi. Construct a sample space for this experiment.

Solution   Here we must select two individuals from among five for promotion. The number of possible outcomes is therefore $C(5, 2) = 10$. A sample space for this experiment is

$$\{AB, AC, AD, AE, BC, BD, BE, CD, CE, DE\},$$

where we have indicated by their initials the pairs of persons to be promoted. ∎

Practice Problem 2   A distributor of business forms has a toll-free phone number on which its out-of-state customers can place orders. The company can handle as many as 15 calls per hour on this line. Construct a sample space for the experiment of observing the number of calls on this toll-free line on a Monday morning between 9 and 10 A.M.

Often we are interested in knowing the probability that one of several outcomes happens. For instance, in rolling a die we may want to know the probability that the die comes up greater than 3. In this case we are interested in finding the probability that one of the outcomes

$$4, \quad 5, \quad \text{or} \quad 6$$

occurs. We call any such set of outcomes an **event**. In other words, *an event is simply a subset of the sample space of an experiment*. Listed below are several other events and the corresponding sets of outcomes for the experiment of rolling a die.

	Description of the event	Set of outcomes
1.	The die comes up odd.	$\{1, 3, 5\}$
2.	The die comes up less than 5.	$\{1, 2, 3, 4\}$
3.	The die comes up 2.	$\{2\}$
4.	The die comes up 7.	$\varnothing$
5.	The die comes up less than 8.	$\{1, 2, 3, 4, 5, 6\}$

We say that an event **occurs** if one of the outcomes in the event happens. Thus the event "the die comes up odd" occurs if the die comes up 1, 3, or 5, and it does not occur otherwise. Note that the fourth event above can never occur. We call the empty subset $\varnothing$ the **impossible** event. On the other hand, the fifth event above must occur, and so we call the sample space itself the **certain** event.

EXAMPLE 6.6
Revisited   For the experiment described in Example 6.6, list the subsets of outcomes corresponding to each of the following events.

(a) Two women are promoted.
(b) Two men are promoted.
(c) Ms. Choi is promoted.
(d) All the men are promoted.

Solution Refer to the sample space in Example 6.6.

(a) Of the five candidates for promotion, only two are women, namely Mrs. Bethea and Ms. Choi. Thus there is only one outcome in which two women are promoted, namely, *BC*. Hence the desired subset of outcomes is {*BC*}.

(b) Of the candidates for promotion, Mr. Adams, Mr. Dunn, and Mr. Eovaldi are the men. Therefore the outcomes in which two men are promoted are the possible pairs of these three candidates; so the desired subset of outcomes is {*AD, AE, DE*}.

(c) An outcome in which Ms. Choi is promoted is a pair containing the letter *C*. Thus the desired set of outcomes is {*AC, BC, CD, CE*}.

(d) Since three of the candidates for promotion are men and only two persons are to be promoted, this subset of outcomes is ∅. Thus promoting all the men is the impossible event. ■

Practice Problem 3 Recall the toll-free telephone number in Practice Problem 2. List the sets of outcomes described by the following events.

(a) Exactly ten calls are received.
(b) At least twelve calls are received.
(c) Fewer than five calls are received.
(d) No calls are received.
(e) More than sixteen calls are received.

THE PROBABILITY OF AN EVENT

We are interested in computing the probability of an event. When the sample space consists of equally likely outcomes, this computation involves counting. For example, in the experiment of rolling a symmetric die, we can easily calculate the probability of the event "the die comes up less than 5." For as we have already seen, this event corresponds to the set of outcomes

$$E = \{1, 2, 3, 4\}.$$

Since the die is symmetric, we assume that each of the six outcomes is equally likely. Hence in the long run we expect that one of the outcomes in E will occur on 4/6 of the rolls. We indicate that the probability of E is 4/6 by writing

$$P(E) = \frac{4}{6} = \frac{2}{3}.$$

This calculation is based on the following fundamental formula.

The Counting Rule ■ ■ ■ ■ ■

Let S be a finite sample space consisting of equally-likely outcomes. Then for any event $E \subseteq S$, the probability of E is given by

$$P(E) = \frac{|E|}{|S|} = \frac{\text{number of outcomes in } E}{\text{number of outcomes in } S}.$$

In order to use the counting rule it is crucial that the sample space consist of equally likely outcomes. The early probabilists usually took this fact for granted on the assumption that two outcomes should be considered equally likely if there is no reason why one will occur rather than the other. Today mathematicians are more cautious. For example, hospital records indicate that the probability of a random child being born male is about .514, not .5 as one might suspect. Even gambling devices may have small but significant biases.* We will use phrases such as "fair die" and "chosen at random" to signify that the possible outcomes are equally likely.

EXAMPLE 6.7 What is the probability that exactly two heads appear in three flips of a fair coin?

Solution Here the experiment consists of flipping a coin three times and observing the sequence of heads and tails that results. Recall that we saw in Example 6.5 that the sample space for this experiment is

$$\{HHH, HHT, HTH, HTT, THH, THT, TTH, TTT\}.$$

Since the coin is fair, this sample space consists of equally likely outcomes. Let E be the set of outcomes for the event "exactly two heads appear." Then

$$E = \{HHT, HTH, THH\}.$$

Hence by the counting rule

$$P(\text{exactly two heads appear}) = P(E) = \frac{|E|}{|S|} = \frac{3}{8}.$$

Therefore the probability of obtaining exactly two heads in three flips of a fair coin is 3/8. ∎

EXAMPLE 6.8 What is the probability of rolling a sum of 7 with a pair of fair dice?

Solution Recall from Example 6.3 that when rolling a pair of dice, the possible outcomes are as follows.

				Second Die			
		1	2	3	4	5	6
	1	(1, 1)	(1, 2)	(1, 3)	(1, 4)	(1, 5)	**(1, 6)**
	2	(2, 1)	(2, 2)	(2, 3)	(2, 4)	**(2, 5)**	(2, 6)
First	3	(3, 1)	(3, 2)	(3, 3)	**(3, 4)**	(3, 5)	(3, 6)
Die	4	(4, 1)	(4, 2)	**(4, 3)**	(4, 4)	(4, 5)	(4, 6)
	5	(5, 1)	**(5, 2)**	(5, 3)	(5, 4)	(5, 5)	(5, 6)
	6	**(6, 1)**	(6, 2)	(6, 3)	(6, 4)	(6, 5)	(6, 6)

---

*In 1947, two students from the California Institute of Technology, A. R. Hibbs and R. Walford, surveyed the roulette wheels in casinos. They found that one out of every four had a bias sufficient to eliminate the casino's expected profits from a player who bet correctly.

Let S denote the sample space of these 36 outcomes. Since the dice are fair, the outcomes in S are equally likely. Now the event "a sum of 7 is rolled" corresponds to the set of outcomes

$$E = \{(6, 1), (5, 2), (4, 3), (3, 4), (2, 5), (1, 6)\}.$$

Therefore by the counting rule

$$P(\text{sum is } 7) = P(E) = \frac{|E|}{|S|} = \frac{6}{36} = \frac{1}{6}.$$ ■

Practice Problem 4 A mouse is placed into the starting room of the maze pictured in Figure 6.5. If the mouse chooses a doorway at random, what is the probability that it will next enter room C?

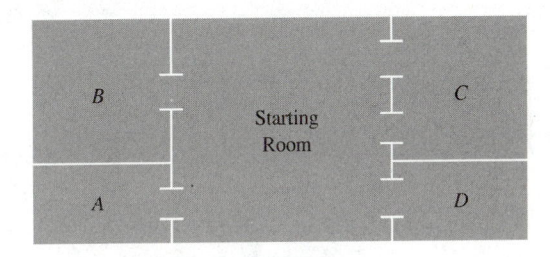

FIGURE 6.5

APPLICATIONS OF COUNTING TECHNIQUES

Let us consider the following problem. A firm has six accountants, three of whom are men (Mr. Allen, Mr. Babcock, and Mr. Casey) and three of whom are women (Ms. Dunbar, Mrs. Edmonds, and Ms. Fish). Two of these accountants are to be chosen at random to attend a financial seminar. What is the probability that the two persons selected are both men?

To solve this problem, we can proceed as before by enumerating the sample space S. In this case S consists of all possible pairs of the accountants, and so

$$S = \{AB, AC, AD, AE, AF, BC, BD, BE, BF, CD, CE, CF, DE, DF, EF\},$$

where we have identified each pair by their initials. Let M denote the event "two men are chosen." Then

$$M = \{AB, AC, BC\}.$$

Thus

$$P(M) = \frac{|M|}{|S|} = \frac{3}{15} = \frac{1}{5}.$$

There is another way to attack this problem, however, that does not require listing all the outcomes. Since S consists of all possible pairs chosen from among six accountants,

$$|S| = C(6, 2) = \frac{P(6, 2)}{2!} = \frac{6 \cdot 5}{2 \cdot 1} = 15.$$

Likewise M consists of all possible pairs chosen from among the three men (Allen, Babcock, and Casey); so

$$|M| = C(3, 2) = \frac{P(3, 2)}{2!} = \frac{3 \cdot 2}{2 \cdot 1} = 3.$$

Hence

$$P(M) = \frac{|M|}{|S|} = \frac{3}{15} = \frac{1}{5},$$

as above.

EXAMPLE 6.7 Revisited What is the probability that exactly two heads appear in three flips of a fair coin?

Solution Rather than enumerate the sample space S as in Example 6.7, we will solve the problem using techniques from Chapter 5. Since there are two possible results when a coin is flipped (heads or tails), the multiplication principle gives the number of outcomes when the coin is flipped three times to be

$$2 \cdot 2 \cdot 2 = 8.$$

Thus $|S| = 8$. To count the event $E =$ "exactly two heads appear," we consider an outcome in E as a listing of two heads and one tail in some order. But then the number of elements in E is the number of ways to choose places for two heads from among three possible positions. Hence

$$|E| = C(3, 2) = 3.$$

Therefore

$$P(E) = \frac{|E|}{|S|} = \frac{3}{8}. \blacksquare$$

Practice Problem 5 Use techniques from Chapter 5 to compute the probability of obtaining two different numbers on a single roll of a pair of fair dice.

When using the counting rule, the size of the sample space may make listing all of the outcomes impossible. In such cases we are forced to use the counting techniques discussed in Chapter 5.

6.1 COMPUTING PROBABILITIES BY COUNTING

EXAMPLE 6.9 A producer of computer software copies its programs onto diskettes to be sold to distributors. To check that the copying is being performed correctly, it randomly tests one fourth of its diskettes. What is the probability that there are 12 faulty diskettes among a lot of 40, and yet none of the faulty diskettes is chosen to be tested?

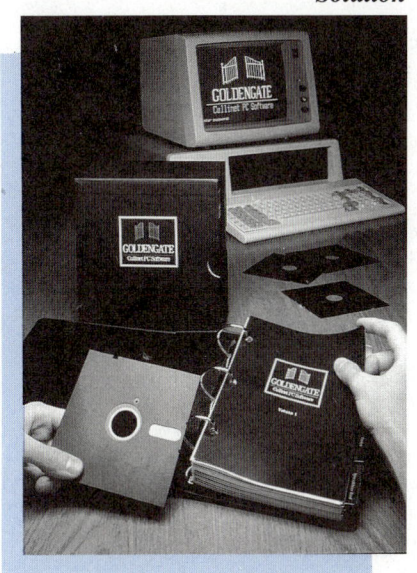

Solution Since the producer checks one fourth of all the diskettes, 10 diskettes from the lot of 40 will be tested. Thus for this experiment we choose as the sample space S the set consisting of all possible samples of 10 diskettes chosen from among the 40 produced. Then

$$|S| = C(40, 10).$$

Note that since the diskettes to be tested are chosen at random, each possible selection of 10 diskettes is equally likely. The event E of interest is the one in which no faulty diskettes are chosen for testing. Any outcome in E is obtained by selecting the 10 diskettes for testing from among the $40 - 12 = 28$ nondefective diskettes. Hence

$$|E| = C(28, 10).$$

Thus by the counting rule

$$P(\text{no faulty diskettes are chosen}) = P(E) = \frac{|E|}{|S|}$$

$$= \frac{C(28, 10)}{C(40, 10)}$$

$$= \frac{13{,}123{,}110}{847{,}660{,}528} \approx .015,$$

and so the probability that all the faulty diskettes go untested is about .015. ∎

EXAMPLE 6.10 The semifinalists in a state lottery consist of 3 men and 6 women. If 4 of these persons are to be randomly selected to win \$1,000,000, what is the probability that 2 men and 2 women are selected?

Solution The sample space S consists of all possible selections of 4 persons from among the 9 semifinalists; so

$$|S| = C(9, 4) = 126.$$

The event E of interest consists of those selections containing 2 men and 2 women. Such a selection can be formed by choosing 2 of the 3 men and 2 of the 6 women. Thus the multiplication principle shows that

$$|E| = C(3, 2) \cdot C(6, 2) = 3 \cdot 15 = 45.$$

Hence by the counting rule

$$P(E) = \frac{|E|}{|S|} = \frac{45}{126} = \frac{5}{14}. \quad \blacksquare$$

Notice that in both Examples 6.9 and 6.10 we are sampling from a population of two types of objects. In Example 6.9 the diskettes are either faulty or nondefective, and we are interested in choosing 0 faulty diskettes and 10 nondefective diskettes. Similarly in Example 6.10 the people are either men or women, and we must select 2 men and 2 women. This type of sampling problem is quite common.

Practice Problem 6 From a committee consisting of 12 representatives of management and 10 representatives of labor, a subcommittee of 5 is to be selected. If members of the subcommittee are chosen at random, what is the probability that 3 representatives of management and 2 representatives of labor are chosen?

EXAMPLE 6.11 Forty persons are attending a seminar discussing financial investments. From among those attending, three *different* persons will be selected at random to receive door prizes.

(a) What is the probability that a particular person at the seminar wins the first door prize?
(b) What is the probability that a particular person at the seminar wins the third door prize?

Solution (a) The probability that someone wins the first door prize can be computed directly from the counting rule:

$$P(\text{someone wins first prize}) = \frac{\text{number of winners}}{\text{number in attendance}} = \frac{1}{40}.$$

(b) In order to compute the probability that someone wins the third door prize, let us regard the winners of the prizes as a permutation of three of the persons attending the seminar. Thus, for instance, the permutation

(Ms. McCauley, Mr. Jeffreys, Mrs. Earls)

will signify that the first prize was won by Ms. McCauley, the second by Mr. Jeffreys, and the third by Mrs. Earls. Let the sample space S consist of all such permutations. Because the prize winners are selected at random, each permutation in S is equally likely. Note that

$$|S| = P(40, 3) = 40 \cdot 39 \cdot 38.$$

In order for Mr. Smith to win the third door prize, his name must appear in the third position in the list of winners. For example,

(Mrs. Clark, Ms. Kennedy, Mr. Smith)

is one configuration in which Mr. Smith receives the third prize. Let E denote the set of all the permutations in S in which Mr. Smith's name occurs in the third position. Since nobody can win more than one prize, each such permutation can be constructed as follows.

1. Choose the name of a first prize winner (other than Mr. Smith).
2. Select the name of a second prize winner (other than Mr. Smith and the first prize winner).
3. Choose Mr. Smith's name.

Hence the multiplication principle shows that

$$|E| = 39 \cdot 38 \cdot 1,$$

and so by the counting rule

$$P(\text{Mr. Smith wins third prize}) = P(E) = \frac{|E|}{|S|}$$

$$= \frac{39 \cdot 38 \cdot 1}{40 \cdot 39 \cdot 38}$$

$$= \frac{1}{40}.$$

Thus we see that the probability of winning the third prize is the same as the probability of winning the first! With the benefit of hindsight, we see that this result is intuitively obvious: We have just as much chance of winning one prize as another. ∎

Practice Problem 7 Six applicants for a particular job, four men and two women, are to be interviewed in a random order. What is the probability that the four men are interviewed before either woman?

EXERCISES 6.1

Describe a sample space for the experiments in Exercises 1–8.

1. One fair coin is flipped.
2. A two-headed coin is flipped.
3. Randomly selected persons are asked the day of the week on which they were born.
4. One marble is selected from an urn containing red, yellow, blue, green, and purple marbles.
5. A randomly selected voter is asked whether she voted for the Republican or Democratic candidate for governor.
6. A die is rolled, and then a coin is flipped.
7. The number of persons riding the subway in New York is counted on a particular day.
8. A coin is flipped until tails appears.
9. Fifteen automobile engines are to be tested and the number of defective units recorded. Describe a sample space for this experiment, and represent each of the following events as a subset of the sample space.
    (a) None of the engines is defective.
    (b) At most two of the engines are defective.
    (c) More than half of the engines are defective.

10. Represent each of the following events as a subset of the sample space in Example 6.6.
    (a) Mr. Adams is promoted.
    (b) Mrs. Bethea and Mr. Dunn are promoted.
    (c) Mr. Eovaldi is not promoted.

11. A coin is to be flipped twice. Describe a sample space for this experiment, and represent each of the following events as a subset of the sample space.
    (a) The coin lands heads on the second flip.
    (b) The coin comes up heads at least once.
    (c) The coin comes up heads twice as many times as it comes up tails.

12. During practice, a basketball player is going to shoot three free throws. Describe a sample space for this experiment, and represent each of the following events as a subset of the sample space.
    (a) The player misses the third shot.
    (b) The player makes exactly two shots.
    (c) The player makes some shot and misses the next.

13. If a fair die is rolled, what is the probability that it will come up
    (a) odd?   (b) less than 3?
    (c) both odd and less than 3?   (d) both odd and greater than 5?

14. American roulette wheels have 38 equal compartments numbered 00, 0, 1, . . ., 36. A ball is spun and comes to rest in one of these compartments. What is the probability that the number obtained is
    (a) the number 29?   (b) in the "second dozen" (13–24)?
    (c) positive and even?   (d) an odd number in the second dozen?

15. Assuming that all birthdays are equally likely, what is the probability that a randomly chosen person will have a birthday
    (a) in March?
    (b) between January 1 and February 15, inclusive?

16. A two-digit number from 00 to 99 is to be randomly generated by a computer. What is the probability that it will be
    (a) less than 47?   (b) positive and divisible by 3?

17. A large assortment of glass beads of uniform shape and size are mixed together. In this mixture, 34% of the beads are clear and blue, 12% are opaque and blue, 10% are clear and green, 18% are opaque and green, and 26% are clear and beige. If a bead is chosen from this mixture in the dark, what is the probability that it will be
    (a) opaque and green?   (b) blue?
    (c) clear?   (d) either blue or beige?

18. There were 250,000 tickets sold in a particular lottery. One ticket will win the first prize of $100,000; ten tickets will win second prizes of $10,000; fifty tickets will win third prizes of $2000; and one hundred tickets will win fourth prizes of $1000. If the winning tickets are selected at random, what is the probability that a random ticket will win
    (a) $100,000?   (b) $1000?
    (c) $2000 or more?   (d) nothing?

19. The registered voters in Baker County are 65% white, 25% black, and 10% Hispanic. Of the white voters, 60% are male, compared to 48% of the black voters and 50% of the Hispanic voters. Registered voters are selected at random for jury duty. What is the probability that someone selected for jury duty will be

   (a) black?   (b) a nonwhite?   (c) an Hispanic male?   (d) a female?

20. From a standard deck of 52 playing cards, one card is selected at random. Determine the probability of obtaining

   (a) the ace of spades   (b) any red card   (c) any diamond   (d) any king.

In Exercises 21–30 use Example 6.3 to determine the probability of the indicated event when two fair dice are rolled.

21. The sum is 5.
22. Both dice show the same number.
23. The white die comes up 6.
24. The sum exceeds 15.
25. Each die is greater than 2.
26. The sum is at least 8.
27. The white die is 3 and the sum is 10.
28. The numbers on both dice are even.
29. The number on the white die exceeds the number on the black die.
30. The number on one die is twice the number on the other die.
31. To win the Illinois State Lottery's daily game, a player must correctly pick a three-digit number, i.e., 000, 001, . . ., 999. What is the probability of winning the daily game?
32. A student guesses the answers to three true-or-false questions. What is the probability that
   (a) all three answers are correct?
   (b) exactly one answer is correct?
33. If a fair coin is flipped four times, find the probability that
   (a) at least three heads appear
   (b) exactly two heads appear
   (c) no two heads occur on successive flips
   (d) tails does not appear on either of the last two flips.
34. The integers 1–7 are written once each on separate slips of paper, placed in a box, and thoroughly mixed. If two slips are chosen simultaneously, find the probability that
   (a) the number 4 is chosen   (b) the number 3 is not chosen
   (c) both 6 and 7 are chosen   (d) some even number is chosen
   (e) two odd numbers are chosen   (f) the sum of the numbers chosen is 15.

Exercises 35–44 require the use of counting techniques from Chapter 5.

35. To play the Illinois State Lottery, a player must pick six numbers which match six numbers chosen at random from the integers 1–44. What is the probability of winning this lottery on a single play?

36. In an 8-horse race, a bettor bet the trifecta, which requires that the first three horses be identified in order of their finish. What is the probability of winning the trifecta by randomly selecting three numbers?

37. A supervisor randomly selected the personnel files of four employees. What is the probability that the files were selected in alphabetical order?

38. In a taste test, three crackers were spread with butter and three with margarine. Someone was asked to identify the three crackers spread with butter. What is the probability of doing so by guessing?

39. If three men and three women are seated at random in a row of six seats, what is the probability that no two people of the same gender are in adjacent seats?

40. To win the second prize in the Illinois State Lottery, a player must pick six numbers and correctly match exactly five of six numbers chosen at random from the integers 1–44. What is the probability of winning the second prize in this lottery on a single play?

41. A committee of four is to be chosen at random from a group of six men and six women. What is the probability that the committee will contain
    (a) no men?   (b) two men and two women?

42. An urn contains 4 red marbles, 6 blue marbles, and 5 green marbles. If two of the marbles are selected at random from the urn, what is the probability that
    (a) both are blue?   (b) neither is green?
    (c) one is red and one is blue?   (d) the two are different colors?

43. In a lot of twenty microcomputer diskettes, exactly four are defective. If the diskettes are packaged in two boxes of ten, what is the probability that
    (a) all of the defective diskettes are packed in a particular box?
    (b) 3 defective diskettes are packed in the same box?
    (c) 2 defective diskettes are packed in each box?

44. A player is dealt five cards from a standard deck of 52 playing cards. What is the probability that the cards form
    (a) a flush (all cards of the same suit)?
    (b) a full house (3 cards of one denomination and 2 of another)?
    (c) a pair (2 cards of one denomination and 1 each of three other denominations)?
    (d) a straight (5 cards of consecutive denominations, where an ace is the highest denomination)?

Answers to Practice Problems

1. (a) The tree diagram is shown at the right.
   (b) $\frac{1}{16}$

2. $\{0, 1, 2, \ldots, 15\}$

3. (a) $\{10\}$ (b) $\{12, 13, 14, 15\}$
   (c) $\{0, 1, 2, 3, 4\}$ (d) $\{0\}$ (e) $\varnothing$

4. $\frac{2}{5}$

5. $\frac{30}{36} = \frac{5}{6}$

6. $\frac{9900}{26,334} = \frac{50}{133}$

7. $\frac{48}{720} = \frac{1}{15}$

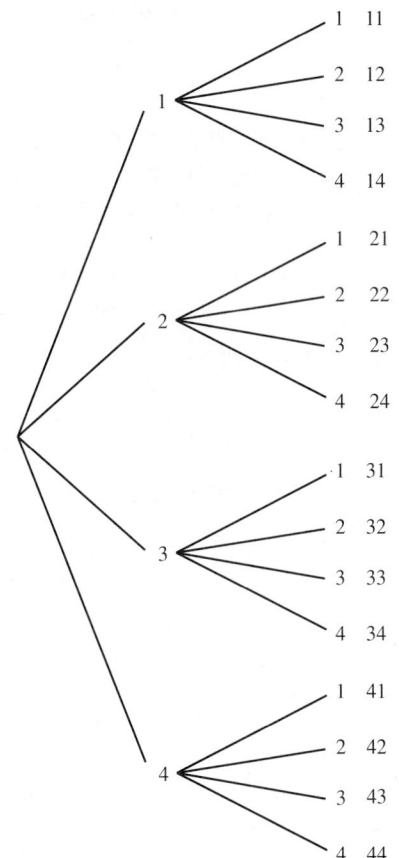

6.2 Assigning Probabilities

In Section 6.1 we learned how to calculate the probability of an event using a sample space consisting of equally likely outcomes. In this section we will extend the computation of probabilities to any finite sample space, whether consisting of equally likely outcomes or not.

PROBABILITY DISTRIBUTIONS

We saw in Section 6.1 that when a fair die is rolled, the theoretical probability of each outcome is 1/6. It is useful to record this information in the table below.

Outcome	Probability
1	1/6
2	1/6
3	1/6
4	1/6
5	1/6
6	1/6

This same type of table can be used even if the outcomes of an experiment are not equally likely. For example, in Example 6.3 we considered the experiment of rolling a pair of fair dice. Proceeding as in Example 6.8, we can compute the probability that the sum of the two dice is any of the possible values from 2 through 12. The results of these computations are shown below.

Outcome	Probability
2	1/36
3	2/36
4	3/36
5	4/36
6	5/36
7	6/36
8	5/36
9	4/36
10	3/36
11	2/36
12	1/36

Notice that each of the probabilities is a number between 0 and 1 and that the sum of all the probabilities is 1.

Generalizing from this example, we define a **probability distribution** for an experiment to be a sample space $\{s_1, s_2, \ldots, s_n\}$ for the experiment together with a set of numbers $p_1, p_2, \ldots, p_n$ that satisfy the following two conditions:

1. Each number is between 0 and 1; that is, $0 \leq p_i \leq 1$ for each i.
2. The sum of the numbers is 1; that is, $p_1 + p_2 + \ldots + p_n = 1$.

The number p_i is the probability of outcome s_i.

The first condition in the definition of a probability distribution requires that the probability of each outcome be a number between 0 and 1 (inclusive). The second condition says that the sum of the probabilities of the outcomes must be 1; this reflects that when the experiment is performed, one of the outcomes $s_1, s_2, \ldots, s_n$ must occur. Notice that nothing is said about *how* the numbers p_i are to be obtained; from the mathematical point of view they are regarded as being given. In practice, it may be difficult to determine an appropriate probability distribution for a particular experiment.

EXAMPLE 6.12 A die was rolled 500 times, and the outcome of each trial was recorded with the results shown in the frequency table below. Use this data to obtain a probability distribution for the experiment of rolling this die.

Outcome	Frequency
1	79
2	83
3	85
4	74
5	91
6	88
	500 Total

Solution If we divide each frequency by the number of rolls (500), we obtain the relative frequency table below.

Outcome	Relative Frequency
1	.158
2	.166
3	.170
4	.148
5	.182
6	.176
	1.000 Total

We call these relative frequencies the *empirical* probabilities of the six outcomes. For instance, a 4 appeared on 74 of the 500 rolls, and so we take the probability of rolling a 4 to be

$$\frac{74}{500} = .148.$$

Since each of the relative frequencies is a number between 0 and 1, and since the sum of all the relative frequencies is 1, the preceding table is a probability distribution for the experiment of rolling the die. ∎

Two questions arise from Example 6.12. Recall that according to the relative frequency interpretation, the probability of an event represents the proportion of the time that the event tends to occur over the long run. How do we know in Example 6.12 that 500 rolls are enough to define "the long run"? We don't. In general, the frequency distribution constructed by this process is only as good as the empirical data on which it is based. Obviously, we would have more confidence in the distribution in Example 6.12 if the die had been rolled 5000 or 10,000 times rather than 500. But no matter how often it is rolled, we can only hope that our data accurately portray the general behavior of the die.

Second, is the probability distribution constructed in Example 6.12 more accurate than the theoretical distribution in which the probability of each outcome is 1/6? Not necessarily. Since the distribution we constructed is only an approximation to the true behavior of the die, we cannot be certain that the theoretical distribution is not a better description of this behavior. In fact, the probabilities in the distribution we constructed are close enough to 1/6 that it is reasonable to use 1/6 as the probability of each outcome. Which probability distribution to use in a particular application is a decision that must be made based on experience, intuition, or personal taste. However, in this book we will use the theoretical distribution whenever appropriate.

Once a probability distribution has been established for an experiment, we can find the probability of an event E by simply adding the probabilities of the outcomes for which E occurs. For example, if E is the event "the die comes up even," then $E = \{2, 4, 6\}$. Thus for the probability distribution in Example 6.12 we find that

$$P(E) = P(2) + P(4) + P(6)$$
$$= .166 + .148 + .176 = .490.$$

Likewise if F is the event "the die comes up less than 5," then

$$P(F) = P(1) + P(2) + P(3) + P(4)$$
$$= .158 + .166 + .170 + .148$$
$$= .642.$$

More generally, suppose that a probability distribution has been established for an experiment with sample space S. If $E = \emptyset$, then we define $P(E) = 0$. In all other cases we define $P(E)$, the probability of event E, to be the sum of the probabilities of the outcomes for which E occurs. Whereas in Section 6.1 we were able to compute probabilities only if we had a sample space consisting of equally likely outcomes, this definition enables us to compute the probabilities of events in any sample space for which we have a probability distribution.

EXAMPLE 6.13 A mail-order electronics supplier has recorded the number of toll-free telephone calls received between noon and 1 P.M. on 250 successive weekdays. These data are recorded below.

Number of calls	Frequency
0	28
1	42
2	59
3	47
4	28
5	19
6	12
7	7
8	5
9	2
10	1
	250 Total

(a) Use these data to create a probability distribution.
(b) Compute the probability that the supplier received fewer than five calls on a weekday between noon and 1 P.M.

Solution (a) First we divide each frequency by the number of observations (250) to obtain the following probability distribution.

Number of calls	Probability
0	.112
1	.168
2	.236
3	.188
4	.112
5	.076
6	.048
7	.028
8	.020
9	.008
10	.004
	1.000 Total

(b) Using this distribution, we see that

$$P(\text{fewer than 5 calls}) = P(0) + P(1) + P(2) + P(3) + P(4)$$
$$= .112 + .168 + .236 + .188 + .112$$
$$= .816.$$

Therefore 81.6% of the time the supplier received fewer than five calls on weekdays between noon and 1 P.M.

Practice Problem 8 A bank tabulated its cashiers' transactions for the week as follows.

Type of Transaction	Frequency
Cashed a check	1592
Made a deposit	1144
Made a withdrawal	936
Paid a bill	256
Requested change	44
Bought a money order	28
	4000 Total

(a) Use these data to create a probability distribution.

(b) Compute the probability that a transaction was either a deposit or withdrawal.

ODDS

Probabilities are often expressed in the language of "odds." For instance, we might say that the odds are 3 to 2 in favor of an event E. In terms of relative frequencies, this means that over the long run E will occur 3 times for every 2 times that it does not occur. In other words, in the long run E will occur 3 times out of 5. Therefore

$$P(E) = \frac{3}{5}.$$

More generally, to say that the **odds in favor of E are m to n** means that

$$P(E) = \frac{m}{m+n}.$$

Conversely, if we write $P(E)$ in this form, then the odds in favor of E are m to n. When the odds in favor of E are m to n, we also say that the **odds against E are n to m**.

Converting odds into probabilities requires nothing more than simple arithmetic. The example below demonstrates the method.

EXAMPLE 6.14 Find the probability of an event E in each of the following cases.

(a) The odds in favor of E are 4 to 1.
(b) The odds in favor of E are 5 to 3.
(c) The odds against E are 13 to 7.

Solution (a) Since the odds in favor of E are 4 to 1,

$$P(E) = \frac{4}{4+1} = \frac{4}{5} = .8.$$

(b) In this case the odds in favor of E are 5 to 3; so

$$P(E) = \frac{5}{5+3} = \frac{5}{8} = .625.$$

(c) Here the odds are 13 to 7 *against* E. This means that the odds *in favor of* E are 7 to 13. Consequently

$$P(E) = \frac{7}{7+13} = \frac{7}{20} = .35. \blacksquare$$

Practice Problem 9   Find the probability of an event E if

(a) the odds in favor of E are 7 to 4
(b) the odds against E are 9 to 5.

It is also easy to convert probabilities into odds; for if

$$P(E) = \frac{a}{b},$$

then

$$P(E) = \frac{a}{a + (b - a)}.$$

Hence if $P(E) = a/b$, then the odds in favor of E are a to $b - a$.

EXAMPLE 6.15   (a) If $P(E) = 11/16$, find the odds in favor of E.
(b) If $P(E) = 2/7$, find the odds against E.
(c) If $P(E) = .76$, find the odds against E.

Solution   (a) Since

$$P(E) = \frac{11}{16},$$

the odds in favor of E are 11 to $16 - 11$, that is, 11 to 5.
(b) Because

$$P(E) = \frac{2}{7},$$

the odds in favor of E are 2 to $7 - 2$, that is, 2 to 5. Thus the odds against E are 5 to 2.
(c) Here we must first write the given probability as a fraction, which we will reduce to lowest terms. In this way we obtain

$$P(E) = .76 = \frac{76}{100} = \frac{19}{25},$$

and so the odds in favor of E are 19 to $25 - 19$, that is, 19 to 6. Therefore the odds against E are 6 to 19. $\blacksquare$

Practice Problem 10 (a) If $P(E) = .75$, find the odds in favor of E.
(b) If $P(E) = 2/9$, find the odds against E.

EXERCISES 6.2

In Exercises 1–8 determine if the given values form a probability distribution for an experiment with sample space $S = \{a, b, c, d\}$.

1.
Outcome	Probability
a	.1
b	.2
c	.3
d	.4

2.
Outcome	Probability
a	1/5
b	1/4
c	1/3
d	1/2

3.
Outcome	Probability
a	.27
b	.53
c	−.22
d	.42

4.
Outcome	Probability
a	3/8
b	1/4
c	5/16
d	1/16

5.
Outcome	Probability
a	.06
b	.14
c	.39
d	.41

6.
Outcome	Probability
a	.6
b	−.1
c	.2
d	.3

7.
Outcome	Probability
a	.31
b	.17
c	.22
d	.25

8.
Outcome	Probability
a	1/4
b	1/4
c	1/4
d	1/4

9. Determine the probability of the given event using the probability distribution below.

Outcome	Probability
a	1/2
b	1/4
c	1/8
d	1/16
e	1/16

(a) $\{b, c\}$
(b) $\{a, d, e\}$
(c) $\{a, b, c, d, e\}$
(d) $\{b, c, d, e\}$

10. Determine the probability of the given event using the probability distribution below.

Outcome	Probability
a	1/5
b	1/5
c	1/5
d	1/5
e	1/5

(a) $\{b, c, e\}$
(b) $\{a, d\}$
(c) $\{b\}$
(d) $\{a, c, d, e\}$

11. Determine the probability of the given event using the probability distribution below.

Outcome	Probability
a	.15
b	.20
c	.25
d	.30
e	.10

(a) ∅
(b) {a, e}
(c) {c, d, e}
(d) {a, b, c, d}

12. Determine the probability of the given event using the probability distribution below.

Outcome	Probability
a	.4
b	.1
c	.2
d	.2
e	.1

(a) ∅
(b) {a, d}
(c) {b, c, e}
(d) {a, b, c, d, e}

13. Of the last 568 bids submitted by a construction company, 93 were accepted. What is the empirical probability that the next bid submitted by this company will be accepted?

14. A basketball player has made 27 free throws out of 35 attempts. What is the empirical probability that she makes her next free throw?

15. A Labor Department statistician interviewed 120 unemployed persons in an inner-city area to determine how many years of high school each had completed. Their responses are shown below.

Number of years	Frequency
0	6
1	24
2	39
3	30
4	21
	120 Total

(a) Use this frequency table to construct a probability distribution for the number of years of high school completed.
(b) What is the probability that someone completed four years of high school?
(c) What is the probability that someone completed no more than one year of high school?
(d) What is the probability that someone completed at least three years of high school?

16. An insurance company checked the records of 80 random motorists to determine how many moving violations each had been cited for during the past five years. The results are shown below.

Violations	Frequency
0	34
1	28
2	12
3	4
4	2
	80 Total

(a) Use this frequency table to construct a probability distribution for the number of moving violations.
(b) What is the probability that someone was guilty of three violations?
(c) What is the probability that someone was guilty of at least two violations?
(d) What is the probability that someone was guilty of no more than one violation?

17. The speeds of 100 cars selected at random from I-74 were distributed as shown below.

Speed s (mph)	Cars
$45 \leq s < 55$	15
$55 \leq s < 65$	31
$65 \leq s < 75$	35
$75 \leq s < 85$	17
$85 \leq s < 95$	2
	100 Total

(a) Use this frequency table to construct a probability distribution for the speed of the cars.
(b) What is the probability that a car was traveling from 55 to 65 mph?
(c) What is the probability that a car was below the 65 mph speed limit?
(d) What is the probability that a car was traveling at least 75 mph?

18. A test of the tensile strength (breaking point under stress) of a sample of 80 cables produced the frequency distribution shown below.

Tensile strength t (lbs)	Number of cables
$1400 \leq t < 1500$	11
$1500 \leq t < 1600$	29
$1600 \leq t < 1700$	26
$1700 \leq t < 1800$	9
$1800 \leq t < 1900$	5
	80 Total

(a) Use this frequency table to construct a probability distribution for the tensile strength of the cables.
(b) What is the probability that a cable had a tensile strength between 1400 and 1500 pounds?
(c) What is the probability that a cable had a tensile strength of at least 1700 pounds?
(d) What is the probability that a cable had a tensile strength of at most 1600 pounds?

19. A baker finds that the number of chocolate cakes requested on a given day has the following distribution.

Requests	Frequency
0	33
1	60
2	96
3	57
4	36
5	15
6	3
	300 Total

(a) Use this frequency table to construct a probability distribution for the number of cakes requested.
(b) What is the probability of selling exactly three cakes?
(c) What is the probability of selling at most 2 cakes?
(d) What is the probability of selling no more than 1 cake?
(e) What is the maximum number of cakes the baker can prepare in order to be at least 50% certain of selling them all?

20. A stockbroker observes that the delivery time for her local mailings has the distribution below.

Delivery time (working days)	Number of items
1	109
2	159
3	136
4	65
5	31
	500 Total

(a) Use this frequency table to construct a probability distribution for the delivery time of local mailings.
(b) What is the probability that an item mailed on Monday will arrive on Wednesday?
(c) What is the probability that an item mailed on Monday will arrive on Wednesday, Thursday, or Friday?
(d) What is the probability that an item mailed on Monday will arrive by Thursday at the latest?
(e) By what day must an item be mailed in order to be at least 80% certain that it will arrive before the weekend?

In Exercises 21–26 complete the missing entries in the table.

	Probability	Odds in favor	Odds against
21.	_____	3 to 1	_____
22.	.85	_____	_____
23.	_____	_____	18 to 7
24.	_____	_____	12 to 13
25.	4/9	_____	_____
26.	_____	9 to 11	_____

In Exercises 27–32 determine the indicated odds when a single die is rolled.

27. the odds in favor of rolling a 5 or 6
28. the odds against rolling a 5 or 6
29. the odds against rolling a number greater than 1
30. the odds in favor of rolling a number less than or equal to 4

31. the odds in favor of rolling a multiple of 3
32. the odds against rolling an even number
33. A lawyer has informed her client that, in her judgment, a stiff sentence is twice as likely as a light sentence and that some sentence (either stiff or light) is twice as likely as receiving probation. What is the probability of receiving probation?
34. A die is weighted in such a way that 1, 2, and 3 are equally likely to appear; 4, 5, and 6 are equally likely to appear; and 1 is 50% more likely to appear than 6. What is the probability of rolling an even number with this die?
35. An observer watching cars at a certain intersection has noticed that they go straight 50% more often than they turn left and 40% more often than they turn right. What is the probability that the next car approaching this intersection will turn left?
36. Two East German swimmers and three Americans are competing in the women's 200 meter freestyle event. One of the East Germans and two of the Americans are given exactly the same chance of winning. The other East German is twice as likely to win as the third American but only half as likely as her East German teammate. What is the probability that an East German will win the event?

Answers to Practice Problems

8. (a) The probability distribution is given below.

Type of Transaction	Probability
Cashed a check	.398
Made a deposit	.286
Made a withdrawal	.234
Paid a bill	.064
Requested change	.011
Bought a money order	.007
	1.000 Total

(b) .520

9. (a) $\dfrac{7}{11}$ (b) $\dfrac{5}{14}$

10. (a) 3 to 1   (b) 7 to 2

6.3 Basic Laws of Probability

When events are logically related to one another, it is reasonable to expect their probabilities to be mathematically related. In this section we will develop two fundamental implications of this type.

OPERATIONS ON EVENTS

Since events are sets, we can combine them using the set operations of union, intersection, and complement that were introduced in Section 5.1. From the Venn diagram in Figure 6.6(a) we see that $A \cup B$ contains those outcomes in

(a) $A \cup B$ is shaded

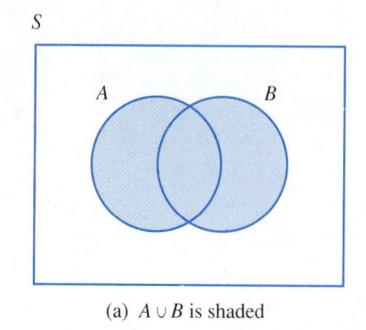

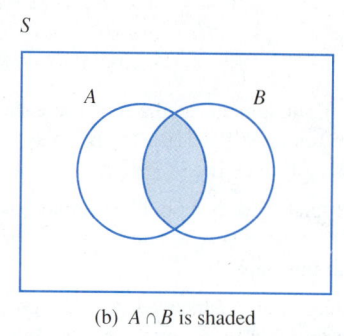

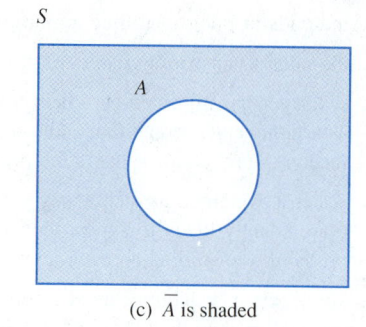

(b) $A \cap B$ is shaded

(c) $\overline{A}$ is shaded

FIGURE 6.6

either A or B or both. Likewise from Figures 6.6(b) and (c) we see that $A \cap B$ contains those outcomes in both A and B, and $\overline{A}$ contains those outcomes not in A. Hence

$A \cup B$ represents the event "*either A* occurs *or B* occurs (or both)," that is, the event "*at least one* of A or B occurs"

$A \cap B$ represents the event "*both A and B* occur"

$\overline{A}$ represents the event "*A* does *not* occur."

Since these set operations will enter into many of our later formulas, the meaning of each of these symbols should be thoroughly understood.

EXAMPLE 6.16 A fair coin is to be flipped three times. Consider the following events:

$$A = \text{"the first flip is heads" and}$$
$$B = \text{"the third flip is tails."}$$

Express each of the events $A \cup B$, $A \cap B$, $\overline{A}$, and $\overline{B}$ in words, and calculate its probability.

Solution The meaning of the events $A \cup B$, $A \cap B$, $\overline{A}$, and $\overline{B}$ is as follows.

$A \cup B$ is "the first flip is heads *or* the third flip is tails."

$A \cap B$ is "the first flip is heads *and* the third flip is tails."

$\overline{A}$ is "the first flip is *not* heads," that is, "the first flip is tails."

$\overline{B}$ is "the third flip is *not* tails," that is, "the third flip is heads."

Recall from Example 6.5 that for this experiment a sample space of equally likely outcomes is

$$\{HHH, HHT, HTH, HTT, THH, THT, TTH, TTT\}.$$

Hence

$$A \cup B = \{HHH, HHT, HTH, HTT, THT, TTT\},$$

and so the counting rule gives

$$P(A \cup B) = \frac{6}{8} = \frac{3}{4}.$$

Similarly we find that

$$A \cap B = \{HHT, HTT\}, \quad \overline{A} = \{THH, THT, TTH, TTT\},$$

and

$$\overline{B} = \{HHH, HTH, THH, TTH\}.$$

Therefore

$$P(A \cap B) = \frac{2}{8} = \frac{1}{4}, \quad P(\overline{A}) = \frac{4}{8} = \frac{1}{2}, \quad \text{and} \quad P(\overline{B}) = \frac{4}{8} = \frac{1}{2}. \blacksquare$$

EXAMPLE 6.17 In the experiment of rolling a pair of fair dice, consider the following events:

$A =$ "the sum rolled is 8" and

$B =$ "each die comes up greater than 2."

Calculate the probability of each of the events $A \cup B$, $A \cap B$, $\overline{A}$, and $\overline{B}$.

Solution The meaning of the events $A \cup B$, $A \cap B$, $\overline{A}$, and $\overline{B}$ is given below.

$A \cup B$ is "the sum rolled is 8 or each die comes up greater than 2."
$A \cap B$ is "the sum rolled is 8 and each die comes up greater than 2."
$\overline{A}$ is "the sum rolled is not 8."
$\overline{B}$ is "1 or 2 comes up on at least one die."

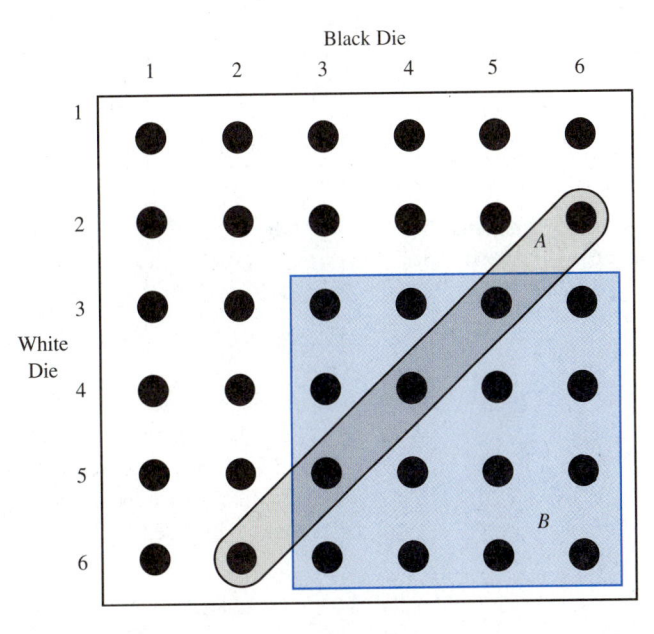

FIGURE 6.7

Recall from Example 6.3 that the possible outcomes are as shown in Figure 6.7. By counting the appropriate dots in this figure, we see that

$$P(A \cup B) = \frac{18}{36} = \frac{1}{2}, \quad P(A \cap B) = \frac{3}{36} = \frac{1}{12},$$
$$P(\overline{A}) = \frac{31}{36}, \quad P(\overline{B}) = \frac{20}{36} = \frac{5}{9}. \blacksquare$$

Practice Problem 11 In the experiment of rolling a pair of fair dice, consider the following events:

$A = $ "the sum rolled is 6" and

$B = $ "at least one die comes up greater than 4."

Calculate the probability of each of the events A, B, $A \cup B$, $A \cap B$, $\overline{A}$, and $\overline{B}$.

THE COMPLEMENT RULE

In Example 6.17 notice that $P(A) = 5/36$ and $P(\overline{A}) = 31/36$. It is not difficult to see that the sum of the probabilities of an event and its complement must always be 1. For let A be any event in an experiment having sample space S, and let

$$A = \{r_1, r_2, \ldots, r_m\} \quad \text{and} \quad \overline{A} = \{s_1, s_2, \ldots, s_n\}$$

be the sets of distinct outcomes corresponding to A and $\overline{A}$. By definition of the complement of a set, we see that $r_1, r_2, \ldots, r_m, s_1, s_2, \ldots, s_n$ must be distinct and comprise all of the sample space S. But in any probability distribution the sum of the probabilities of all the outcomes must be 1, and so we have

$$P(A) + P(\overline{A}) = P(r_1) + \cdots + P(r_m) + P(s_1) + \cdots + P(s_n) = 1.$$

We have obtained the following result.

Theorem 6.1 (the complement rule)

For any event A, we have

$$P(A) + P(\overline{A}) = 1.$$

EXAMPLE 6.18 The American roulette wheel has 38 compartments numbered 0, 00, 1, 2, ..., 36. The numbers 0 and 00 are green, and the remaining 36 numbers are equally divided between red and black. Assuming that each of the 38 numbers is equally likely to appear, what is the probability that someone betting on a red number will lose?

Solution Since the player wins whenever a red number appears, we have

$$P(\text{player wins}) = \frac{18}{38} = \frac{9}{19}.$$

Thus by the complement rule

$$P(\text{player wins}) + P(\text{player loses}) = 1.$$

Hence

$$P(\text{player loses}) = 1 - P(\text{player wins})$$
$$= 1 - \frac{9}{19}$$
$$= \frac{10}{19}. \blacksquare$$

When computing the probability of an event containing a large number of outcomes, the complement rule can often be used to simplify the calculation.

EXAMPLE 6.13 Revisited In Example 6.13 we constructed a probability distribution for the number of toll-free telephone calls received on a weekday between noon and 1 P.M. by a mail-order electronics supplier. What is the probability that this supplier receives at least one call during this time?

Solution The probability distribution from Example 6.13 is reproduced below.

Number of calls	Probability
0	.112
1	.168
2	.236
3	.188
4	.112
5	.076
6	.048
7	.028
8	.020
9	.008
10	.004
	1.000 Total

The event of interest is A = "at least one call is received." Now the set of outcomes corresponding to A is

$$\{1, 2, \ldots, 10\}.$$

Proceeding as in Section 6.2, we see that

$$P(A) = P(1) + P(2) + \cdots + P(10)$$
$$= .168 + .236 + \cdots + .004.$$

There is an easier method of obtaining this answer that uses the complement rule. Note that the complement of $A = $ "at least one call is received" is $\overline{A} = $ "no calls are received." Hence

$$P(A) = 1 - P(\overline{A}) = 1 - .112 = .888.$$

Practice Problem 12 For the probability distribution above, what is the probability that (a) at most 8 calls are received? (b) no more than 9 calls are received?

If the probability of an event seems hard to calculate directly, a good strategy is to consider use of the complement rule.

EXAMPLE 6.19 A subcommittee of four is to be randomly chosen from among the 27 members of the county board. If the board contains 17 men and 10 women, what is the probability that the subcommittee will contain at least one woman?

Solution Here the event of interest is $A = $ "the subcommittee contains at least one woman." Since the number of subcommittees that contain at least one woman is difficult to compute directly, we consider the complementary event $\overline{A} = $ "the subcommittee contains only men." Now

$$P(\overline{A}) = \frac{C(17, 4)}{C(27, 4)} = \frac{2380}{17{,}550},$$

and so

$$P(A) = 1 - \frac{2380}{17{,}500} = \frac{15{,}170}{17{,}550}.$$

Practice Problem 13 A fair die is rolled five times. What is the probability that an even number appears on at least one roll?

THE UNION RULE

The complement rule is analogous to Theorem 5.1(a) in Section 5.2. There is also an analog to Theorem 5.1(b), the principle of inclusion-exclusion.

Theorem 6.2 (the union rule)

For any events A and B,

$$P(A \cup B) = P(A) + P(B) - P(A \cap B).$$

Recall that in Example 6.16 we considered the events

$A = $ "the first flip is heads" and
$B = $ "the third flip is tails."

For these events we found that $P(A \cap B) = 1/4$. Since the probability is 1/2 that a fair coin will land heads or tails on any flip, we see that $P(A) = 1/2$ and $P(B) = 1/2$. Thus by the union rule, we have

$$P(A \cup B) = P(A) + P(B) - P(A \cap B)$$
$$= \frac{1}{2} + \frac{1}{2} - \frac{1}{4} = \frac{3}{4},$$

as we found earlier by use of the counting rule.

EXAMPLE 6.20 A certain construction project requires carpenters and masons. Construction workers say that the probability of a strike by the carpenters is .4, the probability of a strike by the masons is .5, and the probability of a strike by both groups is .2. What is the probability that the project will be delayed by some form of strike?

The probability of a strike by the carpenters is .4. The probability of a strike by the masons is .5.

Solution Let C and M denote the events "the carpenters strike" and "the masons strike," respectively. We are told that

$$P(C) = .4, \quad P(M) = .5, \quad \text{and} \quad P(C \cap M) = .2.$$

The probability of a delay in the project due to a strike is $P(C \cup M)$, which we can compute by the union rule. Since

$$P(C \cup M) = P(C) + P(M) - P(C \cap M) = .4 + .5 - .2 = .7,$$

there is a .7 probability of delay because of a strike. ■

EXAMPLE 6.21 In a survey conducted by the fire department it was found that 62% of all houses had a smoke detector and 35% had a fire extinguisher. Moreover, 24% had both a smoke detector and a fire extinguisher. What percentage of houses do not have either a smoke detector or a fire extinguisher?

Solution Recall that we can regard percentages as probabilities. Therefore if S and F are the events "has a smoke detector" and "has a fire extinguisher," respectively, then $P(S) = .62$, $P(F) = .35$, and $P(S \cap F) = .24$. The event $S \cup F$ is "has a smoke detector or a fire extinguisher," and so we must compute $P(\overline{S \cup F})$. By the union rule

$$P(S \cup F) = P(S) + P(F) - P(S \cap F) = .62 + .35 - .24 = .73.$$

Hence by the complement rule

$$P(\overline{S \cup F}) = 1 - P(S \cup F) = 1 - .73 = .27.$$

Thus 27% of the homes have neither a smoke detector nor a fire extinguisher. ■

Practice Problem 14 A factory requires both aluminum and steel. There is a probability of .06 that it will be short of aluminum, a probability of .05 that it will be short of steel, and a probability of .01 that it will be short of both metals. What is the probability that the factory will have an adequate supply of both metals?

EXAMPLE 6.22 An oil company has undertaken exploratory drilling in northern Canada. Preliminary tests indicate that there is a 90% chance of finding oil and an 85% chance of finding natural gas. What can be said about the probability of finding both oil and natural gas?

Solution By the union rule

$$P(\text{oil or gas}) = P(\text{oil}) + P(\text{gas}) - P(\text{oil and gas}).$$

Since a probability cannot exceed 1, we see that

$$1 \geq P(\text{oil}) + P(\text{gas}) - P(\text{oil and gas}).$$

Hence

$$P(\text{oil and gas}) \geq P(\text{oil}) + P(\text{gas}) - 1 \geq .90 + .85 - 1 = .75.$$

Moreover, since finding both oil and gas is no easier than finding gas, we have

$$P(\text{oil and gas}) \leq P(\text{gas}) = .85.$$

Therefore the probability of finding both oil and gas is between .75 and .85. (We cannot determine this probability exactly without more information.) ■

MUTUALLY EXCLUSIVE EVENTS

An important special case of the union rule arises if A and B are events that cannot happen simultaneously. In this case A and B have no common outcomes, and so $A \cap B = \emptyset$. We call such events *mutually exclusive*. (See Figure 6.8.)

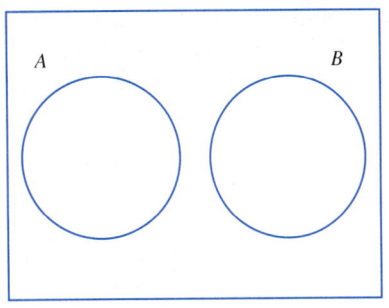

6.3 BASIC LAWS OF PROBABILITY

A and B are mutually exclusive

FIGURE 6.8

If A and B are mutually exclusive, then $P(A \cap B) = 0$. Hence the union rule reduces to the following simpler form.

> If A and B are mutually exclusive events, then
> $$P(A \cup B) = P(A) + P(B).$$

More generally, if all pairs of the events $A_1, A_2, \ldots, A_k$ are mutually exclusive, then we have the following analog of the addition principle (Theorem 5.2).

Theorem 6.3 (union rule for mutually exclusive events)

> If all pairs of the events $A_1, A_2, \ldots, A_k$ are mutually exclusive, then
> $$P(A_1 \cup A_2 \cup \cdots \cup A_k) = P(A_1) + P(A_2) + \cdots + P(A_k).$$

EXAMPLE 6.23 In a particular supermarket, the probabilities of having various numbers of checkout lines open are given below.

Precise Number of Open Checkout Lines	Probability
1	.12
2	.18
3	.21
4	.23
5 or more	.26

What is the probability that at most 3 checkout lines are open?

Solution Let A_1 denote the event "exactly 1 checkout line is open," and let A_2 and A_3 be defined similarly. Now each pair of A_1, A_2, and A_3 is mutually exclusive because the supermarket cannot simultaneously have two different numbers of checkout lines open. Therefore by the union rule for mutually exclusive events

P(at most 3 are open)

$= P(\text{exactly 1 is open}) + P(\text{exactly 2 are open}) + P(\text{exactly 3 are open})$
$= .12 + .18 + .21 = .51.$

Thus 51% of the time there are at most 3 checkout lines open. ∎

EXAMPLE 6.24 From among a group of twelve persons with a certain disease, five are to be randomly selected to receive treatment with a new drug. If eight of the twelve persons are female, what is the probability that a majority of those selected will be female?

Solution In order that a majority of those selected be female, at least three of the five persons chosen must be women. The events "exactly three women are chosen," "exactly four women are chosen," and "exactly five women are chosen" are mutually exclusive. Thus

P(a majority is female)

$= P(\text{exactly 3 women}) + P(\text{exactly 4 women}) + P(\text{exactly 5 women}).$

Now a committee containing exactly three women can be formed by choosing three women (from among eight) and two men (from among four). Hence

$$P(\text{exactly 3 women}) = \frac{C(8, 3) \cdot C(4, 2)}{C(12, 5)} = \frac{56 \cdot 6}{792} = \frac{336}{792}.$$

Likewise

$$P(\text{exactly 4 women}) = \frac{C(8, 4) \cdot C(4, 1)}{C(12, 5)} = \frac{70 \cdot 4}{792} = \frac{280}{792}$$

and

$$P(\text{exactly 5 women}) = \frac{C(8, 5) \cdot C(4, 0)}{C(12, 5)} = \frac{56 \cdot 1}{792} = \frac{56}{792}.$$

It follows that

P(a majority is female)

$= P(\text{exactly 3 women}) + P(\text{exactly 4 women}) + P(\text{exactly 5 women})$

$= \dfrac{336}{792} + \dfrac{280}{792} + \dfrac{56}{792}$

$= \dfrac{672}{792} = \dfrac{28}{33}.$ ∎

EXAMPLE 6.25 In the fourth race at Pimlico the newspaper handicapper has given respective odds of 5 to 3 against, 3 to 1 against, and 7 to 1 against the race being won by the number 4, 2, and 6 horses. What is the probability that one of these three horses wins?

Solution The probabilities that the number 4, 2, and 6 horses will win are 3/8, 1/4, and 1/8, respectively. Since only one horse can win the race, the events "number 4 horse wins," "number 2 horse wins," and "number 6 horse wins" are mutually exclusive. Hence by the union rule for mutually exclusive events the probability that one of these horses will win is

$$P(\text{number 4, 2, or 6 wins}) = P(4 \text{ wins}) + P(6 \text{ wins}) + P(2 \text{ wins})$$
$$= \frac{3}{8} + \frac{1}{4} + \frac{1}{8}$$
$$= \frac{3}{4}.$$ ■

Practice Problem 15 A certain subdivision was developed by the Ark Corporation, a joint venture of the Armstrong Company, Rave Brothers Construction, and Kaisner Construction. In this subdivision 35% of the houses were built by Armstrong, 20% by Rave Brothers, 19% by Kaisner, 11% by Robert Rist, and 15% by others. What percentage of the houses were built by a member of the Ark Corporation?

EXERCISES 6.3

1. In the experiment of observing the U.S. economy for one year, consider the events

   A = "inflation increases"

   B = "unemployment increases."

   Use set operations to symbolize each of the following events in terms of A and B, and illustrate each with a Venn diagram.
   (a) Inflation does not increase.
   (b) Both inflation and unemployment increase.
   (c) Inflation increases but not unemployment.
   (d) Neither inflation nor unemployment increases.

2. In the experiment of marketing three new products, consider the events

   A = "Product 1 is successful"

   B = "Product 2 is successful"

   C = "Product 3 is successful."

Use set operations to symbolize each of the following events in terms of A, B, and C, and illustrate each with a Venn diagram.
(a) All three products are successful.
(b) None of the three products is successful.
(c) Only the second product is successful.
(d) At least two of the three products are successful.

3. In the experiment of selecting a member from a committee consisting of students, faculty, and administrators, let E be the event "a student is chosen" and F be the event "a female is chosen." Describe the following events in words.
 (a) $\overline{E}$ (b) $\overline{F}$ (c) $E \cap F$ (d) $E \cup F$

4. In the experiment of flipping a coin twice, let E be the event "the same side of the coin shows on both flips" and let F be the event "the second flip is tails." Describe the following events in words.
 (a) $\overline{E}$ (b) $\overline{F}$ (c) $E \cap F$ (d) $E \cup F$

5. In the experiment of generating a random three-digit number from 000 to 999, consider the events

$$A = \text{"the number is less than 400"}$$
$$B = \text{"the number is even"}$$
$$C = \text{"the number is at least 200."}$$

Calculate the probabilities of the following events.
 (a) $\overline{A}$ (b) $A \cap C$ (c) $\overline{(A \cup B)}$ (d) $A \cap B \cap C$

6. In the experiment of randomly drawing a card from a standard deck of playing cards, consider the events

$$A = \text{"the card is red"}$$
$$B = \text{"the card is higher than a ten"}$$
$$C = \text{"the card is a spade."}$$

Calculate the probabilities of the following events if an ace is the highest card.
 (a) $\overline{B}$ (b) $A \cap C$ (c) $\overline{(A \cap B)}$ (d) $(A \cup C) \cap B$

7. If the probability that A occurs is .53, the probability that B occurs is .72, and the probability that both occur is .48, what is the probability that
 (a) A does not occur?   (b) B does not occur?
 (c) A or B occurs?   (d) neither A nor B occurs?

8. If the probability that A occurs is .25, the probability that B occurs is .45, and the probability that both occur is .05, what is the probability that
 (a) A does not occur?   (b) B does not occur?
 (c) A or B occurs?   (d) neither A nor B occurs?

9. Suppose that A, B, and C are mutually exclusive events with respective probabilities .23, .42, and .31. What is the probability that
 (a) A does not occur?
 (b) at least one of the three events occurs?
 (c) none of the three events occur?
 (d) more than one of the three events occurs?

10. Suppose that *A*, *B*, and *C* are mutually exclusive events with respective probabilities 1/4, 1/5, and 1/2. What is the probability that
    (a) *B* does not occur?
    (b) at least one of the three events occurs?
    (c) none of the three events occur?
    (d) more than one of the three events occurs?

In Exercises 11–18 determine which pairs of events are mutually exclusive in the experiment of rolling a pair of dice.

11. "the sum is 2, 5, 8, or 11" and "the sum is 9, 10, 11, or 12"
12. "the sum is 7, 8, 9, or 10" and "the sum is 2, 3, 11, or 12"
13. "one die comes up 4" and "one die comes up 2"
14. "one die comes up 2" and "the sum of the dice is 10"
15. "the same number appears on both dice" and "the sum of the dice is 7"
16. "one die comes up 3 or less" and "an even number appears on one die"
17. "the sum of the dice is 2 or 3" and "1 does not appear on either die"
18. "the sum of the dice is a multiple of 3" and "odd numbers appear on both dice"
19. If a fair coin is flipped three times, what is the probability that it will come up heads at least once?
20. If three fair dice are rolled, what is the probability that at least one 6 will appear?
21.* If five letters (not necessarily distinct) are typed at random, what is the probability that at least one will be a vowel?
22.* Four women checked their hats at a restaurant. If the hats are returned at random to these women, what is the probability that someone receives the wrong hat?
23.* A shipment of twenty refrigerators contains five defective units. If three units are chosen from the shipment at random, what is the probability that at least one defective unit will be among them?
24.* If two cards are chosen at random from a standard deck of playing cards, what is the probability that at least one is a jack, queen, or king?
25. In a certain town the probability of reading *The Times* is .51 and the probability of reading *The Herald* is .43. If the probability of reading both papers is .16, what is the probability of reading *The Times* or *The Herald*?
26. In a certain subdivision 67% of the houses have central air conditioning, 32% have a fireplace, and 13% have both central air conditioning and a fireplace. What percentage of these houses have central air conditioning or a fireplace?
27. In a large sample of families it was found that 86% of the husbands and 63% of the wives worked outside the home, while in 54% of the cases both worked outside the home. What is the probability that in a random family from this sample
    (a) at least one spouse works outside the home?
    (b) neither spouse works outside the home?

---

*This exercise requires the use of counting techniques from Chapter 5.

28. During a certain year 53% of the cars sold in a large city were made in America and 63% were economy cars. If 21% of the cars sold in this city were American-made economy cars, what percentage of the cars sold were neither American nor economy cars?

29. According to a computer dating service, the odds are 3 to 1 that Norman will like Sylvia, 7 to 5 that she will like him, and 5 to 7 that they will both like each other. What are the odds that at least one will like the other?

30. Two mechanical systems must both function properly in order for a space probe to land successfully on Venus. According to NASA officials, the probabilities that the systems will function properly are .80 and .75, and the probability that both systems will fail is .05. What are the chances for a successful landing?

31. In the third race, the local handicapper has set the odds against Affirmation at 3 to 1 and the odds against Brobdingnag at 4 to 1. According to this handicapper, what is the probability that one of these two horses will win the race?

32. Three New York critics are screening a new French film. The film's producer rates the chances of favorable reviews from Armstrong, Babcock, and Craft to be 15%, 20%, and 25%, respectively. If no two of these reviewers ever like the same film, what is the probability that all three reviews are unfavorable?

33. An observant commuter noticed that the 5:30 train is usually between one and ten minutes late. Keeping records over the course of several months, she obtained the following probability distribution.

Minutes late	Probability
0	.03
1	.07
2	.12
3	.15
4	.17
5	.16
6	.13
7	.09
8	.05
9	.02
10	.01

What is the probability that on a random day the train will be
(a) at least five minutes late?
(b) at least two minutes late?
(c) at most seven minutes late?

34. In a particular barbershop, the probability distribution for the number of customers waiting for a haircut is as shown below.

Customers	Probability
0	.5
1	.2
2	.2
3	.1
4 or more	0

(a) What is the probability that someone is waiting?

(b) What is the probability that at least 2 are waiting?

(c) What is the probability that no more than two are waiting?

35.* If five people are chosen at random from among four men and six women, what is the probability that a majority of those selected will be women?

36.* If two marbles are chosen at random from an urn containing eight red marbles and four blue marbles, what is the probability that both will be the same color?

37.* A freezer contains three orange and five cherry Popsicles. If four Popsicles are chosen at random from the freezer, what is the probability that

(a) at least two will be cherry?

(b) no more than one will be orange?

38.* A city council consists of four Democrats and six Republicans. If a delegation of three is selected at random, what is the probability that

(a) none are Republicans?

(b) a majority are Republicans?

39. Prove the union rule for a sample space of equally likely outcomes.

40. Prove the union rule for three events: $P(A \cup B \cup C) = P(A) + P(B) + P(C) - P(A \cap B) - P(A \cap C) - P(B \cap C) + P(A \cap B \cap C)$.

Answers to Practice Problems

11. $P(A) = \frac{5}{36}$, $P(B) = \frac{5}{9}$, $P(A \cup B) = \frac{23}{36}$, $P(A \cap B) = \frac{1}{18}$, $P(\overline{A}) = \frac{31}{36}$, $P(\overline{B}) = \frac{4}{9}$

12. (a) .988 (b) .996

13. $\frac{31}{32}$

14. .90

15. 74%

6.4 Conditional Probability, the Intersection Rule, and Independent Events

Our estimate of the probability of an event E reflects the amount of information we have about the event. If we obtain additional information, it may change our opinion of the likelihood that the event occurs. For instance, the probability of snow on a given day in Illinois is higher in January than in July. Thus knowledge of the time of year will cause us to adjust upward or downward our estimate of the probability of snow on a given day in Illinois. In this section we will discuss probabilities when such additional information is available.

---

*This exercise requires the use of counting techniques from Chapter 5.

REDUCED SAMPLE SPACES

Let S be a sample space consisting of equally likely outcomes, and let A be a particular event in S. (See Figure 6.9.) Suppose that we are interested in knowing whether A occurred when the experiment was performed. If we have not yet been told the outcome of the experiment, there are two possible situations.

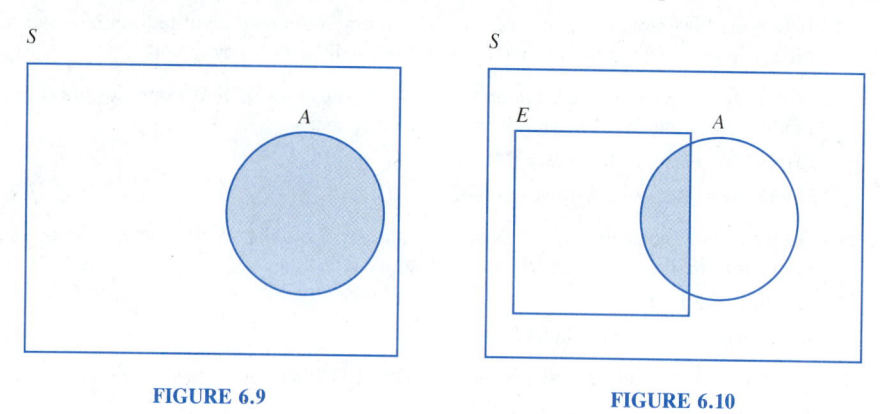

FIGURE 6.9

FIGURE 6.10

In the first case, we have no information whatsoever about the result of the experiment. In this case we are in the same position as before the experiment was performed; namely, as far as we know, all the outcomes in S are equally likely. Thus by the counting rule, the probability that A occurred is

$$P(A) = \frac{|A|}{|S|} = \frac{\text{number of outcomes in } A}{\text{number of outcomes in } S}.$$

In the second case, we are given information that some event E occurred, as illustrated in Figure 6.10. In this case the only possible outcomes for the experiment are those in E. Thus in effect E becomes a reduced sample space for the experiment. The outcomes in E for which A occurs are those in the intersection $A \cap E$, which is shaded in Figure 6.10. Applying the counting rule to the new sample space E, we see that the probability that A occurs given that E occurred is

$$\frac{\text{number of outcomes in } A \text{ and } E}{\text{number of outcomes in } E}.$$

We call the probability that A occurs given that E occurred the *conditional probability of A given E* and denote it by $P(A|E)$.

6.4 CONDITIONAL PROBABILITY, THE INTERSECTION RULE, AND INDEPENDENT EVENTS

Counting Rule for Conditional Probabilities

If A and E are events in a finite sample space consisting of equally likely outcomes, then

$$P(A|E) = \frac{|A \cap E|}{|E|} = \frac{\text{number of outcomes in } A \text{ and } E}{\text{number of outcomes in } E}.$$

Recall that in part (b) of Example 6.11 we considered the following question.

> From among the forty persons attending a seminar discussing financial investments, three different persons will be selected at random to receive door prizes. What is the probability that a particular person at the seminar wins the third door prize?

As we have seen, the answer to this question is 1/40, not 1/38 as some people believe. This question illustrates the first case above. For let us consider the situation at the time when the third door prize is to be awarded. At this time two prizes have already been given, but *we have no information about the winners of the first two prizes*. Thus as far as we know, all forty persons in attendance are equally likely to win the third prize.

On the other hand, if we learn that Mr. Smith did not win either of the first two prizes, then the question becomes an illustration of the second case. With this new information, we can reduce the sample space to the 38 persons who did not win the first or second prize. And in this reduced sample space, we see that the probability of Mr. Smith's receiving the third door prize is 1/38. Thus the information we have about an experiment plays a crucial role in our calculation of probabilities.

EXAMPLE 6.26 In a pharmaceutical research study, half of a group of 500 patients suffering from chronic headaches were given a new painkilling drug, and the other half were given a placebo. The results are shown in the table below.

	Felt better	Felt no better	Total
Given the drug	183	67	250
Given placebo	117	133	250
Total	300	200	500

(a) What is the probability that a random patient felt better?
(b) What is the probability that a random patient felt better, given that he or she received the drug?
(c) What is the probability that a random patient received the placebo, given that he or she felt no better?

Solution (a) Of the 500 patients, 300 felt better. Hence the probability that a random patient felt better is given by the counting rule to be

$$P(\text{felt better}) = \frac{300}{500} = .600.$$

Notice that this is *not* a conditional probability.

(b) In question (b), we are told that the patient received the drug. Hence we restrict our attention to the 250 patients who received the drug. Of these, 183 felt better. Hence by the counting rule for conditional probabilities we see that

$$P(\text{felt better}|\text{given drug}) = \frac{\text{number who felt better given drug}}{\text{number given drug}}$$

$$= \frac{183}{250} = .732.$$

(c) In this case we know that the patient felt no better, and so we restrict our attention to the 200 patients who felt no better. Of this group, 133 received the placebo. Therefore

$$P(\text{placebo}|\text{felt no better}) = \frac{\text{number given placebo who felt no better}}{\text{number who felt no better}}$$

$$= \frac{133}{200} = .665.$$

That is, 66.5% of those who felt no better were given the placebo. ∎

Problem Problem 16 In Example 6.26, what is the conditional probability that a patient felt no better given that the patient received a placebo?

EXAMPLE 6.27 Two fair dice are rolled. What is the probability that the sum is 7 if we are told that each die came up less than 5?

Solution Figure 6.11 shows the sample space S for the experiment of rolling two dice along with the events

$$A = \text{``the sum rolled is 7''} \text{ and}$$
$$E = \text{``each die came up less than 5.''}$$

Since we are told that each die came up less than 5, there are only 16 possible outcomes for the experiment instead of the usual 36. Of these 16, there are only 2 outcomes for which A occurs, namely the pairs (3, 4) and (4, 3). Thus by the counting rule for conditional probabilities, we find that

$$P(A|E) = \frac{|A \cap E|}{|E|} = \frac{2}{16} = \frac{1}{8}.$$

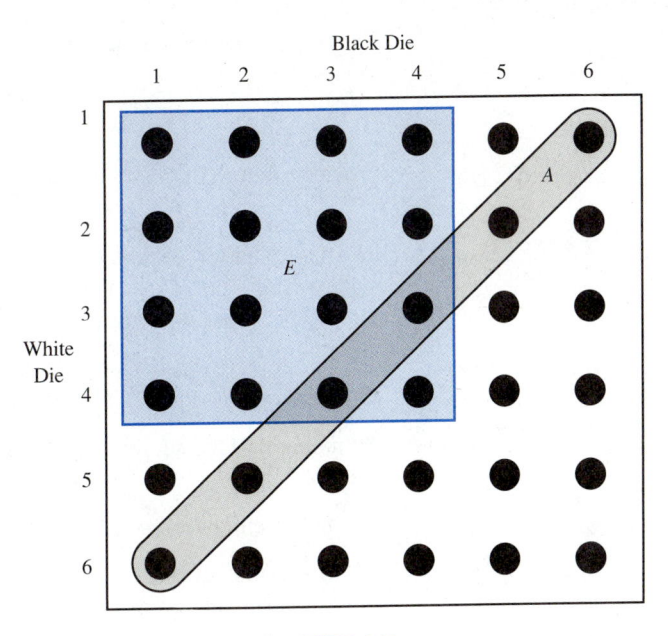

6.4 CONDITIONAL PROBABILITY, THE INTERSECTION RULE, AND INDEPENDENT EVENTS

FIGURE 6.11

Observe that without the information that E occurred, we would have computed the probability of A to be

$$P(A) = \frac{|A|}{|S|} = \frac{6}{36} = \frac{1}{6}.$$

Hence knowing that E occurred decreases the probability that A occurred. ∎

Practice Problem 17   Use the tree diagram in Figure 6.2 on page 273 to determine the probability that, in three flips of a fair coin, the first flip was heads given that two of the three flips were heads.

Recall that the various counting formulas are valid only for sample spaces consisting of equally likely outcomes. Since conditional probabilities may arise in experiments where the outcomes are not equally likely, we must define them in a more general context. Fortunately, a simple modification of the counting rule for conditional probabilities will yield this more general formula. For if we divide the numerator and denominator by $|S|$, the number of elements in the sample space, we obtain

$$\frac{\text{number of outcomes in } A \text{ and } E}{\text{number of outcomes in } E} = \frac{\frac{|A \cap E|}{|S|}}{\frac{|E|}{|S|}} = \frac{P(A \cap E)}{P(E)}.$$

Thus we define the **conditional probability of A given E** by

$$\frac{P(A \cap E)}{P(E)}$$

provided that $P(E) \neq 0$, and denote it, as before, by $P(A|E)$.

EXAMPLE 6.28 In a large survey of its customers, a nationwide retail store found that 20% made major purchases, 48% had a charge account, and 12% fell into both categories. What is the probability that

(a) a customer who makes a major purchase has a charge account?
(b) a customer with a charge account has made a major purchase?

Solution Assuming that the survey was representative of all of the store's customers, we see that

$P(\text{major purchase}) = .20$, $P(\text{charge account}) = .48$, and $P(\text{both}) = .12$.

(a) In the first question we want to know the conditional probability that a random customer has a charge account *given the fact that the customer makes a major purchase*. Using the definition of conditional probability above, we find

$$P(\text{charge account}|\text{major purchase}) = \frac{P(\text{charge account and major purchase})}{P(\text{major purchase})}$$

$$= \frac{.12}{.20} = .6.$$

In other words, among the customers who make major purchases, 60% have charge accounts as compared to 48% of all the store's customers.

(b) Here we want to determine the conditional probability that a random customer has made a major purchase *given the fact that the customer has a charge account*. As above, we obtain

$$P(\text{major purchase}|\text{charge account}) = \frac{P(\text{major purchase and charge account})}{P(\text{charge account})}$$

$$= \frac{.12}{.48} = .25.$$

Thus 25% of the customers with charge accounts make a major purchase, as compared to 20% of all the store's customers. ∎

Note that conditional probabilities can be disguised in ordinary discourse. In Example 6.28, for instance, the phrases "a customer who makes a major purchase" and "a customer with charge account" indirectly give information that limits the type of customers under consideration. For this reason, conditional probabilities are required in this problem.

6.4 CONDITIONAL PROBABILITY, THE INTERSECTION RULE, AND INDEPENDENT EVENTS

Practice Problem 18 An insurance agent has found that, among her clients, 60% have purchased automobile insurance from her, 40% have purchased life insurance from her, and 15% have purchased both types of insurance from her. What is the probability that someone who has purchased life insurance from her has also purchased automobile insurance from her?

THE INTERSECTION RULE

In Section 6.3 we presented formulas for the probability of the complement and union of events. To obtain a similar formula for the probability of the intersection of events, we need only multiply the conditional probability formula

$$P(B|A) = \frac{P(B \cap A)}{P(A)}$$

by $P(A)$.

Theorem 6.4 (the intersection rule)

For any events A and B,

$$P(A \cap B) = P(A) \cdot P(B|A).$$

Thus the probability that two events will *both* occur is equal to the probability that the first occurs times the conditional probability that the second occurs given that the first occurred.

EXAMPLE 6.29 Ms. Caldwell, an insurance agent, often telephones persons at random in hopes of arranging appointments to discuss insurance. She has found that 8% of those who are called agree to an appointment, and 30% of those who agree to an appointment actually buy insurance from her. What is the probability that someone called at random will agree to an appointment and buy insurance?

Solution Let A and B be the following events:

$A =$ "agrees to an appointment" and $B =$ "buys insurance."

We must compute $P(A \cap B)$, the probability that someone will agree to an appointment and buy insurance. By the intersection rule

$$P(A \cap B) = P(A) \cdot P(B|A) = .08(.30) = .024.$$

Thus 2.4% of those who are called agree to an appointment and buy insurance. ∎

Practice Problem 19 Engines on an automobile assembly line are inspected at two different points during the production process. The first inspection catches 70% of the defective units, and the second inspection catches 80% of the defective units that pass the first inspection. What is the probability that a defective engine will pass both inspections?

EXAMPLE 6.30 A firm has six accountants, three men and three women. If two of the accountants are randomly selected to attend an accounting seminar, what is the probability that the two who are chosen are both men?

Solution Instead of regarding the two persons as being chosen simultaneously, we will suppose that two persons are chosen in succession. (Clearly this will not affect the probability of obtaining two men.) Let

$$A = \text{``a man is chosen first''} \quad \text{and} \quad B = \text{``a man is chosen second.''}$$

Then $P(A) = 3/6$ by the counting rule. Moreover, after the first man is selected there are 5 accountants remaining, 2 of whom are men. Thus $P(B|A) = 2/5$. Thus by the intersection rule

$$P(\text{two men are chosen}) = P(A \cap B)$$
$$= P(A) \cdot P(B|A)$$
$$= \frac{3}{6} \cdot \frac{2}{5} = \frac{1}{5}.$$

Compare this solution to that discussed in Section 6.1 after Example 6.8. ∎

Practice Problem 20 In a box of ten transistors, three are defective. What is the probability that if two transistors are chosen at random from this box, both will be defective?

INDEPENDENT EVENTS

In the experiment of rolling two fair dice, one white and one black, consider the events

$$A = \text{``the white die comes up 2, 3, or 4''} \text{ and}$$
$$B = \text{``the black die comes up greater than 4.''}$$

It is intuitively obvious that what happens to the white die does not affect the result of the black die, and vice versa. Thus we expect that the occurrence of B will not change the likelihood that A occurs; that is, the probabilities $P(A)$ and $P(A|B)$ should be equal. Similarly we expect that $P(B) = P(B|A)$. To verify these statements, we can refer to the sample space S in Figure 6.12.

6.4 CONDITIONAL PROBABILITY, THE INTERSECTION RULE, AND INDEPENDENT EVENTS

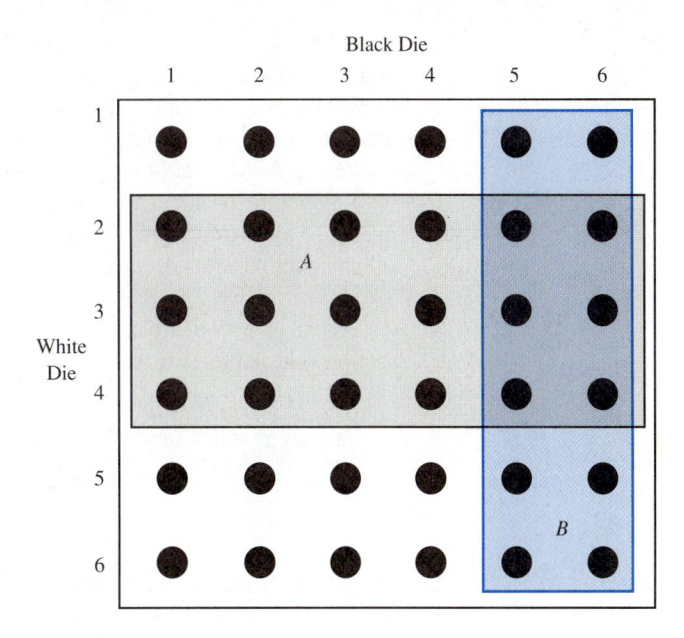

FIGURE 6.12

From this sample space we see that

$$P(A) = \frac{|A|}{|S|} = \frac{18}{36} = \frac{1}{2},$$

$$P(B) = \frac{|B|}{|S|} = \frac{12}{36} = \frac{1}{3},$$

$$P(A|B) = \frac{|A \cap B|}{|B|} = \frac{6}{12} = \frac{1}{2}, \text{ and}$$

$$P(B|A) = \frac{|B \cap A|}{|A|} = \frac{6}{18} = \frac{1}{3}.$$

Hence

$$P(A) = P(A|B) \quad \text{and} \quad P(B) = P(B|A),$$

as we anticipated.

Events A and B are called **independent** if both

$$P(A) = P(A|B) \quad \text{and} \quad P(B) = P(B|A).$$

We also call A and B independent if either of these conditional probabilities is undefined; that is, if either $P(A) = 0$ or $P(B) = 0$. Thus the preceding calculations have shown that the events "the white die comes up 2, 3, or 4" and "the black die comes up greater than 4" are independent.

The example above illustrates the distinction between *independent* events and *mutually exclusive* events. As we have seen, the events "the white die comes up 2, 3, 4" and "the black die comes up greater than 4" are independent, since the occurrence of one does not affect the probability of the other. However, these events are not mutually exclusive because it is possible for both events to occur at the same time. (For instance, the white die can come up 2 and the black die 5.) It can be shown that two independent events A and B can never be mutually exclusive unless $P(A) = 0$ or $P(B) = 0$. (See Exercise 43.)

When A and B are independent events, $P(B|A) = P(B)$. Hence the intersection rule reduces to the following simpler form.

> If A and B are independent events, then $P(A \cap B) = P(A) \cdot P(B)$.

More generally, a collection of events $A_1, A_2, \ldots, A_k$ is independent if the probability of each of the events is unaffected by information concerning the others. In this case the intersection rule can be generalized as follows.

Theorem 6.5 (the intersection rule for independent events)

> For any independent events $A_1, A_2, \ldots, A_k$,
> $$P(A_1 \cap A_2 \cap \cdots \cap A_k) = P(A_1) \cdot P(A_2) \cdots P(A_k).$$

In applying this rule, remember that in order for the intersection of $A_1, A_2, \ldots, A_k$ to occur, *all* of the events must occur. Thus the intersection rule for independent events says that the probability that a collection of independent events all occur is equal to the product of their separate probabilities.

EXAMPLE 6.31 A computer system consists of the computer, a monitor, and a disk drive. Based on past experience, 2% of the computers, 1% of the monitors, and 5% of the disk drives can be expected to fail during the warranty period. If these components are independently produced, what percentage of these computer systems do not fail during the warranty period?

Solution Let C, M, and D be defined as follows:

C = "computer fails," M = "monitor fails," and D = "disk drive fails."

The given information is that

$$P(C) = .02, \quad P(M) = .01, \quad \text{and} \quad P(D) = .05.$$

6.4 CONDITIONAL PROBABILITY, THE INTERSECTION RULE, AND INDEPENDENT EVENTS 321

We are interested in determining the probability that a system does not fail, that is, that none of the components fail. Therefore we must consider the complementary events $\overline{C}$, $\overline{M}$, and $\overline{D}$, for which the complement rule gives

$$P(\overline{C}) = .98, \quad P(\overline{M}) = .99, \quad \text{and} \quad P(\overline{D}) = .95.$$

Since the components of the system are independently produced, we assume that $\overline{C}$, $\overline{M}$, and $\overline{D}$ are independent events. Hence

$$\begin{aligned}
P(\text{system does not fail}) &= P(\text{no component fails}) \\
&= P(\overline{C} \cap \overline{M} \cap \overline{D}) \\
&= P(\overline{C}) \cdot P(\overline{M}) \cdot P(\overline{D}) \\
&= .98(.99)(.95) \\
&= .92169.
\end{aligned}$$

Thus over 92% of the computer systems do not fail during the warranty period. ■

Practice Problem 21 A stereo system consists of a turntable, an amplifier, and two speakers. Suppose that 5% of the turntables, 2% of the amplifiers, and 3% of the pairs of speakers require servicing during the first year. If the turntables, amplifiers, and speakers are manufactured independently, what percentage of these systems do not require servicing during the first year?

The topics discussed in Sections 6.1–6.4 are essential to the understanding of probability. Almost all questions involving probability require use of one or more of the concepts and techniques from these sections. By comparison, the specialized techniques that we will consider in Sections 6.5 and 6.6 are used much less often. Thus the reader should place special importance on understanding the material presented up to this point.

EXERCISES 6.4

1. Suppose that the probability that A occurs is .53, the probability that B occurs is .72, and the probability that both occur is .48. Compute each of the following.
   (a) $P(A|B)$ (b) $P(B|A)$

2. Suppose that the probability that A occurs is 1/3, the probability that B occurs is 1/4, and the probability that both occur is 1/8. Compute each of the following.
   (a) $P(A|B)$ (b) $P(B|A)$

3. Two fair dice are rolled. What is the probability that the sum rolled is at least eight, given
   (a) no information at all?
   (b) that at least one die comes up five?
   (c) that the same number appears on both dice?

4. A card is drawn at random from a standard deck of playing cards. What is the probability that the card is higher than a seven, given
   (a) no information at all?
   (b) that the card is not an ace?
   (c) that the card is a club?
   (d) that the card is a jack, queen, or king?

5. A fair coin is flipped three times. What is the probability that it comes up heads at least once, given
   (a) no information at all?
   (b) that the same side appears on all three flips?
   (c) that it comes up heads at most once?
   (d) that the second flip was tails?

6.* Two letters, not necessarily distinct, were typed at random. What is the probability that both are vowels, given
   (a) no information at all?
   (b) that the two letters are different?
   (c) that at least one is a vowel?

7. The following data from reference [2] shows the number of voters in Elmira, New York, who voted for the Republican candidate (Thomas E. Dewey) in the 1948 presidential election.

	Socioeconomic Status			Total
	High	Middle	Low	
Men	66	48	74	188
Women	60	80	90	230
Total	126	128	164	418

Suppose that one of these voters is selected at random.
   (a) What is the probability that the voter is a man?
   (b) What is the probability that the voter has high socioeconomic status?
   (c) What is the probability that the voter is a woman given that the voter has low socioeconomic status?
   (d) What is the probability that the voter has middle socioeconomic status given that the voter is a man?

8. In the Framingham study (see reference [6]) a group of 828 men in their thirties were classified according to their serum cholesterol level (in milligrams per deciliter). These cases were followed up many years later to see how many developed coronary disease. The results are shown at the top of page 323.

---

*This exercise requires the use of counting techniques from Chapter 5.

Cholesterol Level	Coronary Disease	No Coronary Disease	Total
Under 200	22	293	315
200–219	12	116	128
220–239	29	140	169
240–259	19	78	97
Over 259	35	84	119
Total	117	711	828

Assuming that this data is representative of men in their thirties, what is the probability that

(a) a man in his thirties will develop coronary disease in later life?

(b) a man in his thirties with a cholesterol level between 220 and 239 will develop coronary disease in later life?

(c) a man who develops coronary disease had a cholesterol level of at least 240 when he was in his thirties?

(d) a man without coronary disease had a cholesterol level below 200 when he was in his thirties?

The following data from reference [2] shows the number of voters in Elmira, New York, categorized by religion, who voted for each 1948 presidential candidate.

Religion	Presidential Candidate Voted For Democrat	Republican	Total
Catholic	124	67	191
Protestant	90	345	435
Total	214	412	626

In Exercises 9–18 let C, T, D, and R denote the events of being Catholic, being Protestant, voting Democratic, and voting Republican, respectively.

9. Compute $P(C)$.
10. Compute $P(R)$.
11. Compute $P(C|D)$.
12. Compute $P(R|T)$.
13. Compute $P(R|C)$.
14. Compute $P(D|T)$.
15. Are C and D independent?
16. Are T and R independent?
17. Are C and R independent?
18. Are T and D independent?

19. Among army volunteers it was found that 17% fail the physical exam, 26% fail the written exam, and 8% fail both. What is the probability that a volunteer who fails the physical exam will also fail the written exam?

20. The percentage of voters favoring the incumbent went from 46% on August 1 to 52% on September 1, although only 38% favored him in both polls. What is the probability that a random voter who supported him on August 1 still supported him on September 1?

21. U.S. population life tables indicate that the probability that a white female will survive from birth to age 60 is .864 and the probability that she will survive from birth to age 70 is .733. What is the probability that a 60-year-old white female will survive to age 70?

324 CHAPTER 6 PROBABILITY

22. Oddsmakers say that the odds against Brobdingnag's winning the Kentucky Derby are 3 to 2 and against winning the Triple Crown (the Kentucky Derby and two other races) are 9 to 1. If these odds are correct, what is the probability that if Brobdingnag wins the Kentucky Derby, he will go on to win the Triple Crown?

23. In a certain inner-city high school, 4000 students were asked (i) whether they were regular alcohol users, (ii) whether they were regular marijuana users, and (iii) whether they had ever tried hard drugs. Their responses are shown in the Venn diagram at the right, where U is the set of all students, A is the set of regular alcohol users, M is the set of regular marijuana users, and H is the set of students who have tried hard drugs. A marijuana advocate claimed that these data disproved the claim that marijuana use leads to hard drugs because an even higher percentage of those trying hard drugs were alcohol users (98.2%) than marijuana users (97.4%). Use the data to show that, in fact, marijuana users are more than twice as likely as alcohol users to try hard drugs.

24. An argument similar to the following occurred during a murder trial.
    Prosecutor: Ladies and gentlemen of the jury, you have heard the witnesses testify that the murderer drove a blue GM sedan with a license plate beginning with Z. Out of 735,000 cars registered in this state, there are only five which fit this description and could have been at the scene of the crime. Moreover, Snavely—the defendant—owns one of those five cars! The odds are therefore overwhelming that Snavely is the murderer.
    As Snavely's attorney, how would you defend your client against this "mathematical" attack?

25. A plumber's welds are inspected twice for defects. The first inspection catches 80% of the defective welds, and of those that pass the first inspection, 90% are caught on the second inspection. What is the probability that a defective weld will pass both inspections?

26. A mathematics student would like to know the probability of not giving a wrong answer in class. She will avoid giving a wrong answer if the teacher does not call on her or if she answers correctly when questioned. She estimates that the probability of being questioned by the teacher is .2 and that the probability of answering correctly if questioned is .6. Assuming that she will not be questioned more than once per class, what is the probability that she will not give an incorrect answer during tomorrow's class?

27. Sixty percent of the graduates majoring in accounting at a certain university pass the CPA exam on the first try. Of those who fail on the first try, half pass on the second try; and of those who fail twice, 40% pass on the third attempt. What is the probability that a random graduate majoring in accounting at this university will pass the CPA exam within three tries?

28. The Motor Vehicle Department has found that the probabilities of passing the driver's license test on the first, second, and third tries are .80, .75, and .50, respectively. What are the probabilities of passing the test within two tries and within three tries?

29. A lawyer has told his client that there is a 30% chance of acquittal in the forthcoming trial. If convicted, the lawyer believes that there is a 60% chance of being acquitted when the verdict is appealed to a higher court. What is the probability that the client will eventually be acquitted?

30. A basketball player makes his first free throw 70% of the time. When this happens, he makes the second shot 90% of the time. But if he misses the first shot, then he makes

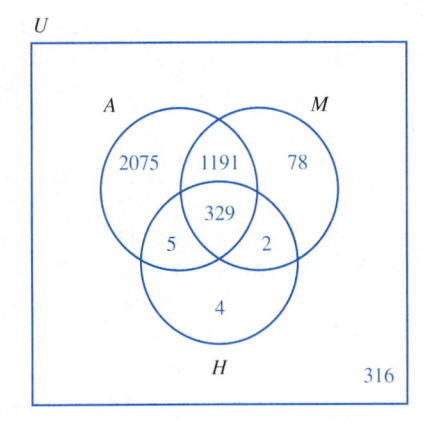

the second shot only 80% of the time. If the player shoots two free throws, what is the probability that he will make

(a) both of them?   (b) at least one of them?

31. From an audience of 15 boys and 10 girls two children are to be selected at random to assist a magician. What is the probability that

   (a) two girls are chosen?
   (b) a boy is chosen first and a girl is chosen second?
   (c) one boy and one girl are chosen?
   (d) a girl is chosen second?

32. A jeweler accidentally dropped two diamonds onto a tray containing three zircons of identical size and shape. In order to find the diamonds he must examine the five gems one by one until both diamonds are found. What is the probability that he will examine

   (a) no more than two gems?   (b) no more than three gems?

33. From an urn containing 10 red, 10 white, and 10 blue marbles, three are to be selected in sequence without replacement. What is the probability that they will be chosen in the order red, white, and blue?

34. From an urn containing 8 red, 5 white, and 6 yellow marbles, two are to be selected in sequence without replacement. What is the probability that two marbles of the same color will be chosen?

35. If $P(E) = .6$ and $P(F) = .2$, compute $P(E \cap F)$ and $P(E \cup F)$ if

   (a) E and F are independent   (b) E and F are mutually exclusive.

36. Let A and B be events for which $P(A) = .85$ and $P(B) = .90$.

   (a) What is the smallest possible value for $P(A \cap B)$?
   (b) Is it possible for A and B to be mutually exclusive?
   (c) If A and B are independent, what is $P(A \cap B)$?

37. The probability that a female over 60 years of age will develop breast cancer is .04 and the probability that she will develop diabetes is .02. Assuming that these events are independent, what is the probability that a female over 60 will develop

   (a) at least one of the diseases?   (b) both of the diseases?

38. Three teams are making independent attempts to climb Mount Everest. If their chances of success are 40%, 45%, and 50%, respectively, what is the probability that at least one team will succeed?

39. During construction of a new office building, a contractor estimates that there is an 8% chance of a shortage of materials, a 10% chance of a strike, and a 20% chance of delays due to bad weather. If these events are independent, what is the probability that at least one of these problems occurs?

40. In blacks the probabilities of having blood types A and B are .30 and .20, respectively. Moreover, the probability that a black has the Rh antigen is .85. Use the fact that the presence or absence of the Rh antigen is independent of the presence or absence of the A and B antigens to compute the probability that a randomly selected black will have the following blood types. (Refer to Example 5.8 in Section 5.1 for a description of human blood types.)

   (a) A+   (b) A−   (c) B+   (d) B−

41. In their past games, chess grandmaster Lipner has defeated grandmaster Traub 40% of the time, lost 10% of the time, and drawn 50% of the time. If they play three games, what is the probability that
  (a) all three games are drawn?
  (b) Lipner will win one game and draw two?

42. An oil company is currently drilling in three different countries, where geologists estimate the chances of finding oil to be 30%, 50%, and 80%, respectively. What is the probability that
  (a) oil will be found at all three sites?
  (b) oil is found at exactly two sites?

43. Prove that if A and B are both independent and mutually exclusive, then $P(A) = 0$ or $P(B) = 0$.

44. Prove the intersection rule for three events A, B, and C:
$$P(A \cap B \cap C) = P(A) \cdot P(B|A) \cdot P(C|A \cap B).$$

45. Prove that if A and B are events such that $P(B) \neq 0$, then
$$P(\overline{A}|B) = 1 - P(A|B).$$

46. Prove that if events A and B are independent, then so are events A and $\overline{B}$.

Answers to Practice Problems

16. $\dfrac{133}{250} = .532$

17. $\dfrac{2}{3}$

18. .375

19. .06

20. $\dfrac{6}{90} = \dfrac{1}{15}$

21. approximately 90.3%

6.5 Bayes's Formula

Much of science is based on *inductive* rather than deductive inference, that is, instead of reasoning from hypotheses to conclusions we work backwards, assigning probabilities to various hypotheses on the basis of known conclusions. For example, a doctor making a diagnosis observes the patient's known symptoms and then assigns probabilities to various diseases that might produce those symptoms. A fundamental tool in the theory of inductive inference is Bayes's formula, which has widespread applications in statistics.

STOCHASTIC DIAGRAMS

Many experiments are performed in stages, each of which may have several possible outcomes. Such a multi-stage experiment is called a **stochastic process,** from the Greek word *stochos* meaning "guess." Such processes arise in many applications. We will begin by considering an example.

6.5 BAYES'S FORMULA

A regional producer of soft drinks fills and caps cola bottles on one of three machines in its plant in Decatur, Georgia. Twenty percent of the total output goes through machine 1, which has a 0.1% defective rate; that is, one out of every 1000 bottles that are processed on this machine is improperly filled or capped. Machine 2 handles 30% of the output with a 0.6% defective rate, and machine 3 processes the rest of the bottles with a 0.4% defective rate. What is the probability that a bottle of this cola is defective?

This problem describes a stochastic process. First, each bottle is sent to one of the three machines, and then it is filled and capped, possibly defectively. Let M_1, M_2, M_3, and D denote the events

$$M_1 = \text{``bottle was processed by machine 1,''}$$
$$M_2 = \text{``bottle was processed by machine 2,''}$$
$$M_3 = \text{``bottle was processed by machine 3,''} \quad \text{and}$$
$$D = \text{``bottle is defective.''}$$

Since 20% of the bottles are processed by machine 1, we have $P(M_1) = .20$. Moreover, since 0.1% of the bottles processed by machine 1 are defective, $P(D|M_1) = .001$. Similarly we have

$$P(M_2) = .30, \; P(D|M_2) = .006, \; P(M_3) = .50, \text{ and } P(D|M_3) = .004.$$

This process can be illustrated by using a tree diagram as in Figure 6.13. Because $\overline{D}$ = "bottle is satisfactory," the complement rule can be used to compute $P(\overline{D}|M_1)$, $P(\overline{D}|M_2)$, and $P(\overline{D}|M_3)$.

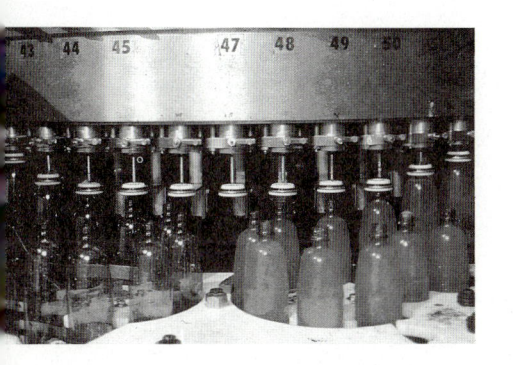

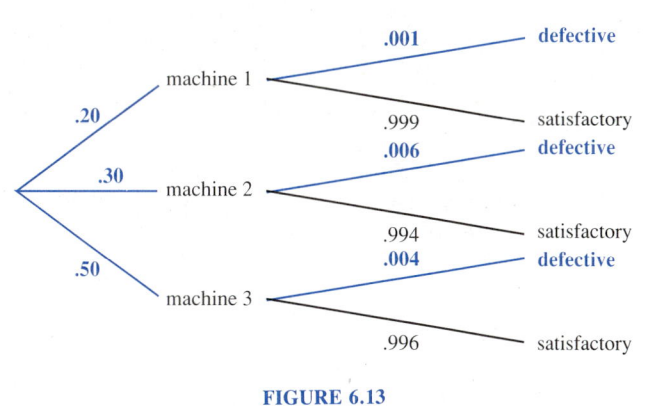

FIGURE 6.13

Notice that the three leftmost branches in this diagram are labeled with the probabilities that a bottle is processed by the corresponding machine, and the six rightmost branches are labeled with the *conditional probabilities* that a defective or satisfactory bottle is produced by the corresponding machine. A tree diagram of this sort, in which each branch is labeled with the appropriate probability or conditional probability, is called a **stochastic diagram**.

The event "bottle is defective" can be broken into three mutually exclusive cases corresponding to the three machines used to fill and cap the bottles. Each of these cases is represented by a different branch in the diagram that ends in "defective." Now the intersection rule states that

$$P(M_1 \cap D) = P(M_1) \cdot P(D|M_1),$$

and so $P(M_1 \cap D)$, the probability that a bottle is defective and was produced by machine 1, can be found by multiplying the probabilities along the corresponding branch of the tree. Thus

$$P(D) = P(M_1 \cap D) + P(M_2 \cap D) + P(M_3 \cap D)$$
$$= P(M_1) \cdot P(D|M_1) + P(M_2) \cdot P(D|M_2) + P(M_3) \cdot P(D|M_3)$$

is the sum of all the branch probabilities that end in "defective." Hence

$$P(D) = (.20)(.001) + (.30)(.006) + (.50)(.004)$$
$$= .0002 + .0018 + .0020 = .0040.$$

Therefore there is a .4% chance that a bottle will be defective.

It turns out that the probability of an event can always be computed in this manner if we have a set of events like M_1, M_2, and M_3. Accordingly we say that a set of events $A_1, A_2, \ldots, A_k$ is a **partition** of a sample space S if *exactly one* of the events occurs whenever the experiment is performed. In set-theoretic terms this means that the intersection of each pair of the events is $\emptyset$ and the union of all k of the events is S. (See Figure 6.14.)

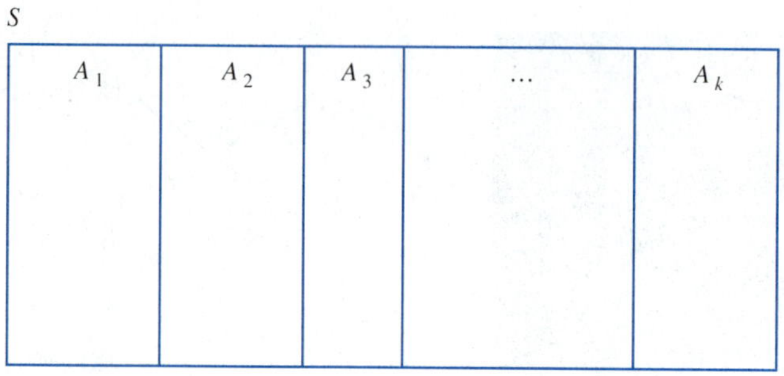

FIGURE 6.14

The effect of a partition on an arbitrary event E is to divide it into mutually exclusive subsets

$$A_1 \cap E, \quad A_2 \cap E, \quad \ldots, \quad A_k \cap E$$

as in Figure 6.15. Since

$$E = (A_1 \cap E) \cup (A_2 \cap E) \cup \cdots \cup (A_k \cap E),$$

we have
$$P(E) = P(A_1 \cap E) + P(A_2 \cap E) + \cdots + P(A_k \cap E)$$
by the union rule for mutually exclusive events. This equation is sometimes called the **law of complete probability.**

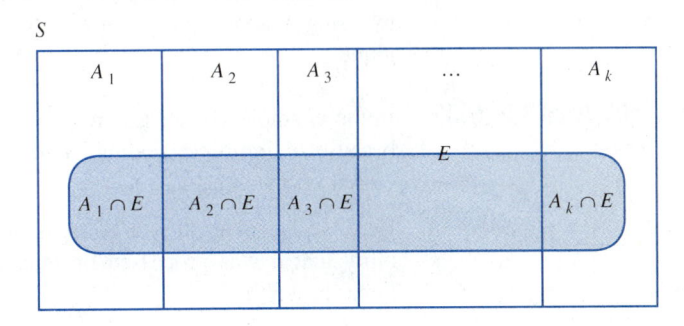

FIGURE 6.15

BAYES'S FORMULA

Given a partition $A_1, A_2, \ldots, A_k$ of a sample space, we can think of each event A_i as a "hypothesis" that has a certain probability of being true. Once an event E has occurred, we can ask how this new information alters the probability of A_i. In other words, what is the probability of A_i given that E has occurred? In the example above, for instance, we can ask the probability that a defectively capped bottle was processed on machine i. Of course, what we are seeking is the conditional probability

$$P(A_i|E) = \frac{P(A_i \cap E)}{P(E)}.$$

If we apply the intersection rule and the law of complete probability to the right side of this equation, we obtain our main result.

Theorem 6.6 (Bayes's formula*)

Let $A_1, A_2, \ldots, A_k$ be a partition of a sample space S. Then for any event $E \subseteq S$ with $P(E) \neq 0$

$$P(A_i|E) = \frac{P(A_i)P(E|A_i)}{P(A_1)P(E|A_1) + P(A_2)P(E|A_2) + \cdots + P(A_k)P(E|A_k)}.$$

---

*Thomas Bayes (1702–1761) was an English clergyman and mathematician. This theorem appeared in a famous paper entitled "An Essay Toward Solving a Problem in the Doctrine of Chances," which was published in 1763.

The significance of Bayes's formula is this: If we know the probability of E under each "hypothesis" A_i, then we can compute the conditional probability of each A_i given E. In other words, we can compute $P(A_i|E)$ in terms of each $P(E|A_i)$. Unfortunately, the notation required to express Bayes's Formula is quite complicated. However, for small values of n we can avoid the notation by using a stochastic diagram. The following example demonstrates this technique.

EXAMPLE 6.32 In the example above, a purchaser of the cola obtained a bottle that tasted flat because of improper sealing. What is the probability that the bottle was processed on machine 1?

Solution Here we are given that a bottle is defective, and we want to know the probability that it was processed on machine 1. Thus we must compute the conditional probability

$$P(\text{machine 1}|\text{defective}).$$

By definition of this conditional probability we have

$$P(\text{machine 1}|\text{defective}) = \frac{P(\text{machine 1 and defective})}{P(\text{defective})}.$$

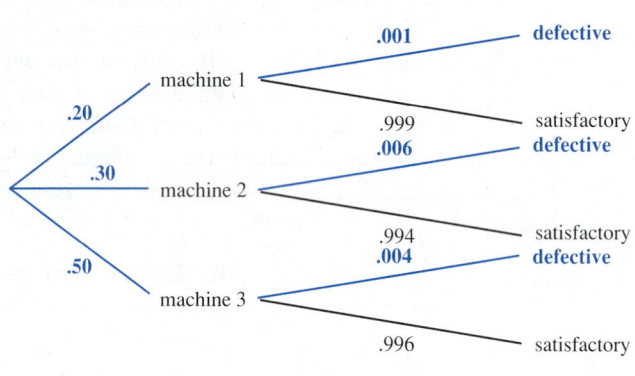

FIGURE 6.16

The right side of this equation can be easily evaluated using the stochastic diagram in Figure 6.16. Indeed, from the diagram we see that

$$P(\text{machine 1 and defective}) = (.20)(.001) = .0002,$$

and as we saw before

$$P(\text{defective}) = (.20)(.001) + (.30)(.006) + (.50)(.004) = .0040.$$

Hence

$$P(\text{machine 1}|\text{defective}) = \frac{.0002}{.0040} = \frac{2}{40} = .05.$$

Similarly we see that

$$P(\text{machine 2}|\text{defective}) = \frac{.0018}{.0040} = \frac{18}{40} = .45.$$

and

$$P(\text{machine 3}|\text{defective}) = \frac{.0020}{.0040} = \frac{20}{40} = .50.$$

Thus if we know that a bottle is defective, the probability that it was processed by machine 1 *decreases* from .20 to .05, the probability that it was processed by machine 2 *increases* from .30 to .45, and the probability that it was processed by machine 3 remains .50. In Bayesian terms, the probability of the events M_1 and M_2 has been altered by the knowledge that D occurred.

Notice that in Example 6.32 we are able to compute

$$P(\text{machine 1}|\text{defective})$$

by computing the probability along the branch containing "machine 1" and "defective" and dividing by the sum of all the branch probabilities that end in "defective." This is exactly what the complicated notation in Bayes's formula means in terms of our stochastic diagram.

EXAMPLE 6.33 A factory employs workers on three shifts. The first shift employs 40% of the workers, the second shift 35%, and the third shift 25%. The absentee rate for workers on the first shift is 1.5%, but increases to 4% on the second shift and 12% on the third. What percentage of the absentees are assigned to the first shift? To the second shift? To the third shift?

Solution In this problem we must compute

$P(\text{first shift}|\text{absent}), \quad P(\text{second shift}|\text{absent}), \quad \text{and} \quad P(\text{third shift}|\text{absent}).$

We are told the conditional probabilities of a worker's being absent given that he or she works on a particular shift. Thus since the three shifts partition the sample space of all workers, Bayes's Theorem will enable us to compute the desired probabilities. As above we will rely on a stochastic diagram, which is shown in Figure 6.17.

To compute $P(\text{first shift}|\text{absent})$, we divide the branch probability through "first shift" and "absent" by the sum of all the branch probabilities that end in "absent." In this way we find that

$$P(\text{first shift}|\text{absent}) = \frac{.40(.015)}{.40(.015) + .35(.04) + .25(.12)}$$

$$= \frac{.006}{.006 + .014 + .030} = \frac{.006}{.050}$$

$$= \frac{6}{50} = .12.$$

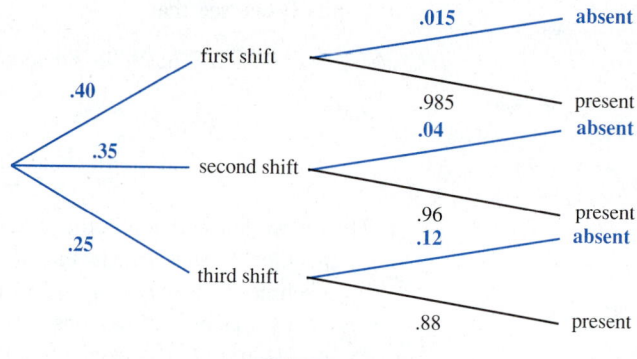

```
                                        .015 ─── absent
                       first shift ─┤
              .40                       .985 ─── present
                                        .04  ─── absent
              .35
         ─┤─── second shift ─┤
                                        .96  ─── present
              .25                       .12  ─── absent
                       third shift ─┤
                                        .88  ─── present
```

FIGURE 6.17

Note that by the law of complete probability the denominator of our first fraction,

$$.40(.015) + .35(.04) + .25(.12) = .050,$$

is the probability of a worker's being absent. This same denominator occurs in our other calculations

$$P(\text{second shift}|\text{absent}) = \frac{.35(.04)}{.050} = \frac{.014}{.050}$$

$$= \frac{14}{50} = .28$$

and

$$P(\text{third shift}|\text{absent}) = \frac{.25(.12)}{.050} = \frac{.030}{.050}$$

$$= \frac{30}{50} = .60.$$

Thus 12% of the absentees work on the first shift, 28% on the second, and 60% on the third. ∎

Practice Problem 22   In a particular county 50% of the registered voters are Democrats, 30% are Republicans, and 20% are independents. During a recent election, 40% of the Democrats voted, 60% of the Republicans voted, and 70% of the independents voted. What is the probability that someone who voted is a Democrat? A Republican? An independent?

6.5 BAYES'S FORMULA

RELIABILITY OF TESTS

Bayes's formula has important applications to the statistical evaluation of tests, particularly in medicine. In one typical situation we want to know how accurately a particular diagnostic test predicts a certain condition, that is, to what extent a positive test result can be believed.

EXAMPLE 6.34 Approximately 8% of all black Americans are carriers of sickle cell anemia, a hereditary form of anemia. A new test to detect this trait was administered to a large group of subjects known to be carriers and to another group known to be noncarriers. The test correctly identified 97% of the known carriers and 95% of the known noncarriers.* If this new test is randomly administered to a black American and the result is positive (indicating the presence of sickle cell anemia), what is the probability that the person really is a carrier?†

Solution We take as our sample space the set of all black Americans and consider the partition formed by the events "person is a carrier" and "person is a noncarrier." The probability that a randomly chosen black American is a carrier is equal to .08, the percentage of carriers in the sample space. Therefore by the complement rule the probability that he or she is a noncarrier is .92. Since the test is positive for 97% of the known carriers, the conditional probability that the test result is positive given that someone is a carrier is .97. (And therefore the conditional probability of a negative test result is .03.) Likewise the conditional probability of a *negative* test result given that someone is a noncarrier is .95, and so the conditional probability of a positive test result is .05 in this case. In this manner we obtain the stochastic diagram in Figure 6.18.

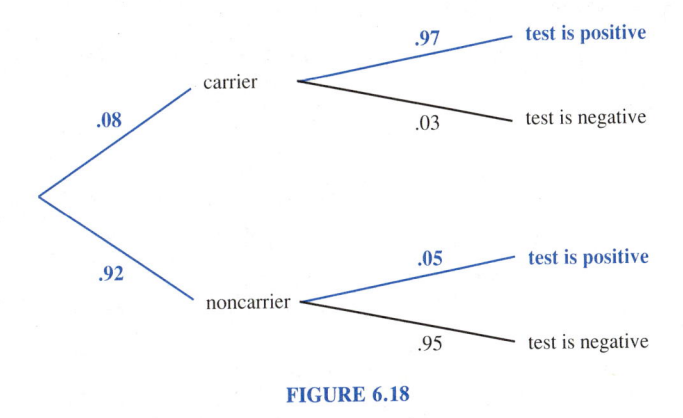

FIGURE 6.18

---

*In the terminology of diagnostic testing this test is said to have 97% *sensitivity* and 95% *specificity*.

†This probability is called the *predictive value* of the test.

What we want to know is *the probability that someone is a carrier given that he or she has a positive test result*. In other words, we want to compute

$$P(\text{carrier}|\text{test is positive}).$$

Since two branches lead to the event "test is positive," we see that

$$P(\text{carrier}|\text{test is positive}) = \frac{.08(.97)}{.08(.97) + .92(.05)}$$

$$= \frac{.0776}{.0776 + .0460}$$

$$= \frac{.0776}{.1236} = \frac{776}{1236} \approx .628.$$

Therefore a positive test result is correct only 62.8% of the time; in over one-third of the cases it is a false alarm! ■

The answer to Example 6.34 is surprising. In general, *the rarer a condition is among the population, the less reliable a diagnostic test will be*. In other words, as the percentage of actual carriers decreases, the percentage of inaccurate test results increases. As this example shows, our intuition can be very misleading in problems of statistical inference.

EXERCISES 6.5

In Exercises 1–8 find the indicated probability using the stochastic diagram shown below.

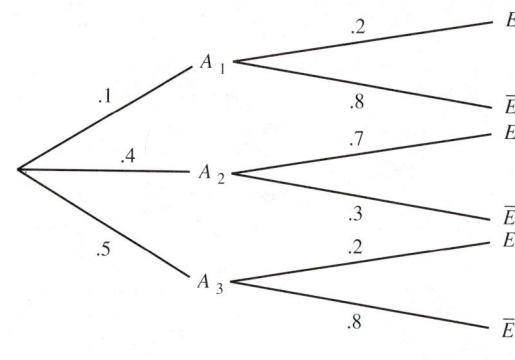

1. $P(E)$
2. $P(\overline{E})$
3. $P(A_1|E)$
4. $P(A_1|\overline{E})$
5. $P(A_2|E)$
6. $P(A_2|\overline{E})$
7. $P(A_3|E)$
8. $P(A_3|\overline{E})$

6.5 BAYES'S FORMULA

Suppose that A_1 and A_2 partition the sample space of an experiment. Find $P(E)$, $P(A_1|E)$, and $P(A_2|E)$ under the conditions in Exercises 9–12.

9. $P(A_1) = .7$, $P(E|A_1) = .4$, and $P(E|A_2) = .1$
10. $P(A_2) = .4$, $P(E|A_1) = .2$, and $P(E|A_2) = .7$
11. $P(A_2) = .8$, $P(E|A_1) = .3$, and $P(E|A_2) = .4$
12. $P(A_1) = .25$, $P(E|A_1) = .32$, and $P(E|A_2) = .12$

Suppose that A_1, A_2, and A_3 partition the sample space of an experiment. Find $P(E)$, $P(A_1|E)$, $P(A_2|E)$, and $P(A_3|E)$ under the conditions in Exercises 13–16.

13. $P(A_1) = .3$, $P(A_2) = .6$, $P(E|A_1) = .7$, $P(E|A_2) = .2$, and $P(E|A_3) = .4$
14. $P(A_1) = .5$, $P(A_3) = .1$, $P(E|A_1) = .4$, $P(E|A_2) = .6$, and $P(E|A_3) = .2$
15. $P(A_2) = .2$, $P(A_3) = .6$, $P(E|A_1) = .35$, $P(E|A_2) = .45$, and $P(E|A_3) = .15$
16. $P(A_1) = .4$, $P(A_2) = .3$, $P(E|A_1) = .5$, $P(E|A_2) = .2$, and $P(E|A_3) = .3$

17. From past experience an oil exploration firm has found that only 10% of the sites it drills yield oil. The firm is considering use of a new prediction method based on infrared analysis. When tested on a large number of randomly chosen drilling sites, the method correctly predicted oil in 82% of the sites where oil was subsequently found, and it correctly predicted no oil in 74% of the sites that subsequently proved to be dry. If the new method predicts oil at a particular site, what is the probability that oil will actually be found there?

18. A teacher found that only 60% of her class completed a particular examination. Of those who completed the examination, 90% passed, but only 75% of the others passed. What percentage of students passed this examination?

19. An automobile insurance company classifies its policyholders as either good risks (class G) or bad risks (class B). Among the company's current policyholders 95% are in class G and 5% are in class B. For drivers in class G the probability of having an accident in a given year is .01, and for drivers in class B the probability of having an accident in a given year is .06. If a policyholder reports that he or she has just had an accident, what is the probability that he or she belongs to class B?

20. An airline has hired a psychologist to determine whether prospective pilots will be able to react well in emergencies. The airline currently employs a large number of pilots, of whom 85% are known to react well. Each of the current pilots was interviewed anonymously and rated as either "satisfactory" or "unsatisfactory." Among the pilots known to react well, 92% were rated "satisfactory," compared to only 12% of the pilots known to react poorly. If a random applicant has a "satisfactory" rating, what is the probability that he or she will react well in emergencies?

21. A television consumer reporter finds that 20% of the complaints she investigates are satisfactorily resolved on her first inquiry to the company in question. The remaining 80% are presented on television. Of these, 70% are satisfactorily resolved after the complaint is aired and 30% remain unresolved. Of the complaints that are ultimately resolved, what percentage are aired on television?

22. A mail order business has an 18% response rate when it uses a mailing list purchased from a market research firm as compared to a 9% response rate for random mailings. If 60% of the items it mails are sent in random mailings, what percentage of the company's responses come from random mailings?

23. An automobile dealer sells Buicks, Oldsmobiles, and Pontiacs. His sales records show that 45% of his new car sales are Buicks, 30% are Oldsmobiles, and 25% are Pontiacs. Of those who buy Buicks, 60% want an AM-FM radio; for purchasers of Oldsmobiles and Pontiacs this percentage drops to 40% and 20%, respectively.

   (a) What is the probability that someone buying a new car from this dealer will want an AM-FM radio?

   (b) What is the probability that a car with an AM-FM radio bought from this dealer is a Buick? An Oldsmobile? A Pontiac?

24. In a particular county 40% of the registered voters are Democrats, 50% are Republicans, and 10% are independents. During a recent election, 50% of the Democrats voted, 60% of the Republicans voted, and 80% of the independents voted.

   (a) What percentage of the eligible voters cast ballots in this election?

   (b) What is the probability that someone who voted is a Democrat? A Republican? An independent?

25. An insurance company finds that 25% of its automobile insurance policies cover drivers under 25 years of age, 15% cover drivers 25–34, 40% cover drivers 35–60, and 20% cover drivers over 60. The probability of an accident during the next year for each age group is as follows: .020 for those under 25, .008 for those aged 25–34, .005 for those aged 35–60, and .012 for those over 60.

   (a) What proportion of this company's policyholders will have an accident during the next year?

   (b) What proportion of the accidents involving this company's policyholders involve drivers of each age group?

26. A physician knows that a patient's peculiar symptoms must be caused by one of four diseases. Research shows that the patient's symptoms occur in 80% of those with the first disease, in 90% of those with the second disease, in 95% of those with the third disease, and in 70% of those with the fourth disease. If these diseases are known to occur in 0.2%, 0.4%, 0.1%, and 0.3% of the population, respectively, what is the most probable cause of the patient's symptoms?

27. Three identical boxes each contain two coins. In the first box both coins are gold, in the second box one is gold and one is silver, and in the third box both coins are silver. One of the boxes is chosen at random, and a coin is randomly selected from this box. If this coin is gold, what is the probability that the other coin in this box is gold?

28. A box contains three coins, two ordinary coins and one two-headed coin. A coin is chosen at random from the box and flipped twice. If it comes up heads both times, what is the probability that the coin is two-headed?

29. Half the students in a certain class read the textbook after each lecture. The percentage of students receiving an A was twice as high among those who read the text after each lecture as among those who did not. If a student in this class got an A, what is the probability that he or she read the text after each lecture? (*Hint:* Let p be the probability that a nonreader got an A.)

30. A sales representative finds that 45% of the buyers she sees are male. Moreover, her success rate with male buyers is three times that with female buyers. If a meeting with a buyer results in a sale, what is the probability that the buyer was male? (*Hint:* Let p be the sales representative's success rate with female buyers.)

31. A do-it-yourself pregnancy test known as Ova II was evaluated in 1976 by the Center for Disease Control in Atlanta. The results, which were reported in reference [1], showed that when the test was administered to pregnant women, it was positive 50% of the time; and when the test was administered to women who were not pregnant, it was negative 41% of the time. Show that according to these findings a positive test result actually *lowers* the probability that a random woman is pregnant! (*Hint:* Let p be the probability of pregnancy.)

32. One percent of the inhabitants of an island suffer from a congenital disease having symptoms that appear only in adults. The disease is associated with the presence of the A and B antigens in the blood of those with the disease. Specifically, among those with the disease 60% have antigen A, 70% have antigen B, and 50% have both; whereas the corresponding rates of occurrence are 12%, 15%, and 9% for the nondiseased population. What is the probability that a child has the disease if his or her blood contains

(a) both antigens?   (b) antigen A but not B?

Answer to Practice Problem   22. $\frac{5}{13}, \frac{9}{26}$, and $\frac{7}{26}$

6.6 Bernoulli Trials

It often happens that an experiment is performed not once but repeatedly, and we are interested in the frequency of occurrence of a particular event E. In this section we will derive a useful formula for the probability that E will occur any specified number of times.

REPEATED INDEPENDENT TRIALS

The relative frequency interpretation of probability, which we encountered in Section 6.1, is based on the notion of an experiment as a *repeatable action* that has various possible outcomes. According to this interpretation, the probability of an event E predicts the proportion of the time E will occur over the long run.

Suppose, for instance, that two fair dice are rolled, and on each roll we are interested only in whether or not a sum of seven is obtained. Since

$$P(\text{sum} = 7) = \frac{6}{36} = \frac{1}{6},$$

over many trials we expect this event to occur one-sixth of the time. Thus out of 3000 rolls, we would expect to obtain a sum of seven approximately 500 times. But what if we obtained no sevens at all? Or what if we obtained a seven on every roll? Such occurrences are unlikely, of course, but they could happen. In general, if an event E is neither certain nor impossible, that is, if $0 < P(E) < 1$, then when the experiment is repeated n times, the number of occurrences of E might be any integer k from 0 to n. Our question now becomes: For a given value of k, what is the probability that E occurs exactly k times out of n repetitions of the experiment?

BERNOULLI PROCESSES

Since we are interested only in whether or not E occurs on each trial, we can simplify the problem by calling an occurrence of E a "success" and an occurrence of $\overline{E}$ a "failure." We call a sequence of repeated trials of an experiment a **Bernoulli process*** if the following three conditions are met:

1. Each trial has only two possible outcomes, success (denoted by S) and failure (denoted by F).
2. The trials are independent.
3. The probability of success is the same for each trial.

The terms "success" and "failure" should not be taken literally. As used here, a trial is a success if the particular outcome of interest occurs. Thus if we are counting the occurrence of adverse side effects produced by a new drug, the occurrence of adverse side effects is called a success, and their absence is called a failure. It is customary to denote the probability of success by p and the probability of failure by q. Note that since success and failure are complementary events, it follows from the complement rule that $p + q = 1$.

For example, suppose that we roll a pair of fair dice three times. If the sum is seven, the outcome is considered a success, otherwise it is considered a failure. The three rolls represent $n = 3$ Bernoulli trials, where for each trial we have

$$p = P(S) = P(\text{sum} = 7) = \frac{6}{36} = \frac{1}{6}$$

$$q = P(F) = 1 - P(S) = 1 - \frac{1}{6} = \frac{5}{6}.$$

This process can be represented by the stochastic diagram in Figure 6.19.

There are $2^3 = 8$ possible outcomes represented by the various paths through the stochastic diagram, but they are not all equally likely. For example, the probability of obtaining three sevens is

$$P(SSS) = p \cdot p \cdot p = p^3 = \left(\frac{1}{6}\right)^3 = \frac{1}{216},$$

whereas the probability of rolling three nonsevens is

$$P(FFF) = q \cdot q \cdot q = q^3 = \left(\frac{5}{6}\right)^3 = \frac{125}{216}.$$

---

*The Swiss mathematician Jacob Bernoulli (1654–1705) was among the first to study problems involving repeated trials. His work the *Ars Conjectandi*, published in 1713, was a landmark in probability theory.

6.6 BERNOULLI TRIALS

```
First        Second       Third   Outcome
Roll         Roll         Roll
                              p — S    SSS
                      p — S <
                              q — F    SSF
              p — S <
                              p — S    SFS
                      q — F <
      S <                     q — F    SFF
  p <                         p — S    FSS
                      p — S <
      q                       q — F    FSF
          F <                 p — S    FFS
                      q — F <
                              q — F    FFF
```

FIGURE 6.19

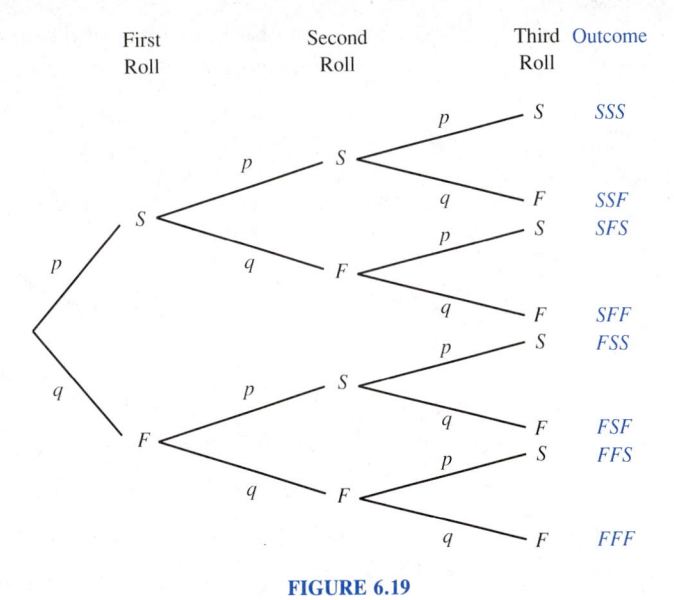

To find the probability of rolling *exactly one* seven, we add the probabilities of the three favorable branches:

$$P(\text{exactly one } 7) = P(SFF) + P(FSF) + P(FFS)$$
$$= pqq + qpq + qqp$$
$$= pq^2 + pq^2 + pq^2$$
$$= 3pq^2 = \frac{75}{216} = \frac{25}{72}.$$

Similarly, the probability of rolling exactly two sevens is

$$P(\text{exactly two } 7s) = P(SSF) + P(SFS) + P(FSS)$$
$$= ppq + pqp + qpp$$
$$= p^2q + p^2q + p^2q$$
$$= 3p^2q = \frac{15}{216} = \frac{5}{72}.$$

Hence for $k = 0, 1, 2, 3$ we have computed the probability of obtaining exactly k sevens in $n = 3$ trials.

Now let us determine the probability of obtaining exactly two sevens in five rolls of a pair of fair dice. If we extend the tree diagram in Figure 6.19 by adding another two levels of branching, we will obtain a sample space for the experiment of rolling a pair of dice five times. Of the $2^5 = 32$ outcomes, there are 10 containing exactly two successes, namely, those listed below.

$$\begin{array}{ccccc} SSFFF & SFSFF & SFFSF & SFFFS & FSSFF \\ FSFSF & FSFFS & FFSSF & FFSFS & FFFSS \end{array}$$

There are 10 such outcomes because the two positions that the successes occupy can be chosen in

$$C(5, 2) = \frac{P(5, 2)}{2!} = \frac{5 \cdot 4}{2 \cdot 1} = 10$$

different ways. Moreover, *all ten of these favorable outcomes have the same probability*, namely p^2q^3. For instance, by independence we have

$$P(SFFSF) = P(S) \cdot P(F) \cdot P(F) \cdot P(S) \cdot P(F) = p \cdot q \cdot q \cdot p \cdot q = p^2q^3,$$
$$P(FSFFS) = P(F) \cdot P(S) \cdot P(F) \cdot P(F) \cdot P(S) = q \cdot p \cdot q \cdot q \cdot p = p^2q^3,$$

and similarly for the other favorable outcomes. Therefore

$$\begin{aligned} P(\text{exactly 2 sevens in 5 rolls}) &= P(SSFFF) + \cdots + P(FFFSS) \\ &= p^2q^3 + \cdots + p^2q^3 \\ &= 10p^2q^3 = 10\left(\frac{1}{6}\right)^2\left(\frac{5}{6}\right)^3 \\ &= \frac{1250}{7776} = \frac{625}{3888} \approx .161. \end{aligned}$$

Consequently, about 16.1% of the time we expect to obtain exactly two sevens in five rolls of a pair of fair dice.

By generalizing the reasoning above, we obtain the following formula for the probability of obtaining exactly k successes in a sequence of n Bernoulli trials.

Theorem 6.7 (Bernoulli's formula)

Let p be the probability of success and $q = 1 - p$ the probability of failure on a single trial of a Bernoulli process. Then the probability of obtaining exactly k successes in a sequence of n Bernoulli trials, where $k = 0, 1, 2, \ldots, n$, is given by

$$P(\text{exactly } k \text{ successes in } n \text{ trials}) = C(n, k)p^kq^{n-k}.$$

Note that the expression $C(n, k)p^kq^{n-k}$ in Bernoulli's formula is a term in the binomial theorem expansion of $(p + q)^n$. (See Section 5.5.) The exponents k and $n - k$ that appear in this term are the number of successes and number of failures, respectively.

Bernoulli's formula agrees with our earlier calculation of the probabilities of obtaining k sevens in three rolls of a pair of fair dice; that is,

$$P(0 \text{ sevens}) = C(3, 0)p^0q^3 = q^3 = \left(\frac{5}{6}\right)^3 = \frac{125}{216},$$

$$P(\text{exactly 1 seven}) = C(3, 1)p^1q^2 = 3pq^2 = 3\left(\frac{1}{6}\right)\left(\frac{5}{6}\right)^2 = \frac{75}{216},$$

$$P(\text{exactly 2 sevens}) = C(3, 2)p^2q^1 = 3p^2q = 3\left(\frac{1}{6}\right)^2\left(\frac{5}{6}\right) = \frac{15}{216},$$

$$P(3 \text{ sevens}) = C(3, 3)p^3q^0 = p^3 = \left(\frac{1}{6}\right)^3 = \frac{1}{216}.$$

APPLICATIONS

Bernoulli's formula can be used in a wide variety of situations when an experiment is repeated under identical conditions.

EXAMPLE 6.35 What is the probability that out of ten flips, a fair coin will come up heads

(a) exactly five times?
(b) either four, five, or six times?

Solution (a) If we define "success" to mean that heads comes up when the coin is flipped, then $p = P(S) = \frac{1}{2}$ and $q = P(F) = \frac{1}{2}$. Hence by Bernoulli's formula

$$P(\text{exactly 5 heads in 10 flips}) = P(\text{exactly 5 successes in 10 trials})$$

$$= C(10, 5)p^5q^5 = 252\left(\frac{1}{2}\right)^5\left(\frac{1}{2}\right)^5$$

$$= \frac{252}{1024} \approx .246.$$

Thus half of the 10 flips are heads less than 25% of the time.

(b) As in part (a) we find that

$$P(\text{exactly 4 heads in 10 flips}) = C(10, 4)p^4q^6 = 210\left(\frac{1}{2}\right)^4\left(\frac{1}{2}\right)^6$$

$$= \frac{210}{1024}$$

and

$$P(\text{exactly 6 heads in 10 flips}) = C(10, 6)p^6q^4 = 210\left(\frac{1}{2}\right)^6\left(\frac{1}{2}\right)^4$$

$$= \frac{210}{1024}.$$

Since obtaining exactly 4, 5, or 6 heads are mutually exclusive events, the union rule for mutually exclusive events gives

$$P(\text{either 4, 5, or 6 heads}) = P(4 \text{ heads}) + P(5 \text{ heads}) + P(6 \text{ heads})$$
$$= \frac{210}{1024} + \frac{252}{1024} + \frac{210}{1024}$$
$$= \frac{672}{1024} = .65625.$$

Hence in 10 flips of a fair coin we obtain 4, 5, or 6 heads over 65% of the time. ■

EXAMPLE 6.36 Sixty percent of the voters in a large metropolitan area favor development of a proposed mass transit system. If seven voters are sampled at random, what is the probability that a majority of them will favor this system?

Solution Each time a voter is questioned we will declare a success if the voter favors the proposed system and a failure otherwise. Since the size of the metropolitan area is very large compared to the size of the sample, the probability of success on each trial remains approximately the same, namely, $p = .60$. Therefore the probability that a majority of the seven voters favor the project is given by

$P(\text{at least 4 successes in 7 trials})$
$= P(4 \text{ successes}) + P(5 \text{ successes}) + P(6 \text{ successes}) + P(7 \text{ successes})$
$= C(7, 4)p^4 q^3 + C(7, 5)p^5 q^2 + C(7, 6)p^6 q^1 + C(7, 7)p^7 q^0$
$= 35(.6)^4(.4)^3 + 21(.6)^5(.4)^2 + 7(.6)^6(.4) + (.6)^7$
$\approx .290 + .261 + .131 + .028 = .710.$

Note that there is an approximately 29% chance that a majority of those sampled will *not* favor development of the proposed mass transit system, and thus this small sample can give a misleading picture of the population at large. This example demonstrates the dangers of a prediction based on too small a sample. ■

Practice Problem 23 A 1987 poll by Louis Harris found that in any given year only 2% of American marriages end in divorce. If this trend continues, what is the probability that there will be exactly 1 divorce among 10 unacquainted married couples during the next year?

EXAMPLE 6.37 A manufacturer of light bulbs finds that 2% of the bulbs produced are defective. If the bulbs are packaged in boxes of four, what percentage of the boxes contains at least one defective bulb?

Solution   Although it is possible to compute the desired probability directly as in the previous example, in this case we can shorten the calculation by using the complement rule. In this way we see that

$$\begin{aligned}P(\text{at least 1 defective}) &= 1 - P(0 \text{ defectives}) \\ &= 1 - C(4, 0)(.02)^0(.98)^4 \\ &= 1 - (.98)^4 \\ &\approx 1 - .922 = .078.\end{aligned}$$

Thus about 7.8% of the boxes will contain a defective bulb. ■

Practice Problem 24   An insurance salesman finds that 10% of her random telephone calls lead to a sale. What is the probability that at least one sale will result from placing calls to 8 persons chosen at random?

We will conclude this section with an application in which Bernoulli's formula is used to determine the number of trials.

EXAMPLE 6.38   From prior experience a manufacturer knows that the probability of making a successful bid on a government contract is .2. What is the smallest number of contracts that must be bid on in order for the manufacturer to have a 90% chance of receiving at least one?

Solution   We will regard each submission of a bid as a Bernoulli trial in which the probability of success (receiving the contract) is .2 and the probability of failure is .8. If n bids are submitted, then the probability of receiving at least one contract is

$$\begin{aligned}P(\text{at least 1 contract received}) &= 1 - P(0 \text{ contracts received}) \\ &= 1 - C(n, 0)(.2)^0(.8)^n \\ &= 1 - (.8)^n.\end{aligned}$$

Since we want the manufacturer to have a 90% chance of receiving at least one contract, we must determine the smallest integer n for which

$$1 - (.8)^n \geq .90,$$

that is, the smallest integer n for which

$$.10 \geq (.8)^n.$$

By trial and error (or the use of logarithms), we see that $n = 11$ is the smallest number of contracts that must be bid on so as to assure a 90% chance of receiving one. ■

EXERCISES 6.6

1. Consider a sequence of $n = 6$ Bernoulli trials, where the probabilities of success and failure on each trial are $p = .7$ and $q = .3$, respectively. Determine the probability of obtaining
   (a) exactly 3 successes
   (b) 6 successes
   (c) at least 5 successes
   (d) at least one failure.

2. Consider a sequence of $n = 10$ Bernoulli trials, where the probabilities of success and failure on each trial are $p = 1/3$ and $q = 2/3$, respectively. Determine the probability of obtaining
   (a) no successes
   (b) exactly 2 successes
   (c) at least one success
   (d) at most two successes.

3. A fair die is rolled three times. What is the probability that the number 4 comes up
   (a) no times?   (b) once?   (c) twice?   (d) three times?

4. A coin is weighted in such a way that the probability of heads is .2 on each flip. If this coin is flipped four times, what is the probability of obtaining
   (a) four tails?
   (b) three heads and one tail?
   (c) two heads and two tails?
   (d) one head and three tails?

5. American roulette wheels have 38 equal compartments numbered 00, 0, 1, . . ., 36. The numbers 00 and 0 are green, and the numbers 1–36 are half red and half black. If a player bets four times in a row that red will come up, what is the probability that
   (a) the player wins all four bets?   (b) the player wins at least three times?

6. An archer has a 75% chance of hitting the target on each shot. If he takes five shots, what is the probability that he will hit the target on
   (a) exactly four of them?   (b) at least four of them?

7. The recovery rate for a certain disease is 75%; that is, on the average, 75% of those with the disease recover. If twelve patients have the disease, what is the probability that exactly nine of the twelve recover?

8. In a certain county, 40% of the registered voters are women. Citizens are selected at random for jury duty from among the registered voters. If twenty persons are called for jury duty, what is the probability that fewer than four are women?

9. Of the items produced on a certain machine, 5% are defective. If ten items are chosen at random from this machine's output, what is the probability that at most one item is defective?

10. A multiple-choice examination consists of ten questions, each having five possible answers, only one of which is correct. If a student randomly guesses the answer to each question, what is the probability that the student answers no more than two questions correctly?

11. A husband and wife are both carriers of a recessive gene for sickle-cell anemia. Although they are not affected by the disease, the probability is 1/4 that each of their children will have the disease. If they have four children, what is the probability that all are free of the disease?

12. Physicist Freeman Dyson, describing British bombing missions during World War II, writes in reference [4]:

    The normal tour of duty for a crew in a regular squadron was thirty missions. The loss-rate during the middle years of the war [that is, the probability of being shot down during a mission] averaged four percent. This meant that a crewman had three chances in ten of completing a normal tour of duty.

    Justify this conclusion.

13. A district attorney has the names of ten persons who witnessed an accident. She knows that in general only 20% of all witnesses can be convinced to testify. If she needs the testimony of at least two witnesses to get a conviction, what are her chances of obtaining a conviction?

14. The famous English diarist Samuel Pepys posed the following question to Isaac Newton in 1693 (see reference [8]). "*A* has six dice in a box, with which he is to fling [at least] one six. *B* has in another box twelve dice, with which he is to fling [at least] two sixes. *C* has in another box eighteen dice, with which he is to fling [at least] three sixes. Question: Whether *B* and *C* have not as easy a taske as *A*, at even luck?" Newton replied that "an easy computation" showed *A* to have the advantage. Justify Newton's conclusion.

15. A workshop has seven lathes and ten full-time employees who work independently of one another. If each worker uses a lathe 60% of the time, what is the probability that at any given moment there will not be enough lathes available for all who need them?

16. A restaurant has found that 20% of those with reservations fail to show up. If it books twelve reservations for ten available tables, what is the probability that everyone who shows up can be seated?

17. A sales representative has a 10% chance of making a sale at any given house. How many houses must be visited in order that the probability of making at least one sale is .50 or higher?

18. What is the smallest number of fair coins that must be flipped in order that the probability of obtaining at least one head is .99 or higher?

19. One of the two gambling problems posed to Pascal by the Chevalier de Méré that led to the development of probability theory was to compare the following probabilities:

    (a) the probability that at least one six appears if a fair die is rolled four times

    (b) the probability that at least one double-six appears if two fair dice are rolled 24 times.

    Calculate these two probabilities.

20. In a certain city the weather bureau takes temperature readings in the center of the city and at the airport thirty miles away. From past experience it is known that the downtown readings are higher 60% of the time. What is the probability that, in ten independent readings, the downtown temperature is higher at least eight times?

21. On Main Street there are four traffic signals that operate independently. Each signal remains green for two minutes and red for one minute. What is the probability that a car traveling on Main Street will pass these four signals without encountering a red light?

22. Two evenly matched teams, *A* and *B,* are playing a best-of-seven series. What is the probability that team *A* will win the series in exactly seven games? (*Hint:* Team *A* must win exactly three of the first six games.)

23. A pharmaceutical company claims that its new drug is 90% effective in curing a certain disease. It is known that half of those with this disease recover without treatment. To test the manufacturer's claim, the drug has been administered to ten patients with the disease, and the claim will be accepted if eight or more recover.
   (a) What is the probability that the drug will fail the test if the claim is true?
   (b) What is the probability that the drug will pass the test even if it has no effect on the disease?

24. In a test of extrasensory perception, an experimenter concentrates on a card chosen at random from a standard deck of playing cards, and a subject attempts to guess its suit (spades, hearts, diamonds, or clubs). If a subject guesses the suit correctly at least four times out of five, he or she is considered to be clairvoyant.
   (a) What is the probability that someone who guesses suits at random will be considered clairvoyant?
   (b) What is the probability that out of 100 subjects, each of whom guesses suits at random, at least one will be considered clairvoyant?

Answers to Practice Problems   **23.** approximately .167
                                    24. approximately .570

CHAPTER 6 REVIEW

IMPORTANT TERMS

- certain event *(6.1)*
- event
- experiment
- impossible event
- outcomes
- probability
- relative frequency
- sample space
- odds against E *(6.2)*
- odds in favor of E
- probability distribution
- mutually exclusive events *(6.3)*
- conditional probability ($P(A|E)$) *(6.4)*
- independent events
- Bayes's formula *(6.5)*
- law of complete probability
- partition of a sample space
- stochastic diagram
- Bernoulli process *(6.6)*

IMPORTANT FORMULAS

The Counting Rule

Let S be a finite sample space consisting of equally-likely outcomes. Then for any event $E \subseteq S$, $P(E) = \dfrac{|E|}{|S|}$.

The Complement Rule

For any event A, $P(A) + P(\overline{A}) = 1$.

The Union Rule

For any events A and B, $P(A \cup B) = P(A) + P(B) - P(A \cap B)$.

The Union Rule for Mutually Exclusive Events

If each pair of the events $A_1, A_2, \ldots, A_k$ is mutually exclusive, then
$$P(A_1 \cup A_2 \cup \cdots \cup A_k) = P(A_1) + P(A_2) + \cdots + P(A_k).$$

The Counting Rule for Conditional Probabilities

If A and E are events in a finite sample space consisting of equally-likely outcomes, then
$$P(A|E) = \frac{|A \cap E|}{|E|}.$$

The Intersection Rule

For any events A and B, $P(A \cap B) = P(A) \cdot P(B|A)$.

The Intersection Rule for Independent Events

If $A_1, A_2, \ldots, A_k$ are independent events, then
$$P(A_1 \cap A_2 \cap \cdots \cap A_k) = P(A_1) \cdot P(A_2) \cdots \cdot P(A_k).$$

Bernoulli's Formula

If p and q are the probability of success and failure, respectively, on a single trial of a Bernoulli process, then $P(\text{exactly } k \text{ successes in } n \text{ trials}) = C(n, k) \, p^k q^{n-k}$.

REVIEW EXERCISES

1. The frequency table below shows the distribution of letter grades awarded over the last few semesters in Professor Gorki's Russian literature course.

Grade	Frequency
A	36
B	63
C	129
D	45
F	27
	300 Total

   Based on this empirical data, what is the probability that a random student in this course will earn
   (a) an A?   (b) at least a B?   (c) at most a C?   (d) a passing grade?

2. If the odds in favor of an event are 18 to 7, what is its probability?

3. If an event E has probability .4, what are the odds against E?

4. If a person is chosen at random, what is the probability that he or she was born on a Tuesday?

5. A large TV station has narrowed its search for a new programming director to three candidates: Anderson, Bertolucci, and Chang. Bertolucci has been told that her chances of being chosen are twice as good as Anderson's but only half as good as Chang's. What is the probability that Bertolucci will get the job?

6. If two fair dice are rolled, what is the probability that the sum will be less than five?

7. If six cards are dealt at random from a standard 52-card deck, what is the probability that four are red and two are black?

8. Adriana and Barbara are classmates. If their class of twenty students is randomly divided into two groups of ten, what is the probability that they will be placed in the same group?

9. Suppose the probability that A occurs is .2, the probability that B occurs is .5, and the probability that both occur is .1. What is the probability that

   (a) A or B occurs?

   (b) neither A nor B occurs?

   (c) B occurs given that A has occurred?

10. In a certain community 40% of the population reads the morning newspaper, 32% reads the afternoon newspaper, and 18% reads both.

    (a) What percentage of the community reads at least one of the papers?

    (b) If someone reads the morning paper, what is the probability that he or she also reads the afternoon paper?

11. Let A, B, and C be mutually exclusive events with probabilities .07, .23, and .12, respectively. What is the probability that none of these three events occurs?

12. A clothes dryer contains twelve identical gray socks and eight identical brown socks. If two socks are removed from the dryer at random, what is the probability of obtaining a matching pair?

13. Of the students entering a certain medical school, 36% will fail out during the first year and 48% will complete the full four-year program. If an entering student survives the first year, what is the probability that he or she will complete the full program?

14. An encyclopedia salesman has found that he has a 20% chance of getting his foot in the door when he calls at a given house. If he does get his foot in the door, then he has a 50% chance of making a sale; but if he does not get his foot in the door, then he has only a 10% chance of making a sale. What is the probability that he will make a sale at a randomly chosen house?

15. Two swimmers are independently attempting to swim from Cuba to Florida. If their chances of success are 20% and 30%, what is the probability that at least one of them succeeds?

16. A certain disease is known to occur in 5% of the population. Among those with the disease, 80% carry a blood antigen called the X-factor, whereas only 10% of those without the disease carry the X-factor. If a random person is found to carry the X-factor, what is the probability that he or she has the disease?

17. The voters in a certain congressional district are 60% white, 15% black, and 25% Hispanic. According to local polls, the Democratic candidate for Congress is supported by 55% of the white voters, 80% of the black voters, and 60% of the Hispanic voters.

(a) According to this poll, what percentage of the district's voters supports this candidate?

(b) What percentage of this candidate's support comes from the black community?

18. Assuming that a newborn kitten is equally likely to be male or female, what is the probability that in a litter of four, there are

    (a) two males and two females?

    (b) at least one kitten of each sex?

19. Each year 2% of the individual tax returns in a certain state are subject to a random audit. What is the probability that a given taxpayer

    (a) will not undergo this audit during the next ten years?

    (b) will undergo this audit at most once in the next ten years?

20. A television network is airing eight new situation comedies this season. If past experience shows that only one of every five new situation comedies is successful, what is the probability that

    (a) at least one of the new sitcoms is successful?

    (b) at least two of the new sitcoms are successful?

REFERENCES

1. Baker, L. D., et al., "Evaluation of a 'do-it-yourself' pregnancy test," *American Journal of Public Health,* vol. 66 (1976), p. 166.

2. Berelson, Bernard R., Paul F. Lazarsfeld, and William N. McPhee, *Voting*. Chicago: University of Chicago Press, 1954.

3. Bourne, Larry S., "Physical Adjustment Process and Land Use Succession: A Conceptual Review and Central City Example," *Economic Geography,* vol. 47 (1971), pp. 1–15.

4. Dyson, Freeman, "Reflections: Disturbing the Universe–I," *The New Yorker,* August 6, 1979, p. 46.

5. Hampton, P. "Regional Economic Development in New Zealand," *Journal of Regional Science,* vol. 8 (1968), pp. 41–51.

6. Kannel, William B., M.D., "Recent Findings of the Framingham Study," *Resident & Staff Physician,* January 1978, p. 57.

7. Prais, S. J., "Measuring Social Mobility," *Journal of the Royal Statistical Society,* vol. 118 (1955), pp. 56–66.

8. Schell, E. D., "Samuel Pepys, Isaac Newton and Probability," *The American Statistician,* October 1960, pp. 27–30.

7

STATISTICS

7.1 Descriptive Statistics

7.2 Grouped Data

7.3 Random Variables: Expected Value and Variance

7.4 Mathematics in Action: Decision Theory

7.5 Probability Distributions

7.6 The Normal Distribution

Chapter Review

7.1 Descriptive Statistics

Statistics is the branch of mathematics concerned with classifying and interpreting numerical data. Because of its many important applications, this subject has become a specialized discipline in its own right. The branch of statistics that deals with collecting, organizing, and summarizing data is called *descriptive statistics,* and the branch that is concerned with drawing conclusions and making predictions is called *inferential statistics*. In this chapter we present a brief survey of some statistical techniques that are related to the discussion of probability in Chapter 6.

7.1 Descriptive Statistics

This section is concerned with the representation of data, which can be classified as either *qualitative* or *quantitative* according to the nature of the categories into which the measurements fall. For example, classifying "countries that produce steel" involves nonnumerical categories and hence is qualitative. On the other hand, classifying "age groups in the United States" involves numerical categories and is therefore quantitative.

A common means for depicting qualitative data like those in Table 7.1 is with a **bar graph** as shown in Figure 7.1. Here the quantity of steel produced by a country is denoted by the length of a horizontal bar. For quantitative data like those in Table 7.2, a somewhat different type of graph is used. (See Figure 7.2.) Here the bars are constructed vertically, and the height of each bar represents the number of persons in the corresponding age bracket. This number is displayed either over the bar (as in Figure 7.2) or within the bar. This type of graph is called a **histogram.** Notice that unlike a bar graph, a histogram has no spaces between its bars.

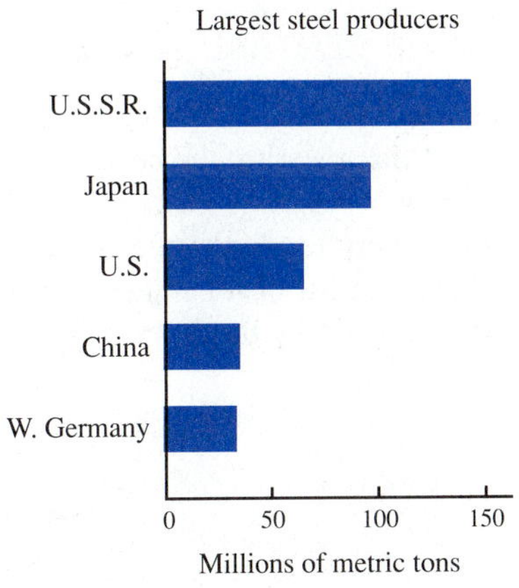

Largest steel producers*
(millions of metric tons)

U.S.S.R.	147.2
Japan	99.5
U.S.	67.7
China	37.1
West Germany	35.9

TABLE 7.1

U.S. population by age†
(millions)

0– 20	69.2
20– 40	69.6
40– 60	45.0
60– 80	29.2
80–100	5.0

TABLE 7.2

*1982 figures
†1980 figures

FIGURE 7.1

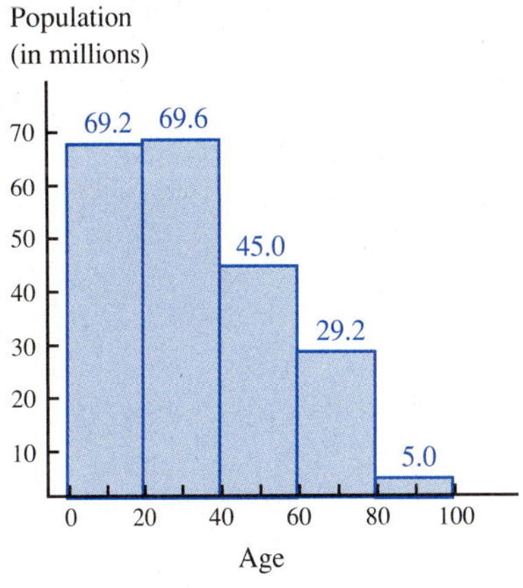

FIGURE 7.2

In working with quantitative data, the numerical categories (such as 0–20, 20–40, 40–60, 60–80, and 80–100 in Table 7.2) are called **class intervals.** It may sometimes happen that a data value falls on the boundary of two different class intervals. For instance, this occurs in Table 7.2 with someone 20 years old. *In this case we will follow the convention of assigning a data value to the higher of the class intervals into which it falls;* so a 20-year-old would be counted in the class interval 20–40 rather than the interval 0–20. This means that the class interval 20–40 in Table 7.2 includes all the ages greater than or equal to 20 and less than 40, and a similar interpretation holds for the other intervals in this table.

The data in Table 7.3 illustrate two common difficulties that occur with quantitative data. Notice that the fifth and sixth class intervals cover unequal income ranges ($5000 and $10,000, respectively). Moreover the last interval ("$50,000 or more") is open-ended, that is, no upper limit is specified. A histogram is normally used only for quantitative data such as in Table 7.2 in which *there are no open-ended intervals and each class interval covers an equal range of values.*

Income of U.S. families*	
Below $5000	4.8%
$5000–$10,000	8.5%
$10,000–$15,000	10.2%
$15,000–$20,000	10.5%
$20,000–$25,000	10.3%
$25,000–$35,000	18.6%
$35,000–$50,000	18.8%
$50,000 or more	18.3%

TABLE 7.3

MEASURES OF CENTRAL TENDENCY

In describing a set of data the first task is to characterize its center or "typical" value. Numbers that measure the center of a set of data are called **measures of central tendency.**

The most widely used measure of central tendency is the mean (or average). The **mean** of a list of n numbers equals their sum divided by n. Thus the mean can be computed by the following formula.

Mean ■ ■ ■ ■ ■

The mean $\bar{x}$ of $x_1, x_2, \ldots, x_n$ is

$$\bar{x} = \frac{x_1 + x_2 + \cdots + x_n}{n}. \tag{7.1}$$

---

*1985 figures

By using the **summation sign** Σ, which is read "sum of", we can express this formula in the more compact form

$$\bar{x} = \frac{\Sigma x}{n}.$$

EXAMPLE 7.1 A student received grades of 84, 94, 87, and 79 on four hourly examinations. Find her mean (average) score.

Solution The student's mean score is

$$\bar{x} = \frac{\Sigma x}{n} = \frac{84 + 94 + 87 + 79}{4} = 86.$$

Practice Problem 1 Find the mean of 26, 38, 17, 32, and 41.

EXAMPLE 7.2 An automobile assembly plant recorded the following numbers of absentees on fifty consecutive working days.

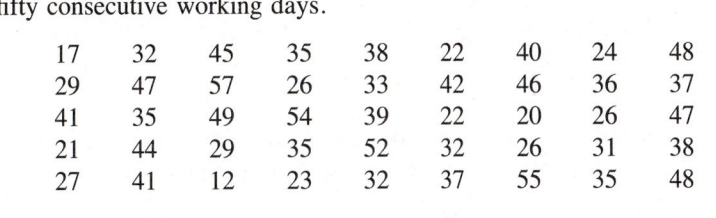

17	32	45	35	38	22	40	24	48	53
29	47	57	26	33	42	46	36	37	16
41	35	49	54	39	22	20	26	47	43
21	44	29	35	52	32	26	31	38	47
27	41	12	23	32	37	55	35	48	46

Find the mean number of absentees from the plant on these fifty days.

Solution Using the fifty values listed above, we find after much calculation that

$$\bar{x} = \frac{\Sigma x}{n} = \frac{17 + 32 + 45 + \cdots + 46}{50}$$

$$= \frac{1810}{50} = 36.2.$$

This is the mean of the fifty observed values. In Section 7.2 we will discuss a method for approximating this number that requires much less calculation.

Practice Problem 2 The following data show the number of tornadoes occurring in the United States during the years 1960–1984.

618	683	658	461	713
898	570	912	660	604
649	888	741	1102	947
920	835	852	788	852
866	783	1046	931	907

Determine the mean number of tornadoes reported in the United States during these years.

Another common measure of central tendency is the **median,** which is the middle value when the data are arranged in increasing order (or the average of the two middle values if there are an even number of data values). For example, to find the median of the seven numbers

$$10, 2, 8, 7, 4, 5, 3,$$

we must arrange them in increasing order

$$2, 3, 4, 5, 7, 8, 10.$$

The middle value, 5, is the median.

If there is another data value, say 9, then the ordered list is

$$2, 3, 4, 5, 7, 8, 9, 10.$$

In this case there are an even number of values, and so instead of a single middle value there are two, 5 and 7. Here the median is the average of these values

$$\frac{5 + 7}{2} = 6.$$

EXAMPLE 7.3 Find the median number of absentees from the automobile assembly line using the data presented in Example 7.2.

Solution If we arrange the raw data in increasing order (a tedious job), we obtain the list below.

12	16	17	20	21	22	22	23	24	26
26	26	27	29	29	31	32	32	32	33
35	35	35	35	36	37	37	38	38	39
40	41	41	42	43	44	45	46	46	47
47	47	48	48	49	52	53	54	55	57

The middle values are 36 and 37. Hence the median is

$$\frac{36 + 37}{2} = 36.5. \blacksquare$$

Practice Problem 3 Determine the median number of tornadoes reported in the United States during the years 1960–1984 using the data given in Practice Problem 2.

When there are a large number of data (as in Example 7.3), the median can be more difficult to calculate than the mean because the data must be ordered. Unlike the mean, however, the median has the advantage of being unaffected by a few atypical large or small values. (See Exercises 15–16.)

MEASURES OF DISPERSION

When Mr. Smith learned that he was being promoted to a position as regional manager in Austin, Texas, he contacted two real estate agencies in order to sell his present house before moving. The first agency claimed that the last six houses comparable to Mr. Smith's that were listed with it had sold in an average of 60 days. The second agency said that the last six houses comparable to Mr. Smith's that were listed with it had sold in an average of 67 days. Which agency should Mr. Smith choose in order to sell his house within 90 days?

With no further information, Mr. Smith should choose the first agency because comparable houses listed recently with the first agency sold one week sooner on the average. In this case, however, Mr. Smith was able to request the data on which the agencies' claims were based. These data are shown below.

Number of days before sale for the first agency	Number of days before sale for the second agency
42	48
99	75
20	83
104	68
65	57
30	71

Looking at these data, Mr. Smith noticed that all six of the houses listed with the second agency had sold within 90 days. On the other hand, only four of the six houses listed with the first agency had sold within 90 days. On this basis Mr. Smith listed his house with the second agency.

In this example, Mr. Smith is concerned more with selling his house within 90 days than in selling it quickly. Thus for this situation the *distribution* of the data is more important than its center. In comparing data, we must be able to describe not only the center of the data but also its distribution. For reasons that will become clear later, it is most useful to measure the distribution of the data around the mean. One common such measure is the variance.

Variance

The **variance*** of $x_1, x_2, \cdots, x_n$ is

$$s^2 = \frac{(x_1 - \bar{x})^2 + (x_2 - \bar{x})^2 + \cdots + (x_n - \bar{x})^2}{n}. \tag{7.2}$$

---

*When the data are a sample of measurements from a larger population, the denominator in (7.2) is often replaced by $n - 1$ for technical reasons. For simplicity we will use only formula (7.2) in this book.

Note that formula (7.2) can be written using the summation sign as

$$s^2 = \frac{\Sigma(x - \bar{x})^2}{n}.$$

Thus the variance is the average of the squares of the distances between the data values and the mean. (The squaring is done to make all the terms in the summation positive. Without squaring, it would always turn out that

$$\frac{\Sigma(x - \bar{x})}{n} = 0$$

because $\bar{x}$ is, by definition, the average of the x values.) We see from the variance formula that when the values of x are close to the mean, the value of s^2 will be small; and when the values of x are far from the mean, the value of s^2 will be large.

Using the agencies' selling times above, we see that the mean selling time for the first agency is

$$\frac{42 + 99 + 20 + 104 + 65 + 30}{6} = \frac{360}{6} = 60$$

days, as it claimed. Hence the variance of these numbers is

$$\frac{(42 - 60)^2 + (99 - 60)^2 + (20 - 60)^2 + (104 - 60)^2 + (65 - 60)^2 + (30 - 60)^2}{6}$$

$$= \frac{324 + 1521 + 1600 + 1936 + 25 + 900}{6} = \frac{6306}{6} = 1051.$$

Because the variance for the first agency is large, its selling times are widely dispersed from the mean. On the other hand, the mean selling time for the second agency is

$$\frac{48 + 75 + 83 + 68 + 57 + 71}{6} = \frac{402}{6} = 67,$$

and so the variance of these numbers is

$$\frac{(48 - 67)^2 + (75 - 67)^2 + (83 - 67)^2 + (68 - 67)^2 + (57 - 67)^2 + (71 - 67)^2}{6}$$

$$= \frac{361 + 64 + 256 + 1 + 100 + 16}{6} = \frac{798}{6} = 133.$$

Because the variance for the second agency is small, its selling times are clustered near the mean.

Since the agencies' data are measured in days, the variances above are measured in square days, a meaningless unit. In order to measure the distribution of the selling times in more meaningful units, we take the square root of the variance. The square root of the variance is called the **standard deviation** and is denoted by s. Thus the standard deviations of the selling times for the two agencies are

$$\sqrt{1051} \approx 32.4 \text{ days} \quad \text{and} \quad \sqrt{133} \approx 11.5 \text{ days},$$

respectively. These numbers show that the selling times for the first agency are considerably more dispersed than those for the second agency.

EXAMPLE 7.4 The following lists contain estimates of the weight of a rock (in pounds). List (a) contains the estimates given by 10 sixth-graders, and list (b) contains the estimates given by 10 ninth-graders.

(a) 1, 1, 1, 2, 2, 4, 4, 5, 5, 5
(b) 2, 2, 2, 3, 3, 3, 3, 4, 4, 4

Find the mean and standard deviation of each set of estimates.

Solution Using formula (7.1), we see that the mean of list (a) is

$$\bar{x} = \frac{\Sigma x}{n} = \frac{1+1+1+2+2+4+4+5+5+5}{10} = \frac{30}{10} = 3$$

pounds, and the mean of list (b) is

$$\bar{x} = \frac{\Sigma x}{n} = \frac{2+2+2+3+3+3+3+4+4+4}{10} = \frac{30}{10} = 3$$

pounds. Thus each list has the same mean.

By formula (7.2) the variance of the values in data set (a) is

$$s^2 = \frac{\Sigma(x-\bar{x})^2}{n}$$

$$= \frac{(1-3)^2 + (1-3)^2 + \cdots + (5-3)^2 + (5-3)^2}{10}$$

$$= \frac{4+4+4+1+1+1+1+4+4+4}{10} = \frac{28}{10} = 2.8.$$

Therefore the standard deviation of the values in data set (a) is

$$s = \sqrt{2.8} \approx 1.67$$

pounds. On the other hand, the variance of the values in data set (b) is

$$s^2 = \frac{\Sigma(x-\bar{x})^2}{n}$$

$$= \frac{(2-3)^2 + (2-3)^2 + \cdots + (4-3)^2 + (4-3)^2}{10}$$

$$= \frac{1+1+1+0+0+0+0+1+1+1}{10} = \frac{6}{10} = 0.6.$$

Therefore the standard deviation of the estimates in data set (b) is

$$s = \sqrt{0.6} \approx 0.77$$

pounds. Thus we see that data set (a) has a greater standard deviation than (b). Hence the estimates of the weight of the rock made by the sixth-graders are more variable than those of the ninth-graders, even though the mean estimate of both groups of students is the same. ∎

It is instructive to write each set of 10 estimates in Example 7.4 in table form.

(a) Estimate	Frequency	(b) Estimate	Frequency
1	3	1	0
2	2	2	3
3	0	3	4
4	2	4	3
5	3	5	0

These distributions are depicted in the histograms of Figure 7.3. Note that in each case the "center" of the distribution is 3, the mean value. The data in set (a), however, are more widely dispersed than those in set (b). This reflects the fact that a ninth-grader is a more accurate estimator than a sixth-grader.

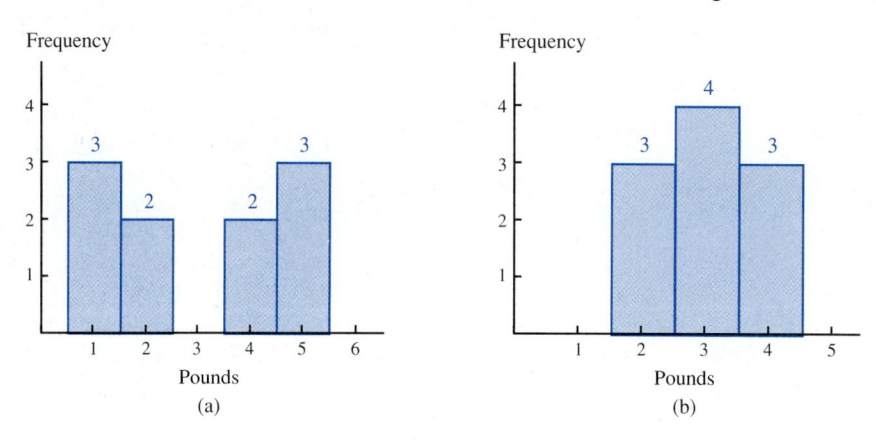

FIGURE 7.3

Practice Problem 4 Determine the standard deviation of the examination scores (84, 94, 87, and 79) for the student in Example 7.1.

For mathematical purposes the mean is the most useful measure of central tendency, and the standard deviation is the most useful measure of dispersion. One reason for their usefulness is that if each entry in a set of data is multiplied by the same factor c, then both the mean and standard deviation are also multiplied by c. As a result, when the data are measurements, a change in the unit of measurement gives the corresponding change in the mean and standard deviation. For example, if the unit of measurement in Example 7.4 is changed from pounds to ounces, then all the data values are multiplied by 16. For these new values, the mean and standard deviation will be 16 times greater than they are when the measurements are given in pounds. We will see applications of the mean and standard deviation throughout this chapter.

EXERCISES 7.1

1. Represent the data in the following table by a bar graph.

   Advertising expenditures (in millions)

1970	$11.9
1975	$15.3
1980	$19.6
1985	$28.2

2. Represent the data in the following table by a bar graph.

   Health expenses per capita, 1977

Hospital care	$297
Physicians	$146
Nursing homes	$57
Drugs	$57
Other	$89

3. A test of the tensile strength (breaking point under stress) of a sample of 80 cables produced the distribution shown below. Represent these data by a histogram.

Tensile strength (lbs)	Number of cables
1400–1500	11
1500–1600	29
1600–1700	26
1700–1800	9
1800–1900	5

4. The speeds of 100 cars selected at random from I–74 were distributed as shown below. Represent these data by a histogram.

Speed (mph)	Cars
40–50	15
50–60	31
60–70	35
70–80	17
80–90	2

For the data in Exercises 5–12 find: (a) the mean, (b) the median, and (c) the standard deviation.

5. 3, 7, 9, 4, 7
6. 3, 5, 7, 3, 7
7. 26, 38, 16, 32
8. 12, 9, 18, 21
9. 4, 10, 9, 4, 6, 9
10. 8, 7, 2, 6, 2, 11
11. 2, 6, 2, 7, 8, 6, 2, 7
12. 4, 12, 7, 9, 6, 5, 8, 13

13. The rates (in dollars per hour) charged by ten moving companies for local moves are as follows:

$$52, 57, 53, 59, 48, 50, 56, 52, 54, \text{ and } 49.$$

Find: (a) the mean, (b) the median, and (c) the standard deviation of these data.

14. Twelve randomly selected owners of General Motors cars were asked how many years they had been driving their present automobiles. Their responses were as follows:

$$2, 8, 5, 11, 7, 6, 1, 14, 4, 5, 3, \text{ and } 6.$$

Find: (a) the mean, (b) the median, and (c) the standard deviation of these data.

15. The salaries of ten randomly chosen major league pitchers are shown below.

$$\begin{array}{ccccc} \$185{,}000 & \$200{,}000 & \$340{,}000 & \$405{,}000 & \$1{,}500{,}000 \\ \$360{,}000 & \$315{,}000 & \$210{,}000 & \$170{,}000 & \$205{,}000 \end{array}$$

Find the mean and median for these salaries. Which value gives a better idea of a representative salary for a major league pitcher?

16. In ten consecutive trips to Las Vegas, a gambler had the following net gains:

$$\begin{array}{ccccc} \$1500 & \$1225 & \$1375 & \$2250 & -\$4725 \\ \$750 & -\$8750 & \$3150 & -\$2725 & \$2150 \end{array}$$

Find the mean and median for these winnings. Which of these measures gives a better idea of the long-term earnings that the gambler can expect?

17. Prove that if the mean and standard deviation of $x_1, x_2, \ldots, x_n$ are $\bar{x}$ and s, respectively, then the mean and standard deviation of $cx_1, cx_2, \ldots, cx_n$ are $c\bar{x}$ and cs, respectively.

18. Prove that the variance of $x_1, x_2, \ldots, x_n$ is equal to

$$\frac{x_1^2 + x_2^2 + \cdots + x_n^2}{n} - \bar{x}^2.$$

This formula simplifies the computation of the variance and standard deviation.

Answers to Practice Problems
1. 30.8
2. 795.36
3. 835
4. $s \approx 5.43$

7.2 Grouped Data

Once we have obtained a set of data, we need to organize it into a form that allows us to comprehend it easily. If the set is small, it is easy to deal with the raw data themselves. A large data set, however, is usually difficult to understand unless it is organized in some systematic way. Consider, for instance, the numbers of absentees from an automobile assembly plant on fifty consecutive working days that were introduced in Example 7.2. (These data are repeated below.)

17	32	45	35	38	22	40	24	48	53
29	47	57	26	33	42	46	36	37	16
41	35	49	54	39	22	20	26	47	43
21	44	29	35	52	32	26	31	38	47
27	41	12	23	32	37	55	35	48	46

In this form, it is very difficult to get a "feel" for the data. In order to understand these numbers better, it is helpful to classify the data in intervals of a uniform width. In this case the number of absentees varies from a low of 12 to a high of 57, for a range of 46 values. Thus all of the data can be accommodated in five class intervals containing ten values each. Of course, there are many different sets of class intervals that can be used to contain the data. Using a small number of wide intervals makes the data more manageable, but having a larger number of narrow intervals is more precise. The general practice is to use at least five and at most fifteen class intervals when grouping data.

Choosing the endpoints of the intervals often requires some thought. Although the intervals 10–20, 20–30, 30–40, 40–50, and 50–60 accommodate all the values in our data set, we encounter a problem with these intervals because values such as 20 or 40 could be placed in two different intervals. Whenever possible, we would like to choose our intervals so that the upper endpoint of one interval equals the lower endpoint of the next interval and each data value falls in exactly one interval. Unfortunately some types of data can (at least theoretically) assume all values in some interval of real numbers. This happens when the data represent ages, as in Table 7.2. In this instance, no matter how the endpoints are chosen, there can be data values that fall in two different intervals. This is not the case for the number of automobile plant absentees, however, and by using the intervals 10.5–20.5, 20.5–30.5, 30.5–40.5, 40.5–50.5, and 50.5–60.5, we can assure that each data value lies in exactly one interval.

In summary, whenever possible we choose the intervals into which the data will be grouped according to the following guidelines.

Guidelines for Choosing the Intervals when Grouping Data

1. The intervals should be of equal width.
2. The number of intervals should be between 5 and 15.
3. The upper endpoint of one interval should equal the lower endpoint of the next interval.
4. Every data value should lie in exactly one interval.

Tallying the number of data values that fall in each of the class intervals produces the **frequency distribution** shown in Table 7.4. Since each data value belongs to exactly one interval, the class frequencies add up to 50, the total number of observations. Note the similarity between a frequency distribution for a set of data and a probability distribution, which we encountered in Section 6.2.

Number of absentees	Tally	Frequency
10.5–20.5	||||	4
20.5–30.5	|||| ||||		11
30.5–40.5	|||| |||| ||||		16
40.5–50.5	|||| |||| ||||	14
50.5–60.5	||||	5

TABLE 7.4

The frequency distribution in Table 7.4 can be represented by a histogram, as in Figure 7.4. This histogram gives us a much better feel for the data than we had by looking at the raw data. For instance, we can see at a glance that the number of absentees is usually between 20.5 and 50.5; in fact, in our 50-

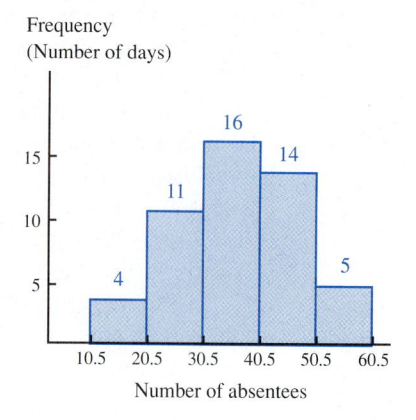

FIGURE 7.4

day sample the number of absentees fell in this range 11 + 16 + 14 = 41 times (82% of the time). The price that we pay for the convenience of grouped data is a loss of information. For example, if we want to know how often the number of absentees was between 25 and 45, our frequency distribution is of no use; we must go back to the original raw data.

Practice Problem 5 The data below show the number of tornadoes occurring in the United States during the years 1960–1984.

618	683	658	461	713
898	570	912	660	604
649	888	741	1102	947
920	835	852	788	852
866	783	1046	931	907

(a) Choose an appropriate set of intervals into which to group these data.

(b) Prepare a frequency distribution and histogram using the class intervals you chose in part (a).

THE MEAN AND STANDARD DEVIATION OF GROUPED DATA

We saw above that by grouping data into a frequency distribution, we could make it easier to understand. In order for a frequency distribution to prove really useful, however, we must be able to use it to estimate the mean and standard deviation without needing to return to the raw data. A simple approximation makes this possible: *For grouped data the values in each class interval are approximated by the midpoint of the interval.* Thus the mean can be estimated as follows.

Mean of Grouped Data

If data values are grouped into class intervals having midpoints $m_1, m_2, \ldots, m_k$ and respective frequencies $f_1, f_2, \ldots, f_k$, then

$$\bar{x} \approx \frac{m_1 f_1 + m_2 f_2 + \cdots + m_k f_k}{n} = \frac{\Sigma mf}{n}, \quad (7.3)$$

where $n = \Sigma f$ is the total number of data values.

EXAMPLE 7.5 Find the mean number of absentees from the automobile assembly plant using the frequency distribution in Table 7.4.

Solution The frequency distribution is reproduced below, showing the midpoint of each class interval. Note that the midpoint is simply the mean of the upper and lower endpoints of the class interval; for instance, the midpoint of the interval 10.5–20.5 is

$$\frac{10.5 + 20.5}{2} = 15.5.$$

Interval	Midpoint m	Frequency f	mf
10.5–20.5	15.5	4	62.0
20.5–30.5	25.5	11	280.5
30.5–40.5	35.5	16	568.0
40.5–50.5	45.5	14	637.0
50.5–60.5	55.5	5	277.5
		50	1825.0 Totals

The last column gives the product mf of the midpoint of each class interval with the corresponding class frequency. The sum of these products is shown at the bottom of the column. Applying formula (7.3), we find that the mean of the grouped data is

$$\bar{x} = \frac{\Sigma\, mf}{n} = \frac{1825}{50} = 36.5.$$

This value agrees fairly well with the true mean of 36.2 computed in Example 7.2. The discrepancy in the two numbers is due to the fact that information is lost when the data are grouped and all the values in each interval are replaced by the midpoint. ∎

Practice Problem 6 Use the frequency distribution in Table 7.5 to estimate the mean number of tornadoes reported in the United States during the years 1953–1977.

Number of Tornadoes	Frequency
399.5– 499.5	1
499.5– 599.5	1
599.5– 699.5	6
699.5– 799.5	4
799.5– 899.5	6
899.5– 999.5	5
999.5–1099.5	1
1099.5–1199.5	1
	25 Total

TABLE 7.5

For grouped data the midpoints $m_1, m_2, \ldots, m_k$ of the class intervals are also used to approximate the actual data in computing the variance and standard deviation. Thus the variance of grouped data can be approximated as follows.

Variance of Grouped Data

If data values are grouped into class intervals having midpoints $m_1, m_2, \ldots, m_k$ and respective frequencies $f_1, f_2, \ldots, f_k$, then

$$s^2 \approx \frac{(m_1 - \bar{x})^2 f_1 + (m_2 - \bar{x})^2 f_2 + \cdots + (m_k - \bar{x})^2 f_k}{n} \quad (7.4)$$

$$= \frac{\Sigma(m - \bar{x})^2 f}{n},$$

where $n = \Sigma f$ is the total number of data values.

EXAMPLE 7.6 Find the standard deviation of the number of absentees from the automobile assembly plant using the grouped data presented in Example 7.5.

Solution It can be shown by a very tedious calculation using formula (7.2) that the actual variance of the fifty observations is 121.84.

Calculating the variance of the grouped data is much simpler. In Example 7.5 we found that the mean of the grouped data is $\bar{x} = 36.5$. The table below shows the frequency distribution from Table 7.4 along with the midpoint of each class interval, the deviation from the mean $m - \bar{x}$, the squared deviation $(m - \bar{x})^2$, and the product $(m - \bar{x})^2 f$.

Interval	Midpoint m	Frequency f	$m - \bar{x}$	$(m - \bar{x})^2$	$(m - \bar{x})^2 f$
10.5–20.5	15.5	4	−21	441	1764
20.5–30.5	25.5	11	−11	121	1331
30.5–40.5	35.5	16	−1	1	16
40.5–50.5	45.5	14	9	81	1134
50.5–60.5	55.5	5	19	361	1805
		50			6050 Totals

Formula (7.4) estimates the variance as

$$s^2 \approx \frac{\Sigma(m - \bar{x})^2 f}{n} = \frac{6050}{50} = 121,$$

and so the standard deviation is

$$s \approx \sqrt{121} = 11.$$

Note that the grouped data have approximately the same variance and standard deviation as the ungrouped data. ■

EXAMPLE 7.7 Use the data in Table 7.2 on page 351 to approximate the mean and standard deviation for the age of someone in the U.S. population.

Solution The frequency distribution in Table 7.2 is reproduced below along with the midpoint of each class interval and the product of each interval's midpoint and frequency.

Interval	Midpoint m	Frequency f	mf
0– 20	10	69.2×10^6	692×10^6
20– 40	30	69.6×10^6	2088×10^6
40– 60	50	45.0×10^6	2250×10^6
60– 80	70	29.2×10^6	2044×10^6
80–100	90	5.0×10^6	450×10^6
		218.0×10^6	7524×10^6	Totals

Thus

$$\bar{x} \approx \frac{\Sigma mf}{n} = \frac{7524 \times 10^6}{218 \times 10^6} \approx 34.5$$

is approximately the average age of an individual in the U.S. population. To find the standard deviation, we extend the table above to include columns for $m - \bar{x}$, $(m - \bar{x})^2$, and $(m - \bar{x})^2 f$.

Interval	Midpoint m	Frequency f	$m - \bar{x}$	$(m - \bar{x})^2$	$(m - \bar{x})^2 f$
0– 20	10	69.2×10^6	-24.5	600.75	$41,571.90 \times 10^6$
20– 40	30	69.6×10^6	-4.5	20.25	$1,409.40 \times 10^6$
40– 60	50	45.0×10^6	15.5	240.25	$10,811.25 \times 10^6$
60– 80	70	29.2×10^6	35.5	1260.25	$36,799.30 \times 10^6$
80–100	90	5.0×10^6	55.5	3080.25	$15,401.25 \times 10^6$
		218.0×10^6			$105,993.10 \times 10^6$	Totals

By formula (7.4) we find the variance to be

$$s^2 \approx \frac{\Sigma (m - \bar{x})^2 f}{n} = \frac{105,993.10 \times 10^6}{218.0 \times 10^6} \approx 486,$$

and so the standard deviation is

$$s \approx \sqrt{486} \approx 22. \blacksquare$$

Practice Problem 7 Determine the standard deviation of the number of tornadoes reported in the United States during the years 1960–1984 using the grouped data in Table 7.5. The mean of this distribution, as computed in Practice Problem 6, is 797.5.

EXERCISES 7.2

1. An insurance company checked the records of 80 motorists to determine how many moving violations each had been cited for during the past five years. The results are shown below.

Violations	Frequency
0	34
1	28
2	12
3	4
4	2
	80 Total

   Use the given frequency distribution to find the mean and standard deviation of the number of moving violations.

2. A Labor Department statistician interviewed 120 unemployed persons in an inner-city area to determine how many years of high school each had completed. Their responses are shown below.

Number of Years	Frequency
0	6
1	24
2	39
3	30
4	21
	120 Total

   Use the given frequency distribution to find the mean and standard deviation of the number of years of high school completed.

3. A test of the tensile strength (breaking point under stress) of a sample of 80 cables produced the frequency distribution shown below.

Tensile strength (lbs)	Number of cables
1400–1500	11
1500–1600	29
1600–1700	26
1700–1800	9
1800–1900	5
	80 Total

   Use the given frequency distribution to find the mean and standard deviation of the tensile strength of the cables.

4. The speeds of 100 cars selected from I–74 were distributed as shown below.

Speed (mph)	Cars
40–50	15
50–60	31
60–70	35
70–80	17
80–90	2
	100 Total

Use this frequency distribution to find the mean and standard deviation of the speed of the cars.

5. The frequency distributions below show the weekly sales records for two realty companies during the past year.

Company A

Number of houses sold	Frequency
0	4
1	25
2	17
3	5
4	1
5	0

Company B

Number of houses sold	Frequency
0	10
1	11
2	12
3	9
4	8
5	2

(a) Find the mean and standard deviation of each company's weekly sales.

(b) Which company had the better sales record? Which company was more consistent?

6. The life spans of 100 light bulbs of two brands were compared with the following results.

Brand S

Life span hours	Frequency
500– 600	12
600– 700	23
700– 800	26
800– 900	21
900–1000	18

Brand W

Life span hours	Frequency
500– 600	5
600– 700	30
700– 800	47
800– 900	16
900–1000	2

(a) Find the mean and standard deviation of each brand's bulb life.

(b) Which brand of bulbs lasts longer, on the average? Which brand is more dependable?

In Exercises 7–14, (a) prepare a frequency distribution for the given data; (b) construct a histogram depicting the frequency distribution in part (a); and (c) approximate the mean and standard deviation of the given data.

7. The daily low temperature readings during the month of November (in degrees Fahrenheit) are shown below for Memphis, Tennessee.

41	43	39	34	38	42
44	47	48	53	58	54
51	59	53	46	47	41
36	31	36	43	43	47
49	48	49	54	57	54

8. The systolic blood pressures (in millimeters of mercury) are shown below for a group of 40 young males.

141	98	132	135	122	117	131	116
139	118	121	128	127	113	138	115
127	142	108	147	117	118	127	138
116	131	119	122	121	94	129	118
119	129	108	123	131	124	131	102

9. In reference [1] Anderson reports the IQ scores of 26 children prior to beginning nursery school. These data are shown below.

108	107	93	128	133	126	125	117	113
115	111	122	100	100	139	108	117	127
119	150	117	108	95	111	127	86	

10. The number of deaths caused by tornadoes in the United States during each year from 1953–1984 is shown below.

515	36	126	83	191	66	58	47
51	28	31	73	296	99	114	131
66	72	156	27	87	361	60	44
43	53	84	28	24	64	34	122

11. The ages of the 39 U.S. presidents at the time of their first inauguration are shown below.

57	57	49	52	50	51	51	55
61	61	64	56	47	56	50	61
57	54	50	46	55	55	62	52
57	68	48	54	54	51	43	69
58	51	65	49	42	54	55	

12. The ages at death of deceased presidents of the U.S. are shown below.

67	80	53	56	56	72	63
90	78	65	66	71	67	88
83	79	74	63	67	57	78
85	68	64	70	58	60	46
73	71	77	49	60	90	64

13. The number of home runs hit by the leading home run hitter in the American League during each season from 1937–1986 is shown below.

46	58	35	41	37	36	34	22	24	44
32	39	43	37	33	32	43	32	37	52
42	42	42	40	61	48	45	49	32	49
44	44	49	44	33	37	32	32	36	32
39	46	45	41	22	39	39	43	40	40

14. The number of points scored by the leading scorer in the National Football League during each season from 1936–1985 is shown below.

73	69	58	68	57	95	138	117	85	110
100	102	110	102	128	102	94	114	114	96
99	77	108	94	176	147	137	113	155	132
119	117	145	129	125	117	128	130	94	138
109	99	118	115	129	121	84	161	131	144

15. We noted in Section 7.1 that if all the raw data are multiplied by the same factor, then the mean and standard deviation are also multiplied by that factor. Show that the same is true when the mean and standard deviation are approximated using the grouped data formulas (7.3) and (7.4).

16. Show that if in formulas (7.3) and (7.4) all the frequencies are multiplied by the same factor, then the mean and variance (and hence the standard deviation) are unchanged.

Answers to Practice Problems

5. (a) There are many acceptable choices of intervals; one is 399.5–499.5, 499.5–599.5, 599.5–699.5, 699.5–799.5, 799.5–899.5, 899.5–999.5, 999.5–1099.5, and 1099.5–1199.5.

   (b) The frequency distribution is given here on the left; it depends on the intervals chosen. The histogram using the intervals in part (a) is shown on the right.

Number of tornadoes	Frequency
399.5– 499.5	1
499.5– 599.5	1
599.5– 699.5	6
699.5– 799.5	4
799.5– 899.5	6
899.5– 999.5	5
999.5–1099.5	1
1099.5–1199.5	1
Total	25

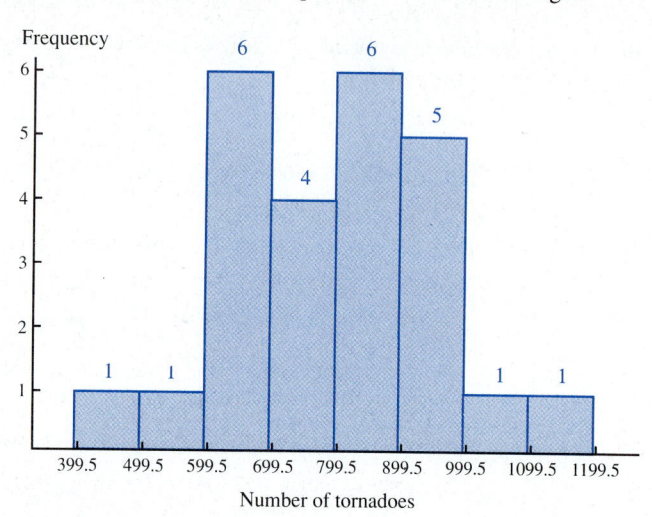

6. 797.5

7. $\sqrt{25696} \approx 160$

7.3 Random Variables: Expected Value and Variance

Quantitative data always represent measurements of some quantity over a certain population, such as the ages of individuals in the United States, the number of absentees at an auto plant on certain days, or the number of hours a light bulb will last before burning out. Note that the value of the quantity being measured depends on which member of the population is chosen. We call a numerical quantity whose value depends on the outcome of an experiment a **random variable**.

7.3 RANDOM VARIABLES: EXPECTED VALUE AND VARIANCE

Consider, for instance, an automobile towing service that has recorded the number of service calls it received each day over a 60-day period. The results are shown in Table 7.6 and depicted in Figure 7.5.

Number of service calls	Frequency
0	6
1	15
2	18
3	12
4	6
5	3
	60 Total

TABLE 7.6

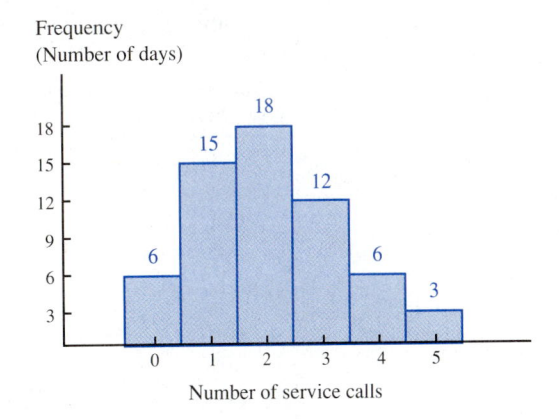

FIGURE 7.5

We can regard Table 7.6 as a frequency distribution for the random variable that counts the number of daily service calls. Recall that if each frequency is divided by the total number of observations (60), we obtain the corresponding relative frequency or empirical probability. This leads to the probability distribution shown in Table 7.7.

Number of service calls	Probability
0	.10
1	.25
2	.30
3	.20
4	.10
5	.05
	1.00 Total

TABLE 7.7

FIGURE 7.6

This distribution can be represented by a probability histogram as in Figure 7.6, where each value is the midpoint of a "class interval" extending one-half unit to the left and right. The height of each bar represents the probability of the corresponding value (the frequency divided by 60). Thus the probability histogram in Figure 7.6 has the same shape as the frequency histogram in Figure 7.5, the only difference being the units of measurement on the vertical axis.

EXAMPLE 7.8 Find the mean and the standard deviation of the number of daily service calls for the towing service described in Table 7.6.

Solution From Table 7.6 we can calculate the products mf of the number of calls times the corresponding frequencies.

Number of calls m	Frequency f	mf	$m - \bar{x}$	$(m - \bar{x})^2$	$(m - \bar{x})^2 f$
0	6	0	−2.1	4.41	26.46
1	15	15	−1.1	1.21	18.15
2	18	36	−0.1	.01	0.18
3	12	36	0.9	.81	9.72
4	6	24	1.9	3.61	21.66
5	3	15	2.9	8.41	25.23
	60	126			101.40 Totals

Adding the products mf and dividing by the number of observations gives

$$\bar{x} = \frac{\Sigma mf}{n} = \frac{126}{60} = 2.10.$$

This is the mean of the number of daily service calls.

Having computed the mean, we can complete columns 4, 5, and 6 in the table above. Dividing the sum of the numbers in the last column by the number of observations gives the variance

$$s^2 = \frac{\Sigma(m - \bar{x})^2 f}{n} = \frac{101.40}{60} = 1.69.$$

Therefore the standard deviation of the number of service calls is

$$s = \sqrt{1.69} = 1.3. \quad \blacksquare$$

THE MEAN AND VARIANCE OF RANDOM VARIABLES

In a similar fashion we can construct frequency and probability distributions for any grouped data. In general we take the values of the random variable to be the midpoints $m_1, m_2, \ldots, m_k$ of the class intervals. The frequency and probability distributions for the random variable are constructed as in Figures 7.5 and 7.6 with the bars centered over the class intervals. If the data fall into the intervals with respective frequencies $f_1, f_2, \ldots, f_k$, then the heights of the bars in the frequency distribution represent $f_1, f_2, \ldots, f_k$ and the heights of the bars in the probability distribution represent the corresponding probabilities $p_i = f_i/n$, where $n = \Sigma f$ is the total number of observations.

Using formulas (7.3) and (7.4), we can estimate the mean and variance of the frequency distribution.

7.3 RANDOM VARIABLES: EXPECTED VALUE AND VARIANCE

$$\bar{x} \approx \frac{\Sigma mf}{n} = \frac{m_1 f_1 + m_2 f_2 + \cdots + m_k f_k}{n}$$

$$= \frac{m_1 f_1}{n} + \frac{m_2 f_2}{n} + \cdots + \frac{m_k f_k}{n}$$

$$= m_1 p_1 + m_2 p_2 + \cdots + m_k p_k$$

$$s^2 \approx \frac{\Sigma (m - \bar{x})^2 f}{n}$$

$$= \frac{(m_1 - \bar{x})^2 f_1 + (m_2 - \bar{x})^2 f_2 + \cdots + (m_k - \bar{x})^2 f_k}{n}$$

$$= \frac{(m_1 - \bar{x})^2 f_1}{n} + \frac{(m_2 - \bar{x})^2 f_2}{n} + \cdots + \frac{(m_k - \bar{x})^2 f_k}{n}$$

$$= (m_1 - \bar{x})^2 p_1 + (m_2 - \bar{x})^2 p_2 + \cdots + (m_k - \bar{x})^2 p_k$$

These calculations express the mean and the variance of the frequency distribution in terms of the probabilities $p_i = f_i/n$ instead of the frequencies f_i. The reader should check that if we compute

$$m_1 p_1 + m_2 p_2 + \cdots + m_k p_k \quad \text{and}$$
$$(m_1 - \bar{x})^2 p_1 + (m_2 - \bar{x})^2 p_2 + \cdots + (m_k - \bar{x})^2 p_k$$

for the data in Table 7.7, then we get exactly the same values for $\bar{x}$ and s^2 that were obtained in Example 7.8.

In view of the above we can generalize the mean and variance to apply to any random variable as follows.

Mean and Variance of a Random Variable

For a random variable taking distinct values $x_1, x_2, \ldots, x_k$ with respective probabilities $p_1, p_2, \ldots, p_k$, the **mean** (or **expected value**) is defined by

$$\mu = \Sigma xp = x_1 p_1 + x_2 p_2 + \cdots + x_k p_k, \qquad (7.5)$$

and the **variance** is defined by

$$\sigma^2 = \Sigma (x - \mu)^2 p \qquad (7.6)$$
$$= (x_1 - \mu)^2 p_1 + (x_2 - \mu)^2 p_2 + \cdots + (x_k - \mu)^2 p_k.$$

As before, the square root of the variance is called the **standard deviation** and is denoted by σ.*

Recall that for grouped data we take the values of the random variable to be the midpoints $m_1, m_2, \ldots, m_k$ of the class intervals. Thus the calculation above shows that the expected average of the random variable over the long run is

$$m_1 p_1 + m_2 p_2 + \cdots + m_k p_k.$$

Hence the term "expected value" is also used in place of "mean."

EXAMPLE 7.9 Determine the mean and standard deviation for the number of heads appearing when three fair coins are flipped.

Solution The number of heads appearing when three fair coins are flipped is a random variable that takes the values 0, 1, 2, and 3. From Example 6.5 in Section 6.1 we see that the probabilities of these outcomes are 1/8, 3/8, 3/8, and 1/8, respectively. Hence the probability distribution is as shown in Figure 7.7. As in Example 7.8 we can perform the necessary calculations to determine the mean and variance in a table.

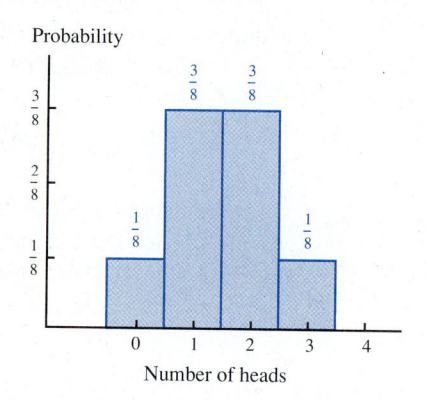

FIGURE 7.7

Number of heads x	Probability p	xp	$x - \mu$	$(x - \mu)^2$	$(x - \mu)^2 p$
0	1/8	0	−3/2	9/4	9/32
1	3/8	3/8	−1/2	1/4	3/32
2	3/8	6/8	1/2	1/4	3/32
3	1/8	3/8	3/2	9/4	9/32
	1.00	12/8			24/32 Totals

---

*The symbols μ and σ (the lower case Greek letters mu and sigma) are used to denote the mean and standard deviation of a random variable because these are *theoretical* quantities. We will continue to use the symbols $\bar{x}$ and s to denote the mean and standard deviation of *empirical* data.

7.3 RANDOM VARIABLES: EXPECTED VALUE AND VARIANCE

The sum of the products xp is the mean, which is

$$\mu = 0\left(\frac{1}{8}\right) + 1\left(\frac{3}{8}\right) + 2\left(\frac{3}{8}\right) + 3\left(\frac{1}{8}\right) = \frac{12}{8} = 1.5.$$

Thus the expected value of the number of heads is 1.5; that is, over the long run we expect the average number of heads appearing to be 1.5. This value agrees with our intuition because we expect heads to appear on half of the flips of a fair coin.

Having computed the expected value, we can complete columns 4, 5, and 6 in the table above. As before, the sum of the numbers in the last column, $24/32 = 0.75$, is the variance; so the standard deviation is

$$\sigma = \sqrt{0.75} \approx 0.87. \quad \blacksquare$$

EXAMPLE 7.10 Find the expected value and standard deviation for the sum obtained when two fair dice are rolled.

Solution The sum obtained when two fair dice are rolled is a random variable that takes the values 2, 3, 4, 5, 6, 7, 8, 9, 10, 11, and 12. Recall from Section 6.2 that it has the probability distribution shown on the left. This distribution is depicted in Figure 7.8.

Sum	Probability
2	1/36
3	2/36
4	3/36
5	4/36
6	5/36
7	6/36
8	5/36
9	4/36
10	3/36
11	2/36
12	1/36

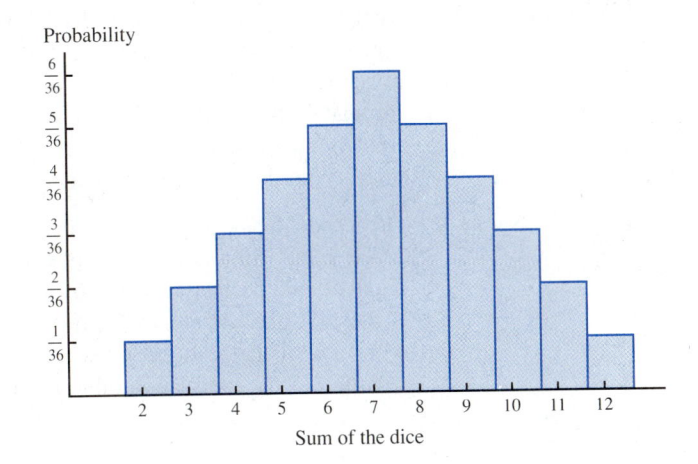

FIGURE 7.8

Thus by formula (7.5) the mean is

$$\mu = \Sigma xp = 2\left(\frac{1}{36}\right) + 3\left(\frac{2}{36}\right) + 4\left(\frac{3}{36}\right) + \cdots + 12\left(\frac{1}{36}\right)$$

$$= \frac{252}{36} = 7.$$

Hence the expected sum is 7. Using formula (7.6), we see that the variance is

$$\sigma^2 = (2-7)^2 \left(\frac{1}{36}\right) + (3-7)^2 \left(\frac{2}{36}\right) + \cdots + (12-7)^2 \left(\frac{1}{36}\right)$$
$$= \frac{210}{36}.$$

Therefore the standard deviation is

$$\sigma = \sqrt{\frac{210}{36}} \approx 2.42. \quad \blacksquare$$

Practice Problem 8   Draw a probability histogram for the number of spots appearing when a fair die is rolled; then determine its mean and standard deviation.

The expected value of a random variable arises naturally in applications involving decision making. We will explore some of these applications in the next section. To close this section, we present several more examples in which the expected value of a random variable occurs.

EXAMPLE 7.11   Kent found a ticket for a raffle in which 1000 tickets were sold. One ticket selected at random will win the first prize of $250, ten tickets will win second prizes of $25 each, and twenty tickets will win third prizes of $10 each. Unfortunately the holder of a winning ticket must be present in order to win, and Kent is about to leave town. Greg has offered to buy Kent's ticket at a fair price. What is a fair price for the ticket?

Solution   Greg should pay as much for the ticket as it is expected to win; that is, the "true value" of the ticket is its expected winnings. The random variable that counts the expected winnings of the ticket can take four possible values: $250 if the first prize is won, $25 if a second prize is won, $10 if a third prize is won, and $0 if the ticket does not win any of the prizes. The probability distribution of the random variable is easily seen to be as follows.

Amount won x	Probability p	xp
$250	.001	$.250
$25	.010	$.250
$10	.020	$.200
$0	.969	$.000
	1.00	$.700 Totals

Thus the expected winnings for each ticket are $\mu = \$0.70$, and so Greg should pay Kent 70¢ for the ticket. $\blacksquare$

7.3 RANDOM VARIABLES: EXPECTED VALUE AND VARIANCE

EXAMPLE 7.12 In the raffle described in Example 7.11, the selling price of each ticket was $1. What are the expected net winnings for someone who buys a single ticket?

Solution The random variable that counts the *net* amount won takes four possible values. If the first prize is received, the net amount won is

$$\$250 - \$1 = \$249,$$

the prize money received minus the cost of the ticket. Likewise the net amount won is $24 if a second prize is received, $9 if a third prize is received, and $-\$1$ if no money is received. Thus the probability distribution for this random variable is as shown in the first two columns of the table below.

Amount won x	Probability p	xp
$249	.001	.249
$24	.010	.240
$9	.020	.180
−$1	.969	−.969
	1.00	−.300 Totals

Thus the expected winnings for each ticket are $\mu = -\$0.30$; in other words, each ticket is expected to *lose* 30¢. Hence the organization holding the raffle will gain an average of 30¢ on each of the 1000 tickets sold, for a net gain of $300. Note that this gain is the difference between the revenue derived from the sale of the tickets, $1000, and the prize money paid out,

$$\$250 + 10(\$25) + 20(\$10) = \$700. \blacksquare$$

EXAMPLE 7.13 An insurance company charges $100 for a life insurance policy that pays $30,000 if a woman 25 years of age dies during the coming year. Mortality tables indicate the probability that a woman of age 25 will survive one more year is .998. What is the company's expected gain on such a policy?

Solution If a woman with this type of policy survives the year, then the company has a net gain of $100, the premium of the policy. If she dies, however, the company keeps the premium but pays out benefits of $30,000, for a net gain of

$$\$100 - \$30,000 = -\$29,900.$$

Thus the random variable that measures the company's gain takes on two values, $100 and $-\$29,900$. The probability distribution for this random variable is as follows.

Gain	Probability
$100	.998
−$29,900	.002

Thus the company's expected net gain is

$$\mu = (\$100)(.998) + (-\$29,900)(.002)$$
$$= \$99.80 - \$59.80 = \$40.$$

Hence over the long run the company will gain $40 on each of these policies that it sells. ∎

Practice Problem 9 Barring complications, a contractor anticipates a $50,000 profit on a certain construction job. Severe winter weather, however, may raise his costs $20,000 more than expected (and lower his profits accordingly). If the probability of a severe winter is .4, what is the contractor's expected profit for this job?

EXERCISES 7.3

1. An insurance company checked the records of 80 motorists chosen at random to determine how many moving violations each had been cited for during the past five years. The results are shown below.

Violations	Frequency
0	34
1	28
2	12
3	4
4	2
	80 Total

   (a) Construct the probability distribution for the number of moving violations.
   (b) Find the mean and standard deviation for the number of moving violations using the probability distribution from part (a).

2. A Labor Department statistician interviewed 120 unemployed persons in an inner-city area to determine how many years of high school each had completed. Their responses are shown below.

Number of years	Frequency
0	6
1	24
2	39
3	30
4	21
	120 Total

   (a) Construct the probability distribution for the number of years of high school completed.
   (b) Find the mean and standard deviation for the number of years of high school completed using the probability distribution from part (a).

3. A test of the tensile strength (breaking point under stress) of a sample of 80 cables produced the frequency distribution shown at the top of the next page.

Tensile strength (lbs)	Number of cables
1400–1500	11
1500–1600	29
1600–1700	26
1700–1800	9
1800–1900	5
	80 Total

(a) Construct the probability distribution for the tensile strength of the cables.
(b) Find the mean and standard deviation for the tensile strength of the cables using the probability distribution from part (a).

4. The speeds of 100 cars selected at random from I–74 were distributed as shown below.

Speed (mph)	Cars
40–50	15
50–60	31
60–70	35
70–80	17
80–90	2
	100 Total

(a) Construct the probability distribution for the speed of the cars.
(b) Find the mean and standard deviation for the speed of the cars using the probability distribution from part (a).

5. A stockbroker observes that the delivery time for her local mailings is a random variable with the distribution below.

Delivery time (working days)	Probability
1	.218
2	.318
3	.272
4	.130
5	.062

Find the mean and standard deviation for the delivery time of the stockbroker's mailings.

6. A baker finds that the number of chocolate cakes requested on a given day has the following distribution.

Requests	Probability
0	.11
1	.20
2	.32
3	.19
4	.12
5	.05
6	.01

Find the mean and standard deviation for the number of chocolate cakes requested on a given day.

7. In a particular lottery, 161 winning tickets will be randomly selected from among the 250,000 tickets sold. One ticket will win $10,000, ten tickets will win $1,000, 50 tickets will win $200, and 100 tickets will win $100.

   (a) What are the expected winnings for each ticket?
   (b) If each ticket cost $1, what is the expected *net* gain for someone who bought a single ticket?

8. A raffle offers a first prize of $2000, 2 second prizes of $500, and 10 third prizes of $100. There were 10,000 tickets sold.

   (a) What are the expected winnings for each ticket?
   (b) If each ticket cost $1, what is the expected *net* gain for someone who bought a single ticket?

9. A bank teller's cash drawer contains 100 one-dollar bills, 50 five-dollar bills, 50 ten-dollar bills, and 25 twenty-dollar bills. What is the expected value of a bill chosen at random from the drawer?

10. American roulette wheels have 38 slots numbered 00, 0, 1, 2, . . . , 36, all of which are equally likely to occur. A person who bets $1 on a given number wins $35 and receives his bet back if that number occurs and loses his bet otherwise. What is the expected value of such a bet?

11. An oil company has found that the cost of drilling for oil on a new site is $5 million and that 10% of the sites that it drills yield oil. When oil is found, a site averages $75 million in revenues. What is the expected net profit from a new site?

12. An investor has bought a tract of land for $50,000. There is a 20% chance that an airport will be built nearby, increasing the value of the property to $250,000. If the airport authority decides not to locate there, however, the value of the land will drop to $35,000. What is the investor's expected net gain?

13. A business venture is expected to earn a $250,000 profit if successful and to lose $100,000 if unsuccessful. If the probability is .4 that this venture is successful, what is its expected value?

14. A farmer expects 110 bushels per acre of corn under normal weather conditions, 125 bushels per acre under excellent conditions, and 75 bushels per acre under poor conditions. If the probabilities of normal weather, excellent weather, and poor weather are .6, .1, and .3, respectively, what is the farmer's expected yield per acre?

15. An insurance company is issuing a new homeowner's policy. Based on their experience with similar policies, the company expects that 1.2% of their policyholders will file a claim each year, and the average amount paid on each claim will be $12,000. How much should the company charge for such a policy in order to earn a $50 profit from each policy sold?

16. A government agency has loaned $25,000 to start a new business. The loan is to be repaid in one year at 10% interest, but there is a 15% chance that the borrower will default, resulting in a $25,000 loss for the agency. What is the agency's expected net gain from such a loan?

17. A basketball player makes her first free throw 80% of the time. If she makes her first shot, her accuracy on the second shot increases to 90%. However if she misses her first shot, her accuracy on the second shot decreases to 70%. What is the expected number of free throws made when she shoots two shots?

18. An urn contains three blue and seven green marbles. Al will choose two marbles at random from the urn. If they are the same color, Bob will pay him $7; but if they are different colors, Al will pay Bob $8. What are Al's expected winnings in this game?

19. An urn contains two red marbles and three white marbles. Marbles are chosen at random from the urn, one after another, until a red marble is selected.
    (a) Find the probability distribution for the number of marbles chosen.
    (b) Find the mean and standard deviation of the number of marbles chosen.

20. Doug has agreed to pay Jennifer for the right to flip a fair coin until either one head or three tails occur. If heads appears on the first, second, or third flip, Jennifer will pay Doug $1, $2, or $4, respectively; but if three tails occur, Doug will win nothing. How much should Doug pay to play this game?

21.* Three prize winners are to be randomly chosen from among six women and two men. What is the expected number of men selected?

22.* A bargaining committee is to be appointed from among six union members and three nonunion workers. If the committee is to consist of four of these persons chosen at random, what is the expected number of union members on the committee?

23.* A university committee of four faculty members and three students intends to form a three-member subcommittee. If the members of the subcommittee are chosen at random, what is the expected number of faculty on the subcommittee?

24.* A planning committee of three is to be selected at random from among the members of the city council. If there are nine Republicans and six Democrats on the city council, what is the expected number of Democrats on the subcommittee?

25. If a random variable takes values $x_1, x_2, \ldots, x_k$ with respective probabilities $p_1, p_2, \ldots, p_k$, prove that its variance equals
$$\Sigma x^2 p - (\Sigma xp)^2.$$

26. Let μ and σ^2 denote the mean and variance of a random variable X. Show that for any constant k, the mean and variance of kX are $k\mu$ and $k^2\sigma^2$, respectively.

27. Let μ and σ^2 denote the mean and variance of a random variable X. Show that for any constant k, the mean and variance of $X + k$ are $\mu + k$ and σ^2, respectively.

Answers to Practice Problems 8. The probability histogram is shown below; $\mu = 3.5$ and $\sigma = \sqrt{\dfrac{70}{24}} \approx 1.71$.

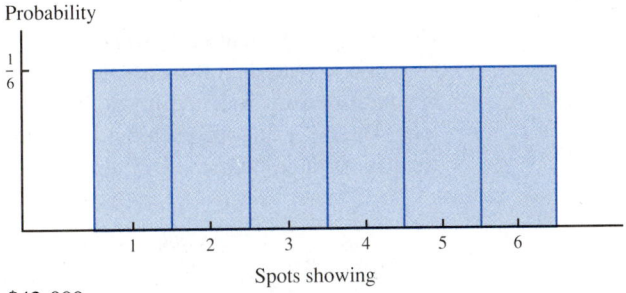

9. $42,000

---

*This exercise requires use of the counting techniques in Chapter 5.

MATHEMATICS IN ACTION

7.4 Decision Theory

In earlier chapters we have seen how mathematical tools such as linear programming and matrix algebra can help us make optimal decisions in a wide range of practical problems. In this section we will use the techniques of probability theory to compare the different options available to a decision-maker.

DECISIONS UNDER UNCERTAINTY

Making a decision amounts to choosing among various possible courses of action. When the outcome of each action can be predicted exactly, we say that we are operating under *conditions of certainty*. In such a case the best decision is simply the one that produces the most desirable outcome, for example, the greatest profit or the least cost.

More often than not, however, we must operate under *conditions of uncertainty*, where the outcome of each action depends in part on factors beyond our control. To model such a situation, we assume that the decision-maker has available a number of possible actions and that one of various possible random events called **states of nature** will occur independently of the action taken. For each state of nature, the consequences of each action can be summarized in a **payoff table,** the entries of which show the decision-maker's net gain.

Consider, for example, an inventor who has patented a new, fuel-efficient carburetor. She would like to produce the device herself, but this is a costly gamble; for if the device is eventually adopted by a major automaker, she will make a profit of $2,000,000, but otherwise she will lose $500,000. A company has offered to finance the project for 70% of any profits; in this case the inventor will either earn $600,000 or break even. Another company has offered to buy the inventor's patent outright for $100,000. What is the inventor's best course of action?

In this situation the inventor has three options available:

1. produce the device herself,
2. retain a 30% interest in the device, or
3. sell the patent.

Her net gain depends on whether or not the device is adopted by a major automaker. Note that the decision as to whether or not the device is adopted is beyond the control of the inventor; these are the states of nature for this problem. The net gains for each option under each state of nature are the entries of the payoff table below.

Action	State of Nature	
	Adopted	Not adopted
Produce device	$2,000,000	−$500,000
Retain 30%	$600,000	$0
Sell patent	$100,000	$100,000

7.4 MATHEMATICS IN ACTION: DECISION THEORY

If the inventor knew that her device would be adopted, she would obviously do best to produce it herself to earn a payoff of $2,000,000. If, on the other hand, she knew that the device would not be adopted, her best course of action would be to sell the patent to earn a payoff of $100,000. Unfortunately her decision must be made under conditions of uncertainty because she does not know whether or not the device will be adopted, that is, she does not know the true state of nature. In this type of situation there are several different criteria for selecting the "best" course of action.

1. The **maximin criterion** chooses the action that *maxi*mizes the *min*imum possible payoff.

   In our example the first action has a minimum payoff of $-\$500,000$, the second action has a minimum payoff of $0, and the third action has a minimum payoff of $100,000. Thus the third action (selling the patent) maximizes the minimum possible payoff, and so this is the action that the inventor will choose if she uses the maximin criterion. This strategy reflects a conservative or pessimistic attitude because it chooses the action that involves the least risk, that is, the action that offers the greatest payoff assuming that the worst happens.

2. The **maximax criterion** chooses the action that *maxi*mizes the *max*imum possible payoff.

   In contrast to the maximin criterion, the maximax criterion is an adventurous or optimistic strategy; it chooses the action that yields the greatest payoff assuming that the best will happen. In our example the maximum possible payoffs for the three actions are $2,000,000, $600,000, and $100,000. Thus if the inventor uses the maximax criterion, she will choose the first action (producing the device herself) because this option offers the greatest opportunity for gain.

   Since the state of nature is determined by external factors, a rational decision strategy should take into account all available information. If we can estimate the probability of each state of nature, then we can compare the options using the notion of expected value. This leads to a third criterion for selecting the "best" course of action.

3. **Bayes's criterion** chooses the action that maximizes the *expected* payoff.

   Suppose that the inventor in our example estimates that there is a 25% chance of her device being adopted by a major automaker. Then the expected payoffs for her three actions are as follows.

$$E_1 = \$2,000,000(.25) + (-\$500,000)(.75) = \$125,000$$
$$E_2 = \$600,000(.25) + \$0(.75) = \$150,000$$
$$E_3 = \$100,000(.25) + (\$100,000)(.75) = \$100,000$$

384 CHAPTER 7 STATISTICS

Thus if the inventor uses Bayes's criterion, she will select the second option (retaining a 30% interest in her device) because this offers the greatest expected payoff of $150,000.

Notice that in this example the three decision criteria (maximin, maximax, and Bayes's) lead to three different decisions. Thus there is no single answer as to which course of action is "best." We will return to this matter shortly, but first let us look at another example.

EXAMPLE 7.14 A Houston-based firm keeps a number of trained crews on call to put out oil fires anywhere in the world. The number of calls it receives has been observed to vary between 0 and 5 per week according to the following distribution.

Number of calls	Probability
0	.1
1	.1
2	.2
3	.3
4	.2
5	.1

The firm charges $50,000 for its services. Each crew can handle one call per week and is paid a weekly salary of $10,000 whether it is used or not. Use Bayes's criterion to determine the optimal number of crews to keep on call.

Solution The firm's net profit depends both on the action it takes (deciding how many crews to keep on call) and the state of nature (the number of calls received). For instance, if the firm has 4 crews on call and 3 calls are received, its profit is

$$\begin{aligned} \text{Profit} &= \text{Revenue} - \text{Cost} \\ &= 3(\$50{,}000) - 4(\$10{,}000) \\ &= \$150{,}000 - \$40{,}000 \\ &= \$110{,}000. \end{aligned}$$

And if there are 4 crews on call and 4 calls are received, the firm's profit is

$$\begin{aligned} \text{Profit} &= \text{Revenue} - \text{Cost} \\ &= 4(\$50{,}000) - 4(\$10{,}000) \\ &= \$200{,}000 - \$40{,}000 \\ &= \$160{,}000. \end{aligned}$$

In this case, the firm's profit is $160,000 if 4 or 5 calls are received, because if there are only 4 crews available, only 4 calls can be accepted. Similar calculations show that the payoff table for this firm is as shown below, where the net profit is given in thousands of dollars per week.

7.4 MATHEMATICS IN ACTION: DECISION THEORY

Action: Number of crews	State of Nature: Number of calls per week					
	0	1	2	3	4	5
1	−10	40	40	40	40	40
2	−20	30	80	80	80	80
3	−30	20	70	120	120	120
4	−40	10	60	110	160	160
5	−50	0	50	100	150	200

Using this table and the given probability distribution for the number of calls per week, we can compute the expected profit for each action. For instance, with 4 crews on call the expected profit is

$$-40(.1) + 10(.1) + 60(.2) + 110(.3) + 160(.2) + 160(.1) = 90.$$

Hence over the long run the firm can expect an average weekly profit of $90,000 by keeping 4 crews on call. Similar calculations produce the expected profits for the other courses of action; these are summarized in the table below.

Number of crews	Expected profit (in thousands)
1	$35
2	$65
3	$85
4	$90
5	$85

Since the maximum expected profit is $90,000 when there are 4 crews on call, 4 is the optimal number of crews according to Bayes's criterion. This is a somewhat surprising result because the expected number of calls per week is only 2.7, as the reader should check.

Observe that if we use the maximin criterion instead of Bayes's criterion, then the optimal number of crews is 1 since this gives the maximum (−$10,000) of the minimum profits for each action. And if the maximax criterion is used, we would keep 5 crews on call since that gives the maximum ($200,000) of the maximum possible profits. ∎

Practice Problem 10 A farmer delivers eggs each Friday to his wholesale and retail customers in the city. From experience he knows that he will sell no fewer than 2 truckloads and no more than 5 truckloads each week; in fact, the probabilities that exactly 2, 3, 4, and 5 truckloads of eggs can be sold in any given week are .2, .4, .3, and .1, respectively. For each truckload that is sold, the farmer receives a profit of $150, but each truckload that is not sold spoils with a resulting loss of $50.

(a) Prepare a payoff table for this problem.

Continuing from part (a), determine the optimal number of truckloads of eggs that the farmer should send to the city each week using

(b) the maximin criterion,
(c) the maximax criterion, and
(d) Bayes's criterion.

What criterion should a rational decision-maker use in a particular situation? When relatively small amounts are at stake, Bayes's criterion makes the most sense because we know that it gives the greatest average payoff over the long run. In Example 7.14, for instance, the fire-fighting firm is not especially concerned with its profit for any single week but rather with the long-term profit if it repeatedly follows a particular course of action. However, in the case of the inventor discussed earlier in this section, it can be argued that the situation is different; here the decision-maker must decide how best to exploit her new invention, a nonrepeatable decision in which there is much at stake. If the inventor needs $80,000 immediately to pay for surgery, she may decide to take the conservative course (selling the patent) in order to be assured of $100,000 no matter what happens. On the other hand, if she is on the verge of bankruptcy and needs $800,000 to survive, she may well decide that there is no alternative but to hope for the largest possible payoff by producing the device herself.

These are not irrational decisions; they merely reflect the different utility of money in different situations. That is, *the utility that a person assigns to a possible gain or loss may not equal its monetary value*. For example, except for the very wealthy, most people agree that "the second $1000 gained is worth less than the first"; in other words, a gain of $2000 is not quite twice as valuable as a gain of $1000. Conversely, a *loss* of $2000 is *more* than twice as painful as a loss of $1000. Thus, given a choice between receiving an outright gift of $10,000 or receiving $20,000 if a fair coin lands heads and nothing otherwise, most people would choose the first option, even though the expected value is the same in each case. Likewise most people prefer a certain loss of $10 to a .001 probability of losing $10,000; indeed, if this were not so, the insurance industry would cease to exist. The average person who occasionally buys a lottery ticket or visits a casino probably feels that the most likely outcome, a small net loss, is more than compensated for by the possibility, however remote, of big winnings. Similarly the individual who buys an insurance policy is not interested in the average return on his investment, but rather in being protected against the worse possible case. In both of these illustrations the person's expected utility may be positive even though the expected net gain is negative.

There have been various attempts to measure subjective utilities, that is, how much a particular gain or loss is "really worth" to a given individual.

The interested reader should see reference [3], in which both the practical and theoretical aspects of utility theory are discussed.

In the case of a private or public corporation with large assets, it is obviously in the best long-term interest of the institution to use Bayes's criterion in making investment decisions. Although some investments will fail and some will succeed, the *overall* net profit will be maximized. Some interesting studies have shown, however, that corporate managers tend to show excessive risk aversion in their decisions, preferring safe courses of action to those promising the greatest expected gain. (See, for instance, reference [4].) The reason for this phenomenon lies in the psychology of the individual, who worries—often justifiably—that his or her performance will be evaluated on the basis of the success or failure of a few key decisions rather than on their inherent rationality. This may explain why the more bureaucratic such institutions become, the more conservative their decisions tend to be, with the result that millions of dollars in potential profits are lost!

DECISION TREES

Often a decision-maker must play a cat-and-mouse game against nature: The decision-maker makes a move and then nature responds, creating a new situation requiring a decision. Such sequential decision problems can be quite involved, but with the aid of a diagram called a **decision tree** they can be systematically analyzed.

Let us consider the case of an aircraft company that is bidding on a contract worth $50 million. One of the company's directors wants to build a $10 million prototype immediately because this will give the company a 50% chance of winning the contract at the preliminary stage of the bidding. Another director wants to submit the bid without the prototype, which will give the company a 20% chance of success at the preliminary stage. If this first-stage bid is rejected, the company will still have the option of building a prototype on a rush basis for $15 million, which will restore a 50% chance of success in the final stage of the bidding. What decision strategy maximizes the company's expected net profit?

The decision tree for this problem is shown in Figure 7.9. The points where branching occurs are called **nodes.** Those points where a decision must be made (nodes A and D) are called **choice-nodes** and are denoted by squares, and those points where nature takes one of several states (B, C, and E) are called **chance-nodes** and are denoted by circles. This tree branches from left to right, each branch finally terminating in an **end-node** (heavy dot).

The initial choice-node A represents the decision whether or not to build the prototype immediately. The decision to build it leads to chance-node B, which branches according to whether the bid is accepted (with probability .5) or rejected (also with probability .5). If the bid is accepted, the company's payoff (in millions of dollars) will be $50 - 10 = 40$, the value of the contract

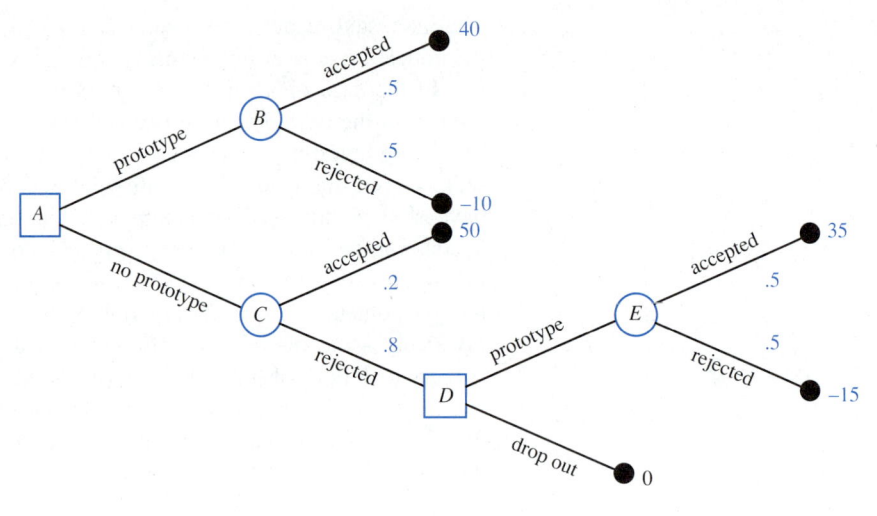

FIGURE 7.9

minus the cost of the prototype. If the bid is rejected, then the company will lose the cost of the prototype, for a payoff of −10.

On the other hand, if the prototype is not built immediately, we arrive at chance-node *C*, which branches according to whether the bid is accepted (with probability .2) or rejected (with probability .8). If the bid is accepted, the payoff is 50; if not, we must decide at choice-node *D* whether to build a rush prototype or drop out of the bidding and take a payoff of 0. If the rush prototype is built (chance-node *E*), the bid will either be accepted at the final stage (with a probability of .5 and a payoff of 50 − 15 = 35) or rejected (with a probability of .5 and a payoff of −15).

Thus Figure 7.9 represents the entire sequential process described in our example. Note that each end-node is labeled with a numerical payoff denoting the company's net profit if that branch of the tree is followed. In order to compare the different strategies available, we work through the tree from right-to-left, assigning a number to each node according to the rules below. The number assigned to a node represents the maximum expected payoff at that point of the decision process.

Rules for Assigning Numbers to the Nodes in a Decision Tree

1. The number assigned to an end-node is the corresponding payoff.

2. The number assigned to a chance-node is the *expected value* of the numbers assigned to the nodes to which it branches.

3. The number assigned to a choice-node is the *maximum* of the numbers assigned to the nodes to which it branches.

In our example, this computation proceeds as follows. (Refer to Figure 7.9. You may wish to label each node with its number as it is assigned.) First, we assign the corresponding payoffs to all of the end-nodes; these numbers are already indicated in Figure 7.9. Second, we assign a number to node E. Since this is a chance-node, we use rule 2 above to assign to node E the expected value of the two end-nodes to which it branches, which is

$$35(.5) + (-15)(.5) = 10.$$

This number represents the expected payoff at this point of the decision problem.

Third, we assign a number to choice-node D in accordance with rule 3. Since node E (building the rush prototype) has been assigned the number 10 and dropping out of the bidding has been assigned the number 0, we assign to D the number 10. Note that at node D we must choose between building the rush prototype and dropping out of the bidding. Since node E (building the rush prototype) has an expected profit of 10 and dropping out of the bidding has an expected profit of 0, we obviously will choose to build the prototype, resulting in an expected profit of 10.

Fourth, we use rule 2 to assign node C the expected value of the numbers at the nodes to which it branches, which is

$$50(.2) + 10(.8) = 18.$$

Fifth, we use rule 2 to assign node B the number

$$40(.5) + (-10)(.5) = 15.$$

Finally, we assign choice-node A the larger of the numbers 15 (at node B) and 18 (at node C) in accordance with rule 3. Thus node A is assigned the number 18. This number represents the company's maximum expected profit (in millions of dollars), which it will realize by initially submitting the bid without the prototype and building a rush prototype later if necessary.

The sort of analysis used in the example above is called **Bayesian analysis** because it is based on Bayes's criterion. It provides the most rational decision method for an institution that wishes to maximize its long-term profits when facing many decisions of this type.

EXAMPLE 7.15 The county agricultural extension service has advised a farmer to apply a pesticide costing $20 per acre to his soybean fields. Depending on the weather, this pesticide might have no effect on the soybean yield per acre, a 10% increase in yield, or a 15% increase in yield. The probability of each effect is shown below.

	Increased Yield		
Weather	*0%*	*10%*	*15%*
Dry	.1	.4	.5
Normal	.2	.6	.2
Wet	.3	.5	.2

Based on experience, the farmer can expect the following probabilities, yields, and soybean prices for the different weather conditions.

Weather	Probability	Expected yield (bushels per acre)	Selling price per bushel
Dry	.3	48	$7.00
Normal	.6	55	$6.20
Wet	.1	60	$5.80

Use a decision tree to decide whether or not the farmer should apply the pesticide.

Solution The farmer has only one decision to make—whether to use the pesticide or not. Thus the decision tree for this problem has only one choice-node, which is node *A*. (See Figure 7.10.) If the farmer does not use the pesticide, we arrive at the chance-node *F*. Here there are three branches according to the type of weather, each leading to an end-node. Things are more complicated, however, if the farmer elects to apply the pesticide. Using the pesticide leads to chance-node *B*, where there are again three branches according to the possible weather conditions. These branches lead to chance-nodes *C*, *D*, and *E*, each of which branches according to the three possible effects from using the pesticide (0% increase in yield, 10% increase in yield, and 15% increase in yield). These branches all terminate at end-nodes.

Computing the payoffs at the end-nodes requires care. If the pesticide is not applied, then the payoff is simply the expected yield for each weather condition multiplied by the corresponding price per bushel. Thus under dry conditions the payoff will be

$$48(7.00) = 336.00$$

dollars per acre. Likewise the payoff under normal conditions is

$$55(6.20) = 341.00$$

dollars per acre, and the payoff under wet conditions is

$$60(5.80) = 348.00$$

dollars per acre.

When the pesticide is used, we must perform a similar calculation in which the yield per acre is adjusted by the appropriate factor to account for the increase in yield. For example, if there is dry weather and the pesticide increases yields by 10%, then the yield per acre is

$$1.10(48) = 52.8$$

bushels per acre. Hence the return per acre in this case will be

$$52.8(7.00) = 369.60.$$

7.4 MATHEMATICS IN ACTION: DECISION THEORY

Figure 7.10 shows a decision tree with choice-node A branching into "pesticide" (leading to node B) and "non-pesticide" (leading to node F).

From node B (pesticide):
- *dry (.3) → node C*
- *normal (.6) → node D*
- *wet (.1) → node E*

From node C:
- *0% (.1) → 316.00*
- *10% (.4) → 349.60*
- *15% (.5) → 366.40*

From node D:
- *0% (.2) → 321.00*
- *10% (.6) → 355.10*
- *15% (.2) → 372.15*

From node E:
- *0% (.3) → 328.00*
- *10% (.5) → 362.80*
- *15% (.2) → 380.20*

From node F (non-pesticide):
- *dry (.3) → 336.00*
- *normal (.6) → 341.00*
- *wet (.1) → 348.00*

FIGURE 7.10

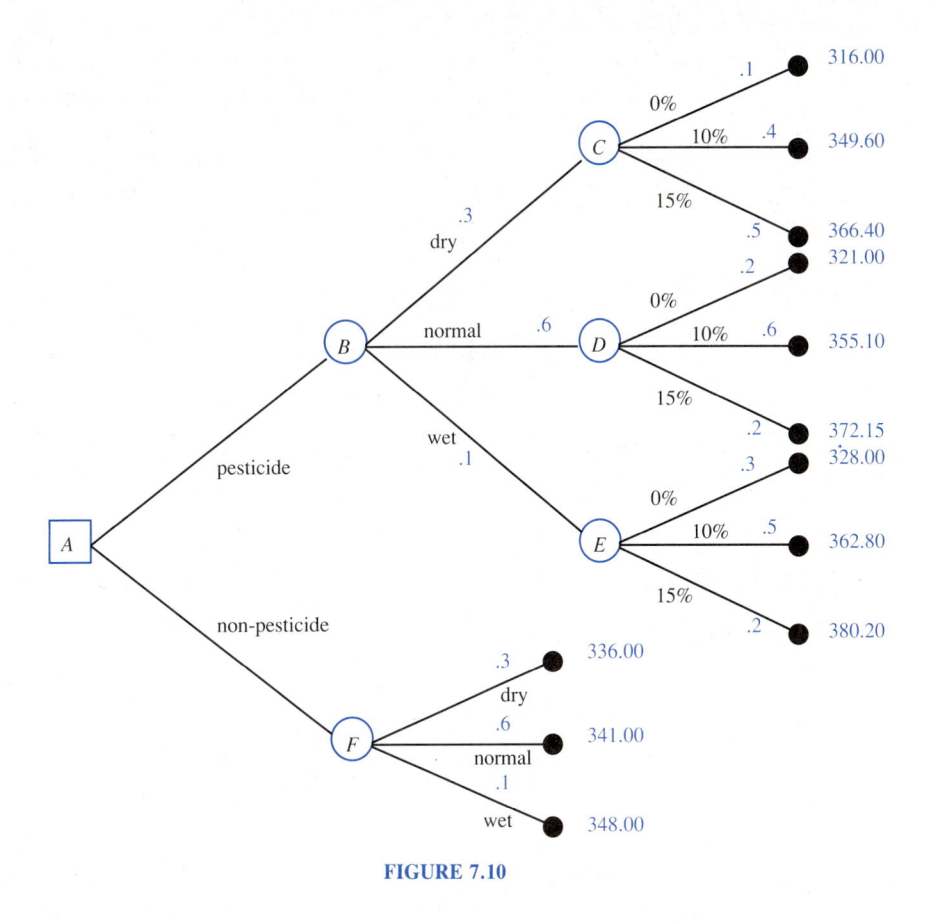

Finally, to obtain the payoff in dry weather when the pesticide increases yields by 10%, we must subtract the pesticide's cost per acre ($20.00) from the return per acre:

$$369.60 - 20.00 = 349.60.$$

Check that the payoffs for the other end-nodes are as shown in Figure 7.10.

We now assign numbers to the nodes using the rules on page 404. To chance-nodes C, D, E, and F, we assign the expected values of the numbers assigned to the end-nodes to which they branch; these expected values are 354.64, 351.69, 355.84, and 340.20, respectively. To chance-node B, we assign the expected value of the numbers assigned to nodes C, D, and E:

$$.3(354.64) + .6(351.69) + .1(355.84) = 352.99.$$

Finally, we assign to choice-node A the maximum of the numbers assigned to nodes B and F, which is $352.99. Thus the farmer's maximum expected gain is $352.99 per acre, and it is obtained by using the recommended pesticide. ∎

Practice Problem 11 A contractor will receive $200,000 in profits for completing a particular construction project. If the project is completed a month early, he will earn an additional $40,000 bonus. However, the only way in which the project can be completed early is to begin in the fall instead of waiting until spring, and if the project is started in the fall, it faces an 80% risk of bad weather. If there is good fall weather, the project will surely be completed early, but bad weather will cost the contractor an additional $20,000 in expenses as well as delaying the project. In this event, the project cannot be finished early unless the contractor chooses to pay his workers an additional $10,000 for overtime, which will give him a 40% chance of finishing early.

(a) Represent this decision process by a decision tree.
(b) Determine the maximum expected payoff at each node.
(c) What strategy should the contractor use to maximize his expected profit?

EXERCISES 7.4

1. Consider the payoff table below.

| | \multicolumn{4}{c}{State of Nature} |
Action	s_1	s_2	s_3	s_4
a_1	20	25	−10	0
a_2	30	15	30	−20
a_3	40	0	50	0

If the probabilities of the four states of nature are .3, .4, .2, and .1, respectively, determine the best course of action using (a) the maximin criterion; (b) the maximax criterion; and (c) Bayes's criterion.

2. Consider the payoff table below.

| | \multicolumn{3}{c}{State of Nature} |
Action	s_1	s_2	s_3
a_1	10	15	25
a_2	−20	30	0
a_3	25	10	−10
a_4	−10	15	45

If the probabilities of the three states of nature are .2, .7, and .1, respectively, determine the best course of action using (a) the maximin criterion; (b) the maximax criterion; and (c) Bayes's criterion.

3. The owner of a newsstand gets a few requests each day for out-of-town papers. One of them, the Baltimore *Sun*, costs the dealer 40¢ and sells for 90¢ per copy. Unsold copies cannot be returned and therefore represent a loss for the dealer. The number of daily requests for the *Sun* has been observed to vary as follows.

Number requested	Probability
0	.1
1	.2
2	.3
3	.3
4	.1

How many copies should the dealer stock to maximize his long-term profit?

4. A fruit grower delivers peaches to a farmers market each Friday. From experience he knows that he can sell between 1 and 4 truckloads each week, as follows.

Number of truckloads	Probability
1	.4
2	.3
3	.2
4	.1

For each truckload of peaches that is sold, the grower will realize a profit of $200, but because of spoilage he will lose $80 for each truckload that is not sold. Under these conditions, how many truckloads should the grower send to the farmers market each week in order to maximize his long-term profit?

5. A fish market observes that the daily demand for fresh salmon varies between 5 and 9 pounds. In fact, the probabilities of selling exactly 5, 6, 7, 8, and 9 pounds of salmon per day are .1, .2, .4, .2, and .1, respectively. The salmon, which sells for $6 per pound, must be ordered the day before at a cost of $3 per pound. The dealer can sell day-old salmon to a local manufacturer of cat food for $2 per pound, and so unsold fish result in a $1 per pound loss to the market. How many pounds of salmon should be ordered each day in order to maximize the market's long-term profit?

6. A baker can sell as many as 3 birthday cakes per day besides those specially ordered. From experience he knows that the probabilities of selling 0, 1, 2, and 3 cakes are .1, .4, .3, and .2, respectively. If each cake that is sold results in a $3 profit and each cake that is not sold results in an $0.80 loss, how many cakes should the baker make in order to maximize his long-term profit?

7. A carnival operator is worried about the possibility of weekend rain. If he cancels the carnival, he will lose $2000 in promotional expenses. On the other hand, if he sets up his equipment, he will make $12,000 if it doesn't rain but will lose $8000 if it does. Assume that he applies Bayes's criterion to decide the best course of action.
   (a) What should he do if the chance of rain is 65%?
   (b) How great must the chance of rain be in order for the operator to decide to cancel the carnival?

8. An insurance company sells single-flight airplane accident policies in major airports. A $10,000 policy costs $10, and a $20,000 policy costs $15. Let p be the probability of an airplane accident.
   (a) If $p = .0003$, which of the two policies is better for the company?
   (b) What must the value of p be in order for the company's expected gain to be the same for both policies?

394 CHAPTER 7 STATISTICS

In Exercises 9–10, find the value of each node in the decision tree and determine the expected payoff if the decision-maker follows an optimal strategy.

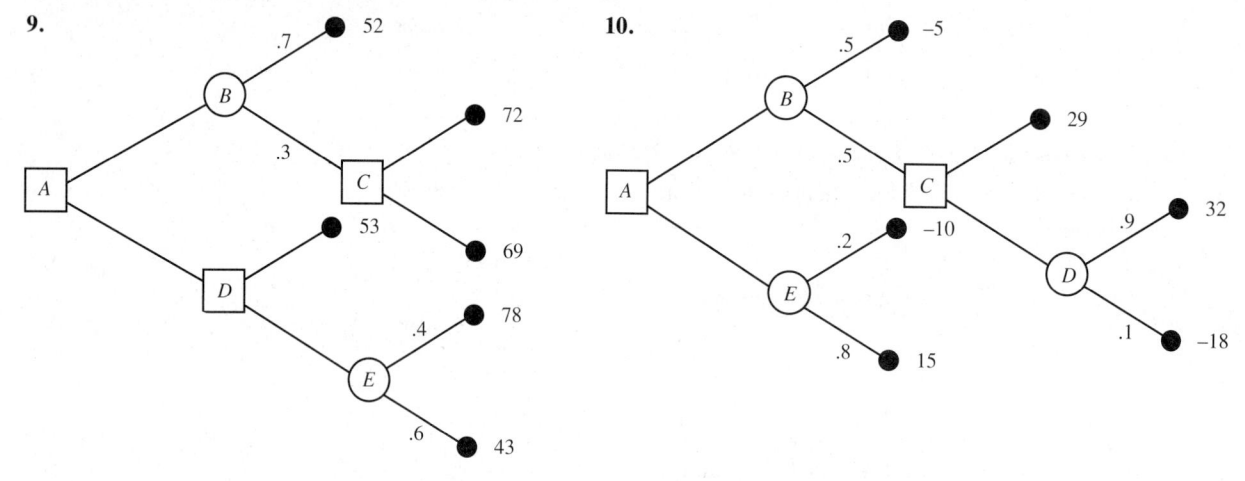

9.

10.

11. A special interest group is trying to decide whether or not to hire a lobbyist at a cost of $250,000 to help pass a key piece of legislation. Hiring the lobbyist increases the bill's chance of passage from 20% to 30%, and passage of the bill is worth $3 million in revenue to the group. What is the group's best course of action, and what is its expected net gain?

12. The director of the San Francisco Opera must decide whether or not to hire a famous tenor for the new season. If the tenor is not hired, the director expects a $2 million profit for the company. On the other hand, if the tenor is hired, the company's profit should increase by $800,000. In this case, however, there is a 50% chance that the company's star soprano will become temperamental and threaten to quit. If so, the director will either have to give her a $500,000 increase in salary (which reduces profits by this amount) or fire her, in which case the net profits will be only $1,900,000. What decision strategy maximizes the company's expected net profit?

13. Elaine often has to park downtown for an hour. She can pay 85¢ at a parking lot or look for a space on the street, which she finds 60% of the time. If she finds a space she can either put 25¢ in the meter or take a 5% risk of getting a $10 ticket. If she can't find a space on the street, she must return to the parking lot, having wasted 50¢ worth of gasoline. What decision strategy minimizes her expected loss?

14. A homeowner is asking $100,000 for his house, which he must sell within two weeks. A buyer has made what she claims is her final offer of $80,000. The seller can either accept this offer or make a counteroffer of $90,000. If he makes the counteroffer, the chances are even that it will be accepted. If it is not, there is an 80% chance that the buyer will lose interest in the house and a 20% chance that the buyer will propose a compromise offer of $85,000, which the seller is prepared to accept. If the buyer loses interest in the house, however, the seller will have to accept a $75,000 offer from another party. What should the seller do to maximize his expected gain from the sale?

15. Mrs. Martin is considering whether or not to file a civil suit against a local roofing company. If she files the suit, there is a 50% chance that she will be offered $2000 to settle

her claim out of court. If no offer is made, or if the offer is made and refused, the case will go to trial, in which case Mrs. Martin has a 70% chance of winning her full $4000 claim. If she loses in court, however, she will have to pay $3000 in court costs. What decision strategy should Mrs. Martin adopt in order to maximize her expected gain?

16. An oil company is trying to decide whether to do shallow or deep drilling at a new site. Shallow drilling costs $5 million and has a 12% chance of finding oil, and deep drilling costs $15 million and has a 23% chance of finding oil. If shallow drilling is unsuccessful, the company can either abandon the project or undertake deep drilling at an extra cost of $15 million, but the chance of finding oil in this case is only 12.5%. If oil is found at a site, the company will realize $90 million in revenues. What decision strategy maximizes the company's expected net profit?

Consider the payoff table below, where the states of nature have respective probabilities of .5, .3, and .2.

Action	s_1	s_2	s_3	Expected payoff
a_1	$4000	$7000	−$1000	$3900
a_2	$5000	$4000	$3000	$4300
a_3	$6000	−$2000	$4000	$3200

According to Bayes's criterion, the best action is a_2 with an expected payoff of $4300. But suppose that the decision-maker knew in advance which state of nature would occur and acted accordingly to maximize the payoff. The expected gain in this case would be

$$\$6000(.5) + \$7000(.3) + \$4000(.2) = \$5900$$

because the maximum payoffs for the three states of nature are $6000, $7000, and $4000, respectively. The difference

$$\$5900 - \$4300 = \$1600$$

is called the **value of perfect information.** *This represents the maximum amount that the decision-maker should pay for information such as market forecasts or opinion polls that can be used to help choose an action.*

17. Find the value of perfect information in Exercise 1.
18. Find the value of perfect information in Exercise 2.
19. Find the value of perfect information in Exercise 3.
20. Find the value of perfect information in Exercise 7(a).

Answers to Practice Problems   **10.** (a) The payoff table is shown below.

Action: Number of truckloads sent	State of Nature: Demand for eggs (number of truckloads)			
	2	3	4	5
2	$300	$300	$300	$300
3	$250	$450	$450	$450
4	$200	$400	$600	$600
5	$150	$350	$550	$750

(b) 2   (c) 5   (d) 4

11. (a) The decision tree is shown below.

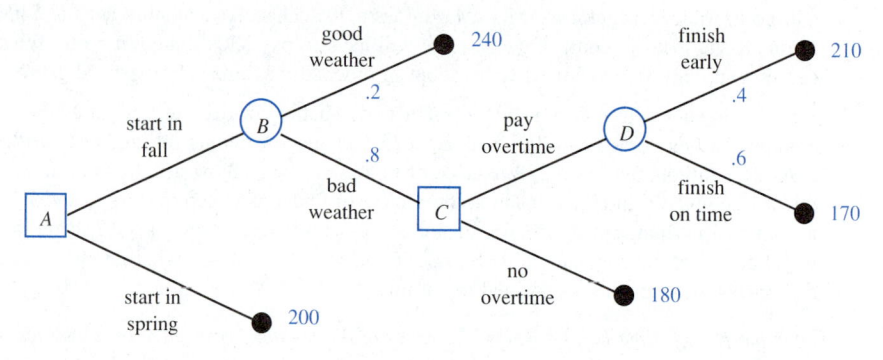

```
                        good weather    240              finish early    210
                          .2                                .4
         start in    B          pay overtime     D
         fall             .8                          .6
    A              bad weather    C                finish on time   170
         start in                   no overtime    180
         spring    200
```

(b) The maximum expected payoffs are 186 at nodes D and C, 196.80 at node B, and 200 at node A.

(c) The contractor should begin construction in the spring.

7.5 Probability Distributions

In Section 7.3 we obtained the following distribution for the random variable that counts the number of daily service calls by an automobile towing service.

Number of service calls	Probability
0	.10
1	.25
2	.30
3	.20
4	.10
5	.05
	1.00 Total

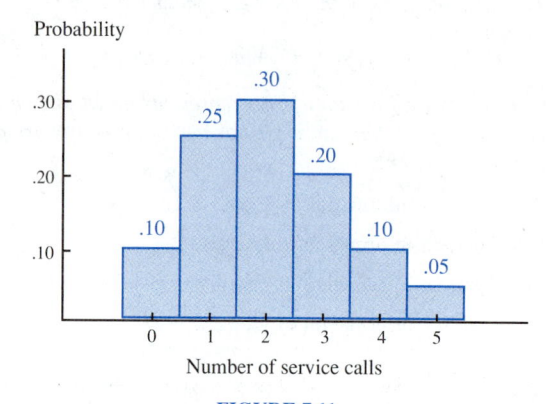

FIGURE 7.11

Recall that we represented this distribution by the histogram in Figure 7.11, where each value of the random variable is the midpoint of a class interval extending one-half unit to the left and right. The height of each bar represents the probability that the random variable equals the midpoint of the class interval corresponding to that bar. This histogram is called the **probability distribution** for the random variable; it is the graph of a function called the **probability density function**.

We can use the probability distribution in Figure 7.11 to find the probability that the towing service receives a given number of service calls on a particular day. For example, the probability that the number of service calls is between 1 and 3, inclusive, is

P(between 1 and 3 calls)

$\quad = P$(exactly 1, 2, or 3 calls)

$\quad = P$(exactly 1 call) $+ P$(exactly 2 calls) $+ P$(exactly 3 calls)

$\quad = .25 + .30 + .20$

$\quad = .75.$

This calculation is depicted in Figure 7.12, where the part of the probability distribution corresponding to the values 1, 2, and 3 is shaded. Since each bar of the histogram has width 1, the total area of the shaded portion is

$$.25(1) + .30(1) + .20(1) = .75,$$

which is the probability in question. Notice that the total area represented by the probability distribution is 1.

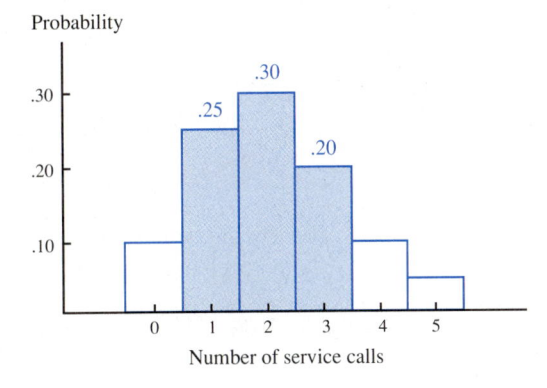

FIGURE 7.12

This example illustrates our first theorem.

Theorem 7.1

Consider a random variable that takes consecutive integer values.* The probability that the variable assumes a value between a and b (inclusive) is equal to the area under its probability distribution between $a - \frac{1}{2}$ and $b + \frac{1}{2}$.

---

*If the values of the variable are not consecutive integers, then Theorem 7.1 must be modified to account for the fact that the bars in its probability distribution do not have width 1. We will not consider this situation.

398 CHAPTER 7 STATISTICS

EXAMPLE 7.16 Use Figure 7.11 to find the probability that on a particular day the towing service receives

(a) between 1 and 4 service calls (inclusive);
(b) at least 3 service calls.

Solution (a) By Theorem 7.1 we see that the probability of receiving between 1 and 4 service calls in Figure 7.11 is

$$P(\text{between 1 and 4 calls}) = \text{area between 0.5 and 4.5} = .85.$$

The corresponding region of the probability distribution is shown in Figure 7.13(a).

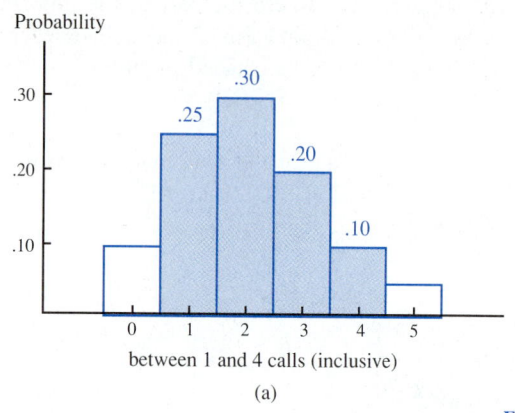

FIGURE 7.13

(b) The probability of receiving at least 3 requests is the same as the probability of receiving between 3 and 5 requests, inclusive. Thus we see that

$$P(\text{at least 3 calls}) = \text{area between 2.5 and 5.5} = \text{area to the right of 2.5} = .35.$$

See Figure 7.13(b). ■

Practice Problem 12 Use Figure 7.11 to find the probability that on a particular day the towing service receives

(a) between 3 and 4 service calls, inclusive;
(b) at most 1 service call.

THE BINOMIAL DISTRIBUTION

Recall that in Section 6.6 we defined a *Bernoulli process* to be a sequence of repeated trials of an experiment in which

1. each trial has only two possible outcomes, called *success* and *failure* (and denoted S and F, respectively);
2. the trials are independent; and
3. the probability of success is the same for each trial.

The principal result concerning Bernoulli processes is Theorem 6.7, which we reformulate below in the language of statistics.

Binomial Distribution

The number of successes in a sequence of n Bernoulli trials is a random variable that assumes the values $0, 1, 2, \ldots, n$. Its probability density function is given by Bernoulli's formula:

$$P(\text{exactly } k \text{ successes in } n \text{ trials}) = C(n, k) p^k q^{n-k} \quad (k = 0, 1, \ldots, n),$$

where p and q denote the probabilities of success and failure, respectively, on each trial.

EXAMPLE 7.17 Find the probability density function and probability distribution for the number of heads obtained when a fair coin is flipped ten times.

Solution We must determine the distribution of the random variable that counts the number of successes in 10 trials, where success means that the coin lands heads. Since the probabilities of success and failure are $p = \frac{1}{2}$ and $q = \frac{1}{2}$ for a fair coin, Bernoulli's formula gives the following probability density function.

Number of heads	Probability
0	1/1024
1	10/1024
2	45/1024
3	120/1024
4	210/1024
5	252/1024
6	210/1024
7	120/1024
8	45/1024
9	10/1024
10	1/1024
	1 Total

For instance, the probability of obtaining exactly 3 heads is computed as follows:

$$P(\text{exactly 3 successes in 10 trials}) = C(10, 3) p^3 q^7$$
$$= 120 \left(\frac{1}{2}\right)^3 \left(\frac{1}{2}\right)^7$$
$$= \frac{120}{1024}.$$

(See also Example 6.35 for the probability of obtaining exactly 5 heads.)

The probability distribution for this density function is shown in Figure 7.14. As we see from this figure, the *most likely* number of heads is 5, which has probability $252/1024 \approx .246$. Using Theorem 7.1 we find that the probability of obtaining between 4 and 6 heads is

$$P(4, 5, \text{ or } 6 \text{ heads}) = \text{area between 3.5 and 6.5}$$
$$= \frac{672}{1024} = .65625.$$

Likewise the probability of obtaining between 3 and 7 heads is

$$P(3, 4, 5, 6, \text{ or } 7 \text{ heads}) = \text{area between 2.5 and 7.5}$$
$$= \frac{912}{1024} = .890625. \blacksquare$$

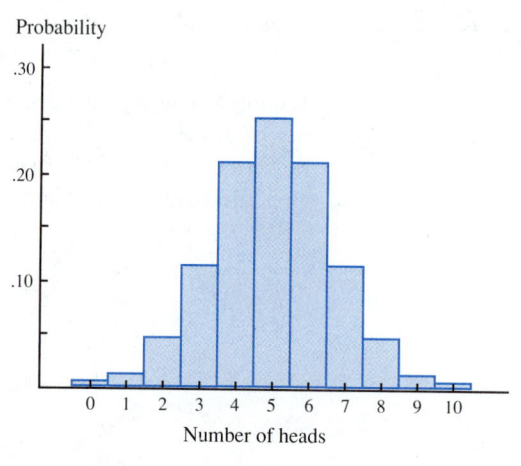

FIGURE 7.14

Practice Problem 13 (a) Find the probability density function for the number of sixes rolled when a fair die is tossed five times.
(b) Determine the probability of obtaining more than one six in five tosses.

Using Formulas (7.5) and (7.6), we can compute the mean and standard deviation for the probability density function in Example 7.17. Fortunately, however, there is a general result about binomial distributions that makes these calculations unnecessary.

Theorem 7.2

Consider the random variable that counts the number of successes in a sequence of n Bernoulli trials in which the probabilities of success and failure on each trial are p and q, respectively. Then the mean of this variable is

$$\mu = np,$$

and its standard deviation is

$$\sigma = \sqrt{npq}.$$

The formula for the mean in Theorem 7.2 makes sense intuitively: If the probability of success on each trial is p, then in n trials we should expect np successes.

EXAMPLE 7.18 Find the mean and standard deviation for the number of heads obtained when a fair coin is flipped three times.

Solution We can solve this problem by applying Theorem 7.2 with $n = 3$ and $p = q = \frac{1}{2}$. The expected number of heads (the mean) is

$$\mu = np = 3(\tfrac{1}{2}) = 1.5,$$

and the standard deviation is

$$\sigma = \sqrt{npq} = \sqrt{3\left(\tfrac{1}{2}\right)\left(\tfrac{1}{2}\right)} = \sqrt{0.75} \approx 0.87.$$

Note that these are the same values obtained in Example 7.9 when we computed the mean and standard deviation directly from the probability density function. ∎

EXAMPLE 7.19 Find the mean and standard deviation for the number of sixes obtained when a fair die is rolled 60 times.

Solution We can solve this problem by applying Theorem 7.2 with $n = 60$, $p = 1/6$, and $q = 5/6$. The expected number of sixes (the mean) is

$$\mu = np = 60\left(\tfrac{1}{6}\right) = 10$$

and the standard deviation is

$$\sigma = \sqrt{npq} = \sqrt{60\left(\tfrac{1}{6}\right)\left(\tfrac{5}{6}\right)} = \sqrt{\tfrac{300}{36}} \approx 2.89.$$

Note, however, that although the expected number of sixes is 10, the probability of obtaining *exactly* 10 sixes is only

$$C(60, 10)\left(\frac{1}{6}\right)^{10}\left(\frac{5}{6}\right)^{50} \approx .137.$$

Thus the probability that the mean actually occurs can be quite small.

Practice Problem 14 From past experience a construction company has found that its probability of making a successful bid on a contract is .2. Assuming that this probability continues, find the mean and standard deviation for the number of contracts obtained if it submits 30 bids per year.

Binomial distributions arise in a wide variety of applications. The next example illustrates their use in testing the effectiveness of drugs.

EXAMPLE 7.20 Suppose that 40% of those having a certain disease recover without treatment. Eight persons with this disease are given an experimental drug, and five of them recover. What is the probability that such a result occurs merely by chance?

Solution To answer this question, we will assume that the experimental drug has no effect and determine the probability that out of eight persons with the disease, five or more recover. Accordingly we will regard this situation as a Bernoulli process with $n = 8$ trials in which the probability of success (recovery) on each trial is $p = .4$. Since the number of successes is binomially distributed, the probability density function for the number of successes is as follows.

Number of successes	Probability
0	.017
1	.090
2	.209
3	.279
4	.232
5	.124
6	.041
7	.008
8	.001
Total	1.000

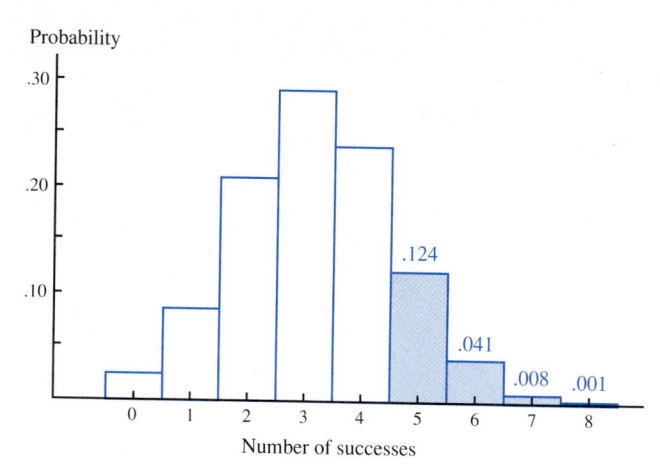

FIGURE 7.15

For instance, the probability that exactly three of the eight persons recover is

$$P(3 \text{ successes in 8 trials}) = C(8, 3)p^3q^5$$
$$= 56(.4)^3(.6)^5$$
$$\approx .279.$$

From the probability histogram in Figure 7.15, we see that

$$P(5 \text{ or more successes}) = \text{area to the right of } 4.5$$
$$= .124 + .041 + .008 + .001$$
$$= .174.$$

Thus the probability is .174 that five or more patients would have recovered even if they had not been given the experimental drug. Because this probability is high, few statisticians would regard this success rate as evidence of the drug's effectiveness. On the other hand, the probability that *six* or more of the eight persons recover is only

$$P(6 \text{ or more successes}) = \text{area to the right of } 5.5 = .05,$$

which many statisticians would accept as evidence of the drug's effectiveness. ■

Practice Problem 15 A graphologist (handwriting expert) claimed to be able to distinguish normal persons from psychotics on the basis of handwriting. A test was devised (see reference [2]) in which the graphologist was presented ten pairs of persons, one normal and one psychotic. In these ten pairs, the graphologist correctly identified the psychotic person six times.

(a) Find the probability of identifying the psychotic person in exactly six of the pairs by chance alone.
(b) Find the probability of identifying the psychotic person in exactly seven of the pairs by chance alone.
(c) Determine the probability of identifying the psychotic person in six or more pairs by chance alone.
(d) Would you accept the graphologist's claim?

When the number of trials is large, the binomial distribution becomes unwieldy. In such cases we can approximate the binomial probability distribution by a smooth curve called the *normal distribution*. We will discuss this curve in the next section.

EXERCISES 7.5

1. The probability distribution for a random variable taking the values 0, 1, 2, 3, and 4 is shown below.

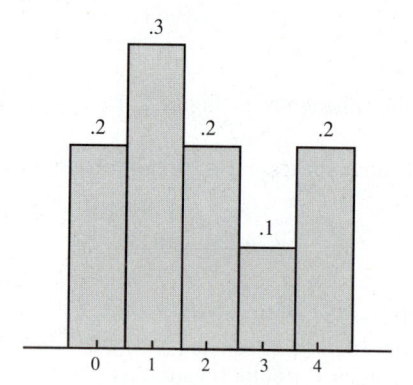

Find the probability that the variable assumes a value
(a) between 1 and 3 (inclusive) (b) at least 2.
(c) less than 3.

2. The probability distribution for a random variable taking the values 0, 1, 2, . . ., 10 is shown below.

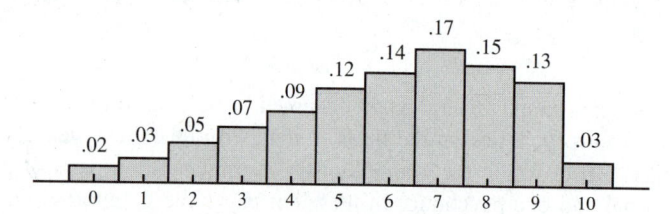

Find the probability that the variable assumes a value
(a) between 4 and 8 (inclusive) (b) greater than 5
(c) no more than 6.

3. An insurance company checked the records of 80 random motorists to determine how many moving violations each had been cited for during the past five years. The results are shown below.

Violations	Frequency
0	34
1	28
2	12
3	4
4	2
	80 Total

(a) Construct the probability density function and the corresponding probability distribution for these empirical data.

(b) Use Theorem 7.1 to determine the probability that someone had at most 2 violations.

(c) Use Theorem 7.1 to determine the probability that someone had between 1 and 3 violations (inclusive).

(d) Use Theorem 7.1 to determine the probability that someone had at least one violation.

4. A Labor Department statistician interviewed 120 unemployed persons in an inner-city area to determine how many years of high school each had completed. Their responses are shown below.

Number of years	Frequency
0	6
1	24
2	39
3	30
4	21
	120 Total

(a) Find the probability density function and the corresponding probability distribution for the number of years of high school completed.

(b) Use Theorem 7.1 to determine the probability that someone completed at most one year of high school.

(c) Use Theorem 7.1 to determine the probability that someone completed at least three years of high school.

(d) Use Theorem 7.1 to determine the probability that someone completed between one and three years of high school (inclusive).

5. A stockbroker observes that the delivery time for her local mailings is a random variable with the distribution below.

Delivery time (working days)	Probability
1	.218
2	.318
3	.272
4	.130
5	.062

(a) Construct the probability distribution for this data.

(b) Use Theorem 7.1 to determine the probability that an item mailed on Monday will arrive on Wednesday, Thursday, or Friday.

(c) Use Theorem 7.1 to determine the probability that an item mailed on Monday will arrive by Thursday at the latest.

6. A baker finds that the number of chocolate cakes requested on a given day has the following distribution.

Requests	Probability
0	.11
1	.20
2	.32
3	.19
4	.12
5	.05
6	.01

(a) Construct the probability distribution for this data.

(b) Use Theorem 7.1 to determine the probability that between 3 and 5 chocolate cakes will be requested on a given day.

(c) Use Theorem 7.1 to determine the probability that at most 2 chocolate cakes will be requested on a given day.

(d) Use Theorem 7.1 to determine the probability that at least 4 chocolate cakes will be requested on a given day.

In Exercises 7–10 determine the probability density function for the random variable that counts the number of successes in a sequence of n Bernoulli trials in which the probability of success on each trial is p.

7. $n = 3$ and $p = .4$

8. $n = 4$ and $p = .8$

9. $n = 5$ and $p = .25$

10. $n = 6$ and $p = .5$

In Exercises 11–18 find the mean and standard deviation of the indicated binomial distribution.

11. Two fair dice are rolled 90 times, and the number of times that a sum of seven occurs is counted.

12. An 80% shooter takes 50 shots at a target, and the number of hits is recorded.

13. A mail solicitation is sent to 10,000 individuals, each of whom has a 9% chance of responding, and the number of responses is counted.

14. A machine produces 10,000 items with a 0.4% defective rate, and the number of defective items is counted.

15. An airline has given 350 reservations for a flight on which there is a 10% chance that a person holding a reservation will fail to show up, and the number of no-shows is counted.

16. On a true-false test consisting of 100 questions, the number of correct answers obtained by random guessing is counted.

17. Of the registered voters in a certain district, 75% favor a tax cut. In a random survey of 200 of these voters, the number of persons favoring the cut is counted.

18. In American roulette, 18 of the 38 possible numbers are red. A player bets on red 95 times in a row, and the number of times he wins is counted.

19. A pharmaceutical house claims to have developed a drug that increases the chance of recovery from a certain disease from 60% to 80%. Suppose that the drug is given to 50 patients with the disease, and the number who recover is counted.
    (a) Find the mean and standard deviation of this random variable if the manufacturer's claim is true.
    (b) Find the mean and standard deviation of this random variable if the drug has no effect on the disease.

20. Suppose that there are 120 infants in a pediatric ward, each needing a nurse's attention 15% of the time. Find the mean and standard deviation of the number needing attention.

Answers to Practice Problems   **12.** (a) .30   (b) .35

13. (a)

Number of sixes	Probability
0	3125/7776
1	3125/7776
2	1250/7776
3	250/7776
4	25/7776
5	1/7776

(b) $\dfrac{1526}{7776}$

14. $\mu = 6$ and $\sigma \approx 2.19$

15. (a) $\dfrac{210}{1024}$

(b) $\dfrac{120}{1024}$

(c) $\dfrac{210}{1024} + \dfrac{120}{1024} + \dfrac{45}{1024} + \dfrac{10}{1024} + \dfrac{1}{1024} = \dfrac{386}{1024} \approx .377$

(d) no

7.6 The Normal Distribution

So far we have considered only random variables that take finitely many values. Probability distributions can also be defined for **continuous** random variables, those with values measured on a continuous scale, such as the set of all nonnegative real numbers. In this section we will discuss the most important case, random variables that are normally distributed.

CONTINUOUS DISTRIBUTIONS

When we measure the length of a javelin throw, the weight of a newborn infant, or the amount of cholesterol in a person's blood, we are dealing with continuous random variables. Although we can approximate such measurements by rounding them off to a finite number of decimal places, the fact remains that in theory their possible values cover an interval of real numbers.

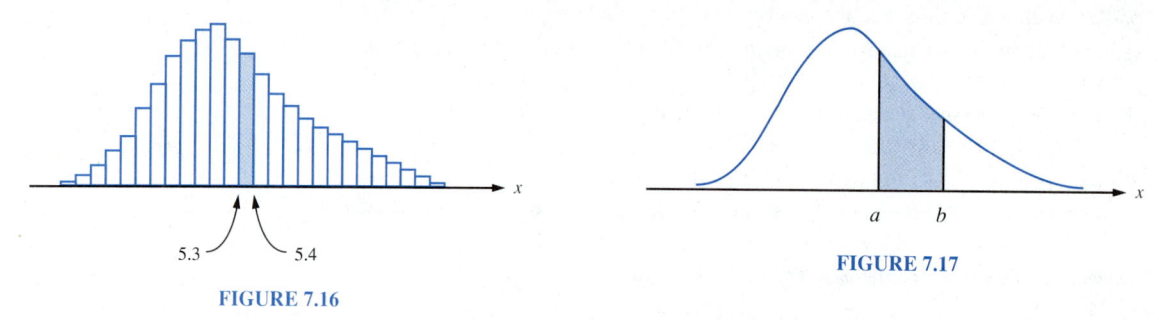

FIGURE 7.16

FIGURE 7.17

To obtain the probability distribution for a continuous random variable, we imagine dividing the interval of possible values into many small subintervals as in Figure 7.16. Over each subinterval (such as 5.3 to 5.4 in the figure) we construct a bar with *area* equal to the probability that the variable assumes a value in that subinterval. Using smaller and smaller subintervals, we find that the resulting histogram approaches a smooth curve, which is shown in Figure 7.17. If we think of the curve as a histogram with "infinitely thin" bars, we see that it has the following basic property.

> The area under the curve between any two values a and b represents the probability that the random variable assumes a value in this interval.

Note that the *total* area under a probability distribution is always 1. Furthermore, if $a = b$ in Figure 7.17, the area between them is zero. Hence *the probability that a continuous random variable takes any particular value is zero.* This is not as unreasonable as it may sound at first. For instance, it is highly unlikely—impossible for all practical purposes—that the length of the next throw of a javelin will be *exactly* 71.6325 meters. In any case this example points out an important difference between finite and continuous random variables: In the former we consider the probability of a given *value,* but for the latter we consider the probability of a given *interval.*

By using concepts from calculus, it is possible to define the mean and standard deviation of a continuous random variable. These numbers have the same significance as in the finite case; that is, the mean represents the "center" of the distribution and the standard deviation measures its dispersion around the mean.

THE NORMAL DISTRIBUTION

In his study of random experimental errors, the French mathematician Abraham De Moivre (1667–1754) discovered that repeated measurements of a physical quantity tend to be distributed according to a precise mathematical pattern called a **normal distribution.** Figure 7.18 shows a typical normal distribution with mean μ and standard deviation σ. There is a unique normal curve* for each value of μ and σ, and it can be shown that every normal curve has the following properties.

Properties of a Normal Curve

1. The curve is bell-shaped with its highest point at $x = \mu$.
2. The curve is symmetric with respect to the vertical line $x = \mu$.
3. The curve extends infinitely far in both directions and comes arbitrarily close to the horizontal axis.
4. The total area under the curve is 1.

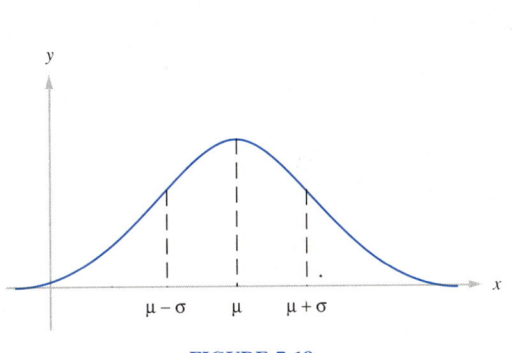

FIGURE 7.18

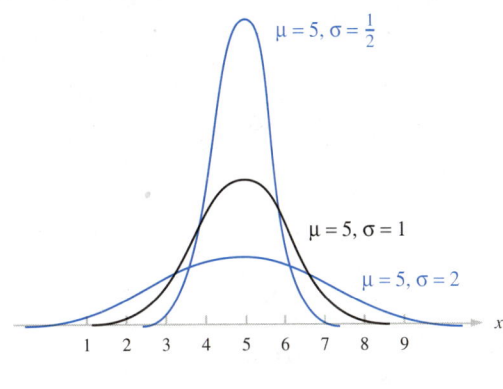

FIGURE 7.19

At the points $x = \mu \pm \sigma$ located at a distance σ on either side of the mean, the normal curve twists or *inflects*. That is, the curve is concave down (like an inverted bowl) between $\mu - \sigma$ and $\mu + \sigma$ and concave up outside this interval.

The mean μ determines the position of the normal curve, and the standard deviation σ determines its shape. Figure 7.19 shows three normal curves, all having the mean $\mu = 5$. Note that a small value of σ gives a tall, narrow curve, indicating that the values are concentrated near the mean. On the other hand, a large value of σ gives a lower, flatter curve, indicating a greater dispersion of the values.

---

*The equation of the normal curve with mean μ and standard deviation σ is
$$y = \frac{1}{\sigma\sqrt{2\pi}} e^{-(x-\mu)^2/2\sigma^2},$$
where $e \approx 2.71828$ and $\pi \approx 3.14159$.

STANDARD UNITS

In any normal distribution approximately 68.26% of the area under the curve lies within one standard deviation of the mean, and approximately 95.44% of the area under the curve lies within two standard deviations of the mean. (See Figure 7.20.) As a result, it is useful to measure distances along the x-axis in terms of standard deviations from the mean. We can convert any x-value into these **standard units** (or **z-values**) by the formula

$$z = \frac{x - \mu}{\sigma}. \tag{7.7}$$

This is merely a change of scale as illustrated in Figure 7.20. That is, the z-value measures the number of standard deviations between the corresponding x-value and the mean. Thus 68.26% of the area under any normal curve lies between $z = -1$ and $z = 1$, and 95.44% of the area lies between $z = -2$ and $z = 2$. More generally, *the area under a normal curve between any two z-values is a constant, independent of the mean and standard deviation of the distribution.*

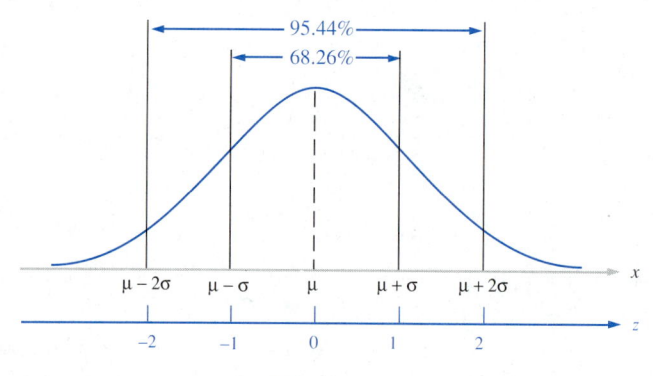

FIGURE 7.20

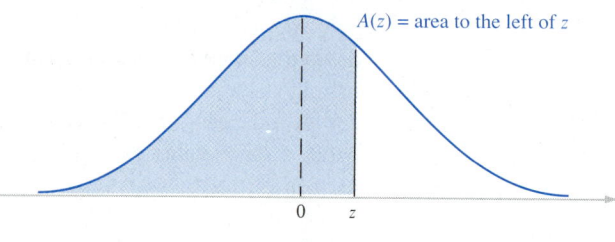

FIGURE 7.21

The table inside the front cover gives the area $A(z)$ under a normal curve to the left of z. (See Figure 7.21.) With this table we can find the area under any normal curve between two given z-values.

EXAMPLE 7.21 A normal distribution has mean 20 and standard deviation 4. Find the area under this curve between 17 and 22.

Solution The region in question is shaded in Figure 7.22. To compute its area, we must first convert the given values to standard units. By formula (7.7) $x = 17$ corresponds to

$$z = \frac{17 - 20}{4} = -.75,$$

and $x = 22$ corresponds to

$$z = \frac{22 - 20}{4} = .50.$$

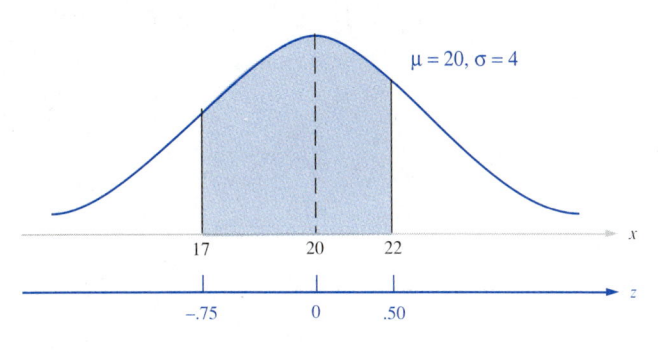

FIGURE 7.22

Thus $x = 17$ lies .75 standard deviations below the mean and $x = 22$ lies .50 standard deviations above the mean. From the table inside the front cover we find that

$$A(-.75) = .2266 \quad \text{and} \quad A(.50) = .6915.$$

Hence the area between $x = 17$ and $x = 22$ is given by

$$A(.50) - A(-.75) = .6915 - .2266$$
$$= .4649.$$

Therefore 46.49% of the total area under the curve lies between the values $x = 17$ and $x = 22$. ■

EXAMPLE 7.22 Find the area under the normal curve with mean 30 and standard deviation 6 that lies

(a) to the right of 36;
(b) between 21 and 24.

Solution (a) The region in question is shaded in Figure 7.23(a). In standard units $x = 36$ becomes

$$z = \frac{36 - 30}{6} = 1.00.$$

That is, $x = 36$ lies one standard deviation above the mean. From the table inside the front cover we find that the area to the *left* of $x = 36$ is $A(1.00) = .8413$. Since the total area is 1, the area to the *right* of $x = 36$ is

$$1 - A(1.00) = 1 - 0.8413$$
$$= 0.1587.$$

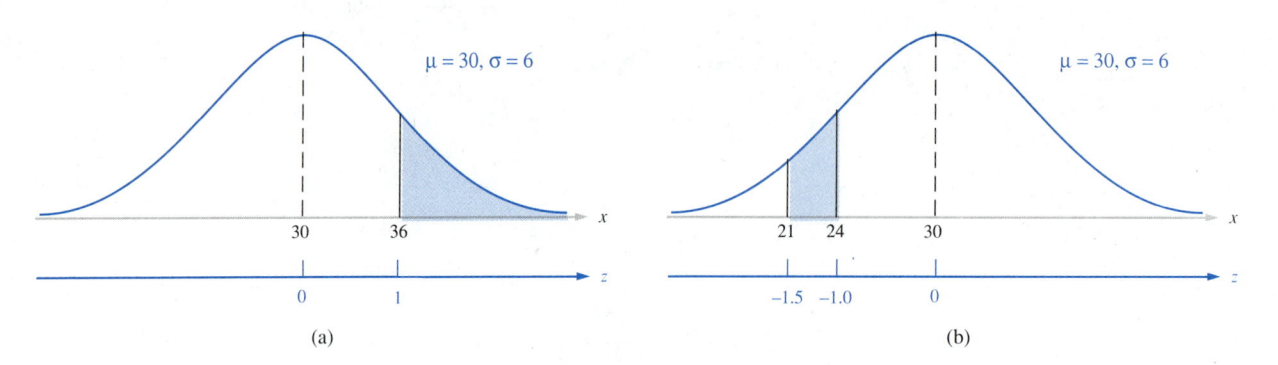

FIGURE 7.23

(b) The desired area is shaded in Figure 7.23(b). As in part (a) we find that $x = 21$ and $x = 24$ convert to $z = -1.50$ and $z = -1.00$, respectively. Thus the desired area is

$$A(-1.00) - A(-1.50) = 0.1587 - 0.0668$$
$$= .0919.$$

Hence 9.19% of the total area lies between $x = 21$ and $x = 24$. ■

Practice Problem 16 Find the area under the normal curve with mean 54 and standard deviation 8 that lies

(a) to the right of 48;
(b) between 68 and 72.

APPLICATIONS

Many of the random variables that arise in statistical applications turn out to be *normally distributed*, that is, distributed as in a normal distribution. In such cases we can calculate probabilities by finding areas under a normal curve.

7.6 THE NORMAL DISTRIBUTION

EXAMPLE 7.23 The annual rainfall in a certain city is normally distributed with mean 48.5 inches and standard deviation 6.2 inches. Find the probability that the rainfall in a given year will be less than 42 inches.

Solution The amount of rainfall per year (in inches) has the normal distribution shown in Figure 7.24. In standard units, $x = 42$ becomes $z = -1.05$. Therefore from the table inside the front cover

$$\begin{aligned} P(\text{rainfall} < 42) &= \text{area to the left of } 42 \\ &= A(-1.05) \\ &= 0.1469. \end{aligned}$$

Hence the probability that the rainfall in a given year will be less than 42 inches is .1469. ∎

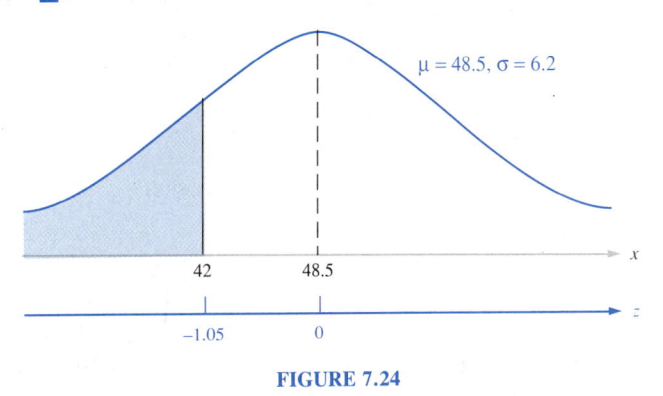

FIGURE 7.24

EXAMPLE 7.24 The mean systolic blood pressure of adult males is normally distributed with a mean of 138 (mm of mercury) and a standard deviation of 9.7. What percentage of adult males have blood pressure between 145 and 150?

Solution The systolic blood pressure of a randomly chosen adult male has the distribution shown in Figure 7.25.

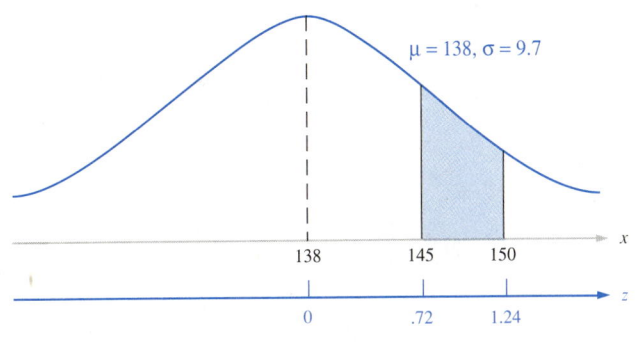

FIGURE 7.25

In standard units $x = 145$ and $x = 150$ become $z = 0.72$ and $z = 1.24$, respectively. Hence

$$P(145 \leq \text{systolic pressure} \leq 150) = A(1.24) - A(.72)$$
$$= .8925 - .7642$$
$$= .1283.$$

Thus approximately 12.83% of all adult males have blood pressure in the given range. ■

THE NORMAL APPROXIMATION TO A BINOMIAL DISTRIBUTION

One of the most important applications of normal curves is based on the fact that they give good approximations to binomial distributions. For instance, the probability distribution for the number of heads obtained when a fair coin is tossed ten times is a binomial distribution with mean

$$\mu = np = 10(\tfrac{1}{2}) = 5$$

and standard deviation

$$\sigma = \sqrt{npq} = \sqrt{10\left(\tfrac{1}{2}\right)\left(\tfrac{1}{2}\right)} \approx 1.58.$$

As we see in Figure 7.26, the binomial histogram for the number of heads is approximated very well by the normal curve with mean 5 and standard deviation 1.58. Consequently we can approximate probabilities for this random variable by computing areas under this normal curve.

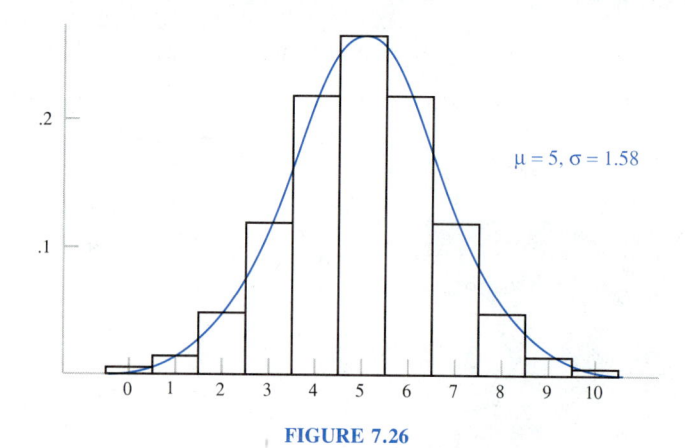

FIGURE 7.26

EXAMPLE 7.25 Use the normal curve with mean 5 and standard deviation 1.58 to approximate the probability that in ten flips of a fair coin

(a) exactly five heads appear;
(b) four, five, or six heads appear.

Solution (a) Consider the random variable that counts the number of heads obtained when a fair coin is flipped ten times. Recall that by Theorem 7.1 the probability of obtaining exactly five heads is equal to the area under the probability distribution for this random variable between 4.5 and 5.5. This area is approximately equal to the area under the normal curve in Figure 7.26 between 4.5 and 5.5. Converting these values to standard units, we obtain

$$z = \frac{4.5 - 5}{1.58} \approx -0.32 \quad \text{for} \quad x = 4.5$$

and

$$z = \frac{5.5 - 5}{1.58} \approx 0.32 \quad \text{for} \quad x = 5.5.$$

Hence by the table inside the front cover

$$P(\text{exactly 5 heads}) \approx A(0.32) - A(-0.32)$$
$$= .6255 - .3745$$
$$= .2510.$$

The true probability was found to be $252/1024 \approx .246$ in Example 7.17.

(b) The probability of obtaining four, five, or six heads equals the area under the probability distribution between 3.5 and 6.5. Since $x = 3.5$ and $x = 6.5$ convert to -0.95 and 0.95 standard units, respectively, the normal approximation gives

$$P(\text{exactly 4, 5, or 6 heads}) \approx A(0.95) - A(-0.95)$$
$$= .8289 - .1711$$
$$= .6578.$$

This value is quite close to the true value of .65625, which we calculated in Example 7.17. ∎

As was true in Example 7.25, a normal curve generally approximates a binomial distribution more accurately when it is used over a larger interval of values. In addition, the approximation improves as the number of trials increases in the binomial distribution. A precise statement of this result, known as the De Moivre-Laplace Limit Theorem, is beyond the scope of this text; however, we can summarize it roughly as follows.

416 CHAPTER 7 STATISTICS

> Consider the random variable that counts the number of successes in n Bernoulli trials, and let p and q denote the probabilities of success and failure, respectively, on each trial. If n is large and neither p nor q is close to zero, then the probability distribution for this random variable can be approximated by the normal curve with mean $\mu = np$ and standard deviation $\sigma = \sqrt{npq}$.

As a rule of thumb, both np and nq should be greater than 5 in order to use a normal curve to approximate a binomial distribution.

EXAMPLE 7.26 Suppose that 40% of those having a certain disease recover without treatment. Assume that 20 individuals with this disease are given an experimental drug, and 13 of them recover. What is the probability that such a result occurs merely by chance?

Solution We proceed as in Example 7.20 by assuming that the drug had no effect. Consider the random variable that counts the number of successes (recoveries) among the 20 patients. Then the patients represent $n = 20$ Bernoulli trials with the probability of success being $p = .4$ on each trial. Moreover this random variable is *binomially distributed,* (that is, its probability distribution is a binomial distribution) with mean

$$\mu = np = 20(.4) = 8$$

and standard deviation

$$\sigma = \sqrt{npq} = \sqrt{20(.4)(.6)} \approx 2.19.$$

Since both $np = 8$ and $nq = 12$ exceed 5, we can approximate the distribution of this random variable by a normal curve with mean 8 and standard deviation 2.19. (See Figure 7.27.) The probability of 13 or more successes in

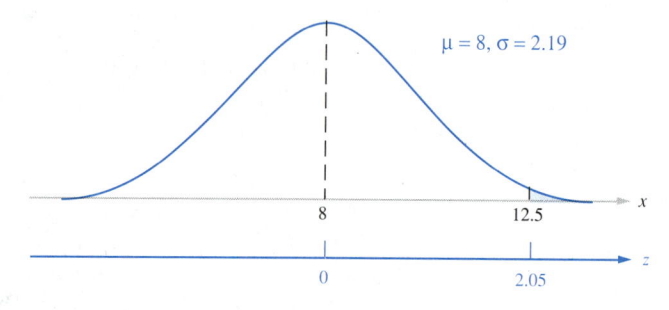

FIGURE 7.27

20 trials is approximately equal to the area under the normal curve to the right of 12.5. Because 12.5 converts to 2.05 standard units, the normal approximation gives

$$P(\text{at least 13 successes}) \approx 1 - A(2.05)$$
$$= 1 - .9798$$
$$= .0202.$$

Thus there is about a 2% probability that the observed number of recoveries could have occurred by chance alone. ■

EXAMPLE 7.27 Actuarial tables indicate an accidental death rate in the United States of 56 persons per 100,000. If an insurance company covers 500,000 individuals for accidental death benefits, what is the probability that the number of claims in a given year will be between 260 and 300 (inclusive)?

Solution The number of claims by the policyholders is binomially distributed with $n = 500,000$, $p = .00056$, and $q = .99944$. The expected number of claims is

$$\mu = np = 280$$

with a standard deviation of

$$\sigma = \sqrt{npq} \approx 16.73.$$

Since both np and nq exceed 5, we can approximate the binomial distribution by a normal curve. (See Figure 7.28.) The probability that the number of claims will be between 260 and 300 is represented by the area under the normal curve between 259.5 and 300.5. Hence

$$P(260 \leq \text{number of claims} \leq 300) \approx A(1.23) - A(-1.23)$$
$$= .8907 - .1093$$
$$= .7814.$$

Thus the number of claims will fall in the given range about 78% of the time. ■

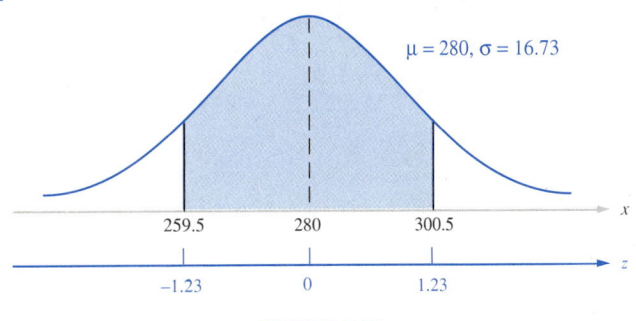

FIGURE 7.28

Practice Problem 17 A particular machine fills and caps cola bottles with a defective rate of 0.5%. That is, five out of every 1000 bottles processed on this machine are filled or capped improperly. What is the probability that more than 20 bottles are defective in a production run of 3000 bottles?

EXERCISES 7.6

In Exercises 1–8 find the area under a normal curve over the given z-interval.

1. to the left of $z = .58$
2. to the right of $z = -.02$
3. between $z = -1.25$ and $z = 1.34$
4. to the left of $z = -.27$
5. to the right of $z = 1.83$
6. between $z = .47$ and $z = 1.63$
7. between $z = -2.09$ and $z = -0.43$
8. between $z = -1.36$ and $z = 0.84$

In Exercises 9–12 find the value of z (standard units) for which the shaded region of a normal curve has the indicated area.

9. Area .1635

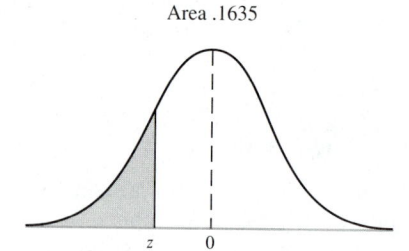

10. Area .1020

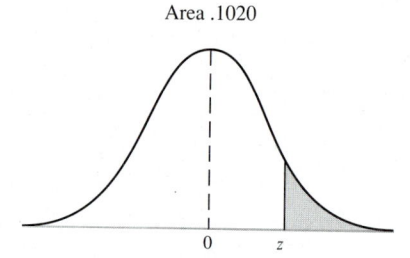

11. Area .3708

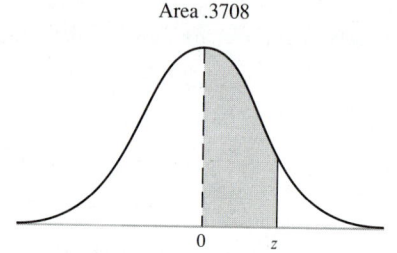

12. Area .6266

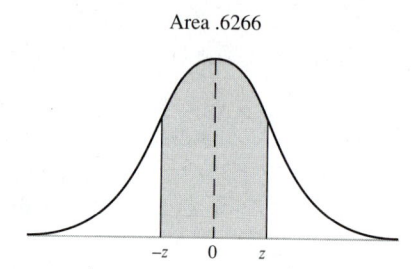

In Exercises 13–20 convert the given values of x into standard units for a normal curve with mean μ and standard deviation σ.

13. $x = 450, \mu = 600, \sigma = 100$
14. $x = 62, \mu = 50, \sigma = 12$
15. $x = 89, \mu = 80, \sigma = 5$
16. $x = 63, \mu = 90, \sigma = 12$

17. $x = 4.36, \mu = 3.5, \sigma = 1$

18. $x = 3.24, \mu = 6, \sigma = 2.4$

19. $x = 13.82, \mu = 17.85, \sigma = 1.55$

20. $x = 8.5, \mu = 6.1, \sigma = 1.5$

In Exercises 21–30 find the area under a normal curve with mean μ and standard deviation σ that lies in the indicated interval.

21. to the left of 99 if $\mu = 120$ and $\sigma = 12$
22. between 25 and 32 if $\mu = 25$ and $\sigma = 10$
23. between 38 and 50 if $\mu = 50$ and $\sigma = 12$
24. to the left of 8.77 if $\mu = 8.75$ and $\sigma = .05$
25. to the right of 54 if $\mu = 75$ and $\sigma = 15$
26. to the right of 90 if $\mu = 72$ and $\sigma = 24$
27. between 44 and 54 if $\mu = 48$ and $\sigma = 5$
28. between 28 and 36 if $\mu = 40$ and $\sigma = 10$
29. between 158 and 168 if $\mu = 150$ and $\sigma = 20$
30. between 50 and 60 if $\mu = 64$ and $\sigma = 8$

31. Scores on the 1960 revision of the Stanford-Binet intelligence test are normally distributed with a mean of 100 and a standard deviation of 16. What percentage of the population has IQ scores in the superior range (120–140)?

32. A slalom skier finds that her times on a certain course are normally distributed with mean 255 seconds and standard deviation 15 seconds. What is the probability that her time will be under 4 minutes on a given run?

33. A certain brand of 100-watt light bulbs has a mean life of 750 hours and a standard deviation of 50 hours. Assuming that the lives of these bulbs are normally distributed, what proportion of these bulbs burn at least 820 hours?

34. Before seeing the doctor, the patients in a certain doctor's office wait an average of 21 minutes with a standard deviation of 7 minutes. Assuming that the waiting times are normally distributed, what percentage of the patients wait between 15 minutes and 30 minutes?

35. A certain machine part must have a diameter between 11.99 and 12.01 cm. If the diameters of these parts are normally distributed with a mean of 12 cm and a standard deviation of 0.004 cm, what percentage of the parts have diameters in the acceptable range?

36. The life of a particular brand of automobile battery is normally distributed with a mean of 39 months and a standard deviation of 6 months. If these batteries are sold with a 36-month warranty, what proportion of them fail before the warranty expires?

In Exercises 37–44 use a normal curve to approximate the indicated binomial distribution. Refer to Exercises 11–19 in Section 7.5 for the mean and standard deviation.

37. If two fair dice are rolled 90 times, what is the probability that a sum of seven occurs between 10 and 20 times (inclusive)?

mean of 255 seconds
standard deviation of 15 seconds

420 CHAPTER 7 STATISTICS

38. What is the probability that an 80% shooter hits a target between 35 and 45 times (inclusive) out of 50 shots?

39. Suppose that a mail solicitation is sent to 10,000 individuals, each of whom has a 9% chance of responding. What is the probability that at least 850 individuals respond?

40. A machine produces 0.4% defective items. What is the probability that in a day's production of 10,000 items, at most 50 are defective?

41. An airline has found that there is a 10% chance that a person holding a reservation will fail to show up. If it gives 350 reservations for a flight with 320 seats, what is the probability that everyone who shows up can be seated?

42. A true-false test consists of 100 questions, each worth one point. What is the probability of scoring 60 or more on this test by guessing at random?

43. In a certain district, 75% of the registered voters favor a tax cut. If 200 voters are surveyed at random, what is the probability that between 70% and 80% (inclusive) will favor the cut?

44. A pharmaceutical house claims to have developed a drug that increases the chance of recovery from a certain disease from 60% to 80%. Suppose that the drug is given to 50 patients with the disease, and 35 of them recover.

    (a) What is the probability of a result this *low* if the maker's claim is true?

    (b) What is the probability of a result this *high* if the drug has no effect?

45. Scholastic Aptitude Test scores are normally distributed with a mean of 500 and a standard deviation of 100.

    (a) What percentage of scores are above 550?

    (b) Approximately what score puts someone in the top 10%?

46. The weekly demand for poultry at a certain market is normally distributed with a mean of 2700 pounds and a standard deviation of 240 pounds.

    (a) If the market stocks 2500 pounds per week, what proportion of the time will all the poultry be sold?

    (b) Approximately how much poultry should the market stock per week in order to be 95% certain of selling it all?

47. The heights of adult males are normally distributed with mean 69 inches and standard deviation 2.5 inches. Approximately what height range contains the middle 50% of adult males?

48. The mean annual salary of beginning engineers is $28,500 with a standard deviation of $2300. Assuming that the salaries are normally distributed, approximately what salary range contains the middle 75% of these engineers?

Answers to Practice Problems **16.** (a) .7734 (b) .0279

17. .1210

CHAPTER 7 REVIEW

IMPORTANT TERMS

- bar graph *(7.1)*
- class intervals
- histogram
- mean
- median
- standard deviation
- summation sign
- variance
- frequency distribution *(7.2)*
- expected value *(7.3)*
- random variable
- Bayes's criterion *(7.4)*
- decision tree
- maximax criterion
- maximin criterion
- payoff table
- states of nature
- binomial distribution *(7.5)*
- probability density function
- probability distribution
- continuous random variable *(7.6)*
- normal distribution
- standard units

IMPORTANT FORMULAS

Mean of ungrouped data

The mean of $x_1, x_2, \ldots, x_n$ is

$$\bar{x} = \frac{x_1 + x_2 + \cdots + x_n}{n} = \frac{\Sigma x}{n}. \tag{7.1}$$

Variance of ungrouped data

The variance of $x_1, x_2, \ldots, x_n$ is

$$s^2 = \frac{(x_1 - \bar{x})^2 + (x_2 - \bar{x})^2 + \cdots + (x_n - \bar{x})^2}{n} = \frac{\Sigma(x - \bar{x})^2}{n}. \tag{7.2}$$

Mean of grouped data

If the class intervals have midpoints $m_1, m_2, \ldots, m_k$ and respective frequencies $f_1, f_2, \ldots, f_k$, then

$$\bar{x} \approx \frac{m_1 f_1 + m_2 f_2 + \cdots + m_k f_k}{n} = \frac{\Sigma mf}{n}, \tag{7.3}$$

where $n = \Sigma f$ is the total number of data values.

Variance of grouped data

If the class intervals have midpoints $m_1, m_2, \ldots, m_k$ and respective frequencies $f_1, f_2, \ldots, f_k$, then

$$s^2 \approx \frac{(m_1 - \bar{x})^2 f_1 + (m_2 - \bar{x})^2 f_2 + \cdots + (m_k - \bar{x})^2 f_k}{n} = \frac{\Sigma(x - \bar{x})^2 f}{n}, \tag{7.4}$$

where $n = \Sigma f$ is the total number of data values.

Mean and variance of a random variable

The mean of a random variable taking distinct values $x_1, x_2, \ldots, x_k$ with respective probabilities $p_1, p_2, \ldots, p_k$ is

$$\mu = \Sigma xp = x_1 p_1 + x_2 p_2 + \cdots + x_k p_k, \tag{7.5}$$

and the variance is

$$\sigma^2 = \Sigma (x - \mu)^2 p = (x_1 - \mu)^2 p_1 + (x_2 - \mu)^2 p_2 + \cdots + (x_k - \mu)^2 p_k. \tag{7.6}$$

Binomial distribution

The distribution of the number of successes in a sequence of n Bernoulli trials is a random variable that assumes the values $0, 1, 2, \ldots, n$. Its distribution is given by Bernoulli's formula:

$$P(\text{exactly } k \text{ successes in } n \text{ trials}) = C(n, k) p^k q^{n-k} \quad (k = 0, 1, 2, \ldots, n),$$

where p and q denote the probabilities of success and failure, respectively, on each trial.

Mean and standard deviation of a binomial distribution

In a binomial distribution of n trials in which the probability of success and failure on each trial are p and q, respectively, the mean is

$$\mu = np$$

and the standard deviation is

$$\sigma = \sqrt{npq}.$$

Standard Units

In a normal distribution with mean μ and standard deviation σ, the number of standard units that a value x lies above or below the mean is

$$z = \frac{x - \mu}{\sigma}.$$

REVIEW EXERCISES

1. Five independent gem experts were asked to estimate the value of a diamond ring. Their appraisals were as follows: $5000, $7000, $6500, $8000, and $7500. Find the mean, median, and standard deviation for these ungrouped data.

2. The percentage price increases in a certain year for ten selected consumer goods are as follows: 7%, 12%, 8%, 10%, 11%, 9%, 13%, 126%, 14%, and 10%. Find the mean and median for these ungrouped data. Which of these two measures better reflects a "typical" increase in consumer prices for the year?

3. The typing speeds (in words per minute) of 80 secretaries are distributed as shown.

    (a) Represent these data by means of a histogram.
    (b) Find the mean and standard deviation for these grouped data.

Speed	Secretaries
40–50	8
50–60	12
60–70	31
70–80	22
80–90	7
	80 Total

4. A class of 25 high school seniors had the following scores on a scholastic aptitude test.

$$\begin{array}{ccccc}
672 & 419 & 523 & 614 & 531 \\
724 & 367 & 431 & 602 & 612 \\
569 & 472 & 763 & 525 & 643 \\
675 & 798 & 664 & 372 & 599 \\
518 & 349 & 617 & 558 & 593
\end{array}$$

(a) Write these data in grouped form.
(b) Represent the grouped data with a histogram.
(c) Find the mean and standard deviation for the grouped data in part (a).

5. A stereo equipment salesman found that the number of tape decks sold in a given day had the distribution shown at the right. Find the mean and standard deviation for the number of units sold per day.

Units sold	Probability
0	.19
1	.35
2	.24
3	.12
4	.09
5	.01

6. An urn contains 10 red marbles, 20 white marbles, and 30 blue marbles. A player chooses a single marble at random from the urn. If it is red, he wins $4; if it is white, he wins $1; and if it is blue, he loses $2. Find the mean and standard deviation for the player's net gain.

7. Find the mean and standard deviation for the number of successes in 15 Bernoulli trials if the probability of success on each trial is 1/3.

8. Thirty percent of the registered voters in a certain area are senior citizens. A pool of 50 voters is chosen at random for jury duty. Find the mean and standard deviation for the number of senior citizens chosen.

9. A botanical research team found that the heights of a certain species of tree at a state park are normally distributed with a mean of 8.44 meters and a standard deviation of 2.4 meters. What percentage of these trees are more than 10 meters tall?

10. The speeds of cars on a two-lane highway were found to be normally distributed with a mean of 58.6 mph and a standard deviation of 3.0 mph. What percentage of cars are observing the 55 mph speed limit on this highway?

11. Estimate the probability that when a fair coin is flipped 100 times, heads will occur between 45 and 55 times (inclusive).

12. Of those having a certain disease, 25% recover without treatment. Suppose that 20 individuals with this disease are given an experimental drug, and 10 of them recover. What is the probability that a recovery rate this high is due merely to chance?

13. An investment analyst anticipates that during the next year the value of a certain stock will decrease by 10% with probability .1, remain unchanged with probability .2, increase by 10% with probability .4, and increase by 20% with probability .3. According to this analyst, what is the expected gain for this stock during the next year?

14. A company that sells solar heating units has found that on any given business day there is a 70% chance of selling no units, a 15% chance of selling one unit, a 10% chance of selling two units, and a 5% chance of selling three units. What is the company's average daily profit from these units if it earns a profit of $1000 on each solar heating unit that it sells?

15. If it rains during the next two weeks, a farmer will make $100,000 profit on his corn crop, but otherwise he will lose $80,000. The chance of rain during the period is estimated at 50%, but it can be increased to 60% by cloud-seeding. If the cloud-seeding costs $15,000, is it worthwhile in terms of the farmer's expected net gain?

16. A textbook publisher estimates that next year's demand for a new calculus text (to the nearest 5000) will be as shown at the right. The publisher must decide whether to print 5, 10, 15, or 20 thousand copies in the first printing. Each printed copy costs the publisher $12, and the book sells for $25. In addition, each unsold copy costs an additional $4 in inventory expenses. How large should the first printing be to maximize the publisher's expected net profit?

Number ordered	Probability
5000	.2
10,000	.5
15,000	.2
20,000	.1

17. Each morning a baker makes a certain number of apple pies to be sold during the day. The number of daily requests for apple pies has been observed to vary as follows.

Number requested	Probability
0	.2
1	.4
2	.3
3	.1

18. A contestant on a quiz show has already won $4200. He can either keep this money, or risk it by trying to answer a jackpot question. If he answers the jackpot question correctly, he will win another $5000, but if he answers incorrectly, he will lose the $4200. In addition, by answering the jackpot question correctly, he becomes eligible for a superjackpot question. As before, he may choose to take the money already won or risk it by trying to answer the superjackpot question. The contestant will win another $20,000 if he answers the superjackpot question correctly, but he will lose all his previous winnings if he answers incorrectly. If the contestant feels that the probability of his answering the jackpot question correctly is .5 and the probability of his answering the superjackpot question correctly is .25, what decision strategy maximizes the contestant's expected net gain?

REFERENCES

1. Anderson, L. Dewey, "A Longitudinal Study of the Effects of Nursery-School Training on Successive Intelligence-Test Ratings," in *Intelligence: Its Nature and Nurture,* Guy M. Whipple, ed. Bloomington, IL: National Society for the Study of Education, 1940.

2. Pascal, Gerald R. and Barbara Suttell, "Testing the Claims of a Graphologist," *Journal of Personality,* vol. 16 (1947), pp. 192–197.

3. Raiffa, Howard, *Decision Analysis.* Reading, MA: Addison-Wesley, 1968.

4. Swalm, Ralph O., "Utility Theory—Insights into Risk Taking," *Harvard Business Review,* November-December 1966, pp. 123–126.

8

MARKOV CHAINS

8.1 Transition Matrices and State Vectors
8.2 Regular Markov Chains
8.3 Absorbing Markov Chains
8.4 Mathematics in Action: Genetics
Chapter Review

In Section 6.6 we discussed Bernoulli trials, in which an experiment consists of a sequence of repeated trials and the outcome of each trial is unaffected by the outcome of any other. In this chapter we will encounter a repeated process called a *Markov chain* in which the outcome of one trial affects the next. Markov chains have been used in many disciplines to model phenomena that change states in some random manner. In Section 8.4 we will consider one such application to the study of genetics.

We will primarily consider two types of Markov chains in this chapter, *regular* and *absorbing* Markov chains. We will begin, however, by examining ideas common to both types.

8.1 Transition Matrices and State Vectors

Consider two basketball players, Al and Bob, who are practicing free throws. Al is an 80% free-throw shooter and Bob a 70% shooter. Each player shoots until he misses, at which time the other begins to shoot.

This is an example of a simple system which is in one of two possible states, according to which person is shooting.

State 1: Al will shoot the next shot.

State 2: Bob will shoot the next shot.

Since Al is an 80% shooter, there is a .8 probability that he will make any free throw and a .2 probability that he will miss. This means that if the system is in state 1, there is a .8 probability that it will remain in state 1 on the next shot and a .2 probability that it will change to state 2. Similarly, since Bob is a 70% shooter, if the system is in state 2, there is a .3 probability that the system will change to state 1 and a .7 probability that it will remain in state 2 on the next shot. Thus we can calculate the probability of the next state of the system if we know its current state. This system is a simple example of a Markov* chain.

Formally, a **Markov chain** is a process consisting of a sequence of trials, each of which results in the process being in one of a finite number of **states.** Moreover, if a given trial results in the process being in state i, there must be a fixed probability that the next trial will result in the process being in state j. This probability is denoted p_{ij} and is called a **transition probability**; because it depends only on the states i and j, it remains the same from trial to trial. Notice that instead of saying "the outcome of the nth trial is i," it is customary to say that "the system is in state i at time n." With this terminology, the transition probability p_{ij} represents the conditional probability that the system will move to state j during the next trial (time period) given that the system is

---

*The Russian mathematician A. A. Markov (1856–1922) developed much of the modern theory of stochastic processes.

currently in state i. Of course, the time period between trials will depend on the context of a particular problem.

Thus the free-throw shooting example is a Markov chain with two states in which each shot is a trial of the experiment. In this case the transition probabilities are $p_{11} = .8$, $p_{12} = .2$, $p_{21} = .3$, and $p_{22} = .7$. All of the necessary information for this Markov chain is conveyed in Figure 8.1, which is called a **transition diagram.** For example, the arrow pointing from state 1 to state 2 indicates that the probability of going from state 1 to state 2 in one trial is .2, that is, $p_{12} = .2$. If some transition probability p_{ij} is 0, then for simplicity we will omit the arrow from state i to state j in the transition diagram.

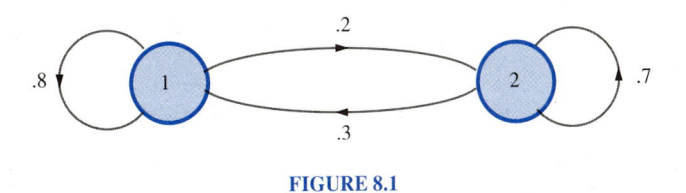

FIGURE 8.1

Another useful way to convey the same information is in a matrix. For a Markov chain with s states, the $s \times s$ matrix $[p_{ij}]$ is called the **transition matrix** of the Markov chain. Thus, for instance, the 2×2 matrix

$$\begin{array}{c} \\ \textbf{Current} \\ \textbf{state} \end{array} \begin{array}{c} \textbf{Next state} \\ \begin{array}{cc} 1 & 2 \end{array} \\ \begin{array}{c} 1 \\ 2 \end{array} \begin{bmatrix} .8 & .2 \\ .3 & .7 \end{bmatrix} \end{array}$$

is the transition matrix for a Markov chain representing the free-throw shooting experiment. Notice that a transition matrix will always consist of nonnegative entries such that the sum of the entries in each *row* is 1. (As the transition matrix above demonstrates, the sum of the entries in each column need not be 1.) The transition matrix plays an important role in the theory of Markov chains because it enables us to use matrix methods to analyze a stochastic process. Before pursuing this idea, however, let us look at another example.

EXAMPLE 8.1 Most of the drivers in a certain community are insured by Allstate or State Farm. Recent data showed that of those who were insured by Allstate in 1987, 70% continued with Allstate in 1988, but 20% switched to State Farm, and 10% switched to another company. Similarly, of those who were insured by State Farm in 1987, 80% continued with State Farm in 1988, 10% switched to Allstate, and 10% switched to another company. Likewise, of those who were insured by another company in 1987, 30% switched to Allstate in 1988, 10% switched to State Farm, and 60% continued with another company. Form a transition diagram and a transition matrix for a Markov chain describing this process.

Solution Since each driver in the community can be classified according to the type of insurance carried, we can represent this situation as a Markov chain with 3 states.

State 1: insured by Allstate
State 2: insured by State Farm
State 3: insured by another company

Here the time period between trials is one year because we are given information about the changes from 1987 to 1988. The transition diagram for this process is shown in Figure 8.2.

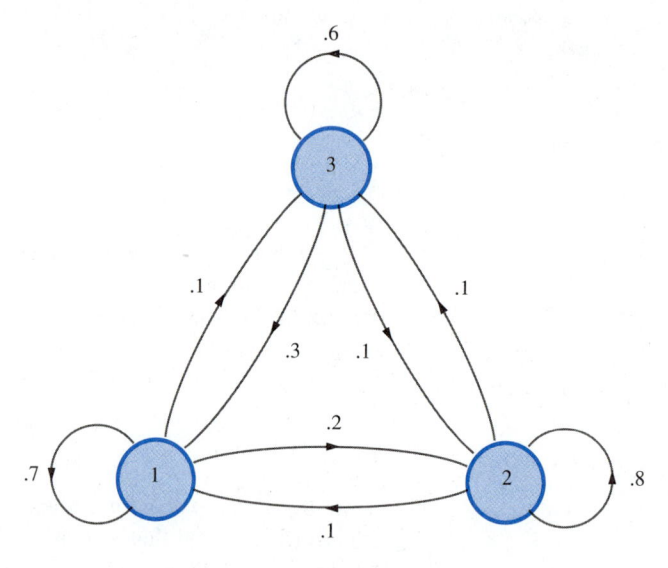

FIGURE 8.2

As the transition diagram shows, the probability of remaining in state 1 during one time period is .7, and the probability of moving from state 1 to state 2 is .2 and from state 1 to state 3 is .1. Since these probabilities are the entries in the first row of the transition matrix, we see that this row is

$$[.7 \quad .2 \quad .1].$$

Likewise in one year the probability of moving from state 2 to state 1 is .1, the probability of remaining in state 2 is .8, and the probability of moving from state 2 to state 3 is .1. Hence the second row of the transition matrix is

$$[.1 \quad .8 \quad .1].$$

Similarly the third row of the transition matrix is

$$[.3 \quad .1 \quad .6].$$

Thus the transition matrix for this Markov chain is

$$\text{Current state} \begin{array}{c} \\ 1 \\ 2 \\ 3 \end{array} \overset{\text{Next state}}{\begin{bmatrix} 1 & 2 & 3 \\ .7 & .2 & .1 \\ .1 & .8 & .1 \\ .3 & .1 & .6 \end{bmatrix}}.$$

Observe again that the entries in each row of the transition matrix are nonnegative numbers that sum to 1. ■

Practice Problem 1 In reference [6], Prais studied class mobility from one generation to the next. Suppose that children of upper-class parents were evenly divided between the upper and middle classes. Among children of middle-class parents, 10% moved to the upper class, 70% remained in the middle class, and 20% moved to the lower class; and children of lower-class parents were evenly divided between the middle and lower classes. Represent this process as a Markov chain and form the transition matrix.

STATE VECTORS

Our main interest in Markov chains centers on the question: *What is the probability that the system will be in a particular state at a given time?* We will denote by $x_i^{(n)}$ the probability that the system is in state i at time n, that is, after n trials. For a Markov chain with s states, the row vector

$$X^{(n)} = [x_1^{(n)} \quad x_2^{(n)} \quad \cdots \quad x_s^{(n)}]$$

is called the **state vector** at time n. Since its entries represent the probabilities of being in each of the respective states $1, 2, \ldots, s$ at time n, their sum will always be 1. Note that the superscripts in this state vector are *not* exponents but merely indicate the time to which the state vector corresponds.

For example, suppose that Al and Bob flip a fair coin to determine who will shoot the first free throw. Then each player has a .5 probability of taking the first shot, and so at time 0 (after no shots) the state vector is

$$X^{(0)} = \overset{1 \quad\; 2}{[.5 \quad .5]}.$$

We call this the **initial state vector** for the Markov chain. To find the corresponding probabilities after the first shot, we represent the first two stages of the process (namely, the coin toss followed by the first shot) by the stochastic diagram in Figure 8.3.

430   CHAPTER 8 MARKOV CHAINS

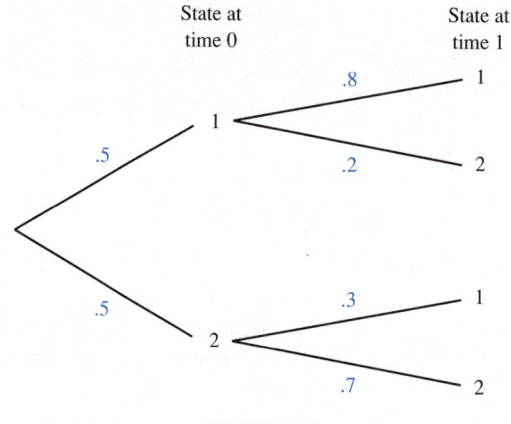

FIGURE 8.3

The probabilities .8, .2, .3, and .7 in this diagram are the entries of the transition matrix

$$P = \begin{array}{c} \\ 1 \\ 2 \end{array} \begin{array}{c} 1 \quad\;\; 2 \\ \begin{bmatrix} .8 & .2 \\ .3 & .7 \end{bmatrix} \end{array}.$$

Since two branches lead to state 1, we see as in Section 6.5 that the probability that the system will be in state 1 at time 1 is

$$.5(.8) + .5(.3) = .40 + .15 = .55.$$

Likewise the probability that the system will be in state 2 at time 1 is

$$.5(.2) + .5(.7) = .10 + .35 = .45.$$

Therefore the state vector at time 1 is

$$X^{(1)} = [.55 \quad .45].$$

Hence the probability is .55 that Al will take the second shot, and the probability is .45 that Bob will take the second shot.

The calculation performed above can be performed very simply by use of matrix multiplication:

$$[.5 \quad .5] \begin{bmatrix} .8 & .2 \\ .3 & .7 \end{bmatrix} = [.55 \quad .45].$$

state vector **transition** **state vector**
at time 0   **matrix**   **at time 1**

We can represent this calculation in symbols as
$$X^{(0)}P = X^{(1)}.$$
More generally, we have the following result.

Theorem 8.1

Let P be the transition matrix for a Markov chain. If the state vector at a given time is X, then the state vector one time period later is XP. In symbols,
$$X^{(n)}P = X^{(n+1)}.$$

We can use Theorem 8.1 to compute subsequent state vectors in our free-throw shooting example. For instance

$$\underset{\substack{\text{state vector}\\\text{at time 1}}}{[.55 \quad .45]} \underset{\substack{\text{transition}\\\text{matrix}}}{\begin{bmatrix} .8 & .2 \\ .3 & .7 \end{bmatrix}} = \underset{\substack{\text{state vector}\\\text{at time 2}}}{[.575 \quad .425]},$$

$$\underset{\substack{\text{state vector}\\\text{at time 2}}}{[.575 \quad .425]} \underset{\substack{\text{transition}\\\text{matrix}}}{\begin{bmatrix} .8 & .2 \\ .3 & .7 \end{bmatrix}} = \underset{\substack{\text{state vector}\\\text{at time 3}}}{[.5875 \quad .4125]},$$

$$\underset{\substack{\text{state vector}\\\text{at time 3}}}{[.5875 \quad .4125]} \underset{\substack{\text{transition}\\\text{matrix}}}{\begin{bmatrix} .8 & .2 \\ .3 & .7 \end{bmatrix}} = \underset{\substack{\text{state vector}\\\text{at time 4}}}{[.59375 \quad .40625]}.$$

From these calculations we see that there is a .59375 probability that Al will take the fifth shot. We will see in Section 8.2 that the state vectors in this example approach the "equilibrium" vector
$$X = [.6 \quad .4]$$
as the process continues. Over the long run, therefore, Al will take approximately 60% of the shots and Bob will take about 40%.

EXAMPLE 8.1 Revisited Suppose we are told that in 1987, 40% of the drivers in the community were insured by Allstate, 30% by State Farm, and 30% by another company. Assuming that the 1987–88 trend continues, what percentage of the drivers will be insured by Allstate and State Farm in 1990?

Solution The distribution for 1987 tells us that the initial state vector is

$$X^{(0)} = [.40 \quad .30 \quad .30].$$

Then the distribution one year later (in 1988) will be given by $X^{(1)} = X^{(0)}P$, where P is the transition matrix for the Markov chain.

$$X^{(1)} = X^{(0)}P = [.40 \quad .30 \quad .30] \begin{bmatrix} .7 & .2 & .1 \\ .1 & .8 & .1 \\ .3 & .1 & .6 \end{bmatrix} = [.40 \quad .35 \quad .25]$$

The distribution for 1989 is given by $X^{(2)} = X^{(1)}P$:

$$X^{(2)} = X^{(1)}P = [.40 \quad .35 \quad .25] \begin{bmatrix} .7 & .2 & .1 \\ .1 & .8 & .1 \\ .3 & .1 & .6 \end{bmatrix} = [.390 \quad .385 \quad .225].$$

Finally the distribution for 1990 is given by $X^{(3)} = X^{(2)}P$:

$$X^{(3)} = X^{(2)}P = [.390 \quad .385 \quad .225] \begin{bmatrix} .7 & .2 & .1 \\ .1 & .8 & .1 \\ .3 & .1 & .6 \end{bmatrix}$$
$$= [.3790 \quad .4085 \quad .2125].$$

Hence in 1990, 37.9% of the drivers will be insured by Allstate and 40.85% by State Farm. ∎

Practice Problem 2 Suppose that the social distribution in Practice Problem 1 on page 429 is currently 20% upper class, 40% middle class, and 40% lower class. Assuming that the birthrate for all three classes is the same, what will be the distribution in two generations?

EXERCISES 8.1

In Exercises 1–6 draw a transition diagram for the Markov chain with the given transition matrix.

1. $\begin{bmatrix} .5 & .5 \\ .6 & .4 \end{bmatrix}$

2. $\begin{bmatrix} \frac{2}{3} & \frac{1}{3} \\ \frac{1}{4} & \frac{3}{4} \end{bmatrix}$

3. $\begin{bmatrix} .2 & .8 \\ 0 & 1 \end{bmatrix}$

4. $\begin{bmatrix} 0 & .6 & .4 \\ .7 & 0 & .3 \\ .4 & .4 & .2 \end{bmatrix}$

5. $\begin{bmatrix} 1 & 0 & 0 \\ \frac{1}{3} & \frac{1}{3} & \frac{1}{3} \\ 1 & 0 & 0 \end{bmatrix}$

6. $\begin{bmatrix} 0 & .5 & .5 \\ 0 & 0 & 1 \\ 1 & 0 & 0 \end{bmatrix}$

In Exercises 7–10 write the transition matrix for the Markov chain with the given transition diagram.

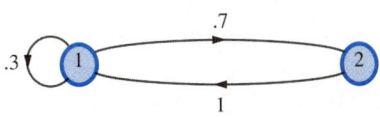

7. [Transition diagram with states 1 and 2: 1→1 with .8, 1→2 with .2, 2→1 with .1, 2→2 with .9]

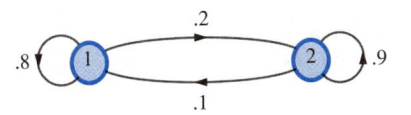

8. [Transition diagram with states 1 and 2: 1→1 with .3, 1→2 with .7, 2→1 with 1]

9. [Transition diagram with states 1, 2, 3: 3→3 with 1, 3→1 with .5, 3→2 with .5, 1→2 with .5, 2→1 with .5]

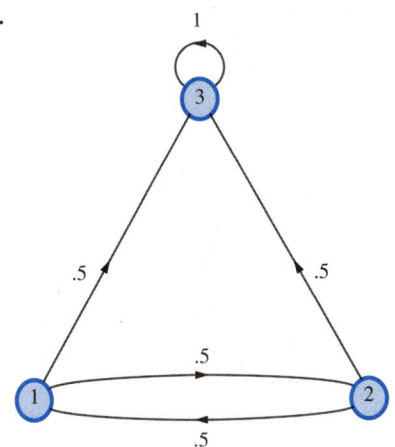

10. [Transition diagram with states 1, 2, 3, 4: various transitions including 4→1 with .5, 2→1 with .2, 1→2 with 1, 2→3 with .8, 3→4 with 1, 4→2 with .4, 4→3 with .1, 3→3 with 1]

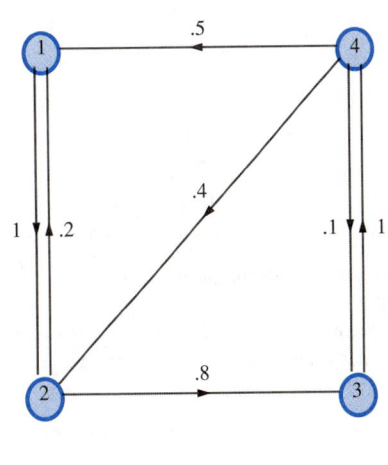

11. A two-state Markov chain has the transition matrix below.

$$\begin{bmatrix} .6 & .4 \\ .1 & .9 \end{bmatrix}$$

Find the state vector at time 3 (that is, after three trials) if the system is initially
(a) in state 1 (b) equally likely to be in state 1 or 2.

12. A two-state Markov chain has the transition matrix below.

$$\begin{bmatrix} .9 & .1 \\ .3 & .7 \end{bmatrix}$$

Find the state vector at time 2 (that is, after two trials) if the system is initially
(a) in state 2 (b) equally likely to be in state 1 or 2.

13. A two-state Markov chain has the transition matrix below.

$$\begin{bmatrix} 1 & 0 \\ .2 & .8 \end{bmatrix}$$

Find the state vector at time 3 (that is, after three trials) if the system is initially
(a) in state 1 (b) in state 2.

14. A three-state Markov chain has the transition matrix below.

$$\begin{bmatrix} .4 & .3 & .3 \\ .2 & .6 & .2 \\ .1 & .1 & .8 \end{bmatrix}$$

Find the state vector at time 2 (that is, after two trials) if the system is initially
(a) in state 2 (b) in state 3.

15. A three-state Markov chain has the transition matrix below.
$$\begin{bmatrix} .5 & 0 & .5 \\ .5 & 0 & .5 \\ 0 & 1 & 0 \end{bmatrix}$$
Find the state vector at time 3 (that is, after three trials) if the system is initially
   (a) in state 1   (b) in state 3.

16. A three-state Markov chain has the transition matrix below.
$$\begin{bmatrix} 0 & 1 & 0 \\ 0 & 0 & 1 \\ \frac{1}{3} & 0 & \frac{2}{3} \end{bmatrix}$$
Find the state vector at time 3 (that is, after three trials) if the system is initially
   (a) in state 2   (b) equally likely to be in state 1, 2, or 3.

17. An actor has found from experience that his performance on a given night affects his performance the following night. A good performance is followed by another good performance 90% of the time, but a bad performance is followed by a good performance only 70% of the time. If the actor performs well on Monday, what is the probability that he will perform well the following Friday?

18. An author has found that retyping his manuscript over and over to eliminate errors yields diminishing returns. If a particular word contains an error, there is a 48% chance it will be corrected on the copy; however, if a word is correct on the original, there is a 2% chance that it will be retyped incorrectly on the copy. If a word is correct initially, what is the probability that it will still be correct after the manuscript has been retyped three times?

19. The sales manager for a metropolitan newspaper has observed an alarming trend: 15% of the subscribers at the beginning of any given month fail to renew their subscriptions the following month, whereas only 5% of all nonsubscribers become subscribers during the same period. If this trend continues, what will happen to the newspaper's current market share of 60% in another three months?

20. There are some recent signs that Booneville's high unemployment rate may be ending. A three-month study revealed that 45% of those unemployed on January 1 were employed on April 1, whereas only 5% of those employed on January 1 were unemployed on April 1. If this trend continues, what will happen to the town's current 20% unemployment rate in the next six months? In the next nine months?

21. Forty percent of an island's population lives in the cities, with the rest dwelling in rural areas. Suppose that, in any given year, 10% of the rural population moves to the cities and 5% of the city dwellers move to rural areas. How long will it be before a majority of the island's population lives in the cities?

22. A sample of American car owners were asked whether their current car was American-made or foreign-made, and which type of car they would buy next. Four out of five owners of foreign cars said that they would buy them again the next time, and two out of five owners of American cars said that they would switch to a foreign car the next time. If foreign cars currently have a 40% share of the market and car owners on the average buy a new car every five years, how long will it be before foreign cars capture 60% of the American market?

23. A mouse is to be placed in a box with three connecting compartments, as shown in the diagram at the right. When the mouse changes rooms, it selects at random one of the doors of the compartment that it is occupying and passes through that door into another compartment. If the mouse is placed in compartment 1 initially, what is the probability that it will still be there after changing rooms four times?

24. A political scientist observed that among children of registered Democrats, 70% become Democrats, 10% become Republicans, and 20% register as independents. For children of Republicans the distribution is 10% Democrats, 60% Republicans, and 30% independents, and for children of independents it is 10% Democrats, 10% Republicans, and 80% independents. If the current voter registration is 40% Democratic, 40% Republican, and 20% independent and if the birthrate for all three groups is the same, what will the distribution be two generations from now?

25. Suppose that 60% of the children of agricultural workers become agricultural workers, 30% become blue-collar workers, and 10% become white-collar workers; 20% of the children of blue-collar workers become agricultural workers, 50% become blue-collar workers, and 30% become white-collar workers; and 10% of the children of white-collar workers become agricultural workers, 30% become blue-collar workers, and 60% become white-collar workers. If the current generation consists of 20% agricultural workers, 50% blue-collar workers, and 30% white-collar workers and if the birthrate for all three groups is the same, what will the distribution be two generations from now?

26. In 1985 a survey of workers in a metropolitan area showed that 20% rode the subway to work, 30% took a bus, and 50% came in automobiles. A follow-up survey two years later showed that 60% of the subway riders in 1985 continued to ride the subway, but 10% rode a bus and 30% used automobiles in 1987. Among the 1985 bus riders, 90% continued to ride the bus and 10% took the subway in 1987. Finally 70% of the 1985 automobile users continued to come by car, but 10% rode the subway and 20% took a bus in 1987. If this trend continues, what percentage of the workers will use each form of transportation in 1993?

27. A 1980 survey showed that 40% of the cars owned by Americans were large cars, 50% were mid-size cars, and 10% were small cars. A follow-up survey in 1985 showed that of those who owned large cars in 1980, 80% still owned large cars, but that 10% owned mid-size cars and 10% owned small cars. Of those who owned mid-size cars in 1980, 30% owned large cars, 60% owned mid-size cars, and 10% owned small cars in 1985. Of those with small cars in 1980, 10% owned mid-size cars and 90% owned small cars in 1985. If this trend continues, what percentage of the cars owned by Americans will be of each size in the year 2000?

28. In 1930, 10% of the land on a certain island was urban, 60% was used for agriculture, and 30% was used for other purposes. Two decades later, 80% of the urban land remained urban, 10% was used for agriculture, and 10% was used for other purposes. Of the land used for agriculture in 1930, 80% continued to be used for agriculture, 10% had become urban, and 10% was used for other purposes in 1950. Of the land used for other purposes in 1930, 20% had become urban and 20% was used for agriculture in 1950. If this trend continues, what percentage of the island's land will be used in each way in the year 2010?

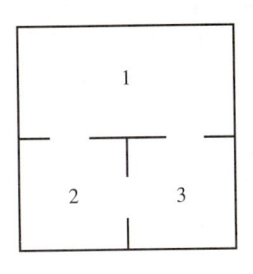

Answers to Practice Problems

1. In this process there are three states.

State 1: belonging to the upper class
State 2: belonging to the middle class
State 3: belonging to the lower class

The time period between trials is a generation, and the transition matrix is

$$\begin{array}{c} \\ \text{Currrent} \\ \text{state} \end{array} \begin{array}{c} \\ 1 \\ 2 \\ 3 \end{array} \overset{\begin{array}{c}\text{Next state}\\ 1 \quad 2 \quad 3\end{array}}{\begin{bmatrix} .5 & .5 & 0 \\ .1 & .7 & .2 \\ 0 & .5 & .5 \end{bmatrix}}.$$

2. In two generations 12.8% of the people will be in the upper class, 61.6% in the middle class, and 25.6% in the lower class.

8.2 Regular Markov Chains

Some Markov chains tend to stabilize over the long run in the sense that their state vectors approach an equilibrium vector as the process continues. For such Markov chains we can predict the proportion of the time over the long run that the system will be in each state.

EQUILIBRIUM VECTORS

Let us consider again the free-throw shooting example presented in Section 8.1. We have seen that if we know the initial state vector $X^{(0)}$, we can determine the state vector at any later time by computing

$$X^{(0)}P = X^{(1)},$$
$$X^{(1)}P = X^{(2)},$$
$$X^{(2)}P = X^{(3)},$$

and so forth. Table 8.1 shows the results of such calculations, rounded off to four decimal places. These calculations are performed for three different ways of beginning the process, according to whether Al shoots the first shot, Bob shoots the first shot, or they flip a fair coin to decide who shoots first.

From Table 8.1 we see that in all three cases, as the process continues the state vectors approach

$$X = [.6 \quad .4].$$

Time	Al Starts	Bob Starts	Coin Flip
0	[1 0]	[0 1]	[.5 .5]
1	[.8 .2]	[.3 .7]	[.55 .45]
2	[.70 .30]	[.45 .55]	[.575 .425]
3	[.650 .350]	[.525 .475]	[.5875 .4125]
4	[.6250 .3750]	[.5625 .4375]	[.5938 .4062]
5	[.6125 .3875]	[.5813 .4187]	[.5969 .4031]
6	[.6063 .3937]	[.5906 .4094]	[.5984 .4016]
7	[.6031 .3969]	[.5953 .4047]	[.5992 .4008]
8	[.6016 .3984]	[.5977 .4023]	[.5996 .4004]
9	[.6008 .3992]	[.5988 .4012]	[.5998 .4002]
10	[.6004 .3996]	[.5994 .4006]	[.5999 .4001]
11	[.6002 .3998]	[.5997 .4003]	[.6000 .4000]
12	[.6001 .3999]	[.5999 .4001]	[.6000 .4000]
13	[.6000 .4000]	[.5999 .4001]	[.6000 .4000]
14	[.6000 .4000]	[.6000 .4000]	[.6000 .4000]
15	[.6000 .4000]	[.6000 .4000]	[.6000 .4000]

TABLE 8.1

If the state vectors reach this vector, they will not change. For if the state vector at a given time is X, then the state vector one step later is

$$XP = [.6 \quad .4]\begin{bmatrix}.8 & .2 \\ .3 & .7\end{bmatrix}$$
$$= [.6(.8) + .4(.3) \quad .6(.2) + .4(.7)]$$
$$= [.48 + .12 \quad .12 + .28]$$
$$= [.6 \quad .4]$$
$$= X.$$

It turns out that *as the process continues, the probability that Al is shooting tends to stabilize at .6, no matter how the process began.* It follows that over the long run, approximately 60% of the shots will be taken by Al and about 40% by Bob.

A vector X is called an **equilibrium vector** for a Markov chain if

$$XP = X,$$

where P is the transition matrix for the Markov chain. Thus the preceding calculation shows that $[.6 \quad .4]$ is an equilibrium vector for the free-throw shooting Markov chain.

Two important questions now arise.

1. Do the state vectors of every Markov chain approach an equilibrium vector?
2. If an equilibrium vector exists, how do we find it?

It is not hard to show that the answer to the first question is negative. (See Exercise 15.) Thus there are Markov chains in which the state vectors fail to stabilize no matter how long the process goes on, and there are others in which the state vectors stabilize at different equilibrium vectors depending on the initial state of the system. However, there is a large class of Markov chains that have a unique equilibrium vector.

A Markov chain is called **regular** if some power of its transition matrix contains only *positive* entries, that is, if at least one of the matrices

$$P, P^2, P^3, P^4, \ldots$$

contains no zero entries. It can be shown that if this does happen for a Markov chain with s states, then it must happen within the first $(s-1)^2 + 1$ powers of P. Thus, for instance, to determine if a three-state Markov chain is regular, we need to check at most the first $(3-1)^2 + 1 = 5$ powers of its transition matrix. The transition matrix P for the free-throw shooting example clearly satisfies this condition, because the transition matrix

$$P = \begin{bmatrix} .8 & .2 \\ .3 & .7 \end{bmatrix}$$

itself contains no zero entries. On the other hand, consider the Markov chain with transition matrix

$$A = \begin{bmatrix} 1 & 0 & 0 \\ 0 & .6 & .4 \\ 0 & .2 & .8 \end{bmatrix}$$

Now

$$A^2 = AA = \begin{bmatrix} 1 & 0 & 0 \\ 0 & .6 & .4 \\ 0 & .2 & .8 \end{bmatrix} \begin{bmatrix} 1 & 0 & 0 \\ 0 & .6 & .4 \\ 0 & .2 & .8 \end{bmatrix} = \begin{bmatrix} 1 & 0 & 0 \\ 0 & .44 & .56 \\ 0 & .28 & .72 \end{bmatrix}$$

and

$$A^3 = AA^2 = \begin{bmatrix} 1 & 0 & 0 \\ 0 & .6 & .4 \\ 0 & .2 & .8 \end{bmatrix} \begin{bmatrix} 1 & 0 & 0 \\ 0 & .44 & .56 \\ 0 & .28 & .72 \end{bmatrix} = \begin{bmatrix} 1 & 0 & 0 \\ 0 & .376 & .624 \\ 0 & .312 & .688 \end{bmatrix}$$

It is now easy to see that the first row of every power of A will be

$$[1 \quad 0 \quad 0],$$

and so this Markov chain is not regular.

Practice Problem 3   If the transition matrix of a Markov chain is

$$A = \begin{bmatrix} 0 & 1 & 0 \\ 0 & 0 & 1 \\ \frac{1}{2} & \frac{1}{2} & 0 \end{bmatrix},$$

is the Markov chain regular?

The next theorem, which we will not prove, guarantees that every regular Markov chain has exactly one equilibrium vector.

Theorem 8.2 ■ ■ ■ ■ ■

> A regular Markov chain has a unique equilibrium vector X, and regardless of the initial state of the system, the state vectors for the Markov chain will approach X as the process continues.

We have already pointed out that the free-throw shooting example is a regular Markov chain and that $[.6 \ \ .4]$ is an equilibrium vector. According to Theorem 8.2, this is the only equilibrium vector for this Markov chain, and, moreover, the probability that Al is shooting tends to stabilize at .6, no matter how the process began.

Fortunately, there is an easy method for computing an equilibrium vector if one exists. To find the equilibrium vector for a Markov chain with transition matrix P, we need only solve the matrix equation $XP = X$. It will prove to be somewhat simpler to solve this equation in the following equivalent form:

$$\begin{aligned} O &= X - XP \\ &= XI - XP \\ &= X(I - P), \end{aligned}$$

where I is an identity matrix of the same size as P.

Consider, for instance, the transition matrix P for the free-throw shooting example. Since this Markov chain has two states, its equilibrium vector X will have two entries. Suppose that

$$X = [x_1 \ \ x_2].$$

Now

$$I - P = I_2 - P = \begin{bmatrix} 1 & 0 \\ 0 & 1 \end{bmatrix} - \begin{bmatrix} .8 & .2 \\ .3 & .7 \end{bmatrix} = \begin{bmatrix} .2 & -.2 \\ -.3 & .3 \end{bmatrix},$$

and so the matrix equation $X(I - P) = O$ becomes

$$[x_1 \ \ x_2] \begin{bmatrix} .2 & -.2 \\ -.3 & .3 \end{bmatrix} = [0 \ \ 0].$$

Multiplying out the left side, we obtain

$$[.2x_1 - .3x_2 \ \ \ -.2x_1 + .3x_2] = [0 \ \ 0],$$

which is equivalent to the following system of linear equations.

$$\begin{aligned} .2x_1 - .3x_2 &= 0 \\ -.2x_1 + .3x_2 &= 0 \end{aligned}$$

We have not yet taken into account that the entries of X must sum to 1. This condition requires that

$$x_1 + x_2 = 1.$$

The resulting system of 3 equations in 2 unknowns

$$.2x_1 - .3x_2 = 0$$
$$-.2x_1 + .3x_2 = 0$$
$$x_1 + x_2 = 1$$

is obviously redundant because the second equation is merely the first equation multiplied by -1; so we may discard the second equation and work with the system

$$.2x_1 - .3x_2 = 0$$
$$x_1 + x_2 = 1$$

instead. This system can be solved either by substitution (see Section 1.3) or by the Gauss-Jordan method (see Sections 2.1–2.2). Applying the substitution method, for example, we first obtain

$$x_2 = 1 - x_1.$$

Now substitute this expression for x_2 in the first equation and simplify.

$$.2x_1 - .3(1 - x_1) = 0$$
$$.2x_1 - .3 + .3x_1 = 0$$
$$.5x_1 = .3$$
$$x_1 = \frac{.3}{.5} = .6$$
$$x_2 = 1 - x_1 = 1 - .6 = .4$$

Therefore the equilibrium vector is

$$X = [.6 \quad .4],$$

as we have already seen.

EXAMPLE 8.2 For the Markov chain in Example 8.1 on page 427, determine the percentage of drivers in the long run who will be insured by each company.

Solution Recall that the states in this Markov chain are: insured by Allstate, insured by State Farm, and insured by another company, respectively. The transition matrix is

$$P = \begin{bmatrix} .7 & .2 & .1 \\ .1 & .8 & .1 \\ .3 & .1 & .6 \end{bmatrix}.$$

Since P contains no zero entries, Theorem 8.2 assures us that the state vectors of this Markov chain approach a unique equilibrium vector

$$X = [x_1 \quad x_2 \quad x_3].$$

The entries of X give the percentage of drivers in each state over the long run.

To compute X, we must solve the matrix equation $X(I - P) = O$, that is,

$$[x_1 \quad x_2 \quad x_3] \begin{bmatrix} .3 & -.2 & -.1 \\ -.1 & .2 & -.1 \\ -.3 & -.1 & .4 \end{bmatrix} = [0 \quad 0 \quad 0].$$

Multiplying out the left side and including the condition that the sum of the unknowns is 1, we obtain the system of equations below.

$$.3x_1 - .1x_2 - .3x_3 = 0$$
$$-.2x_1 + .2x_2 - .1x_3 = 0$$
$$-.1x_1 - .1x_2 + .4x_3 = 0$$
$$x_1 + x_2 + x_3 = 1$$

Although it is not immediately apparent, this system is redundant. (The first equation equals -1 times the sum of the second and third equations.) Solving the system yields the solution

$$x_1 = .35, \quad x_2 = .45, \quad x_3 = .20.$$

Thus over the long run 35% of the drivers in this community will be insured by Allstate, 45% by State Farm, and 20% by another company. ■

EXAMPLE 8.3 Three players are tossing a frisbee. Player 1 always throws to player 2, and player 2 always throws to player 3. Player 3, however, is equally likely to throw to player 1 or to player 2. Over the long run what proportion of the throws will be made by each player?

Solution The three states of this Markov chain are: player 1 has the frisbee, player 2 has the frisbee, and player 3 has the frisbee. The transition matrix is

$$P = \begin{bmatrix} 0 & 1 & 0 \\ 0 & 0 & 1 \\ \frac{1}{2} & \frac{1}{2} & 0 \end{bmatrix}.$$

In Practice Problem 3, we saw that this Markov chain is regular, and so Theorem 8.2 guarantees that it has a unique equilibrium vector

$$X = [x_1 \quad x_2 \quad x_3].$$

The matrix equation $X(I - P) = O$ is

$$[x_1 \quad x_2 \quad x_3]\begin{bmatrix} 1 & -1 & 0 \\ 0 & 1 & -1 \\ -\frac{1}{2} & -\frac{1}{2} & 1 \end{bmatrix} = [0 \quad 0 \quad 0],$$

which is equivalent to the following system of linear equations.

$$x_1 - \tfrac{1}{2}x_3 = 0$$
$$-x_1 + x_2 - \tfrac{1}{2}x_3 = 0$$
$$ -x_2 + x_3 = 0.$$

Including the condition that the sum of the unknowns is 1, we obtain the following system

$$x_1 - \tfrac{1}{2}x_3 = 0$$
$$-x_1 + x_2 - \tfrac{1}{2}x_3 = 0$$
$$ -x_2 + x_3 = 0$$
$$x_1 + x_2 + x_3 = 1.$$

Solving this system, we see that $x_1 = .2$, $x_2 = .4$, and $x_3 = .4$. Thus over the long run player 1 will make 20% of the throws, and players 2 and 3 will each make 40% of the throws. ∎

Practice Problem 4 For the Markov chain in Practice Problem 1 on page 429, determine the percentage of persons in each class over the long run. Recall that its transition matrix is

$$\begin{bmatrix} .5 & .5 & 0 \\ .1 & .7 & .2 \\ 0 & .5 & .5 \end{bmatrix}.$$

Knowledge of the equilibrium vector for a Markov chain can be useful for long-term planning. In reference [1], for instance, Bourne studied changes in land usage in Toronto, Canada, from 1952 to 1962. Land was classified in terms of ten possible uses, resulting in a Markov chain with ten states. When the equilibrium vector for this process was determined, it was found that if the 1952–62 trend continued, then 19% of the city's land would be used for parking and 9.1% would be vacant. Knowledge of these percentages can help the city government to take corrective action to prevent erosion of the city's tax base. A similar study in New Zealand (see reference [3]) showed that about 60% of the country's industry would eventually be located in the Auckland area, supporting the charge of the opposition party leader that the entire south island should be regarded as a depressed area suitable for government subsidies and other development incentives.

EXERCISES 8.2

In Exercises 1–14 determine if the Markov chain having the given transition matrix is regular. If so, find its unique equilibrium vector.

1. $\begin{bmatrix} .2 & .8 \\ 1 & 0 \end{bmatrix}$

2. $\begin{bmatrix} 1 & 0 \\ .6 & .4 \end{bmatrix}$

3. $\begin{bmatrix} .7 & .3 \\ .5 & .5 \end{bmatrix}$

4. $\begin{bmatrix} .1 & .9 \\ .7 & .3 \end{bmatrix}$

5. $\begin{bmatrix} .2 & .8 \\ .4 & .6 \end{bmatrix}$

6. $\begin{bmatrix} .8 & .2 \\ .5 & .5 \end{bmatrix}$

7. $\begin{bmatrix} .7 & .3 \\ 0 & 1 \end{bmatrix}$

8. $\begin{bmatrix} 0 & 1 \\ .5 & .5 \end{bmatrix}$

9. $\begin{bmatrix} \frac{1}{2} & \frac{1}{2} & 0 \\ 0 & 0 & 1 \\ 1 & 0 & 0 \end{bmatrix}$

10. $\begin{bmatrix} .3 & .2 & .5 \\ .3 & .2 & .5 \\ .3 & .2 & .5 \end{bmatrix}$

11. $\begin{bmatrix} .4 & .6 & 0 \\ .5 & .5 & 0 \\ .3 & .3 & .4 \end{bmatrix}$

12. $\begin{bmatrix} .9 & 0 & .1 \\ .2 & .7 & .1 \\ .1 & .3 & .6 \end{bmatrix}$

13. $\begin{bmatrix} .6 & .2 & .2 \\ .1 & .8 & .1 \\ 0 & 1 & 0 \end{bmatrix}$

14. $\begin{bmatrix} 0 & 0 & 1 \\ .3 & .4 & .3 \\ .5 & .5 & 0 \end{bmatrix}$

15. Consider the two-state Markov chain with the transition diagram below.

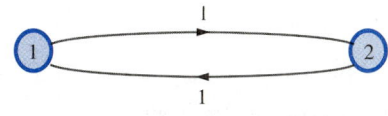

    (a) Does this Markov chain have an equilibrium vector? If so, what is it?
    (b) If the initial state vector is [1 0], do the state vectors approach an equilibrium vector?

16. Consider the three-state Markov chain with the transition diagram below.

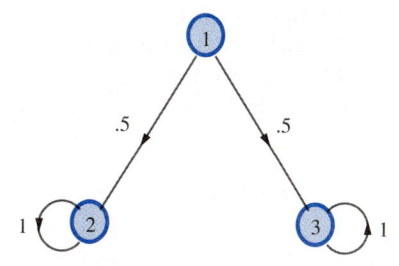

    (a) Is this Markov chain regular?
    (b) Show that, for any nonnegative numbers a and b such that $a + b = 1$, [0 a b] is an equilibrium vector for this Markov chain.

In Exercises 17–28 answer the given question about the corresponding exercise in Section 8.1 on page 434.

17. What proportion of the actor's performances will be good over the long run?
18. What proportion of the words in the manuscript will contain an error after many retypings?
19. What share of the market will the newspaper eventually have?
20. What unemployment rate will Booneville eventually have?
21. What proportion of the population will eventually live in the cities?
22. What share of the market will foreign cars eventually have?
23. What is the probability that the mouse will be in compartment 1 after many room changes?
24. What will the voter distribution be after many generations?
25. What will the distribution of workers be after many generations?
26. What percentage of the workers will eventually use each form of transportation?
27. What percentage of American cars will eventually be large? Mid-size? Small?
28. What proportion of the island's land will eventually be used in each way?
29. Each month that a healthy person remains in the Amazon jungle, there is a 5% risk of contracting an acute infection caused by a local parasite. Once contracted, the disease never completely leaves the body but rather remains present in a dormant state and periodically flares up in acute form. If the disease is acute in a given month, there is a 60% chance that it will still be acute the next month; but if it is dormant during a given month, there is only a 10% chance that it will be acute the next month.

    (a) Represent this process as a three-state Markov chain, and form the transition matrix.
    (b) Is this a regular Markov chain?
    (c) Show that once a person becomes infected, this process can be represented by a two-state Markov chain.
    (d) Show that the Markov chain in part (c) is regular and find its unique equilibrium vector.

30. (a) Apply the Markov chain in Exercise 29(a) to a healthy person to deduce that the probability that he will still be healthy after n months is $(.95)^n$.
    (b) Conclude that in the long run, the probability of being healthy is 0.
    (c) Use part (b) and Exercise 29(d) to show that the disease model has an equilibrium vector.
    (d) Show that the Markov chain in Exercise 29(a) is not regular but has a unique equilibrium vector. (It can be shown that, no matter what the initial state vector is, the state vectors for the Markov chain in Exercise 29(a) approach this equilibrium vector. Thus Theorem 8.2 may hold for a Markov chain that is not regular.)

Answers to Practice Problems 3. Yes, the Markov chain is regular because

$$A^5 = \begin{bmatrix} \frac{1}{4} & \frac{1}{4} & \frac{1}{2} \\ \frac{1}{4} & \frac{1}{2} & \frac{1}{4} \\ \frac{1}{8} & \frac{3}{8} & \frac{1}{2} \end{bmatrix}$$

contains only positive entries.

4. Over the long run 12.5% of the people will be in the upper class, 62.5% in the middle class, and 25% in the lower class.

8.3 Absorbing Markov Chains

In this section we will consider another class of Markov chains that arise frequently in applications. Like regular chains, these processes exhibit a stable behavior in the long run, but here the long range behavior of the system depends on the initial state vector.

ABSORBING STATES

In reference [5] Mosimann presents a simple model for the passage of a phosphorus molecule through a pasture ecosystem having four states: in the soil, in the grass, in cattle, or outside (that is, no longer in the ecosystem). The probabilities that a phosphorus molecule changes between the various pairs of states on a given day are shown in the transition diagram in Figure 8.4.

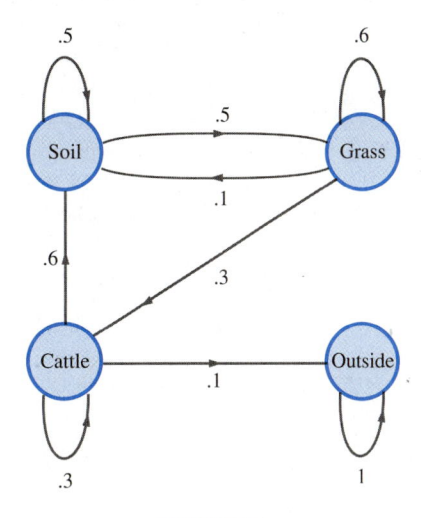

FIGURE 8.4

Thus Mosimann's model can be described by a Markov chain having the transition matrix below.

$$\text{Current state} \begin{array}{c} \text{Soil} \\ \text{Grass} \\ \text{Cattle} \\ \text{Outside} \end{array} \begin{bmatrix} .5 & .5 & 0 & 0 \\ .1 & .6 & .3 & 0 \\ .6 & 0 & .3 & .1 \\ 0 & 0 & 0 & 1 \end{bmatrix}$$

with columns labeled Soil, Grass, Cattle, Outside (Next day's state).

Notice that once a phosphorus molecule leaves the ecosystem, it never returns.

As a result the last row of every power of this transition matrix is

$$[0 \quad 0 \quad 0 \quad 1],$$

and so this Markov chain is not regular.

A state in a Markov chain is called an **absorbing state** if it is impossible to leave that state once it is entered, that is, if the transition probability of moving from that state to itself is 1. A Markov chain is called **absorbing** if

(a) there is at least one absorbing state, and
(b) from every state it is possible to reach an absorbing state in a finite number of trials.

In the model of the pasture ecosystem there is an absorbing state (the state of being outside the ecosystem) and it is possible to reach this state from every other state. (Figure 8.4 shows that the molecule can pass directly to the outside from cattle, from the grass to cattle to the outside, and from the soil to the grass to cattle to the outside.) Thus Mosimann's model of the pasture ecosystem is an absorbing Markov chain.

EXAMPLE 8.4 A laundromat vending machine sells four brands of laundry detergent: brands A, B, C, and D. Those who use brands A and C never change brands. However, of those who buy brand B one week, only 40% continue to buy it the next week; 30% switch to A, 10% to C, and 20% to D. And everyone who tries brand D picks a different brand the next week, with 20% switching to A, 50% to B, and 30% to C. Represent this process as a Markov chain.

Discussion The weekly transition diagram is shown in Figure 8.5, where the states correspond to whichever brand of soap a purchaser bought last.

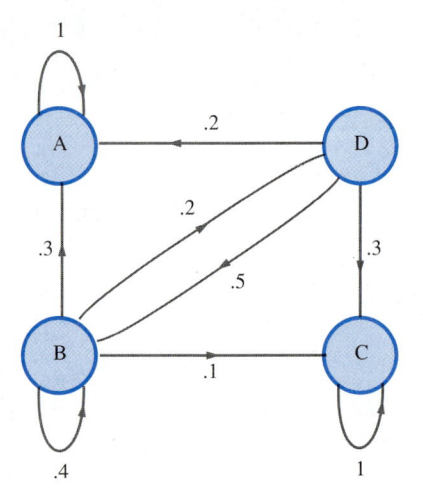

FIGURE 8.5

For instance, if a person bought brand B this week, the probability is .3 that he will buy brand A next week, .4 that he will buy brand B next week, and so forth. Observe that states A and C are absorbing states: Once the process enters either of these states it remains there forever. Moreover, from states B and D it is possible to reach an absorbing state in one trial. Thus this process is an absorbing Markov chain; its transition matrix is

$$\text{Current state} \begin{array}{c} \\ A \\ B \\ C \\ D \end{array} \overset{\text{Next week's state}}{\begin{array}{cccc} A & B & C & D \end{array}} \\ \begin{bmatrix} 1 & 0 & 0 & 0 \\ .3 & .4 & .1 & .2 \\ 0 & 0 & 1 & 0 \\ .2 & .5 & .3 & 0 \end{bmatrix} = P.$$

In Example 8.4 notice that the rows corresponding to the absorbing states A and C have a special form: All their entries are 0 except for a 1 along the main diagonal. Thus the transition probabilities for absorbing state i are such that $p_{ii} = 1$ and $p_{ij} = 0$ for $j \neq i$, which means that once we enter an absorbing state, we are certain to return there on the next trial.

LONG-RUN CONVERGENCE

It is easy to see that a Markov chain with an absorbing state cannot be regular. In fact, if there is more than one absorbing state, then there cannot be a unique equilibrium vector. In Example 8.4, for instance, if the system starts in absorbing state A, it will remain in state A forever; thus each state vector will be

$$[1 \quad 0 \quad 0 \quad 0].$$

Similarly, if the system starts in state C, each state vector will be

$$[0 \quad 0 \quad 1 \quad 0].$$

Hence both of these vectors are equilibrium vectors for this Markov chain. More generally, for any nonnegative numbers a and b such that $a + b = 1$, the vector

$$[a \quad 0 \quad b \quad 0]$$

can be shown to be an equilibrium vector for this Markov chain.

What will happen in Example 8.4 if the system starts in one of the nonabsorbing states (states B or D)? Table 8.2 shows the weekly state vectors for each of these starting states. (Entries are rounded to two decimal places.)

Time	Initial state B	Initial state D
0	[0 1 0 0]	[0 0 0 1]
1	[.30 .40 .10 .20]	[.20 .50 .30 .00]
2	[.46 .26 .20 .08]	[.35 .20 .35 .10]
3	[.55 .14 .25 .05]	[.43 .13 .40 .04]
4	[.61 .08 .28 .03]	[.48 .07 .43 .03]
5	[.64 .05 .30 .02]	[.50 .04 .44 .01]
6	[.66 .03 .31 .01]	[.52 .02 .45 .01]
7	[.67 .02 .31 .01]	[.53 .01 .45 .00]
8	[.67 .01 .32 .00]	[.53 .01 .46 .00]
9	[.68 .01 .32 .00]	[.54 .00 .46 .00]
10	[.68 .00 .32 .00]	[.54 .00 .46 .00]

TABLE 8.2

In each case we see that (to two-decimal place accuracy) the state vectors have stabilized by time 10, but *the equilibrium vector depends on the initial state*. Thus if a person uses brand B initially, there is a 68% chance that he will be using brand A after 10 weeks and a 32% chance that he will be using brand C; whereas if a person uses brand D initially, there is a 54% chance that he will be using brand A after 10 weeks and a 46% chance that he will be using brand C. Note that there is virtually *no* chance that someone will be using brands B or D after 10 weeks. In other words, as the process continues the system is increasingly likely to be in an absorbing state. This result is true in general.

> In an absorbing Markov chain, the long-run probability of being in a nonabsorbing state is zero, regardless of the initial state. Thus over the long run, the probability that the system will reach an absorbing state is 1.

When the system starts out in an absorbing state, the state vectors never change, and so the behavior of the process is completely determined. But when it starts in a nonabsorbing state, three questions can be asked:

1. How many times, on the average, will the system pass through a particular nonabsorbing state?
2. How long will it take the process to enter an absorbing state?
3. What is the probability that the system will enter a particular absorbing state?

The answers to all of these questions involve a matrix called the *fundamental matrix* of the Markov chain.

THE FUNDAMENTAL MATRIX

Recall that the transition matrix in Example 8.4 has the form:

$$\begin{array}{c} \text{Current} \\ \text{state} \end{array} \begin{array}{c} \\ A \\ B \\ C \\ D \end{array} \overset{\begin{array}{c}\text{Next week's state}\\ A \quad B \quad C \quad D\end{array}}{\begin{bmatrix} 1 & 0 & 0 & 0 \\ .3 & .4 & .1 & .2 \\ 0 & 0 & 1 & 0 \\ .2 & .5 & .3 & 0 \end{bmatrix}} = P.$$

If we rearrange the sequence of the four states in Example 8.4 so that the absorbing states come first, we obtain a matrix in which the rows and columns are labeled as shown below.

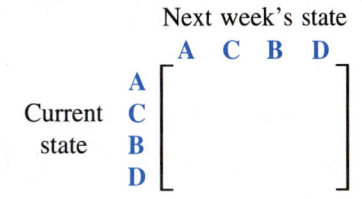

$$\begin{array}{c} \text{Current} \\ \text{state} \end{array} \begin{array}{c} \\ A \\ C \\ B \\ D \end{array} \overset{\begin{array}{c}\text{Next week's state}\\ A \quad C \quad B \quad D\end{array}}{\begin{bmatrix} & & & \\ & & & \\ & & & \\ & & & \end{bmatrix}}$$

The entries of this matrix will be obtained by choosing the entry of the transition matrix P that corresponds to a particular pair of row and column labels. For example, the entry of the matrix lying in the row labeled B and the column labeled C is .1, because that is the entry of P in the row labeled B and the column labeled C. The resulting matrix $P^\star$, called the **canonical form** of P, is shown below.

$$\begin{array}{c} \text{Current} \\ \text{state} \end{array} \begin{array}{c} \\ A \\ C \\ B \\ D \end{array} \overset{\begin{array}{c}\text{Next week's state}\\ A \quad C \quad B \quad D\end{array}}{\left[\begin{array}{cc|cc} 1 & 0 & 0 & 0 \\ 0 & 1 & 0 & 0 \\ \hline .3 & .1 & .4 & .2 \\ .2 & .3 & .5 & 0 \end{array}\right]} = P^\star.$$

The horizontal and vertical lines in $P^\star$ separate the absorbing states from the nonabsorbing states, thereby partitioning it into four submatrices. Note that the matrix in the upper left is an identity matrix, and the matrix in the upper right is a zero matrix.

More generally, consider an absorbing Markov chain with r absorbing states and s nonabsorbing states. The canonical form of its transition matrix will be

$$P^\star = \begin{array}{c} r\{ \\ s\{ \end{array} \overset{\overbrace{\quad\quad}^{r}\;\overbrace{\quad\quad}^{s}}{\left[\begin{array}{c|c} I_r & O \\ \hline R & Q \end{array}\right]},$$

where I_r is the $r \times r$ identity matrix, O is the $r \times s$ zero matrix, and R and Q have dimensions $s \times r$ and $s \times s$, respectively. In the canonical form $P^\star$ of a transition matrix P, we will list the absorbing states in the same order as they occur in P (but preceding all the nonabsorbing states). Likewise in $P^\star$ we will list the nonabsorbing states in the same order as they occur in P (but following all the absorbing states).

The matrix

$$M = (I_s - Q)^{-1},$$

where I_s is the $s \times s$ identity matrix, is called the **fundamental matrix** of the absorbing Markov chain.

EXAMPLE 8.5 Determine the fundamental matrix of the Markov chain described in Example 8.4.

Solution We have seen above that the canonical form of the transition matrix for this Markov chain is

$$\begin{array}{c} \\ A \\ C \\ B \\ D \end{array} \begin{array}{c} A \quad C \quad B \quad D \end{array} \left[\begin{array}{cc|cc} 1 & 0 & 0 & 0 \\ 0 & 1 & 0 & 0 \\ \hline .3 & .1 & .4 & .2 \\ .2 & .3 & .5 & 0 \end{array}\right] = \left[\begin{array}{c|c} I_2 & O \\ \hline R & Q \end{array}\right].$$

To compute the fundamental matrix, we must identify the lower right submatrix, which is

$$Q = \begin{bmatrix} .4 & .2 \\ .5 & 0 \end{bmatrix}.$$

Thus

$$I_2 - Q = \begin{bmatrix} 1 & 0 \\ 0 & 1 \end{bmatrix} - \begin{bmatrix} .4 & .2 \\ .5 & 0 \end{bmatrix} = \begin{bmatrix} .6 & -.2 \\ -.5 & 1 \end{bmatrix}.$$

Applying the technique in Section 2.6, we obtain the fundamental matrix

$$M = (I_2 - Q)^{-1} = \begin{bmatrix} 2 & .4 \\ 1 & 1.2 \end{bmatrix}.$$

The reader should check that this is the correct inverse by verifying that

$$(I_2 - Q)M = I_2. \quad \blacksquare$$

Practice Problem 5 The transition matrix for an absorbing Markov chain is

$$\begin{array}{c} A \\ B \\ C \\ D \end{array} \begin{array}{c} A \quad B \quad C \quad D \end{array} \left[\begin{array}{cccc} .1 & .6 & .1 & .2 \\ 0 & 1 & 0 & 0 \\ .4 & 0 & .4 & .2 \\ 0 & 0 & 0 & 1 \end{array}\right]$$

(a) Determine the canonical form of the transition matrix.
(b) Compute the fundamental matrix of the Markov chain.

8.3 ABSORBING MARKOV CHAINS

ABSORPTION TIMES AND PROBABILITIES

After writing the transition matrix in canonical form and computing the fundamental matrix, the three questions on page 448 can be answered. The following result tells us how to make use of these matrices.

Theorem 8.3

Let

$$P^\star = \left[\begin{array}{c|c} I & O \\ \hline R & Q \end{array}\right]$$

be the canonical form of the transition matrix for an absorbing Markov chain, and let $M = (I - Q)^{-1}$ be its fundamental matrix. Label the rows and columns of M exactly as the rows and columns of Q are labeled in $P^\star$, and label the rows and columns of MR exactly as the rows and columns of R are labeled in $P^\star$. Then:

(a) The entry of M in the row labeled X and the column labeled Y gives the expected number of trials that a system which begins in nonabsorbing state X will be in nonabsorbing state Y.

(b) The sum of the entries in M in the row labeled X gives the expected number of trials before absorption if the system begins in nonabsorbing state X.

(c) The entry of MR in the row labeled X and the column labeled Y gives the probability that a system which begins in nonabsorbing state X will end up in absorbing state Y.

EXAMPLE 8.5 Revisited

For the absorbing Markov chain described in Example 8.4 determine:

(a) The number of weeks that someone can be expected to buy detergent D if he originally buys detergent B.
(b) The expected number of weeks before absorption if someone originally buys detergent B.
(c) The probability that someone will eventually use detergent A if he originally buys detergent B.

Solution Recall from Example 8.5 that the canonical form of the transition matrix for this Markov chain is

$$P^\star = \begin{array}{c} \\ A \\ C \\ B \\ D \end{array} \begin{array}{c} \begin{array}{cccc} A & C & B & D \end{array} \\ \left[\begin{array}{cc|cc} 1 & 0 & 0 & 0 \\ 0 & 1 & 0 & 0 \\ \hline .3 & .1 & .4 & .2 \\ .2 & .3 & .5 & 0 \end{array}\right] \end{array} = \left[\begin{array}{c|c} I_2 & O \\ \hline R & Q \end{array}\right]$$

and that its fundamental matrix is

$$M = (I_2 - Q)^{-1} = \begin{bmatrix} 2 & .4 \\ 1 & 1.2 \end{bmatrix}.$$

When we label the rows and columns of M as the rows and columns of Q are labeled in $P^\star$, the resulting matrix is

$$M = \begin{matrix} & \begin{matrix} B & D \end{matrix} \\ \begin{matrix} B \\ D \end{matrix} & \begin{bmatrix} 2 & .4 \\ 1 & 1.2 \end{bmatrix} \end{matrix}.$$

Thus the rows and columns of the fundamental matrix represent the two non-absorbing states, states B and D.

(a) By Theorem 8.3(a) the number of weeks that someone can be expected to buy detergent D if he originally buys detergent B is .4, the entry of the fundamental matrix in the row labeled B and the column labeled D.

(b) By Theorem 8.3(b) the expected number of weeks before absorption if someone originally buys detergent B is 2.4, the sum of the entries of the fundamental matrix in the row labeled B.

(c) From the canonical form above, we see that

$$R = \begin{bmatrix} .3 & .1 \\ .2 & .3 \end{bmatrix}.$$

Hence

$$MR = \begin{bmatrix} 2 & .4 \\ 1 & 1.2 \end{bmatrix} \begin{bmatrix} .3 & .1 \\ .2 & .3 \end{bmatrix} = \begin{bmatrix} .68 & .32 \\ .54 & .46 \end{bmatrix}.$$

When we label the rows and columns of MR as the rows and columns of R are labeled in $P^\star$, we obtain

$$MR = \begin{matrix} & \begin{matrix} A & C \end{matrix} \\ \begin{matrix} B \\ D \end{matrix} & \begin{bmatrix} .68 & .32 \\ .54 & .46 \end{bmatrix} \end{matrix}.$$

Thus the rows of MR correspond to the nonabsorbing states B and D, whereas its columns correspond to the absorbing states A and C. Hence the probability that someone will eventually use detergent A if he originally buys detergent B is .68, the entry of MR in the row labeled B and the column labeled A. Notice that this answer agrees with the data in Table 8.2. ∎

EXAMPLE 8.6   When obtaining a doctorate at the state university, a graduate student is classified in one of four states:

1. Has not yet passed the qualifying exams;
2. Has passed the qualifying exams, but has no dissertation topic;
3. Has a dissertation topic, but has not received the degree;
4. Has received the degree.

(a) Because a new graduate student begins in state 1, Theorem 8.3(b) tells us that the time before absorption is the sum of the entries in the row of the fundamental matrix labeled 1. Hence we see that on the average a new graduate student will spend 3.75 years in the program before either dropping out or receiving a doctorate.

(b) By Theorem 8.3(c) the probability that a new graduate student will eventually receive a doctorate is an entry of the matrix MR. Computing MR and labeling its rows and columns as the rows and columns of R are labeled in $P^\star$, we obtain

$$MR = \begin{bmatrix} 1.25 & 1.25 & 1.25 \\ 0 & 2.50 & 2.50 \\ 0 & 0 & 5 \end{bmatrix} \begin{bmatrix} .4 & 0 \\ .2 & 0 \\ .1 & .1 \end{bmatrix} = \begin{matrix} 1 \\ 2 \\ 3 \end{matrix} \begin{bmatrix} \overset{0}{.875} & \overset{4}{.125} \\ .750 & .250 \\ .500 & .500 \end{bmatrix}.$$

Thus the probability that a new graduate student (one who is in state 1) will receive a doctorate (be absorbed into state 4) is .125. ■

The theory of absorbing Markov chains can be applied to Bernoulli processes in several interesting ways. The next example illustrates their use in the analysis of *success runs*.

EXAMPLE 8.7 A fair coin is repeatedly tossed until three heads in a row have occurred. On the average, how long will this take?

Solution We will say that the system is in state i at a given time ($i = 0, 1, 2,$ or 3) if the last toss completes a run of i heads; that is, i is the largest number such that the preceding i tosses all were heads. For instance, after the sequence THHT the system is in state 0, and after the sequence THHTH it is in state 1. Because the process ends when we obtain three consecutive heads, we will regard state 3 as an absorbing state. The transition matrix for this process is given below.

$$\text{Current state} \begin{matrix} 0 \\ 1 \\ 2 \\ 3 \end{matrix} \overset{\text{State after next toss}}{\begin{bmatrix} \overset{0}{\tfrac{1}{2}} & \overset{1}{\tfrac{1}{2}} & \overset{2}{0} & \overset{3}{0} \\ \tfrac{1}{2} & 0 & \tfrac{1}{2} & 0 \\ \tfrac{1}{2} & 0 & 0 & \tfrac{1}{2} \\ 0 & 0 & 0 & 1 \end{bmatrix}} = P$$

From state 2, for instance, we will either go to state 3 (if the next toss is heads) or to state 0 (if the next toss is tails).

8.3 ABSORBING MARKOV CHAIN

Since students may drop from the program at any time, we will al[so have] another state:

0. Has dropped out of the program.

The year-to-year transition matrix for this process has been found to [be]

$$\text{Current state} \begin{array}{c} 0 \\ 1 \\ 2 \\ 3 \\ 4 \end{array} \overbrace{\begin{bmatrix} 1 & 0 & 0 & 0 & 0 \\ .4 & .2 & .4 & 0 & 0 \\ .2 & 0 & .6 & .2 & 0 \\ .1 & 0 & 0 & .8 & .1 \\ 0 & 0 & 0 & 0 & 1 \end{bmatrix}}^{\text{Next year's state} \;\; 0 \;\; 1 \;\; 2 \;\; 3 \;\; 4} = P.$$

Thus, for instance, a student who has passed the qualifying exams but [does not] have a dissertation topic has a 20% chance of dropping out, a 60% ch[ance of] remaining without a dissertation topic, and a 20% chance of finding a [disser]tation topic within the next year. Determine:

(a) The number of years that a new student can expect to be in this [doctoral] program before either dropping out or receiving the degree.
(b) The probability that a new graduate student will eventually receive [a doc]torate from the state university.

Solution This process is an absorbing Markov chain with two absorbing states, [0] and 4. To answer the questions posed above, we must rewrite the tra[nsition] matrix in its canonical form:

$$P^\star = \begin{array}{c} 0 \\ 4 \\ 1 \\ 2 \\ 3 \end{array} \overbrace{\begin{bmatrix} 1 & 0 & 0 & 0 & 0 \\ 0 & 1 & 0 & 0 & 0 \\ \hline .4 & 0 & .2 & .4 & 0 \\ .2 & 0 & 0 & .6 & .2 \\ .1 & .1 & 0 & 0 & .8 \end{bmatrix}}^{0 \;\; 4 \;\; 1 \;\; 2 \;\; 3} = \left[\begin{array}{c|c} I_2 & O \\ \hline R & Q \end{array} \right].$$

Next we compute

$$I_3 - Q = \begin{bmatrix} 1 & 0 & 0 \\ 0 & 1 & 0 \\ 0 & 0 & 1 \end{bmatrix} - \begin{bmatrix} .2 & .4 & 0 \\ 0 & .6 & .2 \\ 0 & 0 & .8 \end{bmatrix} = \begin{bmatrix} .8 & -.4 & 0 \\ 0 & .4 & -.2 \\ 0 & 0 & .2 \end{bmatrix}$$

and find its inverse by the method of Section 2.6. The result is the fundame[ntal] matrix for the Markov chain, which we label as the matrix Q is labeled in [...]

$$M = (I_3 - Q)^{-1} = \begin{array}{c} 1 \\ 2 \\ 3 \end{array} \overbrace{\begin{bmatrix} 1.25 & 1.25 & 1.25 \\ 0 & 2.50 & 2.50 \\ 0 & 0 & 5 \end{bmatrix}}^{1 \qquad 2 \qquad 3}.$$

The canonical form of P is obtained by listing absorbing state 3 before nonabsorbing states 0, 1, and 2; it is

$$P^\star = \begin{array}{c} \\ 3 \\ 0 \\ 1 \\ 2 \end{array} \begin{array}{cccc} 3 & 0 & 1 & 2 \\ \left[\begin{array}{c|ccc} 1 & 0 & 0 & 0 \\ \hline 0 & 0 & \frac{1}{2} & \frac{1}{2} \\ 0 & \frac{1}{2} & 0 & \frac{1}{2} \\ \frac{1}{2} & \frac{1}{2} & 0 & 0 \end{array}\right] \end{array} = \left[\begin{array}{c|c} I_1 & O \\ \hline R & Q \end{array}\right].$$

The fundamental matrix for this Markov chain is

$$M = (I_3 - Q)^{-1} = \begin{array}{c} 0 \\ 1 \\ 2 \end{array} \begin{array}{c} 0 \quad 1 \quad 2 \\ \left[\begin{array}{ccc} 8 & 4 & 2 \\ 6 & 4 & 2 \\ 4 & 2 & 2 \end{array}\right] \end{array};$$

its rows and columns are labeled as the rows and columns of Q are labeled in $P^\star$. By Theorem 8.3(b) the time before absorption is the sum of the entries in the row of M labeled 0. Hence it will take 14 tosses on the average to get three heads in succession. ∎

Practice Problem 6 For the transition matrix in Practice Problem 5 on page 450, determine:
(a) The number of trials that the system can be expected to be in state A if it begins in stage C.
(b) The expected number of trials before absorption if the system begins in state A.
(c) The probability that the system will eventually end in state D if it begins in state C.

EXERCISES 8.3

In Exercises 1–4 a transition diagram for a Markov chain is given. Each arrow represents a positive transition probability. Determine if the corresponding Markov chain is absorbing.

1.

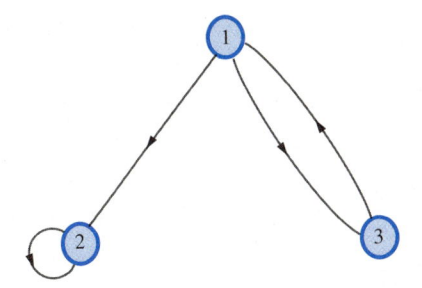

2.

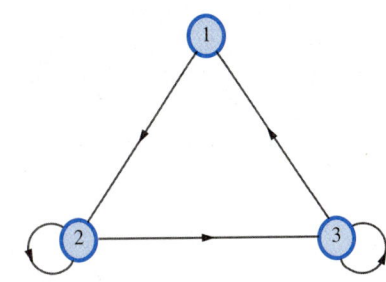

3.

4.

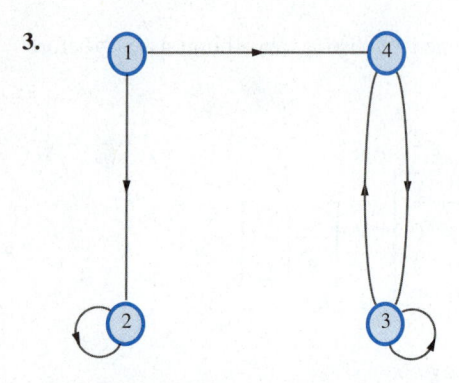

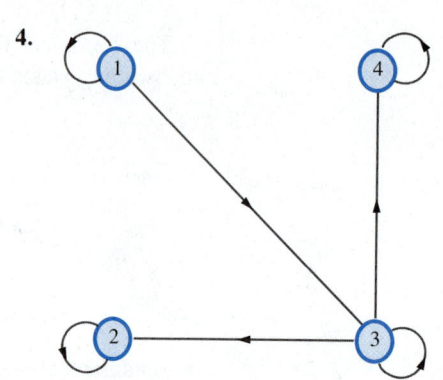

Each of the matrices in Exercises 5–12 is the transition matrix of an absorbing Markov chain. Determine the canonical form $P^\star$ of each matrix and compute the fundamental matrix M of the Markov chain.

5. $\begin{bmatrix} .6 & .2 & .2 \\ .1 & .7 & .2 \\ 0 & 0 & 1 \end{bmatrix}$

6. $\begin{bmatrix} .4 & .5 & .1 \\ 0 & 1 & 0 \\ .5 & 0 & .5 \end{bmatrix}$

7. $\begin{bmatrix} .5 & 0 & .5 \\ 0 & 1 & 0 \\ .1 & .2 & .7 \end{bmatrix}$

8. $\begin{bmatrix} .8 & .1 & 0 & .1 \\ 0 & 1 & 0 & 0 \\ 0 & 0 & 1 & 0 \\ .4 & .1 & .2 & .3 \end{bmatrix}$

9. $\begin{bmatrix} 1 & 0 & 0 & 0 \\ .2 & .5 & .1 & .2 \\ 0 & 0 & 1 & 0 \\ 0 & .5 & .1 & .4 \end{bmatrix}$

10. $\begin{bmatrix} \frac{1}{2} & \frac{1}{8} & \frac{1}{4} & \frac{1}{8} \\ \frac{1}{2} & \frac{1}{8} & \frac{1}{4} & \frac{1}{8} \\ \frac{1}{2} & \frac{1}{8} & \frac{1}{4} & \frac{1}{8} \\ 0 & 0 & 0 & 1 \end{bmatrix}$

11. $\begin{bmatrix} 1 & 0 & 0 & 0 \\ .1 & .2 & .7 & 0 \\ .1 & .4 & .4 & .1 \\ 0 & 0 & 0 & 1 \end{bmatrix}$

12. $\begin{bmatrix} .2 & .4 & .1 & .3 \\ .2 & .4 & .1 & .3 \\ .1 & .2 & .3 & .4 \\ 0 & 0 & 0 & 1 \end{bmatrix}$

13. Suppose that the Markov chain with the transition matrix in Exercise 5 begins in the second state. Determine the expected number of trials that the Markov chain will pass through the first state before absorption.

14. Suppose that the Markov chain with the transition matrix in Exercise 6 begins in the third state. Determine the expected number of trials that the Markov chain will pass through the third state before absorption.

15. Suppose that the Markov chain with the transition matrix in Exercise 7 begins in the first state. Determine the expected number of trials before absorption.

16. Use the transition matrix in Exercise 8 to determine the probability of being absorbed into the third state if the Markov chain begins in the first state.

17. Use the transition matrix in Exercise 9 to determine the probability of being absorbed into the first state if the Markov chain begins in
    (a) the second state   (b) the fourth state.

18. Use the transition matrix in Exercise 10 to determine the expected number of trials before absorption if the Markov chain begins in
    (a) the first state   (b) the second state.

19. Suppose that the Markov chain with the transition matrix in Exercise 11 begins in the third state. Determine the probability of being absorbed into
    (a) the first state   (b) the fourth state.

20. Suppose that the Markov chain with the transition matrix in Exercise 12 begins in the third state. Determine the expected number of trials that it will pass through the second state before absorption.

21. In Mosimann's four-state Markov chain for the passage of a phosphorus molecule through a pasture ecosystem (with the transition diagram in Figure 8.4), compute the expected number of days before a phosphorus molecule leaves the ecosystem if it begins in
    (a) the soil   (b) the grass   (c) cattle.

22. In Mosimann's four-state Markov chain for the passage of a phosphorus molecule through a pasture ecosystem (with the transition diagram in Figure 8.4), suppose that a phosphorus molecule begins in cattle. What is the expected number of days before leaving the ecosystem that the molecule will be in
    (a) the soil?   (b) the grass?   (c) cattle?

23. Consider the absorbing Markov chain with the transition diagram below.

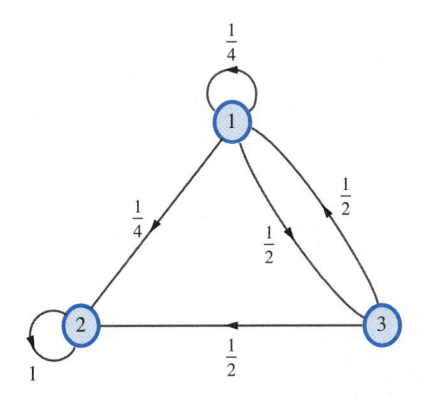

(a) How long will it take, on the average, to reach absorption if the system starts in state 1? In state 3?
(b) What is the probability that the system will end up in state 2 if it starts in state 1? If it starts in state 3?

24. Consider the absorbing Markov chain with the transition diagram below.

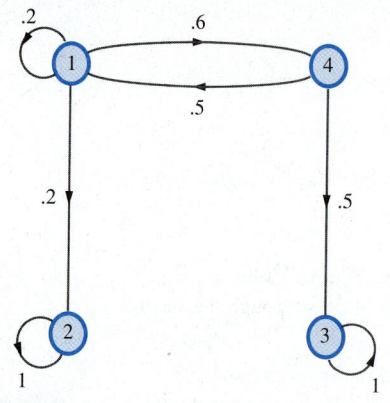

   (a) How long will it take, on the average, to reach absorption if the system starts in state 1? In state 4?
   (b) What is the probability that the system will end up in state 3 if it starts in state 1? If it starts in state 4?

25. Diplomatic personnel from an eastern bloc country are randomly assigned each year as follows: Of those stationed at home, 60% remain there the next year, 30% are transferred to another eastern bloc country, and 10% are assigned to the west. After one year outside the country, all personnel are reassigned to home. Suppose that one of this country's diplomats currently stationed at home intends to defect and never return home as soon as he is assigned to the west. How long should he expect to wait before being assigned to the west?

26. A mouse is placed in a box with four compartments, as shown below.

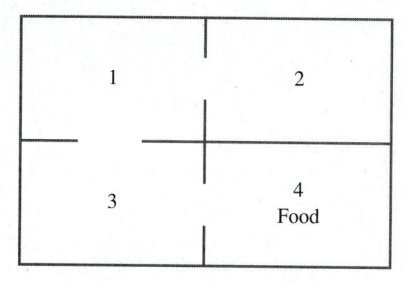

   In each time period the mouse goes through one of the available doors, at random, until it reaches the food in compartment 4. How long will this take, on the average, if the mouse starts in compartment 1? In compartment 2? In compartment 3?

27. A certain childhood disease follows a week-to-week transition pattern: If the patient is sick but has no fever (state S), there is a 40% chance of being in the same state the following week, a 20% chance of developing a fever (state F), and a 40% chance of recovering (state R). If the patient develops a fever, the chances are 60% that he will

still have it the following week, 20% that he will revert to state S, and 20% that he will die (state D). Once recovered from the disease, a patient is immune from contracting the disease again. If a patient is now in state S,

(a) how many weeks will the disease last, on the average?

(b) what is the probability that he will eventually recover?

28. Angela is a slightly better tennis player than Brigit. When the ball is in Angela's court (state A), she returns it to Brigit 80% of the time; the remaining 20% of the time she misses the shot and Brigit wins the point (state L). When the ball is in Brigit's court (state B), she returns it 75% of the time; the remaining 25% of the time she misses the shot and Angela wins the point (state W).

(a) How long will their rally last, on the average, if the ball starts out in Angela's court? In Brigit's court?

(b) What is the probability that Angela will eventually win the point if the ball starts in her court? In Brigit's court?

Angela returns the ball 80% of the time, and Brigit returns the ball 75% of the time.

29. A panel of three judges must reach a unanimous decision either to approve or reject a wiretap request. If the number voting for approval on a particular ballot is 0 or 3, the request is settled. However, if the vote is split (either 1 to 2 or 2 to 1), there is a 20% chance that the dissenter will give in to the majority on the next ballot, a 10% chance he will win exactly one of the others over to his position, and a 70% chance the vote will remain exactly the same on the next ballot. If an initial vote is 1 for approval and 2 against,

(a) how many more ballots will be needed to settle the request, on the average?

(b) what is the probability that the wiretap will eventually be approved?

30. An urn initially contains three marbles, two black and one white. At each step a marble is drawn at random from the urn and replaced by one of the opposite color. This process ends when all three marbles in the urn are the same color.

(a) On the average, how many steps are required before the urn contains three marbles of the same color?

(b) What is the probability that all the marbles in the urn will eventually be black?

31. A fair coin is flipped repeatedly until three heads have occurred (not necessarily in succession). On the average, how many flips will this take?

32. A player rolls two fair dice repeatedly until a sum of seven has occurred twice in succession. On the average, how many rolls will this take?

33. Ann (with $1) agrees to a series of games against Bob, who has $3. In each game Ann has even chances of winning $1 from Bob or losing $1 to him. The series of games will continue until either Ann or Bob goes broke. Thus the number of dollars held by Ann at any time will be 0, 1, 2, 3, or 4, with states 0 and 4 being absorbing states.

(a) On the average, how many games will it take for one of the players to go broke?

(b) What is the probability that Ann will go broke first?

(c) Answer (a) and (b) if Ann and Bob begin with $2 each.

34. Rework Exercise 33 under the assumption that, in each game, the probability is 3/4 that Ann wins $1 from Bob.

Answers to Practice Problems

5. (a) The canonical form of the transition matrix is

$$\begin{array}{c} \\ B \\ D \\ A \\ C \end{array} \begin{array}{cccc} B & D & A & C \end{array} \\ \left[\begin{array}{cc|cc} 1 & 0 & 0 & 0 \\ 0 & 1 & 0 & 0 \\ \hline .6 & .2 & .1 & .1 \\ 0 & .2 & .4 & .4 \end{array} \right]$$

(b) The fundamental matrix of the Markov chain is

$$\begin{bmatrix} 1.2 & 0.2 \\ 0.8 & 1.8 \end{bmatrix}.$$

6. (a) 0.8.   (b) 1.4.   (c) .52

■ ■ ■ ■ ■ ■ ■ ■ *MATHEMATICS IN ACTION*
8.4 Genetics

The principles of probability have important applications to genetics, the science of heredity. In this section we will briefly discuss the Mendelian theory of inheritance and show how Markov chains and other stochastic models can be applied to the study of genetics.

THE MENDELIAN THEORY OF HEREDITY

In a famous series of experiments on garden peas, the Moravian monk Gregor Mendel (1822–1884) deduced several basic principles governing the inheritance of characteristics among living organisms. His account was published in 1886, but it was ignored by the scientific community of his day. In 1900, sixteen years after Mendel's death, his laws were rediscovered by de Vries, Correns, and Tschermak.

In Mendel's theory the basic unit of heredity is the **gene.** An inherited trait is determined by the combination of two genes, one from each parent. The offspring therefore carries two genes for each trait. In garden peas, for example, one pair of genes determines whether the plant is tall or dwarfed, another pair determines whether the seeds are yellow or green, and so forth. The alternative forms of a particular gene are called **alleles.** For the traits studied by Mendel only two alleles are possible, one of which is **dominant,** and the other **recessive.** We will denote the dominant and recessive alleles by G and g, respectively. These alleles can combine in three possible forms called **genotypes** as described below.

Name	Genotype	Description
Pure dominant	GG	Both parents contribute allele G.
Hybrid	Gg	One parent contributes allele G, the other g.
Pure recessive	gg	Both parents contribute allele g.

In the case of a hybrid, the displayed trait is determined by the dominant allele G. In garden peas, for instance, the allele for yellow seed color is dominant over the allele for green seed color. Thus for this gene G denotes yellow seed color and g denotes green seed color, and a pure dominant or hybrid will have yellow seeds, but a pure recessive will have green seeds.

Mendel's first law, called the *law of segregation* states that exactly one gene from each parent's gene-pair is transmitted to the offspring during reproduction; and moreover, each gene has an equal opportunity of being transmitted. Let us see what happens when two individuals with a known genotype reproduce.

Case 1: Neither parent is hybrid. If both parents are pure dominant (genotype GG), the offspring must receive a G allele from each parent. Thus the offspring will also have genotype GG, and so will be a pure dominant. Similarly, if both parents are pure recessive, the offspring will be a pure recessive (genotype gg); and if one parent is pure dominant and the other is pure recessive, the offspring will be a hybrid (genotype Gg).

Case 2: One parent is pure, and the other is hybrid. If one parent is pure dominant (genotype GG) and the other is hybrid (Gg), the offspring must receive a G allele from the first parent but is equally likely to receive a G or g allele from the second parent. Hence the offspring will have genotype GG or Gg, each with probability ½. Similarly, if one parent is pure recessive (genotype gg) and the other is hybrid (Gg), then the offspring will have genotype gg or Gg, each with probability ½.

Case 3: Both parents are hybrid. In this case the offspring is equally likely to receive a G or g allele from each parent, and the allele received from one parent is independent of that received from the other parent. Therefore the probability that the offspring will receive a G allele from both parents is ½ · ½ = ¼. Hence the probability that the offspring will be a pure dominant is ¼, and, likewise, the probability that the offspring will be a pure recessive is also ¼. For the offspring to be hybrid, it must receive a G allele from the first parent and a g allele from the second or a g allele from the first parent and a G allele from the second. Because these two possibilities are mutually exclusive, the probability that the offspring is hybrid is

$$\frac{1}{2} \cdot \frac{1}{2} + \frac{1}{2} \cdot \frac{1}{2} = \frac{1}{4} + \frac{1}{4} = \frac{1}{2}.$$

Thus, on the average, in case 3 one out of four offspring will be pure dominant, two out of four will be hybrid, and one out of four will be pure recessive.

The results of the three cases above are summarized in Table 8.3, where D, H, and R represent the pure dominant, hybrid, and pure recessive geno-

| | Genotype Probabilities for Offspring | | |
Genotypes of Parents	D	H	R
D × D	1	0	0
D × H	½	½	0
D × R	0	1	0
H × H	¼	½	¼
H × R	0	½	½
R × R	0	0	1

TABLE 8.3

types, respectively. Notice that the six combinations in Table 8.3 account for all the possibilities because H × D is the same as D × H, R × D is the same as D × R, and R × H is the same as H × R.

MARKOV CHAIN MODELS

In controlled breeding experiments we are interested in what happens to the genotype of the offspring after several generations. Markov chains are often useful in modeling such experiments.

EXAMPLE 8.8 Albinism is a recessive trait determined by a single pair of genes. Suppose that an albino mates with an individual who does not carry the recessive allele for albinism, then the first offspring mates with another noncarrier, and so on indefinitely. How many generations will it take, on the average, before the recessive allele is eliminated in the firstborn offspring?

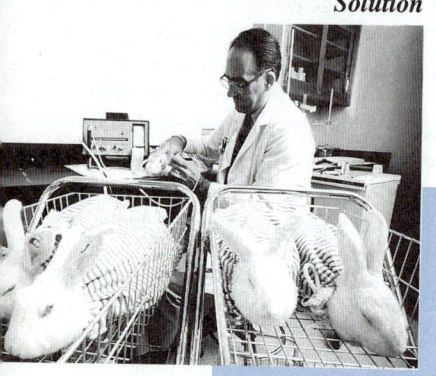

Solution Let g denote the recessive allele for albinism and G denote the normal (dominant) allele for this trait. Then albinos have genotype R = gg, carriers of the trait have genotype H = Gg, and noncarriers have genotype D = GG. We can model this controlled breeding experiment by a three-state Markov chain in which the states are the three possible genotypes (D, H, and R). The transition matrix for this Markov chain is

$$\begin{array}{c} \textit{Genotype of next generation} \\ \text{D} \text{H} \text{R} \end{array}$$

$$\text{Current genotype} \begin{array}{c} \text{D} \\ \text{H} \\ \text{R} \end{array} \begin{bmatrix} 1 & 0 & 0 \\ \tfrac{1}{2} & \tfrac{1}{2} & 0 \\ 0 & 1 & 0 \end{bmatrix} = P.$$

The entries of this matrix are obtained from Table 8.3 by breeding each of the current genotypes with a noncarrier (genotype D). This Markov chain is absorbing and has a single absorbing state, state D. Since the transition matrix is already in canonical form, we see that

$$I_2 - Q = \begin{bmatrix} 1 & 0 \\ 0 & 1 \end{bmatrix} - \begin{bmatrix} \frac{1}{2} & 0 \\ 1 & 0 \end{bmatrix} = \begin{bmatrix} \frac{1}{2} & 0 \\ -1 & 1 \end{bmatrix}.$$

Hence the fundamental matrix is

$$M = (I_2 - Q)^{-1} = \begin{matrix} H \\ R \end{matrix} \begin{bmatrix} \overset{H}{2} & \overset{R}{0} \\ 2 & 1 \end{bmatrix}.$$

From Theorem 8.3(b) we see that the expected number of trials before absorption is 3, the sum of the entries in the second row of the fundamental matrix. Hence it will take three generations, on the average, before the albino's firstborn descendants will become noncarriers. ■

EXAMPLE 8.9 Let us consider an example involving inbreeding. Two plants or animals are mated. Two of their offspring are selected at random and mated. Two of the offspring's offspring are selected at random and mated, and so forth. What will happen in this situation after many generations?

Solution We can take the possible genotypes of the *pair* of offspring as the states of the Markov chain. According to Table 8.3 the six possibilities are DD, DH, DR, HH, HR, and RR, where, for instance, DH represents a pair consisting of one pure dominant and one hybrid. To compute the transition probabilities, we use Table 8.3. If, for example, we start with parents in state DH, we see from Table 8.3 that the probability is ½ that an offspring will be pure dominant and ½ it will be hybrid. Hence we can compute the probabilities that a pair of offspring will be in a particular state:

P(offspring in state DD) $= P$(male offspring is D) $\cdot P$(female offspring is D)
$$= \frac{1}{2} \cdot \frac{1}{2} = \frac{1}{4};$$

P(offspring in state HH) $= P$(male offspring is H) $\cdot P$(female offspring is H)
$$= \frac{1}{2} \cdot \frac{1}{2} = \frac{1}{4};$$

P(offspring in state DH) $= P$(one offspring D and the other H)
$\qquad = P$(male offspring D and female H, or male offspring H and female D)
$\qquad = P$(male offspring D and female H) $+ P$(male offspring H and female D)
$\qquad = P$(male is D) $\cdot P$(female is H) $+ P$(male is H) $\cdot P$(female is D)
$$= \frac{1}{2} \cdot \frac{1}{2} + \frac{1}{2} \cdot \frac{1}{2} = \frac{1}{4} + \frac{1}{4} = \frac{1}{2}.$$

In this manner we obtain the second row of the transition matrix.

$$P = \bordermatrix{
 & DD & DH & DR & HH & HR & RR \cr
DD & 1 & 0 & 0 & 0 & 0 & 0 \cr
DH & \tfrac{1}{4} & \tfrac{1}{2} & 0 & \tfrac{1}{4} & 0 & 0 \cr
DR & 0 & 0 & 0 & 1 & 0 & 0 \cr
HH & \tfrac{1}{16} & \tfrac{1}{4} & \tfrac{1}{8} & \tfrac{1}{4} & \tfrac{1}{4} & \tfrac{1}{16} \cr
HR & 0 & 0 & 0 & \tfrac{1}{4} & \tfrac{1}{2} & \tfrac{1}{4} \cr
RR & 0 & 0 & 0 & 0 & 0 & 1 \cr}$$

(State of parents on left; State of offspring on top.)

It is not difficult to see that this is an absorbing Markov chain in which the absorbing states are DD and RR. Hence in the long run either all the offspring will be pure dominant or else all the offspring will be pure recessive. ∎

Practice Problem 7 In Example 8.9

(a) Determine the canonical form of the transition matrix.
(b) Determine the fundamental matrix.
(c) How many generations, on the average, will it take for absorption to occur if the original parents are in state DR?
(d) If the original parents are in state DH, what is the probability that the offspring will eventually become pure dominant?

OTHER APPLICATIONS

In addition to Markov chains, we can apply Bayes's formula, Bernoulli trials, and other stochastic models to problems in genetics.

EXAMPLE 8.10 Ellen's brother has fibrocystic disease of the pancreas, a recessive condition determined by a single pair of genes. Neither Ellen nor her parents have the disease, however. What is the probability that Ellen carries the recessive allele for this disease?

Solution Since Ellen's brother has the disease, his genotype must be R = gg. Thus each of his parents must carry the recessive allele g. Because they do not have the disease, they must both be of genotype H = Gg.

From Table 8.3 we see that the probabilities that the offspring of two hybrids have each possible genotype are

$$P(D) = \frac{1}{4}, \quad P(H) = \frac{1}{2}, \quad \text{and} \quad P(R) = \frac{1}{4}.$$

Since Ellen does not have the disease, she cannot be of genotype R. Thus the probability that Ellen carries the recessive allele is the conditional probability that her genotype is H, given that it is not R. From the definition of conditional probability (see Section 6.4) we see that

$$P(H|\overline{R}) = \frac{P(H \cap \overline{R})}{P(\overline{R})}$$
$$= \frac{P(H)}{P(\overline{R})}$$
$$= \frac{1/2}{1 - 1/4}$$
$$= 2/3.$$

Thus the probability that Ellen carries the recessive allele is ⅔. ∎

THE HARDY-WEINBERG PRINCIPLE

What happens to a recessive trait after many generations of random mating in a large population? It might seem from the Mendelian theory that such a trait should gradually disappear. Yet many recessive characteristics exist in both plant and animal populations, and these traits seem to persist in fairly constant proportions from one generation to the next. To many biologists in the early 1900s, this fact cast serious doubt on Mendel's theory of inheritance. In 1908, however, the English mathematician G. H. Hardy and the German physician W. Weinberg independently found a simple argument to show that Mendel's theory does not imply that recessive traits will disappear from the population. (See reference [4].)

Consider a trait determined by a single pair of genes with two alleles, G (dominant) and g (recessive). Suppose that the proportions of the genotypes in some population are known to be

p_0 = proportion of pure dominants (GG),

q_0 = proportion of hybrids (Gg), and

r_0 = proportion of pure recessives (gg).

We will assume that these proportions are the same for both males and females, and that genotype has no influence on the choice of a mate or the average number of offspring produced. Under these conditions we can calculate the probability u_0 that a random offspring of parents from this population will receive the G allele from a given parent:

$u_0 = P(\text{offspring receives G})$

$\quad = P(\text{parent is GG and passes G, or parent is Gg and passes G})$

$\quad = P(\text{parent is GG and passes G}) + P(\text{parent is Gg and passes G})$

$\quad = P(\text{parent is GG}) \cdot P(\text{parent passes G}|\text{parent is GG})$
$\quad\quad\quad + P(\text{parent is Gg}) \cdot P(\text{parent passes G}|\text{parent is Gg})$

$\quad = p_0(1) + q_0\left(\dfrac{1}{2}\right)$

$\quad = p_0 + \dfrac{1}{2}q_0.$

Similarly, we find that

$$v_0 = P(\text{offspring receives g})$$
$$= q_0\left(\frac{1}{2}\right) + r_0(1)$$
$$= \frac{1}{2}q_0 + r_0.$$

(Note that $u_0 + v_0 = p_0 + q_0 + r_0 = 1$.) To be pure dominant (genotype GG), an offspring must receive the G allele from *both* parents independently. Hence

$$P(\text{offspring is GG}) = u_0 \cdot u_0 = u_0^2.$$

Likewise

$$P(\text{offspring is gg}) = v_0 \cdot v_0 = v_0^2.$$

To be hybrid (Gg), an offspring must receive a G allele from the father and a g allele from the mother or a G allele from the mother and a g allele from the father. Since these two cases are mutually exclusive, we have

$$P(\text{offspring is Gg}) = u_0 \cdot v_0 + u_0 \cdot v_0 = 2u_0v_0.$$

In a large population these probabilities should approximate the actual genotype proportions in the first generation offspring. Thus if

p_k = proportion of pure dominants (GG) in the kth generation offspring,

q_k = proportion of hybrids (Gg) in the kth generation offspring, and

r_k = proportion of pure recessives (gg) in the kth generation offspring,

then

$$p_1 = u_0^2, \qquad q_1 = 2u_0v_0, \qquad \text{and} \qquad r_1 = v_0^2.$$

Similar statements can be made about each successive generation. Hence in the second generation offspring we expect

$$p_2 = u_1^2, \qquad q_2 = 2u_1v_1, \qquad \text{and} \qquad r_2 = v_1^2,$$

where

$$u_1 = p_1 + \frac{1}{2}q_1 \qquad \text{and} \qquad v_1 = \frac{1}{2}q_1 + r_1.$$

It follows that

$$u_1 = p_1 + \frac{1}{2}q_1 = u_0^2 + \frac{1}{2}(2u_0v_0) = u_0(u_0 + v_0) = u_0(1) = u_0,$$

and similarly $v_1 = v_0$. Thus we find that

$$p_2 = u_1^2 = u_0^2 = p_1,$$
$$q_2 = 2u_1v_1 = 2u_0v_0 = q_1, \text{ and}$$
$$r_2 = v_1^2 = v_0^2 = r_1.$$

Thus *the genotype proportions in the second generation are the same as those in the first generation.*

The Hardy-Weinberg Law

> In a large population with random mating, the theoretical distribution of genotypes will be stable after one generation.

EXAMPLE 8.11 If the initial population is 30% pure dominant (GG), 60% hybrid (Gg), and 10% pure recessive (gg), predict the theoretical distribution of genotypes in succeeding generations.

Solution Using the notation above, we have

$$p_0 = .30, \quad q_0 = .60, \quad \text{and} \quad r_0 = .10.$$

Therefore

$$u_0 = p_0 + \frac{1}{2}q_0 = .30 + \frac{1}{2}(.60) = .60,$$

and

$$v_0 = \frac{1}{2}q_0 + r_0 = \frac{1}{2}(.60) + .10 = .40.$$

Hence

$$p_1 = u_0^2 = (.60)^2 = .36,$$
$$q_1 = 2u_0v_0 = 2(.60)(.40) = .48, \quad \text{and}$$
$$r_1 = v_0^2 = (.40)^2 = .16.$$

Thus in the first generation offspring, and all generations thereafter, the expected genotype distribution will be 36% pure dominant, 48% hybrid, and 16% pure recessive.

Practice Problem 8 If the initial population is 51% pure dominant (GG), 38% hybrid (Gg), and 11% pure recessive (gg), predict the theoretical distribution of genotypes in succeeding generations.

EXAMPLE 8.12 A recessive condition known as phenylketonuria affects one child in 40,000. What proportion of the population are carriers of this disease?

Solution Let G and g denote the dominant and recessive alleles for this disease, respectively. Assuming that the genotype distribution is stable, we must have

$$p_1 = u_0^2,$$
$$q_1 = 2u_0 v_0, \quad \text{and}$$
$$r_1 = v_0^2.$$

Now the proportion of pure recessives, r_1, is the frequency with which the disease appears in the population; so

$$r_1 = \frac{1}{40,000} = .000025.$$

Hence

$$v_0 = \sqrt{r_1} = .005,$$

and therefore

$$u_0 = 1 - v_0 = .995.$$

It follows that

$$q_1 = 2u_0 v_0 = 2(.995)(.005) = .00995,$$

and so $0.995\% \approx 1\%$ of the population carries the recessive allele. ∎

Practice Problem 9 Fibrocystic disease affects one child in 2000. What proportion of the population are carriers of this disease?

The Hardy-Weinberg law is only an approximation to reality because it assumes that mating is completely random and that the genotype proportions in each generation will equal the corresponding theoretical probabilities. The genotype distribution, however, is also affected by factors such as migration, evolutionary selection, differing birth rates among populations, and genetic mutation.

The Mendelian theory has been refined in a number of ways as a result of recent discoveries in molecular biology. For instance, many genes are now known to have more than two alleles, and in such cases the dominance relations among them can be rather complicated. In fact, even when there are only two alleles, it may not be true that one allele is dominant over the other. In snapdragons, for example, the three genotypes for blossom color can all be distinguished: genotype GG produces red blossoms, Gg produces pink blossoms, and gg produces white blossoms.

Furthermore, many traits in which we are interested are determined by more than a single pair of genes, and often the several genes affecting a trait are not inherited independently of one another. For a simple introduction to genetics, see reference [2].

EXERCISES 8.4

Exercises 1–4 refer to a trait determined by a single pair of genes, where one allele is dominant and the other is recessive.

1. Suppose that a hybrid mates with a pure dominant. If they produce two offspring, what is the probability that
   (a) neither is pure dominant?
   (b) exactly one is pure dominant?

2. Two hybrids produce a litter of four offspring. What is the probability that
   (a) none of the four is pure recessive?
   (b) exactly one of the four is pure recessive?

3. Sixty percent of the females in a certain population are pure dominant, 30% are hybrid, and 10% are pure recessive. If a pure recessive male mates with a female chosen at random and their first offspring has the dominant trait, what is the probability that the chosen female is pure dominant?

4. In Example 8.10 it was shown that Ellen has a 2/3 chance of being a carrier and a 1/3 chance of being a noncarrier. Suppose that Ellen marries a healthy man known to be a carrier. If their first offspring is healthy, what is the probability that Ellen is a noncarrier?

5. Consider a population of snapdragons in which 20% have red blossoms (genotype GG), 40% have pink blossoms (genotype Gg), and 40% have white blossoms (genotype gg). Suppose that a plant is chosen at random and crossed with a pink-blossomed plant to produce a single offspring, which is in turn crossed with a pink-blossomed plant to produce a single offspring, and so on.
   (a) Represent this process as a three-state Markov chain in which the state at time n is the genotype of the nth offspring.
   (b) What is the probability that the second offspring will have red blossoms? Pink blossoms? White blossoms?
   (c) What is the long-run probability of each genotype?

6. In Exercise 5, suppose that a snapdragon with pink blossoms is self-pollinated to produce a single offspring, which is in turn self-pollinated to produce a single offspring, and so forth.
   (a) Represent this process as a three-state Markov chain.
   (b) On the average, how many generations will it take before a pure strain (either red or white) is obtained?

7. Consider a large population in which 70% are pure dominant, 20% are hybrid, and 10% are pure recessive. Assuming that mating is random, what will the distribution of genotypes be among the first-generation offspring? Among the second-generation offspring?

8. Consider a large population in which 50% are pure dominant and 50% are hybrid. Assuming that mating is random, what will the distribution of genotypes be among the first-generation offspring? Among the second-generation offspring?

470 CHAPTER 8 MARKOV CHAINS

9. In parts of Africa, 4% of the population has sickle-cell anemia, a recessive blood disease determined by a single pair of genes. Assuming that the genotype distribution is stable and that mating is random, what proportion of the population are carriers?

10. Let p, q, and r be the proportion of GG, Gg, and gg genotypes, respectively, in a given population. Prove that if mating is random, then the next generation has the same genotype distribution if and only if $q^2 = 4pr$.

11. Suppose that a population contains the same proportion of pure dominants as pure recessives. Prove that if mating is random, then the genotype distribution of the next generation will be 25% pure dominants, 50% hybrids, and 25% pure recessives.

12. Consider a large population of rabbits in which mating is random. Suppose that 1% of the rabbits are albinos and that the genotype distribution is stable.
    (a) What is the probability that a random nonalbino rabbit carries the recessive allele?
    (b) If two random nonalbinos produce an offspring, what is the probability that it is an albino?

Exercises 13 and 14 require knowledge of Example 8.10.

13. In the family tree below, males and females are represented by squares and circles, respectively.

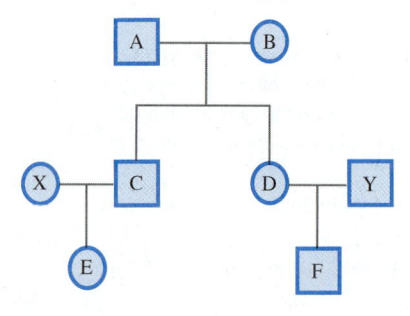

    Thus A and B have a son C and a daughter D who are married to X and Y, respectively. The children (E and F) of these marriages are first cousins. Suppose that C has a certain recessive trait that D does not have and that both X and Y are known to be noncarriers.
    (a) What is the probability that E is a carrier? That F is a carrier?
    (b) In a first-cousin marriage between E and F, what is the probability that their first offspring has the recessive trait?

14. Rework Exercise 13 under the following three assumptions:
    (i) neither A nor B has the recessive trait, but both are known to be carriers,
    (ii) neither C nor D has the recessive trait, and
    (iii) both X and Y are known to be noncarriers.

Answers to Practice Problems   **7.** (a)

$$\begin{array}{c} \\ \begin{array}{c}DD\\RR\\DH\\DR\\HH\\HR\end{array} \end{array} \begin{array}{c}\begin{array}{cccccc}DD & RR & DH & DR & HH & HR\end{array}\\ \left[\begin{array}{cc|cccc}1 & 0 & 0 & 0 & 0 & 0\\0 & 1 & 0 & 0 & 0 & 0\\ \hline \frac{1}{4} & 0 & \frac{1}{2} & 0 & \frac{1}{4} & 0\\0 & 0 & 0 & 0 & 1 & 0\\ \frac{1}{16} & \frac{1}{16} & \frac{1}{4} & \frac{1}{8} & \frac{1}{4} & \frac{1}{4}\\0 & \frac{1}{4} & 0 & 0 & \frac{1}{4} & \frac{1}{2}\end{array}\right]=P^\star.\end{array}$$

(b)
$$M = (I - Q)^{-1} = \begin{array}{c}\\DH\\DR\\HH\\HR\end{array} \begin{array}{c}\begin{array}{cccc}DH & DR & HH & HR\end{array}\\ \left[\begin{array}{cccc}\frac{8}{3} & \frac{1}{6} & \frac{4}{3} & \frac{2}{3}\\ \frac{4}{3} & \frac{4}{3} & \frac{8}{3} & \frac{4}{3}\\ \frac{4}{3} & \frac{1}{3} & \frac{8}{3} & \frac{4}{3}\\ \frac{2}{3} & \frac{1}{6} & \frac{4}{3} & \frac{8}{3}\end{array}\right]\end{array}$$

(c) $6\frac{2}{3}$

(d) $\frac{3}{4}$

8. The expected genotype distribution will be 49% pure dominant, 42% hybrid, and 9% pure recessive.

9. The disease is carried by about 4.4% of the population.

CHAPTER 8 REVIEW

IMPORTANT TERMS

- **Markov chain** *(8.1)*
- **state vector**
- **transition diagram**
- **transition matrix**
- **transition probability**
- **equilibrium vector** *(8.2)*
- **regular Markov chain**
- **absorbing Markov chain** *(8.3)*
- **absorbing state**
- **canonical transition matrix**
- **fundamental matrix**

REVIEW EXERCISES

In Exercises 1–4, find the transition matrix for the Markov chain with the given transition diagram, and determine whether the chain, is regular, absorbing, or neither.

1. 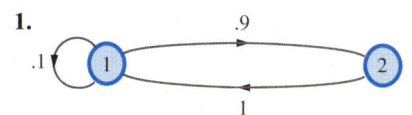
.1 ⟲ (1) →.9→ (2) →1→ (back to 1)

2.
.5 ⟲ (1) →.5→ (2) ⟲ 1

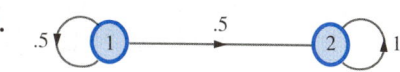

3.

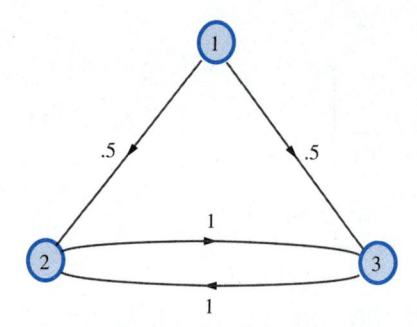

4.

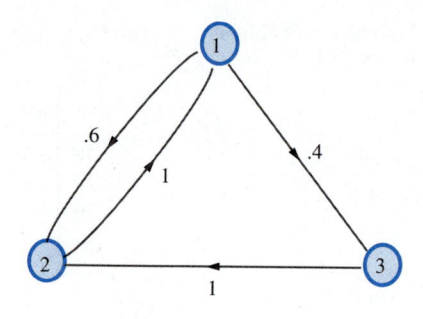

5. A two-state Markov chain has the following transition matrix.

$$\begin{array}{c} \\ 1 \\ 2 \end{array} \begin{array}{c} 1 \quad 2 \\ \begin{bmatrix} .7 & .3 \\ .4 & .6 \end{bmatrix} \end{array}$$

(a) Find the state vector after two steps (at time 2) if the system is initially in state 2.

(b) Find the equilibrium vector of this Markov chain.

6. A three-state Markov chain has the following transition matrix.

$$\begin{array}{c} \\ 1 \\ 2 \\ 3 \end{array} \begin{array}{c} 1 \quad 2 \quad 3 \\ \begin{bmatrix} 0 & 0 & 1 \\ .5 & .5 & 0 \\ .4 & .2 & .4 \end{bmatrix} \end{array}$$

(a) Find the state vector after two steps (at time 2) if the system is initially in state 1.

(b) Show that this Markov chain is regular, and find its equilibrium vector.

7. A three-state Markov chain has the following transition matrix.

$$\begin{array}{c} \\ 1 \\ 2 \\ 3 \end{array} \begin{array}{c} 1 \quad 2 \quad 3 \\ \begin{bmatrix} .8 & .2 & 0 \\ .3 & .5 & .2 \\ .3 & .3 & .4 \end{bmatrix} \end{array}$$

(a) Find the state vector after three steps (at time 3) if the system is initially in state 2.

(b) Show that this Markov chain is regular, and find its equilibrium vector.

8. A four-state Markov chain has the following transition matrix.

$$\begin{array}{c} \\ 1 \\ 2 \\ 3 \\ 4 \end{array} \begin{array}{c} 1 \quad 2 \quad 3 \quad 4 \\ \begin{bmatrix} .2 & .4 & 0 & .4 \\ 0 & 1 & 0 & 0 \\ 0 & 0 & 1 & 0 \\ .4 & .2 & .1 & .3 \end{bmatrix} \end{array}$$

(a) On the average, how long will it take to reach absorption if the system starts in state 1? In state 4?

(b) What is the probability that the system will end up in state 2 if it starts in state 1? If it starts in state 4?

9. Suppose that 52% of the children of doctors become doctors and that 2% of the children of nondoctors become doctors. Assuming that the birthrate is the same for doctors and nondoctors, what proportion of the population will be doctors after many generations?

10. A customer finds that whenever she calls the telephone company with a service complaint, she is put on hold. If she is on hold at any given time, there is an 80% chance that she will still be on hold one minute later, and a 20% chance that she will be speaking to a service representative. If she is talking to a service representative, there is a 50% chance that she will still be speaking to the representative one minute later, a 30% chance that she will be back on hold, and a 20% chance that she will be off the phone. How long can this customer expect an average call to the phone company to last?

11. A barber can cut one person's hair every 15 minutes. He works alone in his shop and never takes a break when customers are waiting. In order to prevent his customers from having to wait for long periods, he does not allow more than two customers to be waiting for a haircut. With this policy he finds that no customer ever leaves his shop before getting a haircut. Suppose that the number of customers arriving at his shop during any 15-minute period has the distribution below.

Number of arrivals	0	1	2	3	4 or more
Probability	.2	.5	.2	.1	0

Assume further that the first customer who arrives during each period arrives at the beginning of the period.

(a) Use the information above to form the transition matrix of a Markov chain in which the states represent the number of customers in the barber shop.

(b) Suppose that every morning when the barber arrives at his shop he is equally likely to find no customers or one customer waiting. What distribution of customers can the barber expect 30 minutes after the shop opens?

(c) Approximately what distribution of customers can the barber expect after many hours of work?

12. Albinism is a recessive trait determined by a single pair of genes. Consider a mating between two nonalbinos, both of whom carry the recessive allele (i.e., both are hybrids). If this mating produces three offspring, what is the probability that

(a) none will be albinos (pure recessives)?

(b) all three will be noncarriers (pure dominant)?

13. Suppose that 9% of the individuals in a large population have a certain recessive trait. Assuming the genotype distribution is stable, what proportion of the population are carriers (i.e., hybrids)?

REFERENCES

1. Bourne, Larry S., "Physical Adjustment Process and Land Use Succession: A Conceptual Review and Central City Example," *Economic Geography,* vol. 47 (1971), pp. 1–15.

2. Emery, Alan E. H., *Heredity, Disease, and Man.* Berkeley, CA: University of California Press, 1968.

3. Hampton, P., "Regional Economic Development in New Zealand," *Journal of Regional Science,* vol. 8 (April 1962), pp. 41–51.

4. Hardy, G. H., "Mendelian Proportions in a Mixed Population," *Science,* new series, vol. 28 (1908), pp. 49–50.

5. Mosimann, J., *Elementary Probability for the Biological Sciences.* Englewood Cliffs, NJ: Prentice–Hall, Inc., 1968.

6. Prais, S. J., "Measuring Social Mobility," *Journal of the Royal Statistical Society,* vol. 118 (1955), pp. 56–66.

GAME THEORY

9.1 Strictly Determined Games
9.2 Games of Strategy
9.3 Simplex Solution of $m \times n$ Games
Chapter Review

Many human activities, such as athletics, politics, and business, involve competition between individuals or groups with opposing interests. In a series of papers written in the 1920s and 1930s, John von Neumann* laid the groundwork for the mathematical theory of conflict situations, which is called **game theory**. In this chapter we will use probability, matrix algebra, and linear programming to find the optimal behavior for the competitors in a certain type of conflict situation. Section 7.3 should be reviewed before reading Section 9.2, and Sections 3.4 and 3.5 should be reviewed before reading Section 9.3.

9.1 Strictly Determined Games

For our purposes a game can be defined as a competition involving two or more parties, called **players**, each of whom has available a number of possible courses of action, called **moves**. In general, the **payoff** to each player depends not only on a player's own move, but also on the moves of others. We will assume that each player must make a move without knowing what the opponents will do and also that the goal of each player is to maximize his or her own payoff.

MATRIX GAMES

The simplest class of games, which are called **matrix games** (or **two-person zero-sum games**), are those in which:
  (i)   there are only two players, and
  (ii) for each choice of moves, the respective payoffs sum to zero.

Condition (ii) means that whatever one player wins, the other loses; thus the outcome of each play will be a payment of a certain amount from one player to the other. We will denote the two players by R (the "row player") and C (the "column player"). Then a matrix game can be completely described by a **payoff matrix**, the entries of which represent the row-player's gain (or equivalently, the column-player's loss).

EXAMPLE 9.1 Two players, R and C, simultaneously show one or two fingers each. If the sum of the fingers shown is even, then R wins that number of dollars from C. On the other hand, if the sum of the fingers shown is odd, then C wins that number of dollars from R.

---

*Von Neumann was mentioned in Section 3.5 in connection with the duality theorem. Together with economist Oskar Morgenstern he also wrote a classic book on game theory entitled *Theory of Games and Economic Behavior.*

Discussion This game is called *two-finger morra*. Here each player has two possible moves, namely, to show one finger or to show two fingers. The payoff matrix for this game is as follows.

$$\text{Player R} \begin{array}{c} \\ 1 \\ 2 \end{array} \begin{array}{c} \text{Player C} \\ \begin{array}{cc} 1 & 2 \end{array} \\ \begin{bmatrix} 2 & -3 \\ -3 & 4 \end{bmatrix} \end{array}$$

The entries of this matrix represent the payoffs from the *row*-player's point of view. Therefore positive entries represent a gain for player R and negative entries represent a loss. Thus if both players show one finger, then R's payoff is 2 (that is, R wins \$2 from C); whereas if R shows one finger and C shows two, then R's payoff is -3 (that is, R loses \$3 to C). Notice that the two players have directly opposing interests. The row-player wants to *maximize* the payoff entry, but the column-player wants to *minimize* it. We will see in Section 9.2 that this game is favorable to the column-player; that is, player C has an advantage if this game is played repeatedly. ■

In general, any $m \times n$ matrix $A = [a_{ij}]$ defines a matrix game by letting the rows of A represent player R's possible moves and the columns represent player C's possible moves. Then the entries a_{ij} denote the payoffs from player C to player R.

EXAMPLE 9.2 Discuss the possible moves in the 3×4 matrix game below.

$$\text{Player R} \begin{array}{c} \\ 1 \\ 2 \\ 3 \end{array} \begin{array}{c} \text{Player C} \\ \begin{array}{cccc} 1 & 2 & 3 & 4 \end{array} \\ \begin{bmatrix} 0 & -4 & 2 & -1 \\ -2 & 5 & 6 & 0 \\ 3 & -1 & 1 & 4 \end{bmatrix} \end{array}$$

Discussion Player R has three possible moves (the rows of the matrix above), and player C has four possible moves (the columns). As before, the entries represent payoffs from player C to player R. Thus, if R chooses row 2 and C chooses column 3, then R will win 6 units from C; but if C chooses column 1 instead, then C will win 2 units from R. ■

Practice Problem 1 Consider the matrix game with the payoff matrix below.

$$\begin{array}{c} \\ 1 \\ 2 \\ 3 \\ 4 \end{array} \begin{array}{c} \begin{array}{ccc} 1 & 2 & 3 \end{array} \\ \begin{bmatrix} 5 & -2 & 3 \\ 0 & 4 & 1 \\ 3 & 1 & -2 \\ -1 & 2 & 2 \end{bmatrix} \end{array}$$

(a) How many moves do players R and C have in this game?

(b) What is the payoff to Player R if R chooses row 2 and C chooses column 2?

(c) What is the payoff to Player R if R chooses row 1 and C chooses column 2?

REDUCED FORM

If we look at the game in Example 9.2 from player R's viewpoint, there appears to be no clear superiority of any move over another. Comparing the first and second rows, for instance, we see that move 1 is better than move 2 if player C chooses column 1, but worse if C chooses column 2, 3, or 4. On the other hand, by comparing player C's four options, we see that *move 2 is definitely preferable to move 3*. The reason is that each entry of column 2 is less than or equal to the corresponding entry in column 3; so player C will always lose a smaller amount by playing column 2 than by playing column 3, *no matter what move the row-player makes*. A rational player C, therefore, should never choose column 3. Removing column 3 from the payoff matrix gives us the reduced game shown below.

$$\begin{array}{c} & 1 & 2 & 4 \\ \begin{matrix}1\\2\\3\end{matrix} & \begin{bmatrix} 0 & -4 & \vdots & -1 \\ -2 & 5 & \vdots & 0 \\ 3 & -1 & \vdots & 4 \end{bmatrix} \end{array}$$

Of course, player R will also realize that a rational opponent will not play column 3. Comparing the options in the reduced game above, R sees that move 3 is preferable to move 1 because each entry in row 3 is now greater than or equal to the corresponding entry in row 1. That is, the payoff for R by playing row 3 will never be worse than the payoff obtained by using row 1, no matter what player C does. We can therefore eliminate row 1 from the payoff matrix, reducing the game still further.

$$\begin{array}{c} & 1 & 2 & 4 \\ \begin{matrix}2\\3\end{matrix} & \begin{bmatrix} \cdots & \cdots & \cdots & \cdots \\ -2 & 5 & \vdots & 0 \\ 3 & -1 & \vdots & 4 \end{bmatrix} \end{array}$$

Comparing columns again in this reduced matrix, player C now sees that each entry in column 1 is less than or equal to the corresponding entry in column 4. Hence move 1 is preferable to move 4, so that column 4 can be deleted. We therefore arrive at the 2 × 2 matrix game below.

$$\begin{array}{c} & 1 & 2 \\ \begin{matrix}2\\3\end{matrix} & \begin{bmatrix} -2 & 5 \\ 3 & -1 \end{bmatrix} \end{array}$$

The 2 × 2 matrix game above is called the **reduced form** of the original game. Note that row 2 is inferior to row 3 if player C chooses column 1, but that row 2 is superior to row 3 if C chooses column 2. Likewise column 1 is superior to column 2 if player R chooses row 2, but column 1 is inferior to column 2 if R plays row 3. Hence neither of the moves for player R or player C is clearly preferable to the other, and so the 2 × 2 game cannot be further simplified.

In general, we can find the reduced form of a matrix game by first comparing the possible moves for the row-player. For the *row-player,* move i will be preferable to move j whenever the entries of row i are all greater than or equal to those in row j. In this case row j can be deleted from the payoff matrix. For the *column-player,* move i will be preferable to move j whenever the entries of column i are all less than or equal to those in column j. In this case column j can be deleted from the payoff matrix. By successively eliminating the inferior moves of each player, we will eventually obtain the reduced form of the game.

EXAMPLE 9.3 Find the reduced form of the matrix game below.

$$\text{Player R} \begin{array}{c} \\ 1 \\ 2 \\ 3 \end{array} \begin{array}{c} \text{Player C} \\ \begin{array}{ccc} 1 & 2 & 3 \end{array} \\ \begin{bmatrix} 2 & 1 & 3 \\ 1 & -2 & 3 \\ 4 & 0 & -1 \end{bmatrix} \end{array}$$

Solution Comparing rows, we see that move 1 is preferable to move 2 for player R. We therefore delete row 2, producing the following matrix.

$$\begin{array}{c} \\ 1 \\ 3 \end{array} \begin{array}{c} \begin{array}{ccc} 1 & 2 & 3 \end{array} \\ \begin{bmatrix} 2 & 1 & 3 \\ \cdots & \cdots & \cdots \\ 4 & 0 & -1 \end{bmatrix} \end{array}$$

Comparing columns, we now see that move 2 is preferable to move 1 for player C, and so column 1 can be deleted.

$$\begin{array}{c} \\ 1 \\ 3 \end{array} \begin{bmatrix} \vdots & 1 & 3 \\ \vdots & \cdots & \cdots \\ \vdots & 0 & -1 \end{bmatrix}$$

Move 1 is now preferable to move 3 for the row player, so the matrix becomes

$$1 \begin{bmatrix} \vdots & 1 & 3 \\ \vdots & \cdots & \cdots \end{bmatrix}$$

9.1 STRICTLY DETERMINED GAMES

Player C will now prefer move 2 to move 3, and so the reduced form of the original game is the 1×1 matrix below.

$$1 \begin{bmatrix} 2 \\ 1 \end{bmatrix}$$

Thus, if both players are rational, the original game is completely determined. In order to maximize their own payoffs, *player R should always play row 1 and player C should always play column 2*. These choices will result in player R receiving a payoff of 1 unit per play. ∎

Practice Problem 2 Find the reduced form of the matrix game below.

$$\begin{array}{c} \\ 1 \\ 2 \\ 3 \\ 4 \end{array} \begin{bmatrix} 1 & 2 & 3 & 4 \\ -1 & 4 & 5 & 3 \\ 3 & -2 & 1 & -3 \\ -1 & 0 & 6 & 2 \\ 2 & 1 & 2 & -4 \end{bmatrix}$$

OPTIMAL MOVES

If the reduced form of a game is a 1×1 matrix as in Example 9.3, then the optimal moves of both players are completely determined. When the reduced matrix is not 1×1, however, we are still left with the problem of determining the best move for each player. Since the row-player has no way to anticipate an opponent's move, one course of action is to take the pessimistic view that *no matter what row is chosen, the column-player will make the best possible countermove*. The safest course is therefore to compare all the available moves by looking at the *smallest* payoff for each row. Likewise the column-player's safest course is to look at the *largest* payoff that might occur for each column.

The safest moves for each player (as described above) are called **optimal moves**. They can be determined as follows.

Optimal Moves ■ ■ ■ ■ ■

(a) To find the optimal move(s) for the row-player, find the minimum entry in each row of the payoff matrix, and choose the row or rows for which this value is largest.

(b) To find the optimal move(s) for the column-player, find the maximum entry in each column of the payoff matrix, and choose the column or columns for which this value is smallest.

EXAMPLE 9.4 Determine the optimal moves for each player in the matrix game below.

$$\begin{array}{c} & \begin{array}{ccc} 1 & 2 & 3 \end{array} \\ \begin{array}{c} 1 \\ 2 \\ 3 \end{array} & \left[\begin{array}{ccc} 3 & -1 & 0 \\ -2 & 2 & 1 \\ 2 & 0 & -3 \end{array} \right] \end{array}$$

Solution This is a 3 × 3 matrix game in reduced form. The minimum entries in rows 1, 2, and 3 are -1, -2, and -3, respectively. These values represent the worst payoffs that player R can receive for each of the three rows in the payoff matrix. Player R's optimal move is therefore row 1, which corresponds to the *largest* of the row minima.

$$\begin{array}{c} & \begin{array}{ccc} 1 & 2 & 3 \end{array} & \text{Row Minimum} \\ \begin{array}{c} 1 \\ 2 \\ 3 \end{array} & \left[\begin{array}{ccc} 3 & -1 & 0 \\ -2 & 2 & 1 \\ 2 & 0 & -3 \end{array} \right] & \begin{array}{l} -1 \leftarrow \textbf{largest of row minima} \\ -2 \\ -3 \end{array} \\ \text{Column Maximum} & \begin{array}{ccc} 3 & 2 & 1 \end{array} \\ & \uparrow \\ & \textbf{smallest of the column maxima} \end{array}$$

The maximum entries in columns 1, 2, and 3 are 3, 2, and 1, respectively. Thus the optimal move for player C is column 3, which minimizes the largest possible payoff to player R. ■

If two or more rows of the payoff matrix have the same minimum entry, it is possible that there may be more than one optimal move for the row player. Similarly, there can be more than one optimal move for the column player.

EXAMPLE 9.5 Determine the optimal moves for each player in the matrix game below.

$$\begin{array}{c} & \begin{array}{cccc} 1 & 2 & 3 & 4 \end{array} \\ \begin{array}{c} 1 \\ 2 \\ 3 \\ 4 \end{array} & \left[\begin{array}{cccc} 3 & 2 & 5 & 4 \\ -1 & 0 & 6 & 2 \\ 4 & 2 & 4 & 3 \\ 1 & 5 & 0 & 2 \end{array} \right] \end{array}$$

Solution The table below shows the row minima and column maxima for the payoff matrix.

$$\begin{array}{c} & \begin{array}{cccc} 1 & 2 & 3 & 4 \end{array} & \text{Row Minimum} \\ \begin{array}{c} 1 \\ 2 \\ 3 \\ 4 \end{array} & \left[\begin{array}{cccc} 3 & 2 & 5 & 4 \\ -1 & 0 & 6 & 2 \\ 4 & 2 & 4 & 3 \\ 1 & 5 & 0 & 2 \end{array} \right] & \begin{array}{l} 2 \leftarrow \textbf{largest of row minima} \\ -1 \\ 2 \leftarrow \textbf{largest of row minima} \\ 0 \end{array} \\ \text{Column Maximum} & \begin{array}{cccc} 4 & 5 & 6 & 4 \end{array} \\ & \uparrow \uparrow \\ & \textbf{smallest of the column maxima} \end{array}$$

Hence player R has two optimal moves, rows 1 and 3; and player C has two optimal moves, columns 1 and 4. ∎

Practice Problem 3   Determine the optimal moves for each player in the matrix game below.

$$\begin{array}{c} \\ 1 \\ 2 \\ 3 \\ 4 \end{array} \begin{array}{ccc} 1 & 2 & 3 \end{array} \\ \left[\begin{array}{ccc} 1 & -4 & 3 \\ -3 & 0 & 4 \\ -1 & 5 & -2 \\ 2 & -3 & 6 \end{array} \right]$$

In Example 9.4 either player could benefit from the knowledge that the opponent intended to use an optimal move. For example, if player R knows that player C will choose column 3, then player R can profit by switching to row 2. (This change will improve the payoff for player R from 0 to 1.) Likewise, if player C knows that player R will choose row 1, then player C can benefit by switching to column 2. (This change will improve the payoff for player C from 0 to 1.)

SADDLE POINTS

Sometimes, however, neither player can benefit by unilaterally switching from an optimal move. Consider, for example, the matrix game below.

$$\begin{array}{c} \\ 1 \\ 2 \\ 3 \end{array} \begin{array}{ccc} 1 & 2 & 3 \end{array} \\ \left[\begin{array}{ccc} 1 & 8 & -2 \\ 3 & 6 & 4 \\ 2 & -5 & 7 \end{array} \right]$$

The minimum entries in rows 1, 2, and 3 are -2, 3, and -5, respectively. Hence row 2 is the optimal move for player R; this move guarantees R a payoff of at least 3 no matter what move player C makes. From the perspective of player C, the maximum entries in columns 1, 2, and 3 are 3, 8, and 7, respectively. Thus the optimal move for player C is column 1, which guarantees that C will lose no more than 3 no matter what move R makes.

In this example neither player can benefit by unilaterally choosing a move other than the optimal move. For if a row other than row 2 is chosen, R's winnings will be less than 3 if C chooses the optimal move (column 1); and if C chooses a column other than column 1, C will lose more than 3 if R chooses the optimal move (row 2). In this game the reason that neither player can benefit by unilaterally changing from the optimal move is that the entry 3 (in row 2, column 1) is both the largest entry in its column and the smallest entry in its row.

An entry *s* of a payoff matrix is called a **saddle point** if it is both the largest entry in its column and the smallest entry in its row; in this case the entry *s* is called the **value** of the game. If the row-player uses an optimal move, then the value of the game is the smallest possible amount that the row player can win each time the game is played. Similarly, if the column-player uses an optimal move, then the value of the game is the largest possible amount that the column player can lose each time the game is played.

Thus the preceding matrix game has a saddle point, and the value of this game is 3. If a matrix game has a saddle point, it is easy to see that the largest of the row minima and the smallest of the column maxima will be the value of the saddle point. Hence the row and column containing the saddle point will be the optimal moves for each player.

Practice Problem 4 (a) Determine the optimal moves for each player in the matrix game below.
(b) Does the matrix game below have a saddle point? If so, what is the value of this game?

$$\begin{array}{c} \\ 1 \\ 2 \\ 3 \end{array} \begin{array}{cccc} 1 & 2 & 3 & 4 \\ \left[\begin{array}{cccc} 0 & 1 & 3 & -2 \\ 1 & -7 & 2 & -3 \\ -5 & 8 & -6 & -4 \end{array}\right] \end{array}$$

It is possible that several entries in a matrix game may be saddle points. In such a case, any pair of optimal moves will determine a saddle point. Moreover, the entries at all such points must have the same value, and so it does not matter which optimal move each player makes.

EXAMPLE 9.6 (a) Determine the optimal moves for each player in the matrix game below.
(b) Does this game have any saddle points? If so, identify them and the value of this game.

$$\begin{array}{c} \\ 1 \\ 2 \\ 3 \\ 4 \end{array} \begin{array}{ccc} 1 & 2 & 3 \\ \left[\begin{array}{ccc} 3 & 2 & 5 \\ -1 & 0 & 6 \\ 4 & 2 & 4 \\ 5 & 1 & 0 \end{array}\right] \end{array}$$

Solution (a) The optimal moves for player R are rows 1 and 3, and the optimal move for player C is column 2.

$$\begin{array}{c} \\ 1 \\ 2 \\ 3 \\ 4 \end{array} \begin{array}{ccc} 1 & 2 & 3 \\ \left[\begin{array}{ccc} 3 & 2 & 5 \\ -1 & 0 & 6 \\ 4 & 2 & 4 \\ 5 & 1 & 0 \end{array}\right] \end{array} \begin{array}{l} \text{Row Minimum} \\ 2 \leftarrow \textbf{optimal move} \\ -1 \\ 2 \leftarrow \textbf{optimal move} \\ 0 \end{array}$$

Column Maximum   5   2   6
                      ↑
                  optimal move

(b) This game has two saddle points, the entries in row 1, column 2 and in row 3, column 2. Note that each of these entries is both the largest in its column and the smallest in its row. The value of this game is 2, the common value of the two saddle points. ∎

APPLICATIONS

In conflict situations between rational opponents, it is often difficult to specify numerical "payoffs" in a meaningful way. When this can be done, however, the concepts of game theory can be used to analyze the available options.

EXAMPLE 9.7 Two competing companies, R and C, are each planning to locate a store in one of the three towns shown in Figure 9.1. The population distribution for the three towns is indicated in this figure. If the companies locate in different towns, the store that is closer to a given town will get all of that town's business. However, if the companies both locate in the same town, company R, which is better known, will get 60% of the total business of the three towns and company C will get the remaining 40%. Where should each company locate?

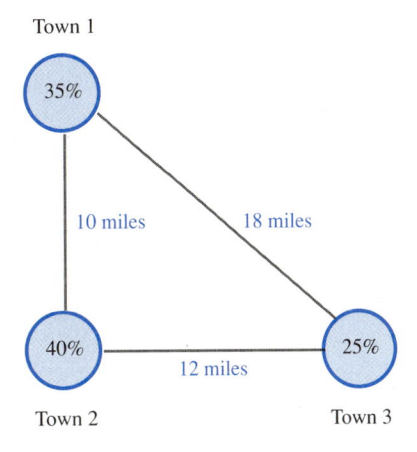

FIGURE 9.1

Town 1: 35%
Town 2: 40%
Town 3: 25%
Town 1 to Town 2: 10 miles
Town 1 to Town 3: 18 miles
Town 2 to Town 3: 12 miles

Solution Since each company can locate in town 1, 2, or 3, players R and C each have three possible moves. Let the payoff to company R be the percentage of the total population whose business R can expect to get. Then the payoff matrix for this situation is as shown below.

$$\begin{array}{c} \text{Company C} \\ \text{Company R} \begin{array}{c} 1 \\ 2 \\ 3 \end{array} \begin{bmatrix} 60 & 35 & 75 \\ 65 & 60 & 75 \\ 25 & 25 & 60 \end{bmatrix} \end{array}$$

For instance, if both companies locate in town 1, then R will receive 60% of the total business of the three towns; but if R locates in town 1 and C in town 3, then R will get the combined business of towns 1 and 2 (75%).*

The row minima and column maxima are calculated below.

$$\begin{array}{c} & & \text{Company C} \\ & & 1 \quad 2 \quad 3 \quad \text{Row Minimum} \\ & 1 & \begin{bmatrix} 60 & 35 & 75 \\ 65 & 60 & 75 \\ 25 & 25 & 60 \end{bmatrix} \begin{array}{l} 35 \\ 60 \leftarrow \textbf{optimal move} \\ 25 \end{array} \\ \text{Company R} & 2 & \\ & 3 & \\ \text{Column Maximum} & & 65 \quad 60 \quad 75 \\ & & \uparrow \\ & & \textbf{optimal move} \end{array}$$

Thus the optimal move for each company is to locate in town 2. Since the entry of the payoff matrix in row 2, column 2 is the largest entry in its column and the smallest in its row, this game has a saddle point. Hence neither company can gain an advantage by unilaterally locating in towns 1 or 3. ∎

A matrix game having a saddle point is called a *strictly determined* game because neither player can benefit by selecting any move other than the optimal move. Note that the game in Example 9.7 is strictly determined because its row 2, column 2 entry is a saddle point. This matrix could also have been reduced to a 1×1 matrix as in Example 9.3. When the reduced form of a matrix game is a 1×1 matrix, the single entry in the reduced form must be a saddle point of the original game. Thus the row 1, column 2 entry of the matrix in Example 9.3 is a saddle point. In Section 9.2 we will consider games that do not have saddle points. We will see there that, unlike strictly determined games, the optimal behavior in such games involves using more than one row or column of the payoff matrix.

One last remark should be made: In Section 7.4 we discussed some decision problems that resemble matrix games. There, however, the opponent was nature rather than a conscious opponent. Recall that when playing a matrix game, we always assume that *each player will choose whichever available move is best from that player's own point of view*. Therefore, unless we are willing to assume that nature will always act in such a way as to produce the worst possible result for a decision-maker, the game-theoretic approach taken in this chapter should not be applied to the decision-making problems in Section 7.4.

---

*Strictly speaking, this is not a zero-sum game but a *constant-sum* game because the payoffs to the two players always add up to 100%. Since one player's gain is the other player's loss, however, the theory of zero-sum games still applies. In fact, this game can be made into a zero-sum game by taking the payoffs to be the amount of business above or below 50% earned by company R.

EXERCISES 9.1

In Exercises 1–8 find the reduced form of the given matrix game.

1. $\begin{bmatrix} 4 & 0 & 3 \\ 3 & 5 & 1 \\ -2 & 4 & 0 \end{bmatrix}$

2. $\begin{bmatrix} 2 & 7 & 3 \\ -1 & 2 & 6 \\ 3 & 5 & 4 \end{bmatrix}$

3. $\begin{bmatrix} -2 & 3 & 1 \\ 4 & 0 & -2 \\ 2 & -5 & -2 \end{bmatrix}$

4. $\begin{bmatrix} 4 & 0 & 3 & -1 \\ 1 & 2 & 2 & 3 \\ -1 & 0 & 4 & 1 \end{bmatrix}$

5. $\begin{bmatrix} 4 & 2 & 3 & 2 \\ 3 & 0 & 2 & -1 \\ 1 & 5 & 6 & 1 \end{bmatrix}$

6. $\begin{bmatrix} 1 & 3 & -2 & 5 \\ 2 & -2 & 4 & -3 \\ 0 & 6 & -3 & 4 \end{bmatrix}$

7. $\begin{bmatrix} 0 & -1 & -3 & 4 \\ -3 & -2 & 2 & -3 \\ -4 & -1 & 1 & -3 \\ 1 & 3 & -2 & 2 \end{bmatrix}$

8. $\begin{bmatrix} 4 & 2 & 0 & -1 \\ 5 & -1 & 4 & 3 \\ 2 & 2 & 2 & 4 \\ -3 & -1 & 2 & -2 \end{bmatrix}$

In Exercises 9–16 determine the optimal moves for each player. Then determine if the game has a saddle point; if so, find the value of the game.

9. $\begin{bmatrix} 2 & 1 \\ -1 & 0 \end{bmatrix}$

10. $\begin{bmatrix} 1 & -2 \\ -3 & 4 \end{bmatrix}$

11. $\begin{bmatrix} 4 & -1 \\ -2 & 3 \\ 1 & 0 \end{bmatrix}$

12. $\begin{bmatrix} -1 & -3 & 3 & 0 \\ 0 & 2 & 1 & 0 \\ -1 & 4 & -3 & -2 \end{bmatrix}$

13. $\begin{bmatrix} -1 & 0 & 4 \\ 2 & 1 & 3 \\ 5 & -1 & -2 \end{bmatrix}$

14. $\begin{bmatrix} 1 & -1 & 2 \\ 0 & 3 & -2 \\ -1 & 2 & 0 \end{bmatrix}$

15. $\begin{bmatrix} -1 & 2 \\ 0 & 1 \\ 1 & 0 \\ 2 & -1 \end{bmatrix}$

16. $\begin{bmatrix} 2 & 5 & 4 & 2 \\ 1 & -1 & 5 & 0 \\ -1 & 7 & -2 & 1 \\ 2 & 6 & 3 & 2 \end{bmatrix}$

In Exercises 17–22, formulate each problem as a matrix game. Then find all the optimal moves and the value of the game.

17. Player R has a $1 bill, a $10 bill, and a $50 bill, and player C has a $5 bill and a $20 bill. At a given signal each player must put one bill on a table. The player who puts down the greater bill wins the amount put down by the other. What should each player do?

18. Player R holds a black 2, a black 4, and a red 6, and player C holds a red 3, a red 5, and a black 7. At a given signal the players must show one of their cards. If they are the same color, then the one of higher rank wins; otherwise the one of lower rank wins. In either case the winner collects from the loser the number of dollars equal to the rank of the loser's card. Which card should each player show?

19. Two competing banks, R and C, are each planning to open a branch in one of two nearby towns. Town 1 contains 60% of their total population, and town 2 contains the remaining 40%. If the banks locate in different towns, the one in each town will get all of that town's business. If they locate in the same town, bank R will get 30% of that town's business (because bank C is better known) but 50% of the other town's business (because it provides a bank-by-mail service). Where should each bank locate?

20. Country R must decide between two costly defense systems against a possible attack by country C. System 1 has a 40% chance of success against land-launched missiles, 50% against sea-launched missiles, and 90% against airborne missiles. For system 2 the corresponding figures are 80%, 60%, and 70%. Although country C wants to minimize the chance of a successful defense by country R, it can afford to develop only one of the three possible delivery systems (land, sea, or air). Which choice is optimal for each side?

21. Candidates R and C are running against each other for a senate seat. The state's 1 million voters are bitterly divided on the issue of nuclear power, with 400,000 in favor and 600,000 opposed. Among those who favor nuclear power, 120,000 prefer R and 280,000 prefer C; whereas among those who oppose nuclear power, 330,000 prefer R and 270,000 prefer C. Each candidate must take a position on the issue at an upcoming debate. If they take different positions, each candidate will receive all the votes of one faction, but if they take the same positions, then the voters who disagree with this position will boycott the election completely. If the goal of each candidate is to obtain the largest possible percentage of those who will vote, what position should each candidate take?

22. The labor union R and management C have reached an impasse. The union demands a 10¢ per hour wage increase, but management claims it cannot afford any increase at all. As a compromise, the federal arbitrator favors an increase of 5¢ per hour. Therefore she has asked the union to submit a final demand of x cents (where $x = 5, 6, 7, 8, 9,$ or 10) and has asked management to submit a final demand of y cents (where $y = 0, 1, 2, 3, 4,$ or 5). She has informed each side that whichever figure is closest to 5 will prevail, and that if both figures are equally close to 5, the increase will be 5¢ per hour. Under these circumstances, what is the best strategy for each side?

23. Prove that in a matrix game every row minimum is less than or equal to every column maximum.

24. Prove that a matrix game has a saddle point if and only if the largest row minimum is equal to the smallest column maximum.

25. Prove that if a matrix game has a saddle point and if row p and column q are optimal moves for players R and C, respectively, then the row p, column q entry of the payoff matrix is a saddle point.

26. Prove that at any two saddle points of a matrix game, the entries are equal.

Answers to Practice Problems **1.** (a) Player R has 4 moves, and player C has 3 moves.
(b) 4 units
(c) −2 units (Thus C wins 2 units from R.)

The union demands a 10¢ per hour wage increase, but management is offering no increase at all.

2. The reduced form of the game is shown below.

$$\begin{array}{c} & 1 & & & 4 \\ 1 & \begin{bmatrix} -1 & \vdots & \vdots & 3 \\ 3 & \vdots & \vdots & -3 \\ \cdots & \vdots & \vdots & \cdots \\ \cdots & \vdots & \vdots & \cdots \end{bmatrix} \\ 2 & \end{array}$$

3. The optimal moves are row 3 for Player R and column 1 for player C.

4. (a) The optimal moves for players R and C are row 1 and column 4.
   (b) The row 1, column 4 entry is a saddle point, and so the value of this game is -2.

9.2 Games of Strategy

In a game having a saddle point, we have seen that a rational player should always make the optimal move. In other games, the element of surprise plays a much greater role.

STRATEGIES

The players in a matrix game always have optimal moves, but when the game lacks a saddle point they may be tempted to make other moves in order to gain an advantage. Consider, for example, the following matrix game.

$$\begin{array}{cc} & \text{Player C} \\ & \begin{array}{cc} 1 & 2 \end{array} \\ \text{Player R} \begin{array}{c} 1 \\ 2 \end{array} & \begin{bmatrix} 1 & -2 \\ -3 & 4 \end{bmatrix} \end{array}$$

The reader should verify that the optimal moves are row 1 for player R and column 1 for player C. Because none of the four entries of this game is a saddle point, however, the players will be tempted to use moves other than their optimal moves.

Suppose, for example, that player R repeatedly chooses his optimal move, row 1. Player C will soon notice this and switch to column 2 in order to give player R a 2-unit loss. If player C does this consistently, however, player R will switch to row 2 to gain a payoff of 4 units. If player R now continues playing row 2, then player C will respond by choosing column 1, whereupon player R will return to row 1. Thus both players will eventually return to the use of their optimal moves!

In the scenario described above, the important observation is that whenever one player's behavior is *predictable,* his opponent can take advantage of it. This suggests that the best strategy for each player might be to choose moves in some *random* manner, so that an opponent cannot anticipate what the player will do next. Player R, for instance, might adopt the strategy

$$P: \begin{cases} \text{Choose row 1 with probability } .5 \\ \text{Choose row 2 with probability } .5, \end{cases}$$

which we will represent by the row vector

$$P = [.5 \quad .5].$$

This means that player R wants to choose moves at random, playing each row half of the time. Player C might adopt the strategy

$$Q: \begin{cases} \text{Choose column 1 with probability } \frac{2}{3} \\ \text{Choose column 2 with probability } \frac{1}{3}, \end{cases}$$

which we will represent by the column vector

$$Q = \begin{bmatrix} \frac{2}{3} \\ \frac{1}{3} \end{bmatrix}.$$

This column vector means that player C wants to choose moves at random, playing the first column two-thirds of the time and the second column one-third of the time. There are various randomizing devices each player can use to implement these strategies. Player R, for instance, may flip a fair coin, playing row 1 if heads appears and row 2 if tails appears. Player C, on the other hand, may roll a fair die, playing column 1 if 1, 2, 3, or 4 is rolled and playing column 2 if 5 or 6 is rolled. By selecting their moves on the basis of some random device, neither player can predict the other's next move on the basis of earlier moves.

We can generalize the notion of a strategy to apply to any matrix game. In an $m \times n$ matrix game, a **strategy for player R** is a row vector

$$P = [p_1 \quad p_2 \quad \cdots \quad p_m],$$

and a **strategy for player C** is a column vector

$$Q = \begin{bmatrix} q_1 \\ q_2 \\ \vdots \\ q_n \end{bmatrix},$$

where in each case the entries are nonnegative numbers that sum to 1. The idea is that player R will choose row i with probability p_i ($i = 1, 2, \ldots, m$), and player C will choose column j with probability q_j ($j = 1, 2, \ldots, n$). Thus in a 2×3 game,

$$P = [.6 \quad .4] \quad \text{and} \quad Q = \begin{bmatrix} .2 \\ .5 \\ .3 \end{bmatrix}$$

are strategies for players R and C, respectively, as are

$$P = [.2 \quad .8] \quad \text{and} \quad Q = \begin{bmatrix} 0 \\ 1 \\ 0 \end{bmatrix}.$$

THE EXPECTED PAYOFF

When a matrix game is played repeatedly, the payoff from player C to player R will vary from one play to the next depending on the actual moves that are made. For a given pair of strategies for each player we can calculate the *expected payoff*, which represents the average amount that the row-player can expect to win per play over the long run.

EXAMPLE 9.8 Find the expected payoff in the matrix game

$$A = \begin{bmatrix} 1 & -2 \\ -3 & 4 \end{bmatrix}$$

when players R and C use the respective strategies

$$P = [.6 \quad .4] \quad \text{and} \quad Q = \begin{bmatrix} .2 \\ .8 \end{bmatrix}.$$

Solution The payoff to the row player is a random variable with four possible values: 1, -2, -3, and 4 (the entries of the payoff matrix). The probability of each payoff is shown in the table below.

Payoff	Probability
1	$(.6)(.2) = .12$
-2	$(.6)(.8) = .48$
-3	$(.4)(.2) = .08$
4	$(.4)(.8) = .32$

For instance a payoff of -2 occurs when player R chooses row 1 (which happens with probability .6) and player C chooses column 2 (which happens with probability .8). Since these choices are made independently, we have

$$\begin{aligned} P(\text{payoff} = -2) &= P(\text{row 1 and column 2 chosen}) \\ &= P(\text{row 1 chosen}) \cdot P(\text{column 2 chosen}) \\ &= (.6)(.8) = .48. \end{aligned}$$

The other probabilities in the table are calculated similarly. The expected payoff is therefore

$$\begin{aligned} E &= 1(.12) + (-2)(.48) + (-3)(.08) + 4(.32) \\ &= .12 - .96 - .24 + 1.28 \\ &= .2. \end{aligned}$$

In other words, player R stands to win .2 units per play if this game is played many times using the given strategies P and Q. ■

It turns out that the expected payoff can be found more simply than in Example 9.8 by multiplying the matrices P, A, and Q in that order. First compute

$$PA = [.6 \quad .4] \begin{bmatrix} 1 & -2 \\ -3 & 4 \end{bmatrix} = [-.6 \quad .4],$$

and then

$$PAQ = (PA)Q = [-.6 \quad .4] \begin{bmatrix} .2 \\ .8 \end{bmatrix} = [.2].$$

Ignoring the brackets around this 1×1 matrix, we see the expected payoff $E = .2$. We leave as an exercise (see Exercise 31) the verification that this technique works for any 2×2 matrix game no matter what strategies the players use. More generally, we have the following rule.

> Let A be the payoff matrix for an $m \times n$ game in which the row and column players use strategies P and Q, respectively. Then the expected payoff to the row player is given by
>
> $$E = PAQ.$$

EXAMPLE 9.9 Suppose that both players make their moves completely at random in the matrix game below. Which side will have the advantage?

$$\begin{array}{cc} & \begin{array}{ccc} & \text{Player C} & \\ 1 & 2 & 3 \end{array} \\ \text{Player R} \begin{array}{c} 1 \\ 2 \end{array} & \begin{bmatrix} 3 & -5 & 4 \\ -2 & 2 & -3 \end{bmatrix} \end{array}.$$

Solution Since player R chooses each row with probability ½ and player C chooses each column with probability ⅓, the strategies for the two players are

$$P = [\tfrac{1}{2} \quad \tfrac{1}{2}] \quad \text{and} \quad Q = \begin{bmatrix} \tfrac{1}{3} \\ \tfrac{1}{3} \\ \tfrac{1}{3} \end{bmatrix},$$

respectively. Hence the expected payoff to player R is

$$E = PAQ = [\tfrac{1}{2} \quad \tfrac{1}{2}] \begin{bmatrix} 3 & -5 & 4 \\ -2 & 2 & -3 \end{bmatrix} \begin{bmatrix} \tfrac{1}{3} \\ \tfrac{1}{3} \\ \tfrac{1}{3} \end{bmatrix}$$

$$= [\tfrac{1}{2} \quad -\tfrac{3}{2} \quad \tfrac{1}{2}] \begin{bmatrix} \tfrac{1}{3} \\ \tfrac{1}{3} \\ \tfrac{1}{3} \end{bmatrix} = -\tfrac{1}{6}.$$

9.2 GAMES OF STRATEGY

Thus player C will have the advantage, with an expected gain of ⅙ unit per play over the long run. ■

Practice Problem 5 What is the expected payoff in the matrix game

$$A = \begin{bmatrix} 0 & -1 & 3 & 9 \\ 1 & -7 & 2 & -3 \\ -5 & 8 & -6 & -4 \end{bmatrix}$$

if players R and C use the respective strategies

$$P = [.2 \quad .5 \quad .3] \quad \text{and} \quad Q = \begin{bmatrix} .1 \\ .2 \\ .3 \\ .4 \end{bmatrix}?$$

Which player has the advantage under these conditions?

OPTIMAL STRATEGIES

For each strategy P of the row-player, there is a best counterstrategy Q for the column-player, namely the one for which the expected payoff $E = PAQ$ is the least. Likewise for each of the column-player's strategies Q, there is a best counterstrategy P for the row-player, namely the one for which the expected payoff is the greatest. A reasonable goal for player R is to try to *maximize* the expected payoff against the best counterstrategy for player C. Player C should similarly try to *minimize* the expected payoff against the best counterstrategy for player R.

Thus we can generalize the definition of "optimal moves" as follows. For each strategy P of the row-player, compute the minimum expected payoff PAQ over all the strategies Q of the column-player. An **optimal strategy** for player R is a strategy P for which this minimum expected payoff is as large as possible. Likewise for each strategy Q of the column-player, compute the maximum expected payoff PAQ over all the strategies P of the row-player. An **optimal strategy** for player C is a strategy Q for which this maximum expected payoff is as small as possible.

In 1928 John von Neumann discovered the following remarkable result.

Theorem 9.1

Let $P^\star$ and $Q^\star$ be optimal strategies for players R and C, respectively, in a matrix game. If R uses strategy $P^\star$, the expected payoff to R will be at least $P^\star A Q^\star$ no matter which strategy C adopts. Similarly, if C uses strategy $Q^\star$, the expected payoff to R will be no more than $P^\star A Q^\star$ no matter which strategy R adopts.

The minimax theorem asserts that an optimal strategy for the row player is the best counterstrategy to an optimal strategy for the column player, and an optimal strategy for the column player is the best counterstrategy to an optimal strategy for the row player. To make this statement more precise, let $P^\star$ and $Q^\star$ denote optimal strategies for players R and C, respectively. The statement that $P^\star$ is the best counterstrategy to $Q^\star$ means that the expected payoff $PAQ^\star$ is greatest when $P = P^\star$. That is,

$$PAQ^\star \leq P^\star AQ^\star \qquad \text{for all strategies } P \text{ of the row-player.}$$

Likewise the statement that $Q^\star$ is the best counterstrategy to $P^\star$ means that

$$P^\star AQ^\star \leq P^\star AQ \qquad \text{for all strategies } Q \text{ of the column-player.}$$

It follows from the minimax theorem that *neither side can benefit unilaterally by adopting a strategy different from an optimal strategy.*

If $P^\star$ and $Q^\star$ are optimal strategies for players R and C, respectively, then the number

$$P^\star AQ^\star$$

is called the **value** of the game. It represents the expected payoff for the row-player every time the game is played with each side using an optimal strategy. If $P^\star AQ^\star = 0$, the game is said to be **fair**, since neither player has an advantage over the long run.

SOLUTION OF 2 × 2 GAMES

We have not yet considered how to solve a matrix game, that is, how to determine its optimal strategies. In the remainder of this section we will discuss a procedure for doing so in a 2 × 2 game. The general case will be considered in Section 9.3.

If the payoff matrix for a 2 × 2 game

$$\begin{bmatrix} a & b \\ c & d \end{bmatrix}$$

has a saddle point, we can determine the optimal strategies as in Section 9.1. For in this case the optimal strategies are just the strategies corresponding to the optimal moves. For example, the row 1, column 2 entry of

$$\begin{bmatrix} 5 & 3 \\ 1 & 2 \end{bmatrix}$$

is a saddle point. Since the optimal moves for players R and C are row 1 and column 2, respectively, the corresponding optimal strategies are

$$P^\star = \begin{bmatrix} 1 & 0 \end{bmatrix} \quad \text{and} \quad Q^\star = \begin{bmatrix} 0 \\ 1 \end{bmatrix}.$$

These strategies mean that player R will use row 1 exclusively, and player 2 will use column 2 exclusively.

The following result enables us to determine the optimal strategies in a 2×2 game having no saddle point.

Theorem 9.2

Consider a 2×2 game with payoff matrix

$$\begin{bmatrix} a & b \\ c & d \end{bmatrix}$$

in which there is no saddle point. Let $s = a - b - c + d$. If

$$p_1^\star = \frac{d-c}{s}, \quad p_2^\star = \frac{a-b}{s},$$

$$q_1^\star = \frac{d-b}{s}, \quad q_2^\star = \frac{a-c}{s},$$

then

$$P^\star = [p_1^\star \ p_2^\star] \quad \text{and} \quad Q^\star = \begin{bmatrix} q_1^\star \\ q_2^\star \end{bmatrix}$$

are the optimal strategies for players R and C, respectively. Moreover, the value of the game is

$$v = \frac{ad - bc}{s}.$$

It can be shown that if the payoff matrix has no saddle point, then the entries on one diagonal of the payoff matrix are both greater than those on the other diagonal. That is, either a and d are both greater than b and c, or vice versa. (See Exercise 30.) Hence s cannot be zero, and the values of $p_1^\star, p_2^\star, q_1^\star$, and $q_2^\star$ in Theorem 9.2 are all positive.

EXAMPLE 9.8 Revisited Solve the 2×2 game with payoff matrix

$$\begin{bmatrix} 1 & -2 \\ -3 & 4 \end{bmatrix}$$

Solution Recall that this payoff matrix has no saddle points, and so its optimal strategies can be found by using Theorem 9.2. First, note that in the notation of Theorem 9.2, we have $a = 1$, $b = -2$, $c = -3$, and $d = 4$. Hence

$$s = a - b - c + d$$
$$= 1 - (-2) - (-3) + 4 = 10.$$

It follows that

$$p_1^\star = \frac{d-c}{s} = \frac{4-(-3)}{10} = \frac{7}{10}, \quad p_2^\star = \frac{a-b}{s} = \frac{1-(-2)}{10} = \frac{3}{10},$$

$$q_1^\star = \frac{d-b}{s} = \frac{4-(-2)}{10} = \frac{6}{10}, \quad q_2^\star = \frac{a-c}{s} = \frac{1-(-3)}{10} = \frac{4}{10}.$$

Thus the optimal strategies for players R and C are

$$P^\star = [.7 \quad .3] \quad \text{and} \quad Q^\star = \begin{bmatrix} .6 \\ .4 \end{bmatrix},$$

respectively, and the value of the game is

$$v = \frac{ad - bc}{s} = \frac{(1)(4) - (-2)(-3)}{10} = -\frac{2}{10} = -.2.$$

Hence over the long run player C can expect to win .2 units per play from player R if both players use their optimal strategies. ■

EXAMPLE 9.10 Solve the game of two-finger Morra (described in Example 9.1 on page 475).

Solution Recall that the payoff matrix for two-finger Morra is

$$\begin{bmatrix} 2 & -3 \\ -3 & 4 \end{bmatrix}.$$

Since none of the entries is a saddle point, we can apply Theorem 9.2. The resulting optimal strategies are

$$P^\star = [\tfrac{7}{12} \quad \tfrac{5}{12}] \quad \text{and} \quad Q^\star = \begin{bmatrix} \tfrac{7}{12} \\ \tfrac{5}{12} \end{bmatrix}.$$

Thus both players should hold up one finger $7/12 \approx 58.3\%$ of the time. The value of this game is $-1/12$, so that over the long run player C can expect to win $1/12$ unit per play from player R. ■

Practice Problem 6 Solve the game with the following payoff matrix.

$$\begin{bmatrix} 2 & -3 \\ -4 & 6 \end{bmatrix}$$

Which player has the advantage in this game?

If the reduced form of a matrix game is 2 × 2, then Theorem 9.2 can be used to solve the game.

EXAMPLE 9.11 A smuggler can cross a mountain border through one of three passes (1, 2, or 3). Normally the chances of getting through undetected are 90%, 80%, and 70%, respectively. When the border police reinforce their patrol, however, the smuggler's chances drop to 50% for pass 1, 70% for pass 2, and 60% for pass 3. Both the smuggler and the police know that there are only enough police to reinforce the patrol at one pass. What is the best strategy for each side?

Solution The smuggler (player R) can attempt to cross through pass 1, 2, or 3, and the border police (player C) can reinforce the patrol at pass 1, 2, or 3. If we let the payoff be the smuggler's chance (in percent) of getting through undetected, then the payoff matrix for this 3 × 3 is as shown below.

$$\text{Smuggler R} \begin{array}{c} \\ 1 \\ 2 \\ 3 \end{array} \overset{\begin{array}{ccc} & \text{Police C} & \\ 1 & 2 & 3 \end{array}}{\begin{bmatrix} 50 & 90 & 90 \\ 80 & 70 & 80 \\ 70 & 70 & 60 \end{bmatrix}}$$

Comparing rows, we see that move 2 is preferable to move 3 for player R. Hence row 3 can be deleted.

$$\begin{array}{c} 1 \\ 2 \end{array} \overset{\begin{array}{ccc} 1 & 2 & 3 \end{array}}{\begin{bmatrix} 50 & 90 & 90 \\ 80 & 70 & 80 \end{bmatrix}}$$

For player C, move 1 (or move 2) is now preferable to move 3, and so the game reduces to the following 2 × 2 game.

$$\begin{array}{c} 1 \\ 2 \end{array} \overset{\begin{array}{cc} 1 & 2 \end{array}}{\begin{bmatrix} 50 & 90 \\ 80 & 70 \end{bmatrix}}$$

Applying Theorem 9.2 to this game, we obtain

$$p_1^\star = \frac{1}{5}, \quad p_2^\star = \frac{4}{5},$$

$$q_1^\star = \frac{2}{5}, \quad q_2^\star = \frac{3}{5}, \quad \text{and}$$

$$v = 74.$$

Because move 3 is inferior for both sides, it follows that the optimal strategies for players R and C are

$$P^\star = [\tfrac{1}{5} \ \ \tfrac{4}{5} \ \ 0] \quad \text{and} \quad Q^\star = \begin{bmatrix} \tfrac{2}{5} \\ \tfrac{3}{5} \\ 0 \end{bmatrix}$$

and the value of the game is 74. Thus, over the long run, if both sides use their optimal strategies, the smuggler can expect to succeed 74% of the time. ■

Practice Problem 7 Consider the 3 × 3 matrix game below.

$$\begin{array}{c c} & \begin{array}{c c c} 1 & 2 & 3 \end{array} \\ \begin{array}{c} 1 \\ 2 \\ 3 \end{array} & \begin{bmatrix} 6 & 8 & -2 \\ -1 & 2 & 3 \\ -2 & -1 & 2 \end{bmatrix} \end{array}$$

(a) Find the reduced form of this game.
(b) Solve the game; that is, determine the optimal strategies for both players and the value of the game.

In general, if the payoff matrix of a game is larger than 2 × 2, we first check whether the game has a saddle point. If so, its solution is given by the optimal moves as in Section 9.1. If there is no saddle point, we check whether its reduced form is 2 × 2, in which case Theorem 9.2 applies. If its reduced form is not 2 × 2, then the optimal strategies can be found using linear programming, a technique that we will discuss in Section 9.3.

EXERCISES 9.2

In Exercises 1–4 determine which vectors are strategies for player R in a 3 × 3 matrix game.

1. [.1 .3 .6]   **2.** [0 1 0]
3. [.4 .2 .7]   **4.** [.5 .5 0]

In Exercises 5–8 determine which vectors are strategies for player C in a 3 × 3 matrix game.

5. $\begin{bmatrix} .9 \\ .1 \\ 0 \end{bmatrix}$ **6.** $\begin{bmatrix} 0 \\ 0 \\ 1 \end{bmatrix}$ **7.** $\begin{bmatrix} \tfrac{1}{2} \\ \tfrac{1}{6} \\ \tfrac{1}{3} \end{bmatrix}$ **8.** $\begin{bmatrix} 0 \\ \tfrac{1}{2} \\ \tfrac{1}{4} \end{bmatrix}$

9. Suppose that in the game with payoff matrix

$$\begin{bmatrix} 4 & -3 & 6 \\ -1 & 7 & -4 \end{bmatrix}$$

the row player uses the strategy [.7 .3]. Which of the following strategies is best for the column player?

(a) $\begin{bmatrix} .4 \\ .1 \\ .5 \end{bmatrix}$ (b) $\begin{bmatrix} .8 \\ .2 \\ 0 \end{bmatrix}$ (c) $\begin{bmatrix} .3 \\ .5 \\ .2 \end{bmatrix}$

10. For the matrix game in Exercise 9, suppose that the column-player uses the strategy

$$\begin{bmatrix} .3 \\ .1 \\ .6 \end{bmatrix}$$

Which of the following strategies is best for the row player?
(a) [.8 .2]   (b) [.5 .5]   (c) [.1 .9]

11. In the matrix game

$$\begin{bmatrix} -5 & 2 \\ 7 & -4 \end{bmatrix}$$

suppose that: (i) player R rolls a fair die, playing row 1 if 1 appears and playing row 2 otherwise; (ii) player C flips a fair coin, playing column 1 if it lands heads and playing column 2 otherwise. Under these conditions, which side will have the advantage?

12. In the matrix game

$$\begin{bmatrix} 1 & -6 & 7 \\ -3 & 6 & -5 \end{bmatrix}$$

suppose that: (i) player R flips a fair coin twice, playing row 1 if heads appears on both tosses and playing row 2 otherwise; (ii) player C chooses a card at random from a standard deck, playing column 1 if it is a spade, column 2 if it is a club, and column 3 otherwise. Which side will have the advantage?

In Exercises 13–20, solve the given matrix game by finding the value of the game and the optimal strategies $P^\star$ and $Q^\star$ for players R and C.

13. $\begin{bmatrix} 2 & -2 \\ -1 & 3 \end{bmatrix}$

14. $\begin{bmatrix} 0 & 2 \\ 3 & 0 \end{bmatrix}$

15. $\begin{bmatrix} -2 & 3 \\ 1 & 2 \end{bmatrix}$

16. $\begin{bmatrix} -5 & 1 \\ 4 & -2 \end{bmatrix}$

17. $\begin{bmatrix} 3 & -2 & -3 \\ -2 & 3 & 2 \end{bmatrix}$

18. $\begin{bmatrix} 0 & 2 & 5 & -1 \\ 3 & -4 & -1 & 2 \end{bmatrix}$

19. $\begin{bmatrix} 2 & -3 & -5 \\ -1 & 4 & 4 \\ -3 & 2 & 3 \end{bmatrix}$

20. $\begin{bmatrix} -5 & 3 & -2 \\ 4 & 2 & 1 \\ 5 & -4 & 0 \end{bmatrix}$

21. At a given signal player C puts either a nickel or a dime on a table, and simultaneously player R guesses which coin it is. If the guess is correct, player R wins the coin; otherwise player C keeps the coin and wins its value from R. Find the optimal strategy for each player. Is this game fair?

22. Player R holds a black ace (having value 1), a red 3, and a black 5; and player C holds a black 2 and a red 4. At a given signal each player shows one card. If the cards are the same color, then the one of higher rank wins; otherwise the one of lower rank wins. In either case, the winner collects from the loser the value of the loser's card. Find the optimal strategy for each player. Is this game fair?

23. Competing TV stations R and C always schedule a movie or a sporting event on Tuesday nights. The schedule is announced one month in advance and cannot be changed. When both stations offer the same type of program, R gets 60% of the audience and C gets the rest. When one offers a movie and the other offers a sporting event, however, the movie gets 55% of the audience and the sporting event gets 45%. Find the best strategy for each station and their expected shares of the audience.

24. Ambassador R always travels in a convoy of two identical cars to confuse guerrilla group C. She knows that if the lead car is attacked, then she will be captured if she is in it but will have a 60% chance of escape if she is in the second car. On the other hand, if the second car is attacked, then she will be captured if she is in it but will definitely escape if she is in the lead car. Under these circumstances find the best strategy for each side and the ambassador's expected chance of escaping an attack.

25. A band of outlaws conduct regular monthly raids on either village 1 or village 2. The local sheriff can defend only one of the two villages, as both he and the outlaws know. An average raid yields $9000 worth of spoils from village 1 and $11,000 worth of spoils from village 2. When the outlaws raid a village that is defended, they must retreat, however, with no gain. Find the best strategy for each side and the band's expected spoils.

26. On first down the Redskins always call either a draw play or a pass play against their rivals, the Cowboys. The average gain depends on whether the Cowboys use man-to-man coverage, a zone defense, or a pass rush. The Redskin coach knows that on the average a draw play will gain 4 yards against man-to-man coverage, 4 yards against a zone defense, and 7 yards against a pass rush. For a pass play the corresponding figures are 6 yards, 5 yards, and 3 yards. Find the best strategy for each team and the expected gain for the Redskins.

27. Prove that for $p_1^\star$, $p_2^\star$, $q_1^\star$, and $q_2^\star$ as defined in Theorem 9.2, we have $p_1^\star + p_2^\star = 1$ and $q_1^\star + q_2^\star = 1$.

28. Prove that for $P^\star$ and $Q^\star$ as defined in Theorem 9.2, the value of the game is $(ad - bc)/s$.

In Exercises 29–32, let

$$A = \begin{bmatrix} a & b \\ c & d \end{bmatrix}$$

be the payoff matrix for a 2 × 2 matrix game.

29. Prove that if $a = b$, then the game has a saddle point.

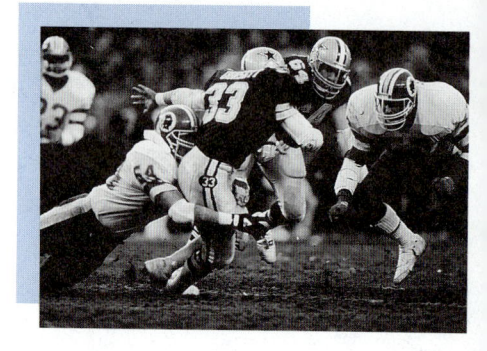

30. Prove that if the game has no saddle points, then either a and d are both greater than b and c, or else both b and c are greater than a and d. (*Hint:* Use Exercise 29 and consider whether $a > b$ or $a < b$.)

31. Prove that if players R and C use the respective strategies

$$P = [p_1 \ p_2] \quad \text{and} \quad Q = \begin{bmatrix} q_1 \\ q_2 \end{bmatrix},$$

then the expected payoff for player R is PAQ.

32. Prove that if the entry a is a saddle point, then

$$P^\star = [1 \ 0] \quad \text{and} \quad Q^\star = \begin{bmatrix} 1 \\ 0 \end{bmatrix}$$

are optimal strategies for A. (*Hint:* Verify that for all strategies P and Q we have $PAQ^\star \leq P^\star AQ^\star \leq P^\star AQ$.)

Answers to Practice Problems

5. The expected payoff from player R to player C is .78 unit per play, and so player C has the advantage.

6. The optimal strategies are

$$P^\star = [\tfrac{2}{3} \ \tfrac{1}{3}] \quad \text{and} \quad Q^\star = \begin{bmatrix} \tfrac{3}{5} \\ \tfrac{2}{5} \end{bmatrix}.$$

Since the value of the game is 0 (that is, the game is fair), neither player has an advantage.

7. (a) The reduced form of this game is

$$\begin{array}{c c} & \begin{array}{cc} 1 & 3 \end{array} \\ \begin{array}{c} 1 \\ 2 \end{array} & \left[\begin{array}{cc} 6 & -2 \\ -1 & 3 \end{array} \right]. \end{array}$$

(b) The optimal strategies for players R and C in the given game are

$$P^\star = [\tfrac{1}{3} \ \tfrac{2}{3} \ 0] \quad \text{and} \quad Q^\star = \begin{bmatrix} \tfrac{5}{12} \\ 0 \\ \tfrac{7}{12} \end{bmatrix},$$

respectively, and the value of the game is $\tfrac{4}{3}$.

9.3 Simplex Solution of $m \times n$ Games

In this section we present a method for using linear programming to solve any matrix game.

GAMES WITH POSITIVE VALUE

We will see shortly that the problem of finding optimal strategies in a matrix game can be expressed as a linear program provided that the game has a positive value v. If at least one row of the payoff matrix has only positive entries, this will surely be the case since player R can choose this row repeatedly and be assured of a positive payoff no matter what player C does. If no such row exists, we can simply add a suitably large constant k to each entry of the payoff matrix. Although this change will increase the value of the game by k, it will not change the optimal strategies $P^\star$ and $Q^\star$, for the expected payoff corresponding to each pair of strategies P and Q will merely be increased by k.

EXAMPLE 9.12 Convert the following game into one having a positive value and find its optimal strategies.

$$\begin{array}{c} \text{Player C} \\ \begin{array}{cc} 1 & 2 \end{array} \\ \text{Player R} \begin{array}{c} 1 \\ 2 \end{array} \left[\begin{array}{cc} -2 & 1 \\ 3 & -4 \end{array} \right] \end{array}$$

Discussion If we add $k = 3$ to each entry of the payoff matrix, we obtain the matrix A below, which has a positive row.

$$A = \left[\begin{array}{cc} -2+3 & 1+3 \\ 3+3 & -4+3 \end{array} \right] = \left[\begin{array}{cc} 1 & 4 \\ 6 & -1 \end{array} \right]$$

This new game with payoff matrix A is obviously favorable to player R because R will receive a payoff of at least 1 by playing the first row no matter what player C does. In fact, using Theorem 9.2 we see that the original game has value $-1/2$ and the new game has value

$$\frac{5}{2} = -\frac{1}{2} + 3 = -\frac{1}{2} + k.$$

The reader should verify that the optimal strategies are the same for both games, namely

$$P^\star = [.7 \quad .3] \quad \text{and} \quad Q^\star = \left[\begin{array}{c} .5 \\ .5 \end{array} \right]. \quad \blacksquare$$

In the same way, any $m \times n$ matrix game can be converted into a game with a positive value v in which the optimal strategies remain unchanged.

Practice Problem 8 Determine a constant k to add to each entry of the payoff matrix A below in order to obtain a matrix B having the same optimal strategies as A. Then write the matrix B.

$$A = \left[\begin{array}{cccc} 4 & -1 & 2 & 1 \\ -2 & 2 & 1 & 0 \\ 6 & 0 & -1 & 1 \end{array} \right]$$

MATRIX GAMES AND LINEAR PROGRAMS

To explain the connection between matrix games and linear programs, we will consider the simplest case, a 2×2 matrix game

$$A = \begin{bmatrix} a & b \\ c & d \end{bmatrix}$$

with positive value v. We will first analyze this game from the *column*-player's viewpoint. For each possible strategy Q, player C is interested in minimizing v_Q, the expected payoff against R's best counterstrategy. Now v_Q equals the maximum value of PAQ over all strategies P. Consider arbitrary strategies

$$P = [p_1 \; p_2] \quad \text{and} \quad Q = \begin{bmatrix} q_1 \\ q_2 \end{bmatrix}$$

for players R and C, respectively. Then

$$\begin{aligned} PAQ &= [p_1 \; p_2] \begin{bmatrix} a & b \\ c & d \end{bmatrix} \begin{bmatrix} q_1 \\ q_2 \end{bmatrix} \\ &= [p_1 \; p_2] \begin{bmatrix} aq_1 + bq_2 \\ cq_1 + dq_2 \end{bmatrix} \\ &= p_1(aq_1 + bq_2) + p_2(cq_1 + dq_2). \end{aligned}$$

It is not hard to show that as P ranges over all strategies, this expression takes on all values between $aq_1 + bq_2$ and $cq_1 + dq_2$. (See Exercise 21.) Hence the maximum value of PAQ over all strategies P is just the larger of the numbers $aq_1 + bq_2$ and $cq_1 + dq_2$; that is,

$$v_Q = \max\{aq_1 + bq_2, cq_1 + dq_2\}.$$

When Q is an optimal strategy, v_Q assumes its minimum value v. Thus the value of the game is the least number v for which

$$\begin{aligned} aq_1 + bq_2 &\leq v \\ cq_1 + dq_2 &\leq v \end{aligned}$$

for all strategies Q. In other words, an optimal strategy Q for player C is a solution of the following linear program.

Minimize v
subject to $\;aq_1 + bq_2 \leq v$
$\phantom{\text{subject to }}\;cq_1 + dq_2 \leq v$
$\phantom{\text{subject to }}\;q_1 + q_2 = 1$
$\phantom{\text{subject to }}\;q_1 \geq 0, q_2 \geq 0.$

It will prove to be more convenient to work with this linear program in a modified form. First, divide the inequalities by v (which is positive) to obtain

$$\frac{aq_1}{v} + \frac{bq_2}{v} \leq 1$$

$$\frac{cq_1}{v} + \frac{dq_2}{v} \leq 1.$$

Second, because $q_1 + q_2 = 1$, $q_1 \geq 0$, and $q_2 \geq 0$, we see that

$$\frac{q_1}{v} + \frac{q_2}{v} = \frac{1}{v}, \qquad \frac{q_1}{v} \geq 0, \quad \text{and} \quad \frac{q_2}{v} \geq 0.$$

Now we introduce new variables

$$y_1 = \frac{q_1}{v}, \qquad y_2 = \frac{q_2}{v}, \quad \text{and} \quad F = \frac{1}{v},$$

and observe that minimizing v is the same as maximizing $F = 1/v$. Thus the linear program above becomes:

$$\begin{aligned}
\text{Maximize} \quad & F = y_1 + y_2 \\
\text{subject to} \quad & ay_1 + by_2 \leq 1 \\
& cy_1 + dy_2 \leq 1 \\
& y_1 \geq 0, \, y_2 \geq 0.
\end{aligned} \tag{9.1}$$

Note that the coefficients of y_1 and y_2 in this linear program are simply the entries in the payoff matrix for the game. This linear program can be solved for y_1, y_2, and F, and the values q_1, q_2, and v can then be obtained from the relations

$$v = \frac{1}{F}, \quad q_1 = vy_1, \quad \text{and} \quad q_2 = vy_2. \tag{9.2}$$

EXAMPLE 9.13 For the game with payoff matrix

$$A = \begin{bmatrix} 1 & 4 \\ 6 & -1 \end{bmatrix}$$

use linear programming to find an optimal strategy $Q^\star$ and the value of the game v.

Solution Since the matrix has a positive row (row 1), we know that $v > 0$. The associated linear program is

$$\begin{aligned}
\text{Maximize} \quad & F = y_1 + y_2 \\
\text{subject to} \quad & y_1 + 4y_2 \leq 1 \\
& 6y_1 - y_2 \leq 1 \\
& y_1 \geq 0, \, y_2 \geq 0.
\end{aligned}$$

Introducing nonnegative slack variables s and t, we can write the program as

$$\text{Maximize} \quad F = y_1 + y_2$$
$$\text{subject to} \quad y_1 + 4y_2 + s = 1$$
$$\phantom{\text{subject to} \quad} 6y_1 - y_2 + t = 1$$
$$y_1 \geq 0, y_2 \geq 0, s \geq 0, t \geq 0.$$

The initial simplex tableau for this program is

$$\begin{array}{c} \\ s \\ t \\ \end{array} \begin{array}{c} y_1 \quad y_2 \quad s \quad t \\ \left[\begin{array}{cccc|c} 1 & 4 & 1 & 0 & 1 \\ 6 & -1 & 0 & 1 & 1 \\ \hline 1 & 1 & 0 & 0 & F \end{array} \right]. \end{array}$$

Notice that the payoff matrix A appears in the first two rows and columns and that the indicators for y_1 and y_2 are both 1. Now we apply the simplex algorithm as described in Section 3.4.

$$\begin{array}{c} \\ s \\ t \\ \end{array} \begin{array}{c} y_1 \quad y_2 \quad s \quad t \\ \left[\begin{array}{cccc|c} 1 & 4 & 1 & 0 & 1 \\ \boxed{6} & -1 & 0 & 1 & 1 \\ \hline 1^* & 1 & 0 & 0 & F \end{array} \right] \end{array} \begin{array}{l} 1/1 = 1 \\ 1/6^* \end{array}$$

$$\begin{array}{c} \\ s \\ y_1 \\ \end{array} \begin{array}{c} y_1 \quad y_2 \quad s \quad t \\ \left[\begin{array}{cccc|c} 0 & \boxed{\tfrac{25}{6}} & 1 & -\tfrac{1}{6} & \tfrac{5}{6} \\ 1 & -\tfrac{1}{6} & 0 & \tfrac{1}{6} & \tfrac{1}{6} \\ \hline 0 & \tfrac{7}{6}* & 0 & -\tfrac{1}{6} & F - \tfrac{1}{6} \end{array} \right] \end{array} \begin{array}{l} (5/6)/(25/6) = 1/5^* \\ \text{(negative quotient)} \end{array}$$

$$\begin{array}{c} \\ y_2 \\ y_1 \\ \end{array} \begin{array}{c} y_1 \quad y_2 \quad s \quad t \\ \left[\begin{array}{cccc|c} 0 & 1 & \tfrac{6}{25} & -\tfrac{1}{25} & \tfrac{1}{5} \\ 1 & 0 & \tfrac{1}{25} & \tfrac{4}{25} & \tfrac{1}{5} \\ \hline 0 & 0 & -\tfrac{7}{25} & -\tfrac{3}{25} & F - \tfrac{2}{5} \end{array} \right] \end{array}$$

From the tableau above we can read the optimal solution:

$$\text{Maximum } F = \frac{2}{5} \quad \text{when} \quad y_1 = \frac{1}{5} \quad \text{and} \quad y_2 = \frac{1}{5}.$$

Relations (9.2) now give

$$v = \frac{1}{F} = \frac{5}{2},$$

$$q_1 = vy_1 = \left(\frac{5}{2}\right)\left(\frac{1}{5}\right) = \frac{1}{2}, \quad \text{and}$$

$$q_2 = vy_2 = \left(\frac{5}{2}\right)\left(\frac{1}{5}\right) = \frac{1}{2}.$$

Therefore the value of this game is $v = 5/2$, and the column player's optimal strategy is

$$Q^\star = \begin{bmatrix} \frac{1}{2} \\ \frac{1}{2} \end{bmatrix},$$

which agrees with the result obtained in Example 9.12. ∎

It can be shown that the row-player's optimal strategy can be obtained from the solution of the dual linear program to (9.1). Thus the duality theorem shows that when we compute the column-player's optimal strategy, we are also finding the row-player's optimal strategy. In the final tableau in Example 9.13

$$\begin{array}{c c} & \begin{array}{cccc} y_1 & y_2 & s & t \end{array} \\ \begin{array}{c} y_2 \\ y_1 \\ \\ \end{array} & \left[\begin{array}{cccc|c} 0 & 1 & \frac{6}{25} & -\frac{1}{25} & \frac{1}{5} \\ 1 & 0 & \frac{1}{25} & \frac{4}{25} & \frac{1}{5} \\ \hline 0 & 0 & -\frac{7}{25} & -\frac{3}{25} & F - \frac{2}{5} \\ & & \uparrow & \uparrow & \\ & & -x_1 & -x_2 & \end{array} \right], \end{array}$$

notice that the negatives of the indicators in the columns corresponding to the slack variables are

$$x_1 = \frac{7}{25} \quad \text{and} \quad x_2 = \frac{3}{25},$$

respectively. Then the row-player's optimal strategy $[p_1 \ p_2]$ is such that $p_1 = vx_1$ and $p_2 = vx_2$, where v is the value of the game. Thus in Example 9.13 we have

$$v = \frac{1}{F} = \frac{5}{2},$$

$$p_1 = vx_1 = \left(\frac{5}{2}\right)\left(\frac{7}{25}\right) = \frac{7}{10} = .7, \quad \text{and}$$

$$p_2 = vx_2 = \left(\frac{5}{2}\right)\left(\frac{3}{25}\right) = \frac{3}{10} = .3.$$

Hence the row-player's optimal strategy is

$$P^\star = [.7 \quad .3],$$

which agrees with our answer in Example 9.12.

SIMPLEX SOLUTION OF $m \times n$ MATRIX GAMES

The method just illustrated for 2×2 matrix games generalizes to any matrix game. The steps are summarized as follows.

Simplex Solution of an $m \times n$ Matrix Game

■ ■ ■ ■ ■

Step 1. If necessary, add a suitable constant k to each entry of the payoff matrix to obtain a matrix

$$A = \begin{bmatrix} a_{11} & a_{12} & \cdots & a_{1n} \\ a_{21} & a_{22} & \cdots & a_{2n} \\ \vdots & \vdots & & \vdots \\ a_{m1} & a_{m2} & \cdots & a_{mn} \end{bmatrix}$$

in which all the entries of some row are positive.

Step 2. Apply the simplex algorithm to the linear program

Maximize $\quad F = y_1 + y_2 + \cdots + y_n$

subject to $\quad a_{11}y_1 + a_{12}y_2 + \cdots + a_{1n}y_n \leq 1$

$\qquad\qquad\quad a_{21}y_1 + a_{22}y_2 + \cdots + a_{2n}y_n \leq 1$

$\qquad\qquad\qquad\qquad\qquad\vdots$

$\qquad\qquad\quad a_{m1}y_1 + a_{m2}y_2 + \cdots + a_{mn}y_n \leq 1$

$\qquad\qquad\quad y_1 \geq 0, y_2 \geq 0, \cdots, y_n \geq 0.$

Step 3. Read the solution $y_1, y_2, \ldots, y_n$, the maximum value F, and $x_1, x_2, \ldots, x_m$, the negatives of the indicators corresponding to the slack variables.

Step 4. Let $v = 1/F$. Then

$$P^\star = [vx_1 \quad vx_2 \quad \cdots \quad vx_m] \quad \text{and} \quad Q^\star = \begin{bmatrix} vy_1 \\ vy_2 \\ \vdots \\ vy_n \end{bmatrix}$$

are optimal strategies for the row and column players, and v is the value of the game A.

Step 5. Subtract k from v to obtain the value of the original game.

Although the procedure on page 505 works even for a strictly determined or nonreduced game, these cases should be eliminated in advance to avoid unnecessary work.

EXAMPLE 9.14 Solve the matrix game below.

$$\begin{array}{c} \text{Player C} \\ \begin{array}{cc} & \begin{array}{ccc} 1 & 2 & 3 \end{array} \\ \text{Player R} & \begin{array}{c} 1 \\ 2 \end{array} \begin{bmatrix} 0 & 1 & 2 \\ 1 & 0 & -2 \end{bmatrix} \end{array} \end{array}$$

Solution After checking that the game has no saddle point and is in reduced form, we proceed as described on page 505.

Step 1 Since neither row is positive, we add $k = 1$ to each entry of the payoff matrix to obtain the game below, which has a positive value (because all of the entries in row 1 are positive).

$$A = \begin{bmatrix} 1 & 2 & 3 \\ 2 & 1 & -1 \end{bmatrix}$$

Step 2 Apply the simplex algorithm to the linear program below.

$$\text{Maximize} \quad F = y_1 + y_2 + y_3$$

$$\text{subject to} \quad y_1 + 2y_2 + 3y_3 \leq 1$$

$$2y_1 + y_2 - y_3 \leq 1$$

$$y_1 \geq 0, y_2 \geq 0, y_3 \geq 0$$

Thus we obtain the following sequence of tableaus.

$$\begin{array}{c} \begin{array}{cccccc} y_1 & y_2 & y_3 & s & t & \end{array} \\ \begin{array}{c} s \\ t \end{array} \left[\begin{array}{ccccc|c} 1 & 2 & 3 & 1 & 0 & 1 \\ ② & 1 & -1 & 0 & 1 & 1 \\ \hline 1* & 1 & 1 & 0 & 0 & F \end{array} \right] \begin{array}{l} 1/1 = 1 \\ 1/2* \end{array} \end{array}$$

$$\begin{array}{c} \begin{array}{cccccc} y_1 & y_2 & y_3 & s & t & \end{array} \\ \begin{array}{c} s \\ y_1 \end{array} \left[\begin{array}{ccccc|c} 0 & \frac{3}{2} & ⑦/② & 1 & -\frac{1}{2} & \frac{1}{2} \\ 1 & \frac{1}{2} & -\frac{1}{2} & 0 & \frac{1}{2} & \frac{1}{2} \\ \hline 0 & \frac{1}{2} & \frac{3}{2}* & 0 & -\frac{1}{2} & F - \frac{1}{2} \end{array} \right] \begin{array}{l} (1/2)/(7/2) = 1/7* \\ \text{(negative quotient)} \end{array} \end{array}$$

$$\begin{array}{c} \begin{array}{cccccc} y_1 & y_2 & y_3 & s & t & \end{array} \\ \begin{array}{c} y_3 \\ y_1 \end{array} \left[\begin{array}{ccccc|c} 0 & \frac{3}{7} & 1 & \frac{2}{7} & -\frac{1}{7} & \frac{1}{7} \\ 1 & \frac{5}{7} & 0 & \frac{1}{7} & \frac{3}{7} & \frac{4}{7} \\ \hline 0 & -\frac{1}{7} & 0 & -\frac{3}{7} & -\frac{2}{7} & F - \frac{5}{7} \end{array} \right] \\ \uparrow \uparrow \\ -x_1 -x_2 \end{array}$$

Step 3 From the final tableau we read the optimal solution

$$y_1 = \frac{4}{7}, \quad y_2 = 0, \quad y_3 = \frac{1}{7}, \quad F = \frac{5}{7}$$

and the negatives of the indicators corresponding to the slack variables

$$x_1 = \frac{3}{7} \quad \text{and} \quad x_2 = \frac{2}{7}.$$

Step 4 Letting $v = 1/F = 7/5$, we obtain the optimal strategies

$$P^\star = [vx_1 \quad vx_2] = [\tfrac{3}{5} \quad \tfrac{2}{5}] \quad \text{and} \quad Q^\star = \begin{bmatrix} vy_1 \\ vy_2 \\ vy_3 \end{bmatrix} = \begin{bmatrix} \tfrac{4}{5} \\ 0 \\ \tfrac{1}{5} \end{bmatrix}.$$

Step 5 The value of the original game is

$$v - k = \frac{7}{5} - 1 = \frac{2}{5}. \quad \blacksquare$$

Practice Problem 9 Solve the game with the payoff matrix below.

$$\begin{bmatrix} 2 & -2 & -1 \\ -1 & 0 & 1 \end{bmatrix}$$

NONZERO-SUM AND *n*-PERSON GAMES

In this chapter we have considered only zero-sum games, those in which one player's loss is the other's gain. For nonzero-sum games the concept of "rational behavior" is much harder to characterize. A famous game-theoretic paradox created by A. W. Tucker illustrates this problem.

EXAMPLE 9.15 Prisoners R and C are both implicated in a crime. The district attorney informs them that if neither confesses, they will each get a 1-year sentence on a minor charge. If one confesses but the other doesn't, the one who cooperates will go free and his accomplice will receive a 10-year sentence. If both confess, they will each get reduced sentences of 5 years. The prisoners are separated and given 24 hours to respond.

If neither confesses, each receives a 1-year sentence. If only one confesses, he goes free and the other receives a 10-year sentence. If both confess, each receives a 5-year sentence.

Discussion This problem is known as the *prisoner's dilemma*. Each of the prisoners has two moves, "confess" or "don't confess." If we interpret a sentence of n years as a payoff of $-n$, then for each choice of moves there will be *two* payoffs, one for player R and one for player C. These pairs of payoffs are shown in the table below.

	Prisoner C Confess	Prisoner C Don't confess
Prisoner R — Confess	$(-5, -5)$	$(0, -10)$
Prisoner R — Don't confess	$(-10, 0)$	$(-1, -1)$

The first number in each pair denotes the payoff to R and the second number denotes the payoff to C. If, for instance, R confesses but C does not, then the payoff to R is 0 and the payoff to C is -10.

No matter what prisoner C does, the payoff to prisoner R is always greater in row 1 than row 2 (-5 compared to -10, and 0 compared to -1). In other words, disregarding any moral qualms or fears about cooperating with the district attorney, R will be better off by confessing. Likewise prisoner C will be better off by confessing, no matter what R does, because the payoffs in column 1 are always greater than the corresponding payoffs in column 2. Thus if both prisoners make the "rational" move, they will each receive 5-year sentences, whereas they would receive only 1-year sentences by both making the "irrational" move of not confessing. ∎

Of course, if the prisoners were permitted to communicate, they might promise each other to remain silent, since it is in both their interests to do so. Such a pact would be inherently unstable, however, because each side could gain by double-crossing the other. Many real-life negotiations on issues such as disarmament and tariff reduction involve the same features as this example. Thus the best solution for both sides may require mutual trust.

Games with more than two players also arise in applications. For example, several countries may be competing for the same resources or several corporations competing for the same market. The theory of n-person games is complicated by the large number of possible coalitions that can be formed among the players, and at present this theory is far less satisfactory than the theory of two-person games. Despite the early enthusiasm for game theory in the 1940s and 1950s, there still seems to be a wide gap between what the theory can handle and the types of problems that actually occur in most conflict situations.

EXERCISES 9.3

In Exercises 1–4, solve the given 2 × 2 game using the simplex method. Check your answer using Theorem 9.2 in Section 9.2.

1. $\begin{bmatrix} 1 & -1 \\ -1 & 1 \end{bmatrix}$

2. $\begin{bmatrix} 1 & 0 \\ 0 & 3 \end{bmatrix}$

3. $\begin{bmatrix} 1 & -2 \\ -3 & 4 \end{bmatrix}$

4. $\begin{bmatrix} 3 & -1 \\ -3 & 1 \end{bmatrix}$

5. Use the simplex method to solve the strictly determined game below.
$$\begin{bmatrix} 2 & 1 & 3 \\ 4 & 0 & 2 \end{bmatrix}$$
Check your answer by using the techniques in Section 9.1.

6. Use the simplex method to solve the nonreduced game below.
$$\begin{bmatrix} 1 & -2 & 2 \\ -3 & 0 & 3 \end{bmatrix}$$
Check your answer by using the techniques in Section 9.2.

Use the simplex method to solve the matrix games in Exercises 7–16.

7. $\begin{bmatrix} 3 & -1 & 0 \\ -2 & 1 & -1 \end{bmatrix}$

8. $\begin{bmatrix} -3 & 1 & 3 \\ 2 & 0 & -2 \end{bmatrix}$

9. $\begin{bmatrix} 1 & -2 & 2 \\ 0 & 5 & -1 \end{bmatrix}$

10. $\begin{bmatrix} 2 & 1 & 3 \\ 1 & 5 & 0 \end{bmatrix}$

11. $\begin{bmatrix} 2 & -3 \\ -2 & 3 \\ -1 & 1 \end{bmatrix}$

12. $\begin{bmatrix} -1 & 3 \\ 0 & -1 \\ 3 & -2 \end{bmatrix}$

13. $\begin{bmatrix} 0 & 1 & 0 \\ 1 & -1 & -1 \\ -1 & -1 & 3 \end{bmatrix}$

14. $\begin{bmatrix} 1 & 0 & 0 \\ 0 & -2 & 1 \\ 0 & 1 & 0 \end{bmatrix}$

15. $\begin{bmatrix} -1 & 0 & 2 \\ 0 & -2 & -1 \\ 1 & 2 & -1 \end{bmatrix}$

16. $\begin{bmatrix} -1 & 1 & 1 \\ 2 & -2 & 2 \\ 3 & 3 & -2 \end{bmatrix}$

17. An inmate R regularly breaks out of the state prison. When he heads for the desert (*D*), his chance of escape is 30%, 20%, or 10% depending on whether the warden sends 0, 1, or 2 search teams to that area. When the inmate heads for the mountains (*M*), his chance of escape is 50%, 40%, or 25% depending on whether 0, 1, or 2 search teams are sent to that area. The warden has only two search teams, and they can both be sent to the desert (*DD*), both to the mountains (*MM*), or one to each area (*DM*). Find the best strategy for each side and the inmate's expected chance of escape.

18. Army R plans to attack one of two outposts held by army C. Outpost *A* is worth 3 points and outpost *B* is worth 5 points. Army C can deploy both of its divisions at *A* (*AA*), one at *A* and one at *B* (*AB*), or both at *B* (*BB*). The commander of army R

knows that he can capture either outpost, but he estimates that there will be losses worth 2 or 3 points, respectively, if the outpost he attacks is defended by 1 or 2 divisions, respectively. Find the best strategy for each side.

19. In the game of *three-finger morra*, two players (R and C) simultaneously show one, two, or three fingers. If the sum of the numbers shown is even, then R wins that many dollars from C. If the sum is odd, then C wins that many dollars from R. Show that this game is fair, and find the best strategy for each side.

20. In the well-known children's game *scissors-paper-stone*, each of two players simultaneously shows either two fingers (scissors), a palm (paper), or a fist (stone). If they each make the same move, then the game is a tie; otherwise the winner is determined by the rule: "Scissors cut paper, paper covers stone, and stone breaks scissors." Suppose that the loser of any game must pay a penny to the winner. Show that this game is fair, and find the best strategy for each player.

21. Prove that as $P = [p_1 \ p_2]$ ranges over all strategies, the quantity $p_1 x_1 + p_2 x_2$ takes on all values between x_1 and x_2, inclusive. (*Hint:* p_1 and p_2 are nonnegative numbers that sum to 1.)

22. Write the dual linear program to program (9.1) on page 502.

23. Let
$$\begin{bmatrix} a & b \\ c & d \end{bmatrix}$$
be a 2 × 2 game having a positive value v, and let $P = [p_1 \ p_2]$ be the row-player's optimal strategy. If
$$x_1 = \frac{p_1}{v}, \quad x_2 = \frac{p_2}{v}, \quad \text{and} \quad G = \frac{1}{v},$$
prove that x_1, x_2, G is the optimal solution of the linear program
$$\text{Minimize} \quad G = x_1 + x_2$$
$$\text{subject to} \quad ax_1 + cx_2 \geq 1$$
$$bx_1 + dx_2 \geq 1$$
$$x_1 \geq 0, x_2 \geq 0.$$

24. Prove that if a constant k is added to each entry of the payoff matrix, then the expected payoff corresponding to each pair of strategies P and Q for the row and column players is increased by k.

Answers to Practice Problems 8. We can choose $k = 2$; the resulting matrix B is
$$B = \begin{bmatrix} 6 & 1 & 4 & 3 \\ 0 & 4 & 3 & 2 \\ 8 & 2 & 1 & 3 \end{bmatrix}.$$

9. The optimal strategies for players R and C are
$$P^\star = [.2 \ .8] \quad \text{and} \quad Q^\star = \begin{bmatrix} .4 \\ .6 \\ 0 \end{bmatrix},$$
and the value of the game is -0.4.

CHAPTER 9 REVIEW

IMPORTANT TERMS

- matrix game *(9.1)*
- optimal move
- payoff matrix
- reduced form
- saddle point
- expected payoff *(9.2)*
- optimal strategy
- strategy
- value of a game

IMPORTANT FORMULAS

Let
$$\begin{bmatrix} a & b \\ c & d \end{bmatrix}$$
be a 2 × 2 game having no saddle point, and let $s = a - b - c + d$. If

$$p_1^\star = \frac{d-c}{s}, \quad p_2^\star = \frac{a-b}{s},$$
$$q_1^\star = \frac{d-b}{s}, \quad q_2^\star = \frac{a-c}{s},$$

then

$$P^\star = [p_1^\star \; p_2^\star] \quad \text{and} \quad Q^\star = \begin{bmatrix} q_1^\star \\ q_2^\star \end{bmatrix}$$

are the optimal strategies for players R and C, respectively. Moreover, the value of the game is

$$\frac{ad - bc}{s}.$$

REVIEW EXERCISES

In Exercises 1–4 find the reduced form of the given matrix game.

1. $\begin{bmatrix} 3 & -1 & 2 \\ -1 & 4 & -3 \end{bmatrix}$

2. $\begin{bmatrix} -2 & 3 & 2 & 0 \\ 3 & 2 & 5 & -1 \end{bmatrix}$

3. $\begin{bmatrix} 2 & -1 & 5 & 7 \\ 3 & 2 & -6 & -2 \\ 0 & 4 & 1 & 3 \\ 3 & 1 & -8 & 2 \end{bmatrix}$

4. $\begin{bmatrix} 9 & -5 & 2 \\ 1 & 7 & -4 \\ 3 & 2 & 1 \\ -5 & 0 & 5 \end{bmatrix}$

In Exercises 5–8 determine the optimal moves for each player. If the game has a saddle point, also find the value of the game.

5. $\begin{bmatrix} 3 & 2 \\ -1 & 5 \\ 4 & 0 \end{bmatrix}$

6. $\begin{bmatrix} 6 & 0 & -3 \\ -2 & -1 & 7 \\ 4 & 2 & 5 \end{bmatrix}$

7. $\begin{bmatrix} -1 & 4 & -3 \\ 0 & -2 & 5 \\ 1 & 3 & 2 \end{bmatrix}$

8. $\begin{bmatrix} 1 & -2 & 5 & -1 \\ 2 & 4 & 3 & 2 \\ 0 & 6 & -1 & 1 \end{bmatrix}$

9. Consider the game with the payoff matrix at the right.
   $$\begin{bmatrix} 50 & -90 \\ -20 & 40 \end{bmatrix}$$
   (a) Suppose that each player flips a fair coin to determine which move to make. Who will have the advantage?
   (b) What is the optimal strategy for each player? If they use these strategies, who will have the advantage?

10. Consider the game with the payoff matrix at the right.
    $$\begin{bmatrix} 5 & 0 & -1 \\ -3 & 4 & 1 \end{bmatrix}$$
    (a) Suppose that player R chooses each row half the time and that player C chooses the first column 20% of the time, the second column 30% of the time, and the third column 50% of the time. Who will have the advantage?
    (b) What is the optimal strategy for each player? If they use these strategies, who will have the advantage?

Use the simplex method to solve the matrix games in Exercises 11 and 12.

11. $\begin{bmatrix} 1 & -5 & 9 \\ -1 & 0 & -2 \end{bmatrix}$
12. $\begin{bmatrix} 3 & -1 \\ -2 & 2 \\ -1 & 0 \end{bmatrix}$

13. Two players, R and C, each simultaneously put 1, 2, or 3 dollars into a pot. If the total amount is even, then R wins the amount put in by C. If the total amount is odd, then C wins the amount put in by R. Find the best strategy for each player. Is this game fair?

14. Army C must send a supply convoy from point A to point B along one of the three mountain passes shown at the right. The convoy will take three days along route 1, four days along route 2, and six days along route 3. Army R can plan to ambush the convoy along any of the three routes. If the ambush is planned along the same route taken by the convoy, then army R will be able to attack the convoy during each day of travel. If, however, the ambush is planned along a different route than that taken by the convoy, exactly two days of attack will be lost by army R. Find the best strategy for each side and the expected number of days that army R will be able to attack the convoy.

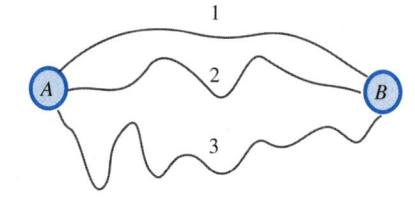

REFERENCES

1. Bernard, Jessie, "The Theory of Games as a Modern Sociology of Conflict," *American Journal of Sociology*, vol. 59 (1954), pp. 411–424.
2. Brams, Steven J., *Game Theory and Politics*. New York: Free Press, 1975.
3. Haywood, O. G. Jr., "Military Decision and the Mathematical Theory of Games," *Air University Quarterly Review*, vol. 4 (1958), pp. 699–709.
4. Haywood, O. G. Jr., "Military Decision and Game Theory," *Operations Research*, vol. 2 (1954), pp. 365–385.
5. Luce, R. Duncan and Howard Raiffa, *Games and Decisions*. New York, John Wiley and Sons, 1957.
6. Shubik, Martin, "The Uses of Game Theory in Management," *Management Science*, vol. 2 (1955), pp. 40–54.
7. Von Neumann, John and Oscar Morgenstern, *Theory of Games and Economic Behavior*. Princeton, NJ: Princeton University Press, 1947.

TABLE 1 Compound Interest $(1 + i)^n$

n \ i	1%	1½%	2%	3%	4%	5%	6%	8%
1	1.01000	1.01500	1.02000	1.03000	1.04000	1.05000	1.06000	1.08000
2	1.02010	1.03023	1.04040	1.06090	1.08160	1.10250	1.12360	1.16640
3	1.03030	1.04568	1.06121	1.09273	1.12486	1.15763	1.19102	1.25971
4	1.04060	1.06136	1.08243	1.12551	1.16986	1.21551	1.26248	1.36049
5	1.05101	1.07728	1.10408	1.15927	1.21665	1.27628	1.33823	1.46933
6	1.06152	1.09344	1.12616	1.19405	1.26532	1.34010	1.41852	1.58687
7	1.07214	1.10984	1.14869	1.22987	1.31593	1.40710	1.50363	1.71382
8	1.08286	1.12649	1.17166	1.26677	1.36857	1.47746	1.59385	1.85093
9	1.09369	1.14339	1.19509	1.30477	1.42331	1.55133	1.68948	1.99900
10	1.10462	1.16054	1.21899	1.34392	1.48024	1.62889	1.79085	2.15892
11	1.11567	1.17795	1.24337	1.38423	1.53945	1.71034	1.89830	2.33164
12	1.12683	1.19562	1.26824	1.42576	1.60103	1.79586	2.01220	2.51817
13	1.13809	1.21355	1.29361	1.46853	1.66507	1.88565	2.13293	2.71962
14	1.14947	1.23176	1.31948	1.51259	1.73168	1.97993	2.26090	2.93719
15	1.16097	1.25023	1.34587	1.55797	1.80094	2.07893	2.39656	3.17217
16	1.17258	1.26899	1.37279	1.60471	1.87298	2.18287	2.54035	3.42594
17	1.18430	1.28802	1.40024	1.65285	1.94790	2.29202	2.69277	3.70002
18	1.19615	1.30734	1.42825	1.70243	2.02582	2.40662	2.85434	3.99602
19	1.20811	1.32695	1.45681	1.75351	2.10685	2.52695	3.02560	4.31570
20	1.22019	1.34686	1.48595	1.80611	2.19112	2.65330	3.20714	4.66096
21	1.23239	1.36706	1.51567	1.86029	2.27877	2.78596	3.39956	5.03383
22	1.24472	1.38756	1.54598	1.91610	2.36992	2.92526	3.60354	5.43654
23	1.25716	1.40838	1.57690	1.97359	2.46472	3.07152	3.81975	5.87146
24	1.26973	1.42950	1.60844	2.03279	2.56330	3.22510	4.04893	6.34118
25	1.28243	1.45095	1.64061	2.09378	2.66584	3.38635	4.29187	6.84848
26	1.29526	1.47271	1.67342	2.15659	2.77247	3.55567	4.54938	7.39635
27	1.30821	1.49480	1.70689	2.22129	2.88337	3.73346	4.82235	7.98806
28	1.32129	1.51722	1.74102	2.28793	2.99870	3.92013	5.11169	8.62711
29	1.33450	1.53998	1.77584	2.35657	3.11865	4.11614	5.41839	9.31727
30	1.34785	1.56308	1.81136	2.42726	3.24340	4.32194	5.74349	10.06266
31	1.36133	1.58653	1.84759	2.50008	3.37313	4.53804	6.08810	10.86767
32	1.37494	1.61032	1.88454	2.57508	3.50806	4.76494	6.45339	11.73708
33	1.38869	1.63448	1.92223	2.65234	3.64838	5.00319	6.84059	12.67605
34	1.40258	1.65900	1.96068	2.73191	3.79432	5.25335	7.25103	13.69013
35	1.41660	1.68388	1.99989	2.81386	3.94609	5.51602	7.68609	14.78534
36	1.43077	1.70914	2.03989	2.89828	4.10393	5.79182	8.14725	15.96817
37	1.44508	1.73478	2.08069	2.98523	4.26809	6.08141	8.63609	17.24563
38	1.45953	1.76080	2.12230	3.07478	4.43881	6.38548	9.15425	18.62528
39	1.47412	1.78721	2.16474	3.16703	4.61637	6.70475	9.70351	20.11530
40	1.48886	1.81402	2.20804	3.26204	4.80102	7.03999	10.28572	21.72452
41	1.50375	1.84123	2.25220	3.35990	4.99306	7.39199	10.90286	23.46248
42	1.51879	1.86885	2.29724	3.46070	5.19278	7.76159	11.55703	25.33948
43	1.53398	1.89688	2.34319	3.56452	5.40050	8.14967	12.25045	27.36664
44	1.54932	1.92533	2.39005	3.67145	5.61652	8.55715	12.98548	29.55597
45	1.56481	1.95421	2.43785	3.78160	5.84118	8.98501	13.76461	31.92045
46	1.58046	1.98353	2.48661	3.89504	6.07482	9.43426	14.59049	34.47409
47	1.59626	2.01328	2.53634	4.01190	6.31782	9.90597	15.46592	37.23201
48	1.61223	2.04348	2.58707	4.13225	6.57053	10.40127	16.39387	40.21057
49	1.62835	2.07413	2.63881	4.25622	6.83335	10.92133	17.37750	43.42742
50	1.64463	2.10524	2.69159	4.38391	7.10668	11.46740	18.42015	46.90161

TABLE 2 Present Value

$$\frac{1}{(1+i)^n}$$

n \ i	1%	1½%	2%	3%	4%	5%	6%	8%
1	.99010	.98522	.98039	.97087	.96154	.95238	.94340	.92593
2	.98030	.97066	.96117	.94260	.92456	.90703	.89000	.85734
3	.97059	.95632	.94232	.91514	.88900	.86384	.83962	.79383
4	.96098	.94218	.92385	.88849	.85480	.82270	.79209	.73503
5	.95147	.92826	.90573	.86261	.82193	.78353	.74726	.68058
6	.94205	.91454	.88797	.83748	.79031	.74622	.70496	.63017
7	.93272	.90103	.87056	.81309	.75992	.71068	.66506	.58349
8	.92348	.88771	.85349	.78941	.73069	.67684	.62741	.54027
9	.91434	.87459	.83676	.76642	.70259	.64461	.59190	.50025
10	.90529	.86167	.82035	.74409	.67556	.61391	.55839	.46319
11	.89632	.84893	.80426	.72242	.64958	.58468	.52679	.42888
12	.88745	.83639	.78849	.70138	.62460	.55684	.49697	.39711
13	.87866	.82403	.77303	.68095	.60057	.53032	.46884	.36770
14	.86996	.81185	.75788	.66112	.57748	.50507	.44230	.34046
15	.86135	.79985	.74301	.64186	.55526	.48102	.41727	.31524
16	.85282	.78803	.72845	.62317	.53391	.45811	.39365	.29189
17	.84438	.77639	.71416	.60502	.51337	.43630	.37136	.27027
18	.83602	.76491	.70016	.58739	.49363	.41552	.35034	.25025
19	.82774	.75361	.68643	.57029	.47464	.39573	.33051	.23171
20	.81954	.74247	.67297	.55368	.45639	.37689	.31180	.21455
21	.81143	.73150	.65978	.53755	.43883	.35894	.29416	.19866
22	.80340	.72069	.64684	.52189	.42196	.34185	.27751	.18394
23	.79544	.71004	.63416	.50669	.40573	.32557	.26180	.17032
24	.78757	.69954	.62172	.49193	.39012	.31007	.24698	.15770
25	.77977	.68921	.60953	.47761	.37512	.29530	.23300	.14602
26	.77205	.67902	.59758	.46369	.36069	.28124	.21981	.13520
27	.76440	.66899	.58586	.45019	.34682	.26785	.20737	.12519
28	.75684	.65910	.57437	.43708	.33348	.25509	.19563	.11591
29	.74934	.64936	.56311	.42435	.32065	.24295	.18456	.10733
30	.74192	.63976	.55207	.41199	.30832	.23138	.17411	.09938
31	.73458	.63031	.54125	.39999	.29646	.22036	.16425	.09202
32	.72730	.62099	.53063	.38834	.28506	.20987	.15496	.08520
33	.72010	.61182	.52023	.37703	.27409	.19987	.14619	.07889
34	.71297	.60277	.51003	.36604	.26355	.19035	.13791	.07305
35	.70591	.59387	.50003	.35538	.25342	.18129	.13011	.06763
36	.69892	.58509	.49022	.34503	.24367	.17266	.12274	.06262
37	.69200	.57644	.48061	.33498	.23430	.16444	.11579	.05799
38	.68515	.56792	.47119	.32523	.22529	.15661	.10924	.05369
39	.67837	.55953	.46195	.31575	.21662	.14915	.10306	.04971
40	.67165	.55126	.45289	.30656	.20829	.14205	.09722	.04603
41	.66500	.54312	.44401	.29763	.20028	.13528	.09172	.04262
42	.65842	.53509	.43530	.28896	.19257	.12884	.08653	.03946
43	.65190	.52718	.42677	.28054	.18517	.12270	.08163	.03654
44	.64545	.51939	.41840	.27237	.17805	.11686	.07701	.03383
45	.63905	.51171	.41020	.26444	.17120	.11130	.07265	.03133
46	.63273	.50415	.40215	.25674	.16461	.10600	.06854	.02901
47	.62646	.49670	.39427	.24926	.15828	.10095	.06466	.02686
48	.62026	.48936	.38654	.24200	.15219	.09614	.06100	.02487
49	.61412	.48213	.37896	.23495	.14634	.09156	.05755	.02303
50	.60804	.47500	.37153	.22811	.14071	.08720	.05429	.02132

TABLE 3   Amount of an Annuity

$$s_{\overline{n}|i} = \frac{(1+i)^n - 1}{i}$$

n \ i	1%	1½%	2%	3%	4%	5%	6%	8%
1	1.00000	1.00000	1.00000	1.00000	1.00000	1.00000	1.00000	1.00000
2	2.01000	2.01500	2.02000	2.03000	2.04000	2.05000	2.06000	2.08000
3	3.03010	3.04523	3.06040	3.09090	3.12160	3.15250	3.18360	3.24640
4	4.06040	4.09090	4.12161	4.18363	4.24646	4.31013	4.37462	4.50611
5	5.10101	5.15227	5.20404	5.30914	5.41632	5.52563	5.63709	5.86660
6	6.15202	6.22955	6.30812	6.46841	6.63298	6.80191	6.97532	7.33593
7	7.21354	7.32299	7.43428	7.66246	7.89829	8.14201	8.39384	8.92280
8	8.28567	8.43284	8.58297	8.89234	9.21423	9.54911	9.89747	10.63663
9	9.36853	9.55933	9.75463	10.15911	10.58280	11.02656	11.49132	12.48756
10	10.46221	10.70272	10.94972	11.46388	12.00611	12.57789	13.18079	14.48656
11	11.56683	11.86326	12.16872	12.80780	13.48635	14.20679	14.97164	16.64549
12	12.68250	13.04121	13.41209	14.19203	15.02581	15.91713	16.86994	18.97713
13	13.80933	14.23683	14.68033	15.61779	16.62684	17.71298	18.88214	21.49530
14	14.94742	15.45038	15.97394	17.08632	18.29191	19.59863	21.01507	24.21492
15	16.09690	16.68214	17.29342	18.59891	20.02359	21.57856	23.27597	27.15211
16	17.25786	17.93237	18.63929	20.15688	21.82453	23.65749	25.67253	30.32428
17	18.43044	19.20136	20.01207	21.76159	23.69751	25.84037	28.21288	33.75023
18	19.61475	20.48938	21.41231	23.41444	25.64541	28.13238	30.90565	37.45024
19	20.81090	21.79672	22.84056	25.11687	27.67123	30.53900	33.75999	41.44626
20	22.01900	23.12367	24.29737	26.87037	29.77808	33.06595	36.78559	45.76196
21	23.23919	24.47052	25.78332	28.67649	31.96920	35.71925	39.99273	50.42292
22	24.47159	25.83758	27.29898	30.53678	34.24797	38.50521	43.39229	55.45676
23	25.71630	27.22514	28.84496	32.45288	36.61789	41.43048	46.99583	60.89330
24	26.97346	28.63352	30.42186	34.42647	39.08260	44.50200	50.81558	66.76476
25	28.24320	30.06302	32.03030	36.45926	41.64591	47.72710	54.86451	73.10594
26	29.52563	31.51397	33.67091	38.55304	44.31174	51.11345	59.15638	79.95442
27	30.82089	32.98668	35.34432	40.70963	47.08421	54.66913	63.70577	87.35077
28	32.12910	34.48148	37.05121	42.93092	49.96758	58.40258	68.52811	95.33883
29	33.45039	35.99870	38.79223	45.21885	52.96629	62.32271	73.63980	103.96594
30	34.78489	37.53868	40.56808	47.57542	56.08494	66.43885	79.05819	113.28321
31	36.13274	39.10176	42.37944	50.00268	59.32834	70.76079	84.80168	123.34587
32	37.49407	40.68829	44.22703	52.50276	62.70147	75.29883	90.88978	134.21354
33	38.86901	42.29861	46.11157	55.07784	66.20953	80.06377	97.34316	145.95062
34	40.25770	43.93309	48.03380	57.73018	69.85791	85.06696	104.18375	158.62667
35	41.66028	45.59209	49.99448	60.46208	73.65222	90.32031	111.43478	172.31680
36	43.07688	47.27597	51.99437	63.27594	77.59831	95.83632	119.12087	187.10215
37	44.50765	48.98511	54.03425	66.17422	81.70225	101.62814	127.26812	203.07032
38	45.95272	50.71989	56.11494	69.15945	85.97034	107.70955	135.90421	220.31595
39	47.41225	52.48068	58.23724	72.23423	90.40915	114.09502	145.05846	238.94122
40	48.88637	54.26789	60.40198	75.40126	95.02552	120.79977	154.76197	259.05652
41	50.37524	56.08191	62.61002	78.66330	99.82654	127.83976	165.04768	280.78104
42	51.87899	57.92314	64.86222	82.02320	104.81960	135.23175	175.95054	304.24352
43	53.39778	59.79199	67.15947	85.48389	110.01238	142.99334	187.50758	329.58301
44	54.93176	61.68887	69.50266	89.04841	115.41288	151.14301	199.75803	356.94965
45	56.48107	63.61420	71.89271	92.71986	121.02939	159.70016	212.74351	386.50562
46	58.04589	65.56841	74.33056	96.50146	126.87057	168.68516	226.50812	418.42607
47	59.62634	67.55194	76.81718	100.39650	132.94539	178.11942	241.09861	452.90015
48	61.22261	69.56522	79.35352	104.40840	139.26321	188.02539	256.56453	490.13216
49	62.83483	71.60870	81.94059	108.54065	145.83373	198.42666	272.95840	530.34274
50	64.46318	73.68283	84.57940	112.79687	152.66708	209.34800	290.33590	573.77016

TABLE 4 Present Value of an Annuity

$$a_{\overline{n}|i} = \frac{1 - (1+i)^{-n}}{i}$$

n \ i	1%	1½%	2%	3%	4%	5%	6%	8%
1	.99010	.98522	.98039	.97087	.96154	.95238	.94340	.92593
2	1.97040	1.95588	1.94156	1.91347	1.88609	1.85941	1.83339	1.78326
3	2.94099	2.91220	2.88388	2.82861	2.77509	2.72325	2.67301	2.57710
4	3.90197	3.85438	3.80773	3.71710	3.62990	3.54595	3.46511	3.31213
5	4.85343	4.78264	4.71346	4.57971	4.45182	4.32948	4.21236	3.99271
6	5.79548	5.69719	5.60143	5.41719	5.24214	5.07569	4.91732	4.62288
7	6.72819	6.59821	6.47199	6.23028	6.00205	5.78637	5.58238	5.20637
8	7.65168	7.48593	7.32548	7.01969	6.73274	6.46321	6.20979	5.74664
9	8.56602	8.36052	8.16224	7.78611	7.43533	7.10782	6.80169	6.24689
10	9.47130	9.22218	8.98259	8.53020	8.11090	7.72173	7.36009	6.71008
11	10.36763	10.07112	9.78685	9.25262	8.76048	8.30641	7.88687	7.13896
12	11.25508	10.90751	10.57534	9.95400	9.38507	8.86325	8.38384	7.53608
13	12.13374	11.73153	11.34837	10.63496	9.98565	9.39357	8.85268	7.90378
14	13.00370	12.54338	12.10625	11.29607	10.56312	9.89864	9.29498	8.24424
15	13.86505	13.34323	12.84926	11.93794	11.11839	10.37966	9.71225	8.55948
16	14.71787	14.13126	13.57771	12.56110	11.65230	10.83777	10.10590	8.85137
17	15.56225	14.90765	14.29187	13.16612	12.16567	11.27407	10.47726	9.12164
18	16.39827	15.67256	14.99203	13.75351	12.65930	11.68959	10.82760	9.37189
19	17.22601	16.42617	15.67846	14.32380	13.13394	12.08532	11.15812	9.60360
20	18.04555	17.16864	16.35143	14.87747	13.59033	12.46221	11.46992	9.81815
21	18.85698	17.90014	17.01121	15.41502	14.02916	12.82115	11.76408	10.01680
22	19.66038	18.62082	17.65805	15.93692	14.45112	13.16300	12.04158	10.20074
23	20.45582	19.33086	18.29220	16.44361	14.85684	13.48857	12.30338	10.37106
24	21.24339	20.03041	18.91393	16.93554	15.24696	13.79864	12.55036	10.52876
25	22.02316	20.71961	19.52346	17.41315	15.62208	14.09394	12.78336	10.67478
26	22.79520	21.39863	20.12104	17.87684	15.98277	14.37519	13.00317	10.80998
27	23.55961	22.06762	20.70690	18.32703	16.32959	14.64303	13.21053	10.93516
28	24.31644	22.72672	21.28127	18.76411	16.66306	14.89813	13.40616	11.05108
29	25.06579	23.37608	21.84438	19.18845	16.98371	15.14107	13.59072	11.15841
30	25.80771	24.01584	22.39646	19.60044	17.29203	15.37245	13.76483	11.25778
31	26.54229	24.64615	22.93770	20.00043	17.58849	15.59281	13.92909	11.34980
32	27.26959	25.26714	23.46833	20.38877	17.87355	15.80268	14.08404	11.43500
33	27.98969	25.87895	23.98856	20.76579	18.14765	16.00255	14.23023	11.51389
34	28.70267	26.48173	24.49859	21.13184	18.41120	16.19290	14.36814	11.58693
35	29.40858	27.07559	24.99862	21.48722	18.66461	16.37419	14.49825	11.65457
36	30.10751	27.66068	25.48884	21.83225	18.90828	16.54685	14.62099	11.71719
37	30.79951	28.23713	25.96945	22.16724	19.14258	16.71129	14.73678	11.77518
38	31.48466	28.80505	26.44064	22.49246	19.36786	16.86789	14.84602	11.82887
39	32.16303	29.36458	26.90259	22.80822	19.58448	17.01704	14.94907	11.87858
40	32.83469	29.91585	27.35548	23.11477	19.79277	17.15909	15.04630	11.92461
41	33.49969	30.45896	27.79949	23.41240	19.99305	17.29437	15.13802	11.96723
42	34.15811	30.99405	28.23479	23.70136	20.18563	17.42321	15.22454	12.00670
43	34.81001	31.52123	28.66156	23.98190	20.37079	17.54591	15.30617	12.04324
44	35.45545	32.04062	29.07996	24.25427	20.54884	17.66277	15.38318	12.07707
45	36.09451	32.55234	29.49016	24.51871	20.72004	17.77407	15.45583	12.10840
46	36.72724	33.05649	29.89231	24.77545	20.88465	17.88007	15.52437	12.13741
47	37.35370	33.55319	30.28658	25.02471	21.04294	17.98102	15.58903	12.16427
48	37.97396	34.04255	30.67312	25.26671	21.19513	18.07716	15.65003	12.18914
49	38.58808	34.52468	31.05208	25.50166	21.34147	18.16872	15.70757	12.21216
50	39.19612	34.99969	31.42361	25.72976	21.48218	18.25593	15.76186	12.23348

ANSWERS TO SELECTED EXERCISES

TO THE STUDENT

If you want further help with this course, you may want to obtain a copy of the *Student's Solutions Manual* that accompanies this textbook. This manual provides detailed step-by-step solutions to the odd-numbered exercises in the textbook and can help you study and understand the course material. Your college bookstore either has this manual or can order it for you.

■ Chapter 1   Section 1.1 (page 8)

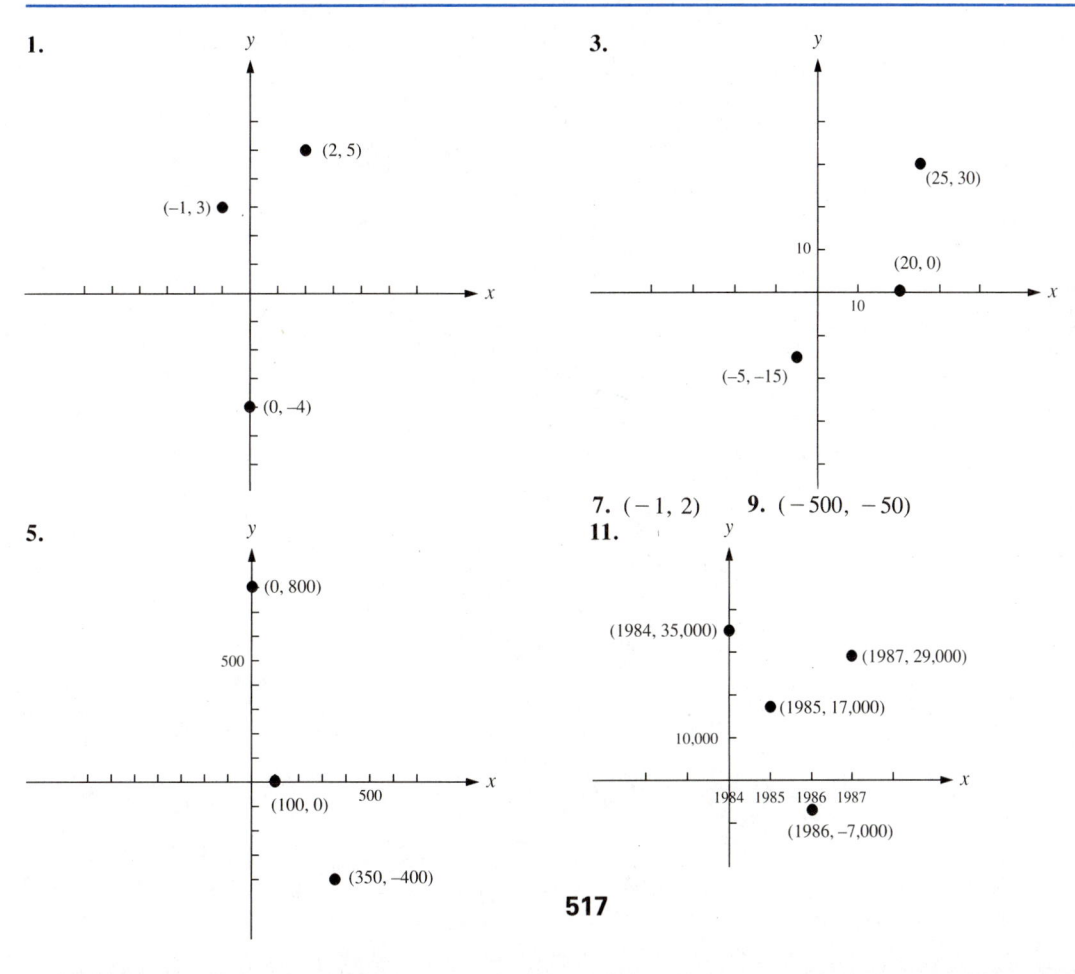

1. Points plotted: (2, 5), (−1, 3), (0, −4)

3. Points plotted: (25, 30), (20, 0), (−5, −15)

5. Points plotted: (0, 800), (100, 0), (350, −400)

7. (−1, 2)   **9.** (−500, −50)

11. Points plotted: (1984, 35,000), (1985, 17,000), (1986, −7,000), (1987, 29,000)

518 ANSWERS TO SELECTED EXERCISES

13.

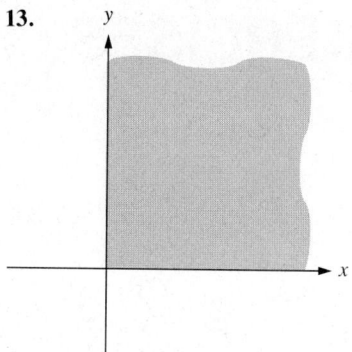

15.

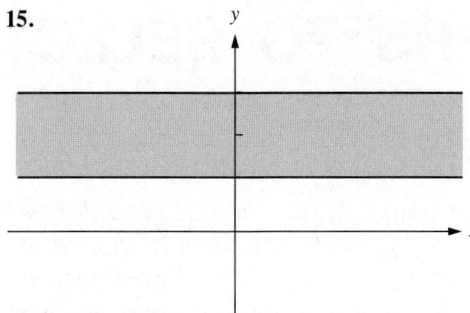

17.

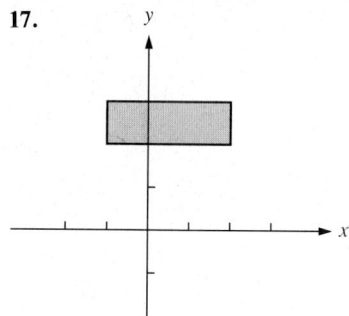

19. $-5x + y = 7$ **21.** $x + 4y = -4$

23. (0, 4), (6, 0)

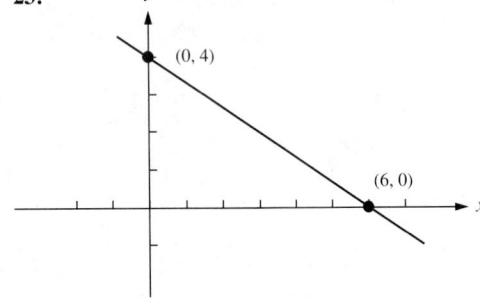

25. (6, 0), (0, −4)

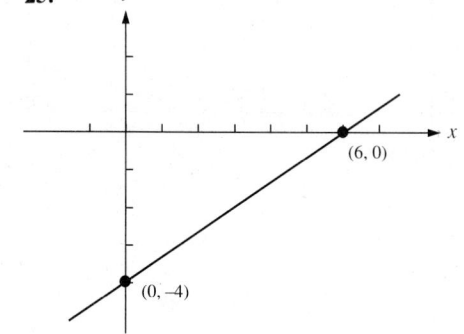

27. $(3\tfrac{1}{3}, 0)$, (0, −2)

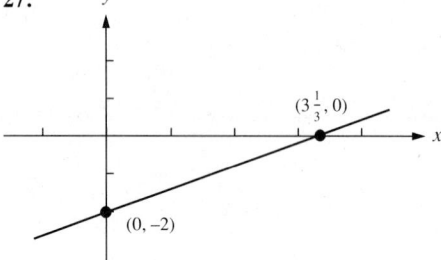

29.

(Graph showing line through (−6, 0) and (0, −4))

31.

(Graph showing line through (−10, 0) and (0, 4))

33.

(Graph showing line through (0, 150) and (90, 0))

35.

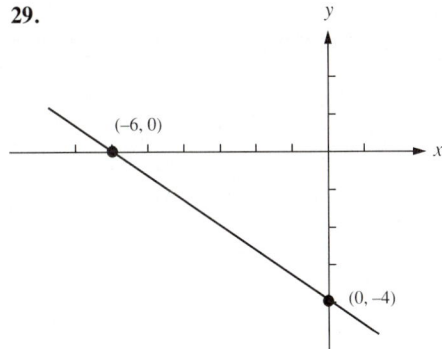

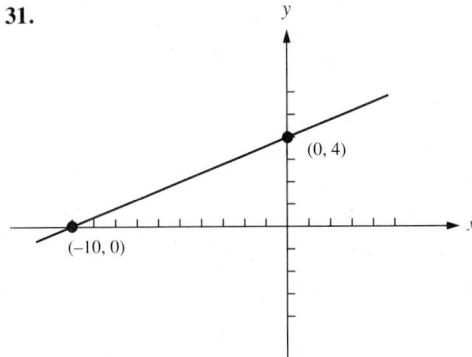

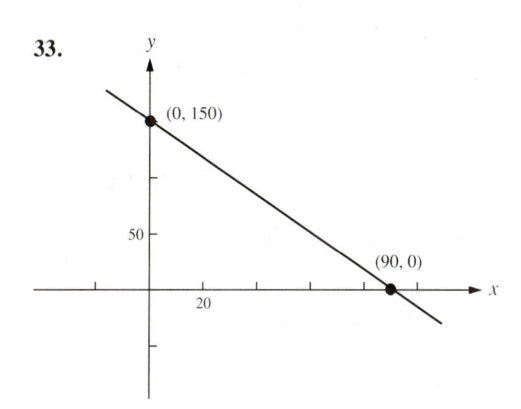

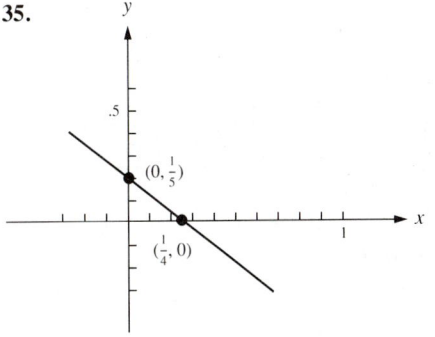

(Graph showing line through $(0, \frac{1}{5})$ and $(\frac{1}{4}, 0)$)

■ Section 1.2 (page 15)

1. $y = -\frac{3}{4}x + \frac{9}{2}$ **3.** $y = \frac{1}{3}x - \frac{5}{3}$ **5.** $y = 0x - \frac{7}{4}$ **7.** $-\frac{7}{8}, \left(0, \frac{21}{4}\right)$ **9.** $0, \left(0, -\frac{5}{3}\right)$
11. $2, \left(0, -\frac{3}{2}\right)$ **13.** no slope or y-intercept   **15.** 2   **17.** $-\frac{3}{4}$ **19.** undefined   **21.** $y = 4x - 3$
23. $y = -3x - 1$ **25.** $y = x - 2$ **27.** $x = 7$ **29.** $y = -x + 7$ **31.** $y = -\frac{8}{3}x + \frac{1}{3}$
33. $y = 0x + 2$ **35.** (a) 325,000 (b) $3,700,000   **37.** 47.5
39. (a) $V = -500t + 10{,}000$ (b) $8500, $7500   (c) 12 years

■ Section 1.3 (page 24)

1. (2, −2)   **3.** (23, 10)   **5.** (2, 1)   **7.** (6, −3)   **9.** (3, 4)   **11.** (64, 18)   **13.** (5.5, 2.5)
15. no solutions   **17.** (30, 5)   **19.** (8, −2)   **21.** 20 oz Food A, 5 oz Food B
23. (a) 1440 lbs Blend A, 1260 lbs Blend B   (b) no   **25.** 80 liters Solution A, 20 liters Solution B
27. 200 rings, 80 bracelets   **29.** $80 for stock A, $50 for stock B

Section 1.4 (page 33)

1. 11,200 units

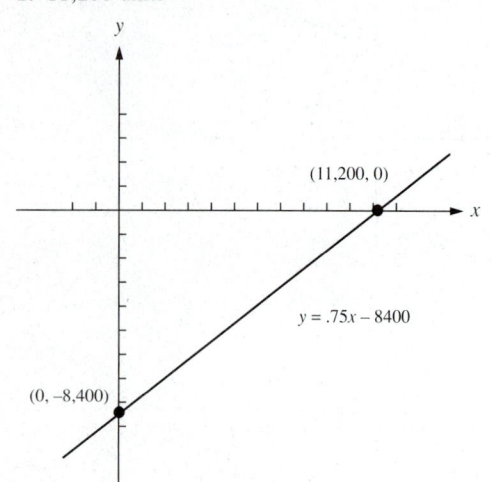

3. 1500 chips

5. (a) 19,200 (b) 16,800 (c) 33,600

7. (a) 5250 (b) 4000 (c) 5000

9. (a) 11,200 units, 12,000 units
(b) Method 2 is better for $x > 16,000$.

11. (a) 48 cars, 65 cars
(b) When > 100 cars are made.

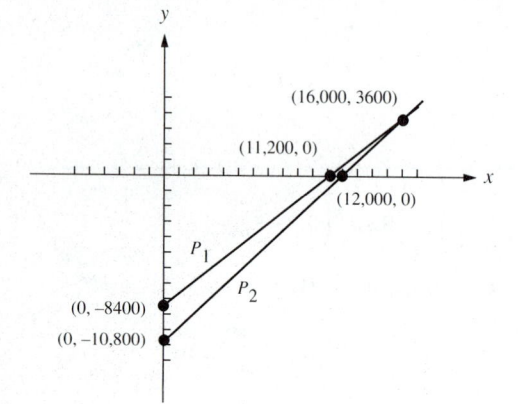

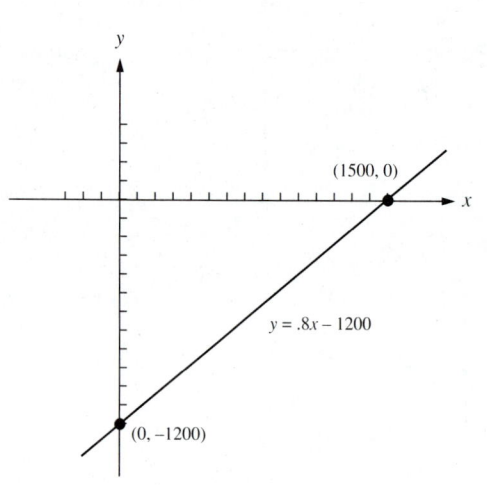

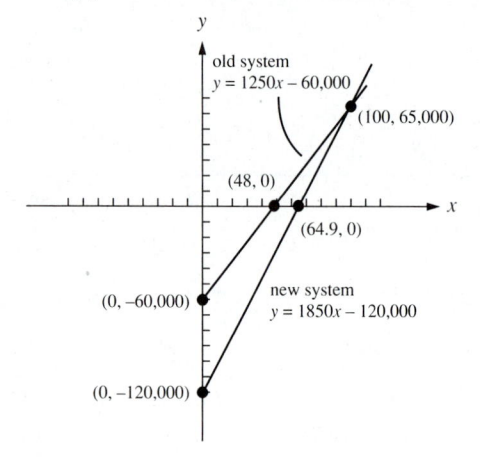

13. Printer 2 is better for > 3000 copies.

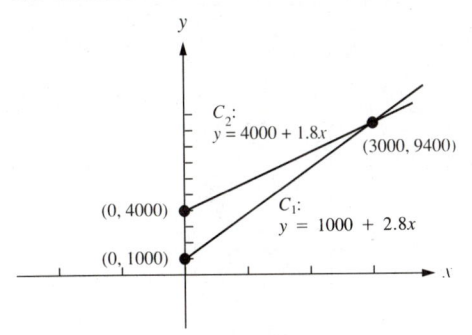

15. Use Red Cab for $x \leq 4$, White Cab for $4 \leq x \leq 10$, and Blue Cab for $x \geq 10$.

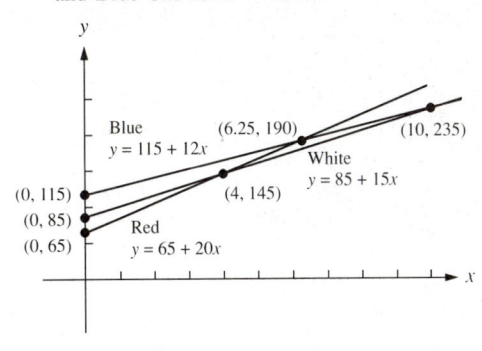

17. First company is better for $x < \$4.5$ million, second for $x > \$4.5$ million.

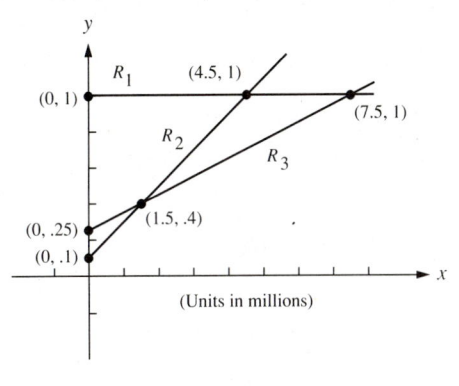

■ Section 1.5 (page 40)

1. \$6, 1500 **3.** \$7.50, 750 **5.** \$7, 800 **7.** \$13.40, 1288 **9.** \$5.63, 1406, \$791.02 **11.** \$6.67, 667, \$1111.11 **13.** \$11.47, 864, \$2973.60 **15.** \$5.50, 1375, \$1031.25 **17.** \$6.90, 760, \$380 **19.** \$12.90, 1177, \$1059.41 **21.** \$10.40, 880 **23.** \$7, 437 **25.** The factory gets \$7.38 for each box, which sells for \$6.38, and 646 boxes are sold.

Chapter 1 Review Exercises (page 42)

1.

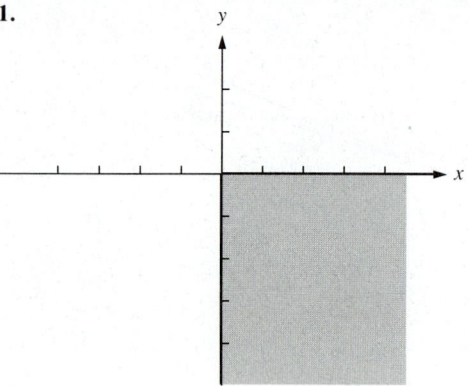

2.

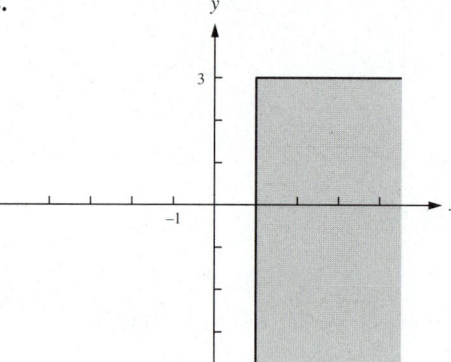

3.

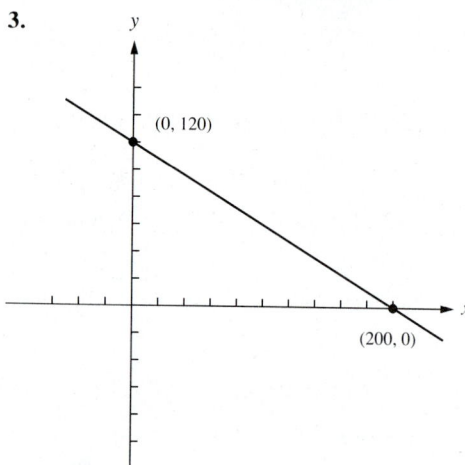

4.

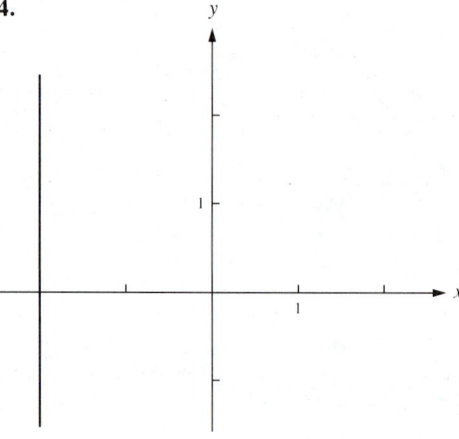

5.

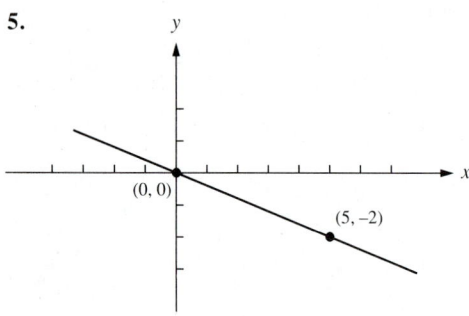

6.

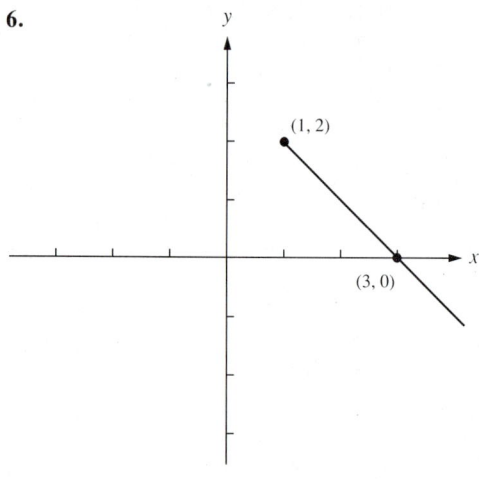

7. $-\frac{3}{2}$, $(1, 0)$, $\left(\frac{2}{3}, 0\right)$ **8.** $y = -8x + 13$ **9.** 25 **10.** $x = 2.5$, $y = -3$ **11.** $a = .1$, $b = .2$
12. 12 Fun-Packs, 1 Economy Box **13.** \$4000 in investment A and \$6000 in investment B
14. (a)

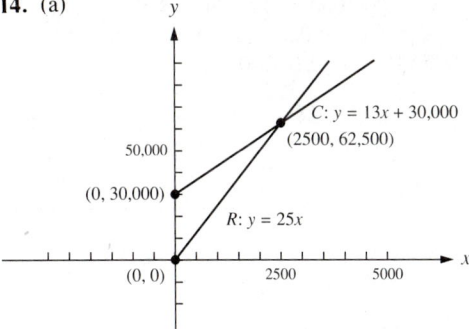

Graph showing lines C: y = 13x + 30,000 and R: y = 25x, intersecting at (2500, 62,500). Points labeled (0, 30,000), (0, 0), 2500, 5000, 50,000.

15. (a)

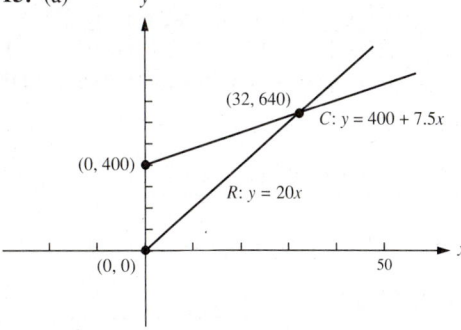

Graph showing lines C: y = 400 + 7.5x and R: y = 20x, intersecting at (32, 640). Points labeled (0, 400), (0, 0), 50.

(b) 2500   (c) 3500   (d) 3750   (e) 5000   (b) 32   (c) 46   (d) 40   (e) 50

16. \$3.75, 550   **17.** \$3.64, 519   **18.** \$3.60, 508

■ Chapter 2   Section 2.1 (page 54)

1. $\begin{bmatrix} 2 & 3 & | & 12 \\ 1 & -\frac{1}{4} & | & \frac{3}{4} \end{bmatrix}$ **3.** $\begin{bmatrix} 1 & -2 & \frac{1}{2} & | & 4 \\ 0 & \frac{1}{2} & -1 & | & -5 \\ 0 & 4 & -\frac{5}{3} & | & -2 \end{bmatrix}$ **5.** $\begin{bmatrix} 2 & -1 & 1 & 0 & | & 6 \\ 0 & 1 & \frac{5}{2} & -2 & | & 4 \\ \frac{3}{4} & -1 & 1 & 2 & | & 2 \\ -1 & 0 & 1 & 1 & | & -5 \end{bmatrix}$ **7.** $\begin{bmatrix} 14 & 0 & | & 21 \\ -4 & 1 & | & -3 \end{bmatrix}$

9. $\begin{bmatrix} 2 & -4 & 1 & | & 8 \\ 2 & -\frac{7}{2} & 0 & | & 3 \\ \frac{10}{3} & -\frac{8}{3} & 0 & | & \frac{34}{3} \end{bmatrix}$ **11.** $\begin{bmatrix} 2 & -1 & 1 & 0 & | & 6 \\ \frac{3}{8} & 0 & \frac{7}{4} & 0 & | & 3 \\ \frac{3}{8} & -\frac{1}{2} & \frac{1}{2} & 1 & | & 1 \\ -\frac{11}{8} & \frac{1}{2} & \frac{1}{2} & 0 & | & -6 \end{bmatrix}$ **13.** $x = 8$, $y = 5$ **15.** $x = 2$, $y = 1$, $z = 4$

17. $x = 0$, $y = 1$, $z = 0$ **19.** $x = 1$, $y = 3$, $z = 2$ **21.** $x = -1$, $y = 0$, $z = 1$, $w = 2$
23. 9 gal. of the 1st, 5 of the 2nd, 2 of the 3rd   **25.** 50 slices pie, 90 shortcakes, 20 baked Alaska
27. Bus 200 whites and 350 blacks from Eastern to Western, and bus 300 whites and 150 blacks from Western to Eastern.
29. $(1/a)R_i$ **31.** $R_i + R_j$, $(-1)R_j$, $R_j + R_i$, $R_i + (-1)R_j$

■ Section 2.2 (page 65)

1. $(x, y, z) = (3z, 2 - 3z, z)$ **3.** $(x, y, z) = (2 - 2y, y, -1)$ **5.** $(w, x, y, z) = (1 - 3y - 2z, -1 + 5y + 5z, y, z)$ **7.** $(u, v, w) = (1, 2, 3)$ **9.** $(r, s, t) = (8/7 + (11/7)t, -3/7 + (2/7)t, t)$
11. inconsistent   **13.** $(x, y, z, u, v) = (-5 + 5z + 2u + 4v, 3 - 2z - u - 3v, z, u, v)$ **15.** $(x, y, z) = \left(-\frac{3}{5} - \frac{1}{5}z, -\frac{4}{5} + \frac{7}{5}z, z\right)$ **17.** (a) $(x, y, z) = (4 + z, -1 - 2z, z)$ (b) $(x, y, z) = \left(\frac{7}{2} - \frac{1}{2}y, y, -\frac{1}{2} - \frac{1}{2}y\right)$
(c) $(x, y, z) = (x, 7 - 2x, -4 + x)$ **19.** $x = -50 + \frac{1}{2}z + w$, $y = 250 - \frac{3}{2}z - 2w$
21. no. lbs peanuts $= \frac{88}{9} + \frac{5}{9}$ (no. lbs almonds), no. lbs cashews $= \frac{272}{9} - \frac{14}{9}$ (no. lbs almonds)

Section 2.3 (page 74)

1. Mine 1, 12 days; mine 2, 16 days; mine 3, 3 days   **3.** 450 lbs Type A, 100 lbs Type B, 50 lbs Type C
5. (a) $180,000 + 20x + 25y + 40z + 50w = 25(12,000 + x + y + z + w)$; $2,700,000 + 125x + 100y + 75z + 25w = 150(12,000 + x + y + z + w)$ (b) $x = 3z + 5w - 24,000$, $y = -3z - 5w + 30,000$ (c) Those such that $z \geq 0$, $w \geq 0$, and $24,000 \leq 3z + 5w \leq 30,000$ **7.** $8000, $11,000, $5000, $10,800, $10,700
9. $40,000, $60,000, $20,000, $42,000, $36,000, $56,000   **11.** 200

Section 2.4 (page 84)

1. $\begin{bmatrix} 9 & 2 & 11 & 12 \\ -1 & 4 & 2 & 4 \end{bmatrix}$ **3.** undefined   **5.** $\begin{bmatrix} 3 & 4 & 11 & 4 \\ 5 & -6 & -2 & 2 \end{bmatrix}$ **7.** $\begin{bmatrix} 18 & 9 & 33 & 24 \\ 6 & -3 & 0 & 9 \end{bmatrix}$

9. $\begin{bmatrix} 12 & 1 & 11 & 16 \\ -4 & 9 & 4 & 5 \end{bmatrix}$ **11.** $\begin{bmatrix} 24 & 2 & 22 & 32 \\ -8 & 18 & 8 & 10 \end{bmatrix}$ **13.** $\begin{bmatrix} 0 & 5 & 11 & 0 \\ 8 & -11 & -4 & 1 \end{bmatrix}$

15. $\begin{bmatrix} \frac{15}{2} & \frac{15}{2} & 22 & 10 \\ \frac{17}{2} & -\frac{19}{2} & -3 & \frac{9}{2} \end{bmatrix}$ **17.** $\left[\begin{array}{ccc|c} 1 & 3 & 1 & 5 \\ 1 & 7 & -1 & 0 \end{array}\right]$ **19.** $\left[\begin{array}{ccc|c} 1 & 3 & 0 & 9 \\ 1 & 0 & 7 & 12 \end{array}\right]$ **21.** $\left[\begin{array}{cccc|c} 1 & 0 & 2 & 0 & -3 \\ 0 & 1 & 0 & -1 & 5 \\ 1 & -1 & 4 & 0 & 7 \end{array}\right]$

23. $[3\ 3\ 3\ 3\ 3]$ **25.** $\begin{bmatrix} 5 & 5 \\ 5 & 5 \\ 5 & 5 \end{bmatrix}$ **27.** $300   **29.** $E - F$ **31.** $3E + 2F$

Section 2.5 (page 91)

1. undefined   **3.** $[15]$ **5.** $\begin{bmatrix} 5 \\ 15 \\ 16 \end{bmatrix}$ **7.** $\begin{bmatrix} 8 \\ 35 \\ 15 \end{bmatrix}$ **9.** $\begin{bmatrix} 9 \\ 26 \end{bmatrix}$ **11.** $\begin{bmatrix} 17 & -4 & -11 \\ 8 & 9 & -9 \\ 17 & 3 & -16 \end{bmatrix}$ **13.** $\begin{bmatrix} -7 & 32 \\ 12 & -2 \\ 7 & 18 \end{bmatrix}$

15. undefined   **17.** $\begin{bmatrix} 0 & 12 & -12 \\ -10 & -4 & -16 \\ 30 & 3 & 57 \end{bmatrix}$ **19.** $[18\ 1\ 10]$ **21.** $\begin{bmatrix} 12 & -8 & 8 \\ 9 & -6 & 6 \\ 0 & 0 & 0 \end{bmatrix}$ **23.** $\begin{bmatrix} 101 & -58 \\ 59 & -72 \\ 116 & -102 \end{bmatrix}$

25. $AB = \begin{bmatrix} 34.5 \\ 77 \\ 47 \end{bmatrix}$ cups sugar, cups flour, number eggs   **27.** $CAB = \begin{bmatrix} 12.43 \\ 12.56 \end{bmatrix}$ Chicago cost, Detroit cost   **29.** $ZY = \begin{bmatrix} 190 \\ 285 \end{bmatrix}$ wholesale, retail

31. $WXY = [47.75]$

Section 2.6 (page 100)

1. $I_3 = \begin{bmatrix} 1 & 0 & 0 \\ 0 & 1 & 0 \\ 0 & 0 & 1 \end{bmatrix}$ **3.** $A = \begin{bmatrix} 1 & 2 & 1 \\ 2 & 0 & 2 \\ 0 & 3 & -1 \end{bmatrix}$, $X = \begin{bmatrix} x \\ y \\ z \end{bmatrix}$, $B = \begin{bmatrix} 3 \\ 5 \\ 0 \end{bmatrix}$ **5.** $A = \begin{bmatrix} 1 & 1 & 1 \\ 0 & 1 & 1 \end{bmatrix}$, $X = \begin{bmatrix} r \\ s \\ t \end{bmatrix}$, $B = \begin{bmatrix} 5 \\ 8 \end{bmatrix}$

7. $A = \begin{bmatrix} 1 & 0 & 1 \\ 1 & 1 & -1 \\ 1 & 0 & 1 \end{bmatrix}$, $X = \begin{bmatrix} x_1 \\ x_2 \\ x_3 \end{bmatrix}$, $B = \begin{bmatrix} 5 \\ 0 \\ 0 \end{bmatrix}$ **9.** $2x - 2y = 0$, $3x + 5y = 2$

11. $u + 2v = 0$, $2u + v = 2$, $3v = 5$ **13.** $\begin{bmatrix} -5 & 3 \\ 2 & -1 \end{bmatrix}$ **15.** $\begin{bmatrix} 1 & -2 & 0 \\ -1 & 3 & 0 \\ -1 & 0 & 1 \end{bmatrix}$ **17.** $\begin{bmatrix} 1 & \frac{1}{2} & -\frac{3}{2} \\ 0 & \frac{1}{2} & -\frac{3}{2} \\ -1 & 0 & 1 \end{bmatrix}$

19. no inverse   **21.** $\begin{bmatrix} 1 & -1 & 0 & 0 \\ 0 & 1 & 0 & 0 \\ 1 & 0 & 1 & -1 \\ -1 & 0 & 0 & 1 \end{bmatrix}$ **23.** $x = -8$, $y = 3$, $z = 11$ **25.** $x = -16$, $y = 11$, $z = -11$

27. $u = 4$, $v = -2$, $w = -7$ **29.** $x_1 = 16$, $x_2 = -3$, $x_3 = -5$

■ Section 2.7 (page 109)

1. nonmetals, services   **3.** $300,000, $1.65 million, $225,000, $825,000   **5.** agriculture, machinery   **7.** $5 billion, $0, $400 million, $1 billion, $1.4 billion

9. $\begin{bmatrix} .3 & .2 \\ .4 & .1 \end{bmatrix} \begin{matrix} mfg \\ engy \end{matrix}$ $\begin{matrix} mfg & engy \end{matrix}$ **11.** (a) $\begin{bmatrix} 1250 \\ 1800 \end{bmatrix}$ (b) $\begin{bmatrix} 169 \\ 204 \end{bmatrix}$ (c) $\begin{bmatrix} 1081 \\ 1596 \end{bmatrix}$ (d) diagram with 1st sector and 2nd sector, arrows labeled $.02x_1$, $.12x_1$, $.08x_2$, $.03x_2$

13. (a) $10,600 goods, $11,600 services   (b) $34,000 goods, $24,000 services
15. 40 units housing, 40 units clothing, 35 units food   **17.** 440 units electricity, 540 units oil, 360 units coal

■ Chapter 2 Review Exercises (page 112)

1. $x = -1$, $y = 2$, $z = 3$ **2.** $(x, y, z, w) = (1.2 + .4z - .6w, .6 + .2z + .2w, z, w)$ **3.** no solutions

4. 300 lbs 5-15-10, 100 lbs 20-5-5, 200 lbs 10-10-10   **5.** $\begin{bmatrix} -4 & -1 & 1 \\ 0 & -2 & 3 \\ -1 & 5 & -6 \end{bmatrix}$ **6.** $\begin{bmatrix} 0 & -1 & 2 \\ 3 & 0 & 3 \\ -5 & -2 & -1 \end{bmatrix}$

7. $\begin{bmatrix} 0 \\ -1 \\ 3 \end{bmatrix}$ **8.** undefined   **9.** $\begin{bmatrix} 1 & 1 & 1 \\ 2 & 1 & 1 \\ -3 & 1 & 1 \end{bmatrix}$ **10.** $\begin{bmatrix} 2 & 1 & 0 \\ 3 & -1 & 5 \end{bmatrix}$ **11.** $\begin{bmatrix} 1 & -1 & -1 \\ 3 & -3 & -2 \\ -2 & 3 & 2 \end{bmatrix}$ **12.** no inverse

13. $A = \begin{bmatrix} 3 & 1 & 1 \\ 1 & -1 & 1 \\ -1 & 4 & -2 \end{bmatrix}$, $X = \begin{bmatrix} x \\ y \\ z \end{bmatrix}$, $B = \begin{bmatrix} 2 \\ 0 \\ 3 \end{bmatrix}$

$A = \begin{bmatrix} 2 & 1 & -1 & 1 \\ 1 & -2 & 0 & 1 \\ -1 & 7 & -1 & -2 \end{bmatrix}$, $X = \begin{bmatrix} x \\ y \\ z \\ w \end{bmatrix}$, $B = \begin{bmatrix} 3 \\ 0 \\ 3 \end{bmatrix}$

526 ANSWERS TO SELECTED EXERCISES

$$A = \begin{bmatrix} 1 & 2 & -3 & 1 \\ 2 & -1 & 1 & -1 \\ 4 & 3 & -5 & 1 \end{bmatrix}, X = \begin{bmatrix} a \\ b \\ c \\ d \end{bmatrix}, B = \begin{bmatrix} 6 \\ -1 \\ 2 \end{bmatrix}$$

14. (a) $(x, y, z) = (-4.5, -2.5, 7.5)$

(b) $(x, y, z) = (-175, -125, 525)$ **15.** (a) $S = \begin{bmatrix} 60 \\ 80 \end{bmatrix}$ (b) 9400 units goods, 5800 units services

Chapter 3   Section 3.1 (page 123)

1. Maximize $x + 1.25y$,
subject to $8x + 4y \leq 1000$
$8x + 12y \leq 1500$
$x \geq 0, y \geq 0$

3.

	Big Barrel	Supersack	Needed
White pieces	10	7	100
Dark pieces	10	4	80
Cost	$9	$6	
Number bought	x	y	

Minimize $9x + 6y$
subject to $10x + 7y \geq 100$
$10x + 4y \geq 80$
$x \geq 0, y \geq 0$

5. Maximize $110x + 120y$
subject to $90x + 80y \leq 72{,}000$
$10x + 20y \leq 10{,}000$
$x \geq 0, y \geq 0$

7. Maximize $3.5x + 4y$
subject to $.25x + .6y \leq 300$
$.5x + .2y \leq 250$
$.25x + .2y \leq 150$
$x \geq 0, y \geq 0$

9. Let x amethysts and y beryls be finished.
Maximize $75x + 100y$
subject to $x + 2y \leq 40$
$x + 5y \leq 40$
$2x + 2y \leq 40$
$x \geq 0, y \geq 0$

11. Let x rolls from Company A and y rolls from Company B be ordered.
Minimize $100x + 300y$
subject to $2x + 5y \geq 1000$
$35x + 25y \leq 10{,}500$
$x \geq 0, y \geq 0$

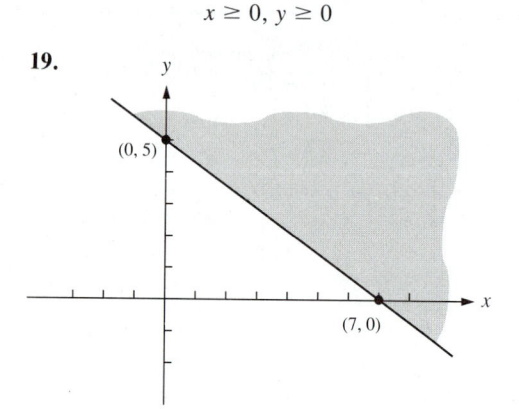

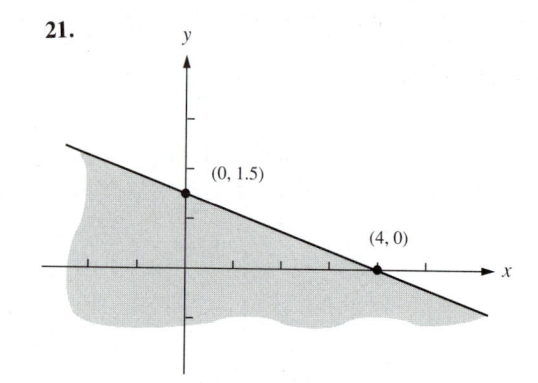

13. $(3, 5)$ **15.** $(5, 1)$ **17.** $(1, 4, 0)$

19.

21.

23.

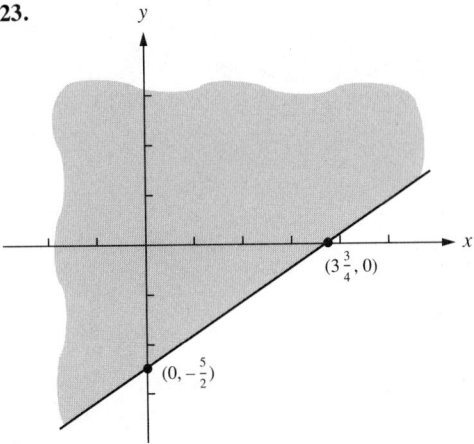

Points: $(3\tfrac{3}{4}, 0)$, $(0, -\tfrac{5}{2})$

25.

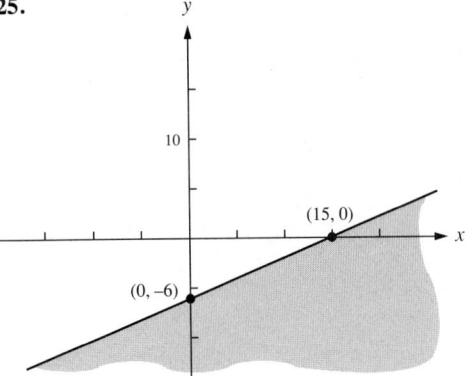

Points: $(15, 0)$, $(0, -6)$

27.

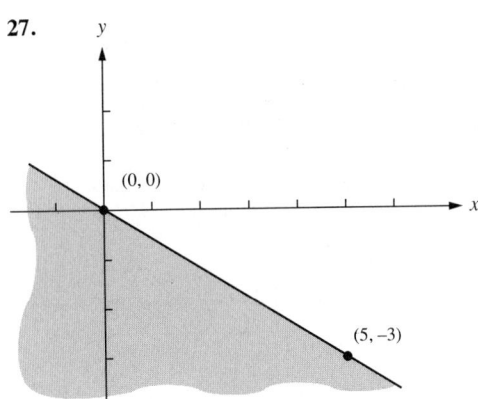

Points: $(0, 0)$, $(5, -3)$

29.

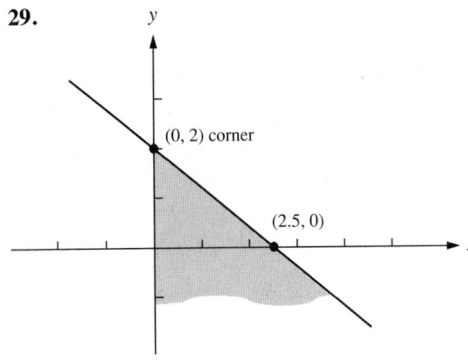

$(0, 2)$ corner, $(2.5, 0)$

31.

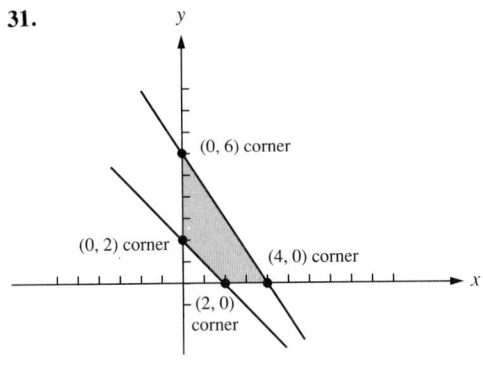

$(0, 6)$ corner, $(0, 2)$ corner, $(4, 0)$ corner, $(2, 0)$ corner

33.

$(0, 2)$ corner, $(3.5, 0)$ corner, $(0, 0)$ corner

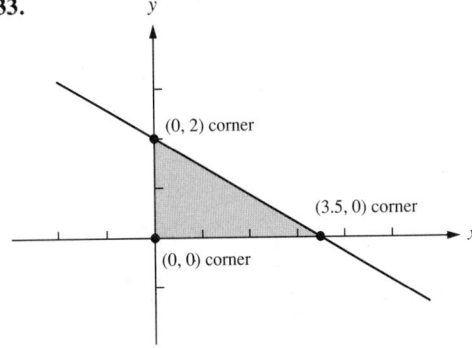

35.

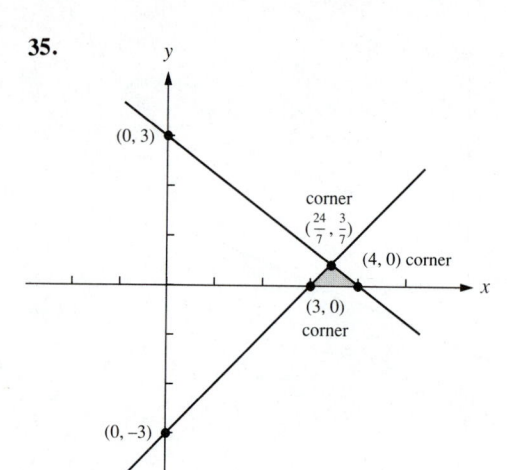

(0, 3), (0, −3), (3, 0) corner, (4, 0) corner, corner $\left(\frac{24}{7}, \frac{3}{7}\right)$

■ Section 3.2 (page 136)

1. 100 at (20, 20) **3.** $18\frac{8}{19}$ at $\left(\frac{45}{19}, \frac{20}{19}\right)$ **5.** 90 at (10, 40) **7.** 92,000 at (640, 180) **9.** 2550 at (300, 375)
11. 1625 at (15, 5) **13.** 55,600 at (220, 112) **15.** 24 animals, using 12 rats and 12 mice
17. 1800 calories, using 6 oz of Food 1 and 4 oz of Food 2 **19.** $242,400 with Albany 96 days, Boston 12 days

■ Section 3.3 (page 147)

1. $\begin{bmatrix} 1 & 3 & 1 & 0 & | & 200 \\ 2 & 5 & 0 & 1 & | & 500 \\ \hline 6 & 4 & 0 & 0 & | & R \end{bmatrix}$ **3.** $\begin{bmatrix} 2 & 4 & 1 & 0 & | & 1200 \\ 1 & 3 & 0 & 1 & | & 1000 \\ \hline 3 & 8 & 0 & 0 & | & R \end{bmatrix}$ **5.** $\begin{bmatrix} 3 & 4 & 1 & 0 & 0 & | & 300 \\ 1 & 2 & 0 & 1 & 0 & | & 120 \\ 1 & 0 & 0 & 0 & 1 & | & 30 \\ \hline 5 & 12 & 0 & 0 & 0 & | & R \end{bmatrix}$

7. $\begin{bmatrix} 2 & 1 & 1 & 1 & 0 & | & 30 \\ 0 & 1 & 3 & 0 & 1 & | & 40 \\ \hline 5 & 4 & 3 & 0 & 0 & | & R \end{bmatrix}$ **9.** $\begin{array}{c} \\ x\\ t\\ \end{array}\begin{bmatrix} x & y & s & t & \\ 1 & \frac{4}{3} & \frac{1}{3} & 0 & | & \frac{7}{3} \\ 0 & \frac{2}{3} & -\frac{1}{3} & 1 & | & \frac{23}{3} \\ \hline 0 & -2 & -1 & 0 & | & R-7 \end{bmatrix}$
$(x, y) = (7/3, 0)$

11. $\begin{array}{c} \\ s\\ x\\ \end{array}\begin{bmatrix} x & y & s & t & \\ 0 & -2 & 1 & -3 & | & -23 \\ 1 & 2 & 0 & 1 & | & 10 \\ \hline 0 & -4 & 0 & -3 & | & R-30 \end{bmatrix}$
$(x, y) = (10, 0)$

13. $\begin{array}{c} \\ y\\ t\\ \end{array}\begin{bmatrix} x & y & s & t & \\ \frac{3}{2} & 1 & \frac{1}{2} & 0 & | & 4 \\ -4 & 0 & -2 & 1 & | & -4 \\ \hline 1 & 0 & -1 & 0 & | & R-8 \end{bmatrix}$
$(x, y) = (0, 4)$

15. $\begin{array}{c} \\ x\\ y\\ \end{array}\begin{bmatrix} x & y & s & t & \\ 1 & 0 & \frac{1}{2} & -\frac{1}{4} & | & 1 \\ 0 & 1 & -\frac{1}{4} & \frac{3}{8} & | & \frac{5}{2} \\ \hline 1 & 0 & -1 & 0 & | & R-8 \end{bmatrix}$
$(x, y) = (1, 5/2)$

17. $R = 1200$ at $(x, y) = (200, 0)$ **19.** $R = 2400$ at $(x, y) = (0, 300)$ **21.** $R = 720$ at $(x, y) = (0, 60)$
23. $R = 120$ at $(x, y, z) = (0, 30, 0)$

25. $\begin{bmatrix} 6 & 5 & 1 & 0 & | & 600 \\ 3 & 4 & 0 & 1 & | & 240 \\ \hline 4.5 & 3.5 & 0 & 0 & | & R \end{bmatrix}$

Pivoting on the 3 in row 2, column 1 shows $R = 360$ at $(x, y) = (80, 0)$.

■ Section 3.4 (page 156)

1. row 2, column 1   **3.** row 1, column 1   **5.** row 1, column 1   **7.** row 1, column 2   **9.** no pivot
11. $R = 13, x = 0, y = 7$ **13.** $R = 22, x = 7, y = 0, z = 5$

15.

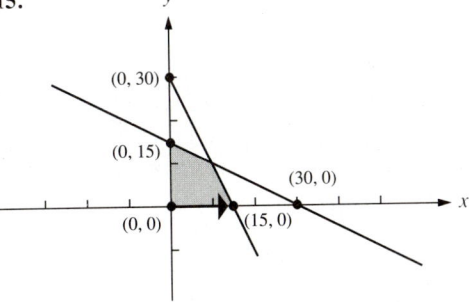

(0, 30)
(0, 15)
(30, 0)
(0, 0) (15, 0)

$R = 75$ at $(x, y) = (15, 0)$

17.

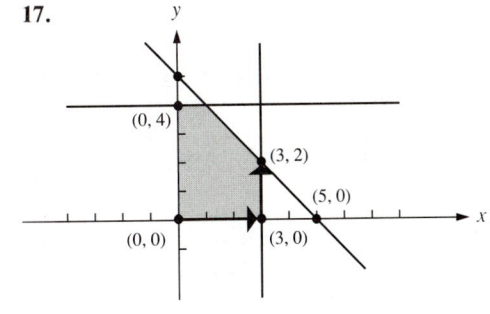

(0, 4)
(3, 2)
(5, 0)
(0, 0) (3, 0)

$R = 40$ at $(x, y) = (3, 2)$

19. $R = 40$ at $(x, y) = (20, 10)$ **21.** $R = 120$ at $(x, y, z) = (60, 0, 0)$ **23.** $R = 7$ at $(x, y, z) = (3, 1, 0)$
25. $R = 10$ at $(x, y, z, w) = (2, 4, 0, 0)$ **27.** 500 acres of asparagus, 500 acres of brussels sprouts, no cauliflower
29. $P = 1980$ at $(x, y, z) = (100, 0, 140)$

■ Section 3.5 (page 167)

1. Maximize $-5X + 20Y + 15Z = R$ **3.** Maximize $40X + 40Y + 25Z = R$ **5.** Maximize $5X + 20Z = R$
  subject to $X + Y + Z \leq 5$ subject to $X + 4Y + Z \leq 6$ subject to $-Y + 5Z \leq 10$
               $-X + 2Y + Z \leq 4$ $4X + Y + Z \leq 2$ $X + Y + 2Z \leq 20$
               $X \geq 0, Y \geq 0, Z \geq 0$ $X \geq 0, Y \geq 0, Z \geq 0$ $X \geq 0, Y \geq 0, Z \geq 0$

7. $C = 65$ at $(x, y) = (5, 10)$ **9.** $C = 70$ at $(x, y) = (5, 20)$ **11.** $C = 120$ at $(x, y) = (2, 5)$ **13.** $C = 110$ at $(x, y, z) = (10, 0, 20)$ **15.** $C = 100$ at $(x, y, z, w) = (0, 40, 60, 0)$ **17.** $1775 for 5 tons oil, 5 tons coal
19. $66,000 with 6 Type A and 12 Type B planes   **21.** $3.30 with 70 g of the first food, 30 g of the second food

■ Section 3.6 (page 183)

1. $\begin{bmatrix} x & y & s & t & a & \\ 1 & 1 & 1 & 0 & 0 & | & 20 \\ 3 & 1 & 0 & -1 & 1 & | & 36 \\ \hline 3 & 5 & 0 & 0 & 0 & | & R \\ 0 & 0 & 0 & 0 & -1 & | & Q \end{bmatrix}$ **3.** $\begin{bmatrix} x & y & s & t & a & \\ 1 & 2 & 1 & 0 & 0 & | & 200 \\ 2 & 1 & 0 & -1 & 1 & | & 250 \\ \hline 3 & 1 & 0 & 0 & 0 & | & R \\ 0 & 0 & 0 & 0 & -1 & | & Q \end{bmatrix}$

5.
$$\begin{bmatrix} x & y & s & t & a & \\ 2 & 1 & 1 & 0 & 0 & 75 \\ 1 & 1 & 0 & -1 & 1 & 60 \\ \hline -3 & -6 & 0 & 0 & 0 & R \\ 0 & 0 & 0 & 0 & -1 & Q \end{bmatrix}$$

7.
$$\begin{bmatrix} x & y & s & t & a & \\ 1 & 2 & 1 & 0 & 0 & 150 \\ 2 & 3 & 0 & -1 & 1 & 240 \\ \hline -6 & -4 & 0 & 0 & 0 & R \\ 0 & 0 & 0 & 0 & -1 & Q \end{bmatrix}$$

9. $R = -5$ at $(x, y) = (0, 5)$ **11.** $R = 84$ at $(x, y) = (8, 12)$ **13.** $R = 600$ at $(x, y) = (200, 0)$
15. $C = 315$ at $(x, y) = (15, 45)$ **17.** $C = 420$ at $(x, y) = (30, 60)$ **19.** $R = 800$ at $(x, y) = (0, 400)$
21. $C = 320$ at $(x, y) = (160, 0)$ **23.** $C = 95$ at $(x, y, z) = (35, 0, 5)$ **25.** Profit is $216 with 22 of 1st style, none of 2nd style, and 25 of 3rd style. **27.** Cost is $650 for 50 lbs of mixture A and 75 lbs of mixture B.
29. Maximum $\frac{11}{3}$ at $(x, y, z) = (\frac{7}{3}, \frac{4}{3}, 0)$ **31.** Minimum 4 at $(x, y, z) = (4, 0, 0)$

■ Chapter 3 Review Exercises (page 186)

1. Let x bears and y dogs be made.
Maximize $3x + 3.2y$
subject to $15x + 18y \leq 300$
$6x + 5y \leq 120$
$12x + 10y \leq 240$
$x \geq 0, y \geq 0$

2. Let x cups of snap beans, y cups of kidney beans, and z cups of baked beans with pork and molasses be used.
Minimize $30x + 230y + 385z$
subject to $63x + 74y + 161z \geq 350$
$2x + 15y + 16z \geq 50$
$x \geq 0, y \geq 0, z \geq 0$

3. *[graph showing region with vertices $(-\frac{3}{2}, 3)$, $(\frac{11}{2}, 3)$, $(2, \frac{2}{3})$, $(0, -\frac{2}{3})$, $(3, 0)$]*

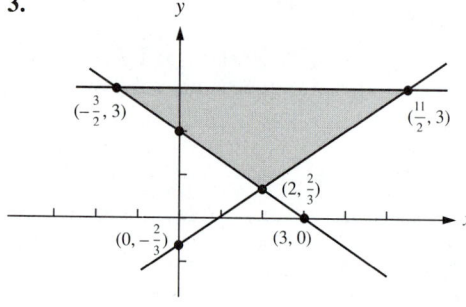

4. *[graph showing region with points $(20, 20)$, $(0, 10)$, $(20, -10)$, $(0, -20)$]*

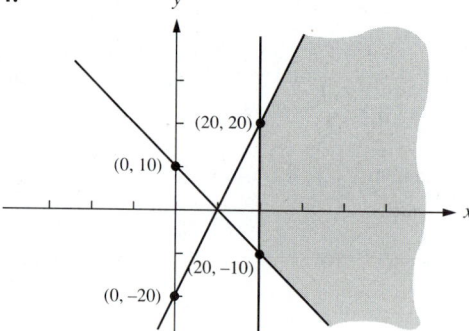

5. $C = 16$ for $(x, y) = (1, 2.4)$ **6.** No solution. Critical region unbounded for decreasing C. **7.** No solution. Critical region is empty. **8.** $R = 32/3$ at $(x, y) = (\frac{44}{15}, \frac{14}{15})$ **9.** $R = 60$ at $(x, y, z) = (0, 12, 0)$ **10.** $R = 12$ at $(x, y, z) = (0, \frac{14}{3}, \frac{4}{3})$ **11.** $C = 13$ at $(x, y, z) = (0, 4, 5)$ **12.** $C = 50$ at $(x, y, z) = (25, 0, 25)$ **13.** $R = 400$ at $(x, y) = (0, 200)$ **14.** $R = 720$ at $(x, y) = (180, 0)$ **15.** 13 tons of titanium with 100 tons from Mine A and 400 tons from Mine C. **16.** $184 with 4 super economy books and 20 "B" tickets

■ Chapter 4 Section 4.1 (page 194)

1. 3, 14, 69, 344, 1719, 8594   neither   **3.** 0, 1.5, 3, 4.5, 6, 7.5   arithmetic progression   **5.** 625, -1125, 2025, -3645, 6561, -11809.8 geometric progression   **7.** 5000, 5600, 6260, 6986, 7784.6, 8663.06   neither
9. $x_n = 6n + 2$; $x_{10} = 62$; $x_{20} = 122$ **11.** $x_n = -2.5n + 35$; $x_{10} = 10$; $x_{20} = -15$ **13.** $x_n = (1.5)^n$;

ANSWERS TO SELECTED EXERCISES 531

$x_{10} \approx 57.665$; $x_{20} \approx 3325.257$ **15.** $x_n = 100(0.9)^n$; $x_{10} \approx 34.868$; $x_{20} \approx 12.158$ **17.** $525   **19.** $24,000
21. $1440   **23.** $4200   **25.** $8228; $1428   **27.** 20%   **29.** 16%   **31.** 16%

■ Section 4.2 (page 200)

1. $7321.14   **3.** $323.21   **5.** $862.30   **7.** $2827.79   **9.** 6.17%   **11.** 7.25%   **13.** $261.42
15. (a) $739.96   (b) $743.23   (c) $744.18   (d) $744.65   **17.** $1,068,843.68   **19.** $144,892.92
21. $91.74   **23.** $3297.19   **25.** 10.97%   **27.** $121.67   **29.** about 3.74%   **31.** about 3.42%

■ Section 4.3 (page 209)

1. $x_n = 7(3^n) - 2$; $x_{10} = 413,341$ **3.** $x_n = -5(-0.6)^n + 15$; $x_{10} \approx 4.97$ **5.** $x_n = 6 - 5n$; $x_{10} = -44$
7. $x_n = 2600(1.01)^n - 2500$ **9.** $2682.42   **11.** $3933.61   **13.** $16,573.64   **15.** $8516.06
17. $49,054.30   **19.** $391.81   **21.** (a) $B_{n+1} = 1.02B_n + 50$ (b) $B_n = $3000(1.02)^n - 2500
(c) $1957.84   **23.** $49,723.17   **25.** $25,168.33   **27.** $98,225.80   **29.** $271,545.95   **31.** $13,131.70
33. not replacing the machinery   **35.** $14,214.37   **37.** $x_n = x_0 + nb$ **39.** $18,033.42

■ Section 4.4 (page 216)

1. $278.83   **3.** $233.02   **5.** $6832.89   **7.** $199.29   **9.** $138.39   **11.** $1022.74   **13.** $21,254.37
15. (a) $947.90   (b) $194,370   **17.** $9307.30   **19.** $288,652.40   **21.** $252.34   **23.** $44,201.35
25. $1179.20   **27.** $5964.77
29.

Month	Old Balance	Interest	Payment	New Balance
1	$1000.00	$15.00	$175.53	$839.47
2	$839.47	$12.59	$175.53	$676.53
3	$676.53	$10.15	$175.53	$511.15
4	$511.15	$7.67	$175.53	$343.29
5	$343.29	$5.15	$175.53	$172.91
6	$172.91	$2.59	$175.50	$0

31. $1029.34   **33.** $2218.10

35. (a) $B_{k+1} = (1 + i)B_k - p$ (b) $B_k = \left(L - \dfrac{p}{i}\right)(1 + i)^k + \dfrac{p}{i}$

■ Chapter 4   Review Exercises (page 220)

1. $x_n = 2n - 1$; $x_8 = 15$ **2.** $x_n = 125(0.8)^n$; $x_8 \approx 20.97$ **3.** $x_n = 3(4^n) - 3$; $x_8 = 196,605$
4. $x_n = 254(0.5)^n + 2$; $x_8 \approx 2.99$ **5.** $x_n = 3(-2)^n$; $x_8 = 768$ **6.** $x_n = -1500(1.02)^n + 2500$; $x_8 \approx 742.51$
7. $x_n = 7(-\frac{2}{3})^n + 1$; $x_8 \approx 1.27$ **8.** $x_n = 0.75n + 0.5$; $x_8 = 6.5$ **9.** (a) $81,000   (b) $6000   **10.** 18%
11. (a) $2970.25   (b) $2987.08   (c) $2991.03   **12.** 16.075%   **13.** $13,112.30   **14.** $12,093.83
15. (a) $B_{n+1} = 1.007B_n - 700 (b) $x_n = -$75,000(1.007)^n + $100,000$ (c) $3589.97   **16.** $184.46
17. $85,111.81   **18.** $12,416.81   **19.** $17,289,588.07   **20.** $16,299.52   **21.** $420,162.64
22. $44,399.00   **23.** $1488.89   **24.** $2051.75

ANSWERS TO SELECTED EXERCISES

Chapter 5 Section 5.1 (page 229)

1. {3, 4, 5, 6, 7, 8} **3.** {5, 6} **5.** {1, 2, 5} **7.** {1, 2, 5, 10} **9.** ∅ **11.** {u, y} **13.** ∅
15. {u, v, w, y, z} **17.** {z} **19.** {u, v, w, x, y} **21.** true **23.** true **25.** true **27.** false **29.** false
31. true **33.** (a) {a}, {b}, {c}, {d} (b) {a, b}, {a, c}, {a, d}, {b, c}, {b, d}, {c, d}
(c) {a, b, c}, {a, b, d}, {a, c, d}, {b, c, d}

35. (a) $M \cap \overline{F}$ (b) $M \cap F$ 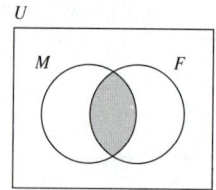 (c) $(M \cap \overline{F}) \cup (\overline{M} \cap F)$ 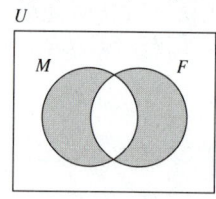 (d) $\overline{M} \cap \overline{F}$

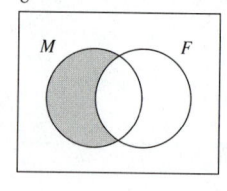

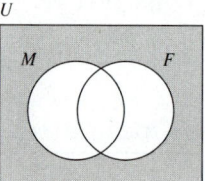

37. (a) $F \cap \overline{S} \cap E$ 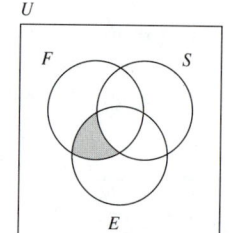 (b) $\overline{F} \cap S \cap \overline{E}$ 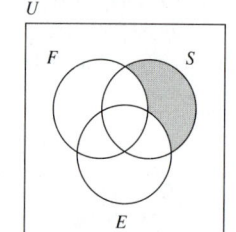 (c) $F \cap S$ 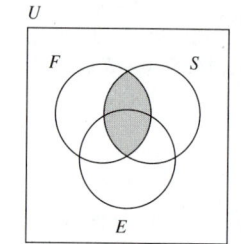 (d) $S \cup E$

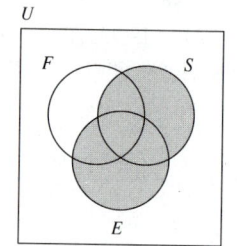

Section 5.2 (page 237)

1.

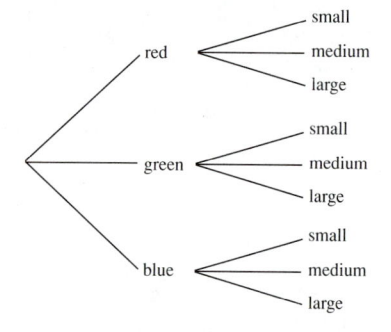

Color	Size
red	small, medium, large
green	small, medium, large
blue	small, medium, large

3. First flip, Second flip, Third flip — H/T branches producing HHH, HHT, HTH, HTT, THH, THT, TTH, TTT

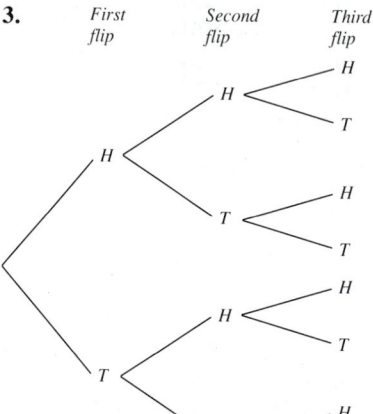

ANSWERS TO SELECTED EXERCISES 533

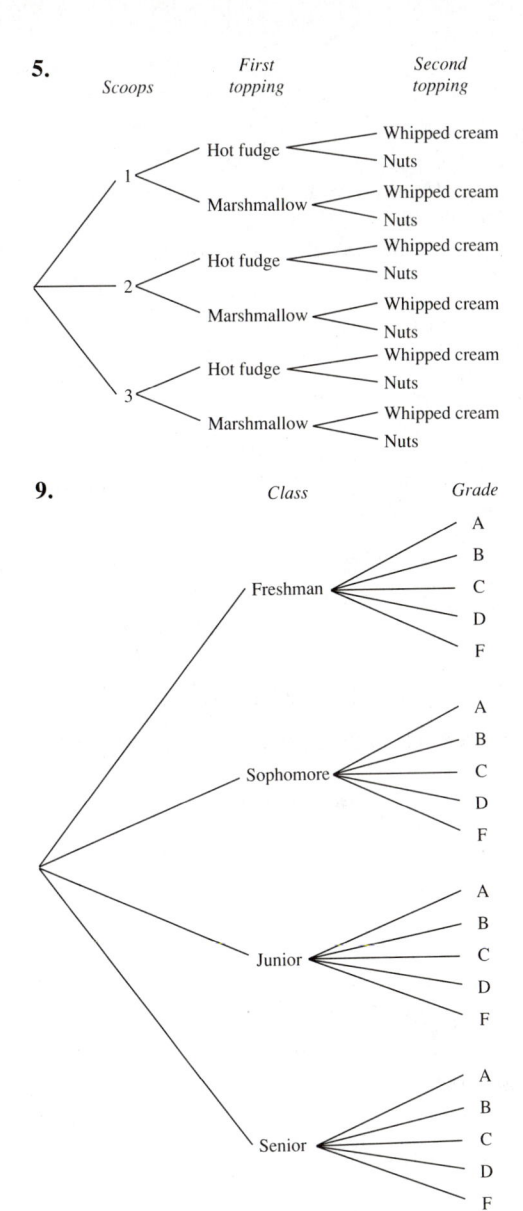

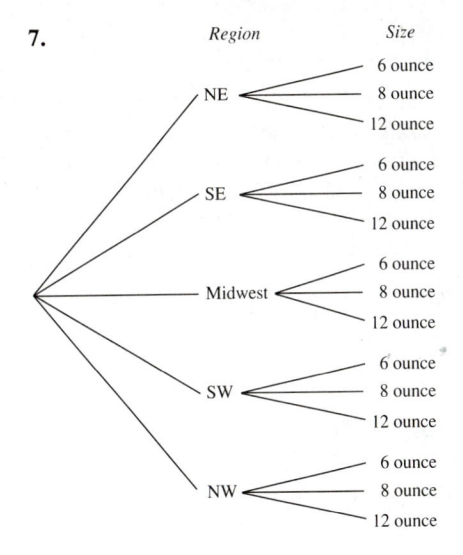

5. [tree diagram: Scoops (1, 2, 3) → First topping (Hot fudge, Marshmallow) → Second topping (Whipped cream, Nuts)]

7. [tree diagram: Region (NE, SE, Midwest, SW, NW) → Size (6 ounce, 8 ounce, 12 ounce)]

9. [tree diagram: Class (Freshman, Sophomore, Junior, Senior) → Grade (A, B, C, D, F)]

11. yes **13.** 896 **15.** (a) 7 (b) 5 **17.** (a) 167 (b) 1059 **19.** 17 are AB+, 3 are AB−, 102 are A+, 18 are A−, 68 are B+, 12 are B−, 153 are O+, and 27 are O−. **21.** The data indicates that there are 910 new policyholders, but the company claims to have only 900. **23.** $S = \varnothing$ **25** (a) 64 (b) 39 (c) 74 **27.** 430 husbands and 215 wives

■ **Section 5.3** (page 249)

1. 70 **3.** 23 **5.** 15 **7.** 21 **9.** 24 **11.** 15 **13.** 1320 **15.** 4096 **17.** 60
19. (a) 4096 (b) 244,140,625 **21.** 384 **23.** (a) 6,400,000 (b) 1,024,000,000 **25.** 1267
27. (a) 720 (b) 144 (c) 48 (d) 72 (e) 72 **29.** (a) 196 (b) 182 (c) 63
31. (a) 192 (b) 48 (c) 240 **33.** (a) 90,000 (b) 3125 (c) 120 (d) 20,000 (e) 37,512

Section 5.4 (page 260)

1. 2520 **3.** 3360 **5.** 24 **7.** 40,320 **9.** 84 **11.** 70 **13.** 1 **15.** 1
17. (a) 20; $ab, ac, ad, ae, ba, bc, bd, be, ca, cb, cd, ce, da, db, dc, de, ea, eb, ec, ed$
(b) 10; $\{a, b\}, \{a, c\}, \{a, d\}, \{a, e\}, \{b, c\}, \{b, d\}, \{b, e\}, \{c, d\}, \{c, e\}, \{d, e\}$ **19.** (a) 1,860,480 (b) 15,504
21. 720 **23.** 15,120 **25.** 56 **27.** (a) 38,760 (b) 10,010 **29.** 27,720 **31.** (a) 126 (b) 40
(c) 60 (d) 45 **33.** $C(2170, 1438) \cdot C(732, 522) \cdot C(210, 210)$ **35.** 11 **37.** (a) 2^n (b) $C(n, k)$
(c) Permutations can only be used for sequences of *distinct* objects.
39. $r \cdot C(n, r) = r \cdot \dfrac{n!}{r!(n - r)!} = \dfrac{n!}{(r - 1)!(n - r)!} = n\dfrac{(n - 1)!}{(r - 1)!(n - r)!} = n \cdot C(n - 1, r - 1)$
41. (a) 1728 (b) 576 **43.** $(n - 1)!$

Section 5.5 (page 266)

1. 1 9 36 84 126 126 84 36 9 1 **3.** $x^6 + 6x^5y + 15x^4y^2 + 20x^3y^3 + 15x^2y^4 + 6xy^5 + y^6$
5. $x^7 - 7x^6y + 21x^5y^2 - 35x^4y^3 + 35x^3y^4 - 21x^2y^5 + 7xy^6 - y^7$ **7.** $a^4 + 16a^3b + 96a^2b^2 + 256ab^3 + 256b^4$
9. $243u^5 - 405u^4v + 270u^3v^2 - 90u^2v^3 + 15uv^4 - v^5$ **11.** $64x^6 - 96x^5 + 60x^4 - 20x^3 + \dfrac{15}{4}x^2 - \dfrac{3}{8}x + \dfrac{1}{64}$
13. $1 - 8x + 28x^2 - 56x^3 + 70x^4 - 56x^5 + 28x^6 - 8x^7 + x^8$ **15.** 125,970 **17.** $-119{,}759{,}850$ **19.** 319,770
21. $-347{,}373{,}600$ **23.** 1,188,096 **25.** $-\dfrac{3003}{32}$ **27.** 92 **29.** Take $x = y = 1$ in the binomial theorem.
31. $C(n, k)$ is the number of k-element subsets of a set with n elements. So the left side of the equation counts all the subsets of a set with n elements; there are 2^n such subsets.
33. $C(n, r - 1) + C(n, r) = \dfrac{n!}{(r - 1)!(n - r + 1)!} + \dfrac{n!}{r!(n - r)!} = \dfrac{r(n!)}{r!(n - r + 1)!} + \dfrac{(n - r + 1)n!}{r!(n - r + 1)!}$
$= \dfrac{(n + 1)n!}{r!(n + 1 - r)!} = \dfrac{(n + 1)!}{r!(n + 1 - r)!} = C(n + 1, r)$

Chapter 5 Review Exercises (page 269)

1. (a) $\{1, 2, 3, 4, 5, 6, 8\}$ (b) $\{2, 5, 6\}$ (c) $\{1, 3, 4, 7, 9, 10\}$ (d) $\{2, 5, 6, 7, 8, 9, 10\}$ **2.** (a) true
(b) true (c) false (d) true (e) false (f) true (g) true (h) true **3.** 138 **4.** (a) 38 (b) 12
(c) 260

SOLUTIONS TO PRACTICE PROBLEMS

■ Chapter 1

Problem 1 If $3x - 4(0) = 6$, then
$$3x = 6$$
$$x = 2,$$
so the x-intercept is $(2, 0)$.
If $3(0) - 4y = 6$, then
$$-4y = 6$$
$$y = -\tfrac{6}{4} = -\tfrac{3}{2},$$
so the y-intercept is $(0, -3/2)$.

Problem 2 Since $m = -2$ and $(x_1, y_1) = (5, 3)$, we have
$$y - 3 = -2(x - 5)$$
$$y - 3 = -2x + 10$$
$$y = -2x + 13.$$

Problem 3 Let $(x_1, y_1) = (1, 3)$ and $(x_2, y_2) = (2, -5)$. Then
$$m = \frac{-5 - 3}{2 - 1} = -8.$$

Thus the equation of the line is
$$y - 3 = -8(x - 1)$$
$$y - 3 = -8x + 8$$
$$y = -8x + 11.$$

Problem 4 Multiplying the first equation by -3 gives

$$-3x - 9y = 15$$
$$3x - 2y = 7.$$

Adding yields

$$-11y = 22$$
$$y = -2.$$

Substituting this in the first equation gives

$$x + 3(-2) = -5$$
$$x - 6 = -5$$
$$x = 1.$$

The point of intersection is $(1, -2)$.

Problem 5 Suppose the man buys x bags of apples and y bags of pears. Then the total weight is

$$6x + 4y = 112,$$

and the total cost is

$$4x + 3y = 79.$$

We multiply the first equation by 2 and the second by -3.

$$12x + 8y = 224$$
$$-12x - 9y = -237$$

Adding gives

$$-y = -13$$
$$y = 13.$$

We substitute this in the first equation.

$$6x + 4(13) = 112$$
$$6x + 52 = 112$$
$$6x = 60$$
$$x = 10$$

He bought 10 bags of apples and 13 bags of pears.

Problem 6 (a) We have

$$R = .50x$$
$$C = .10x + 44,$$

where x is the number of glasses sold. To break even we need $R = C$, or

$$.50x = .10x + 44$$
$$.40x = 44$$
$$x = \frac{44}{.40} = 110 \text{ glasses}.$$

(b) The profit is

$$P = R - C = .50x - (.10x + 44)$$
$$= .40x - 44.$$

(c) Setting $P = \$100$ gives

$$.40x - 44 = 100$$
$$.40x = 144$$
$$x = \frac{144}{.40} = 360 \text{ glasses}.$$

Problem 7 Let x wigs be sold per week. Then

$$R = 60x$$
$$C = 25x + 21{,}000.$$

The company breaks even when $R = C$, or

$$60x = 25x + 21{,}000$$
$$35x = 21{,}000$$
$$x = \frac{21{,}000}{35} = 600 \text{ wigs}.$$

Problem 8 (a) If x_0 is the equilibrium price, then

$$10{,}000x_0 - 7000 = -8000x_0 + 20{,}000$$
$$18{,}000x_0 = 27{,}000$$
$$x_0 = \frac{27{,}000}{18{,}000} = 1.5 = \$1.50.$$

(b) The number of chips supplied at $\$1.50$ is

$$10{,}000(1.50) - 7000 = 8000.$$

Problem 9 Under a 10% sales tax the price paid by the consumer will be $x + .10x$, and so the demand will be

$$D_1 = 1200 - 150(x + .10x) = 1200 - 165x.$$

Under a 20¢ stamp tax the consumer will pay $x + .20$, and so the demand will be

$$D_2 = 1200 - 150(x + .20) = 1170 - 150x.$$

Chapter 2

Problem 1

$$\begin{bmatrix} 1 & 2 & 5 & | & 0 \\ 0 & ③ & -6 & | & 9 \\ 0 & -1 & -1 & | & 2 \end{bmatrix}$$

$$\begin{bmatrix} 1 & 2 & 5 & | & 0 \\ 0 & ① & -2 & | & 3 \\ 0 & -1 & -1 & | & 2 \end{bmatrix} \quad (\tfrac{1}{3}R_2)$$

$$\begin{bmatrix} 1 & 0 & 9 & | & -6 \\ 0 & 1 & -2 & | & 3 \\ 0 & 0 & -3 & | & 5 \end{bmatrix} \quad \begin{array}{l}(R_1 + (-2)R_2)\\ \\ (R_3 + R_2)\end{array}$$

Problem 2

$$\begin{bmatrix} 3 & 0 & 1 & | & 0 \\ ① & 1 & 1 & | & 4 \\ 0 & 3 & -1 & | & 3 \end{bmatrix}$$

$$\begin{bmatrix} 0 & ① & \tfrac{2}{3} & | & 4 \\ 1 & 1 & 1 & | & 4 \\ 0 & 3 & -1 & | & 3 \end{bmatrix} \quad (-\tfrac{1}{3}R_1)$$

$$\begin{bmatrix} 0 & ⊖3 & -2 & | & -12 \\ 1 & 1 & 1 & | & 4 \\ 0 & 3 & -1 & | & 3 \end{bmatrix} \quad (R_1 + (-3)R_2)$$

$$\begin{bmatrix} 0 & 1 & \tfrac{2}{3} & | & 4 \\ 1 & 0 & \tfrac{1}{3} & | & 0 \\ 0 & 0 & ⊖3 & | & -9 \end{bmatrix} \quad \begin{array}{l}(R_2 + (-1)R_1)\\ (R_3 + (-3)R_1)\end{array}$$

$$\begin{bmatrix} 0 & 1 & \tfrac{2}{3} & | & 4 \\ 1 & 0 & \tfrac{1}{3} & | & 0 \\ 0 & 0 & ① & | & 3 \end{bmatrix} \quad (-\tfrac{1}{3}R_3)$$

$$\begin{bmatrix} 0 & 1 & 0 & | & 2 \\ 1 & 0 & 0 & | & -1 \\ 0 & 0 & 1 & | & 3 \end{bmatrix} \quad \begin{array}{l}(R_1 + (-\tfrac{2}{3})R_3)\\ (R_2 + (-\tfrac{1}{3})R_3)\end{array}$$

$$\begin{bmatrix} 1 & 0 & 0 & | & -1 \\ 0 & 1 & 0 & | & 2 \\ 0 & 0 & 1 & | & 3 \end{bmatrix} \quad \begin{array}{l}(R_2)\\ (R_1)\end{array}$$

Thus $x = -1, y = 2, z = 3$.

Problem 3 Let there be x toasters, y blenders, and z mixers. Then

$$x + y + z = 12$$
$$6x + 5y + 3z = 57$$
$$20x + 15y + 15z = 205.$$

We give the result of each pivot.

$$\begin{bmatrix} ① & 1 & 1 & | & 12 \\ 6 & 5 & 3 & | & 57 \\ 20 & 15 & 15 & | & 205 \end{bmatrix} \quad \begin{bmatrix} 1 & 1 & 1 & | & 12 \\ 0 & ⊖1 & -3 & | & -15 \\ 0 & -5 & -5 & | & -35 \end{bmatrix}$$

$$\begin{bmatrix} 1 & 0 & -2 & | & -3 \\ 0 & 1 & 3 & | & 15 \\ 0 & 0 & ⑩ & | & 40 \end{bmatrix} \quad \begin{bmatrix} 1 & 0 & 0 & | & 5 \\ 0 & 1 & 0 & | & 3 \\ 0 & 0 & 1 & | & 4 \end{bmatrix}$$

Thus $x = 5, y = 3,$ and $z = 4$.

SOLUTIONS TO PRACTICE PROBLEMS   551

Problem 4
$$\begin{bmatrix} ① & 2 & -1 & | & 3 \\ 2 & 0 & 1 & | & 2 \\ 4 & 4 & -1 & | & 6 \end{bmatrix} \quad \begin{bmatrix} 1 & 2 & -1 & | & 3 \\ 0 & -4 & 3 & | & -4 \\ 0 & -4 & 3 & | & -6 \end{bmatrix} \quad \begin{matrix} (R_2 + (-2)R_1) \\ (R_3 + (-4)R_1) \end{matrix}$$

$$\begin{bmatrix} 1 & 2 & -1 & | & 3 \\ 0 & ① & -\frac{3}{4} & | & 1 \\ 0 & -4 & 3 & | & -6 \end{bmatrix} \quad (-\tfrac{1}{4}R_2) \quad \begin{bmatrix} 1 & 0 & \frac{1}{2} & | & 1 \\ 0 & 1 & -\frac{3}{4} & | & 1 \\ 0 & 0 & 0 & | & -2 \end{bmatrix} \quad \begin{matrix} (R_1 + (-2)R_2) \\ \\ (R_3 + 4R_2) \end{matrix}$$

Since the last row has the form $0\ 0\ \ldots\ 0\ |\ a$, with $a = -2 \neq 0$, there is no solution.

Problem 5
$$\begin{bmatrix} 1 & 2 & 1 & | & 3 \\ 0 & ① & -1 & | & -2 \end{bmatrix} \quad \begin{bmatrix} 1 & 0 & 3 & | & 7 \\ 0 & 1 & -1 & | & -2 \end{bmatrix} \quad (R_1 + (-2)R_2)$$

The last tableau corresponds to
$$\begin{aligned} x \quad\quad + 3z &= 7 \\ y - z &= -2. \end{aligned}$$

Thus $(x, y, z) = (-3z + 7, z - 2, z)$, where z is any real number.

Problem 6 Suppose she buys x standard, y large, and z jumbo rolls.
$$\begin{aligned} x + y + z &= 30 \\ 9x + 12y + 18z &= 357 \end{aligned}$$

$$\begin{bmatrix} ① & 1 & 1 & | & 30 \\ 9 & 12 & 18 & | & 357 \end{bmatrix} \quad \begin{bmatrix} 1 & 1 & 1 & | & 30 \\ 0 & ③ & 9 & | & 87 \end{bmatrix} \quad (R_2 + (-9)R_1)$$

$$\begin{bmatrix} 1 & 1 & 1 & | & 30 \\ 0 & ① & 3 & | & 29 \end{bmatrix} \quad (\tfrac{1}{3}R_2) \quad \begin{bmatrix} 1 & 0 & -2 & | & 1 \\ 0 & 1 & 3 & | & 29 \end{bmatrix} \quad (R_1 + (-1)R_2)$$

The corresponding equations are $x - 2z = 1$, $y + 3z = 29$. Thus she must buy $1 + 2z$ standard rolls and $29 - 3z$ large rolls.

Problem 7 (a) Calories provided $= 70x + 80y = 1090$,
grams of protein provided $= 2x + 6y = 46$.
(b)
$$\begin{bmatrix} 70 & 80 & | & 1090 \\ ② & 6 & | & 46 \end{bmatrix} \quad \begin{bmatrix} 70 & 80 & | & 1090 \\ ① & 3 & | & 23 \end{bmatrix} \quad (\tfrac{1}{2}R_2)$$

$$\begin{bmatrix} 0 & -130 & | & -520 \\ 1 & 3 & | & 23 \end{bmatrix} \quad (R_1 + (-70)R_2) \quad \begin{bmatrix} 0 & ① & | & 4 \\ 1 & 3 & | & 23 \end{bmatrix} \quad (-\tfrac{1}{130}R_1)$$

$$\begin{bmatrix} 0 & 1 & | & 4 \\ 1 & 0 & | & 11 \end{bmatrix} \quad (R_2 + (-3)R_1)$$

The solution $y = 4$, $x = 11$ tells us that 11 slices of bread and 4 eggs should be provided per week.

Problem 8 (a) From the first two rows of the table

$$x_1 = 400 + 0.2x_1 + 0.1x_2$$
$$x_2 = 210 + 0.2x_1 + 0.2x_2.$$

(b)

$$0.8x_1 - 0.1x_2 = 400$$
$$-0.2x_1 + 0.8x_2 = 210$$

$$\begin{bmatrix} .8 & -.1 & | & 400 \\ -.2 & .8 & | & 210 \end{bmatrix}$$

$$\begin{bmatrix} 0 & 3.1 & | & 1240 \\ 1 & -4 & | & -1050 \end{bmatrix} \quad \begin{bmatrix} 0 & 1 & | & 400 \\ 1 & 0 & | & 550 \end{bmatrix}$$

We see that $x_2 = 400$ and $x_1 = 550$. The equations corresponding to (2.11) are

$$x_3 = 500 + 0.2x_1 + 0.4x_2,$$
$$x_4 = 600 + 0.4x_1 + 0.3x_2.$$

Substituting the above values for x_2 and x_1 gives $x_3 = 770$ and $x_4 = 940$. Thus $770 should be alloted to Direct Sales and $940 to Mail Order.

Problem 9 Let u, v, w, x, y, and z denote the number of cars per hour on certain sections of streets, and let A, B, C, D, and E denote the intersections shown on the diagram.

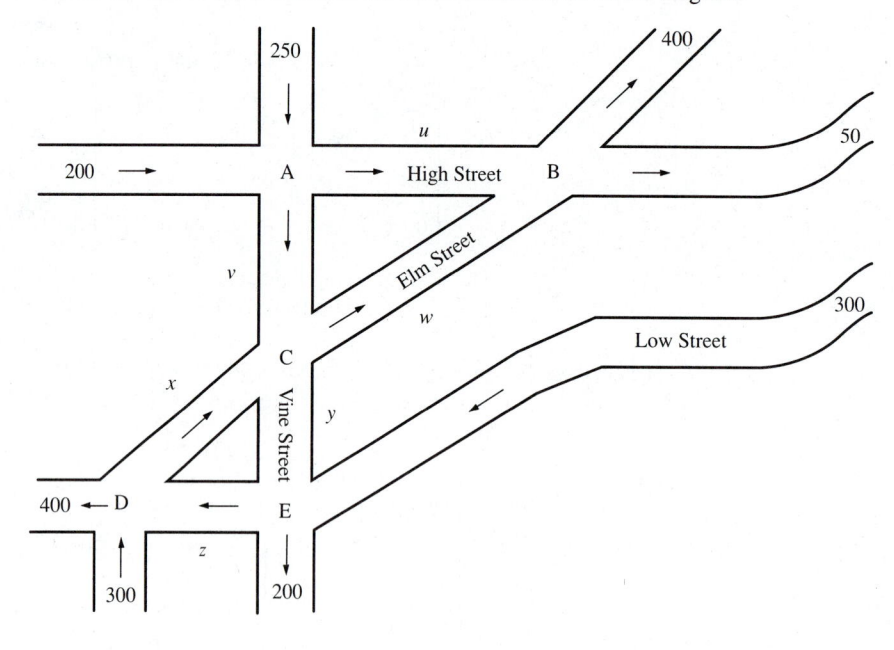

SOLUTIONS TO PRACTICE PROBLEMS 553

$$A: 250 + 200 = u + v$$
$$B: u + w = 400 + 50$$
$$C: v + x = w + y$$
$$D: 300 + z = 400 + x$$
$$E: 300 + y = 200 + z$$

or

$$u + v = 450$$
$$u + w = 450$$
$$ v - w + x - y = 0$$
$$ x - z = -100$$
$$ y - z = -100.$$

$$\begin{bmatrix} 1 & 1 & 0 & 0 & 0 & 0 & | & 450 \\ 1 & 0 & 1 & 0 & 0 & 0 & | & 450 \\ 0 & 1 & -1 & 1 & -1 & 0 & | & 0 \\ 0 & 0 & 0 & 1 & 0 & -1 & | & -100 \\ 0 & 0 & 0 & 0 & 1 & -1 & | & -100 \end{bmatrix}$$

Gauss-Jordan elimination produces

$$\begin{bmatrix} 1 & 0 & 1 & 0 & 0 & 0 & | & 450 \\ 0 & 1 & -1 & 0 & 0 & 0 & | & 0 \\ 0 & 0 & 0 & 1 & 0 & -1 & | & -100 \\ 0 & 0 & 0 & 0 & 1 & -1 & | & -100 \\ 0 & 0 & 0 & 0 & 0 & 0 & | & 0 \end{bmatrix}$$

which gives the equations

$$u = 450 - w$$
$$v = w$$
$$x = -100 + z$$
$$y = -100 + z.$$

Since no variable can be negative, we have $w \leq 450$ and $z \geq 100$. In particular, the number of cars per hour between Vine and Elm on Low must be at least 100.

Problem 10

$$\begin{bmatrix} 3 & 0 \\ -1 & 5 \\ -2 & 1 \end{bmatrix} + \begin{bmatrix} 2 & 2 \\ 6 & 4 \\ 5 & -3 \end{bmatrix} = \begin{bmatrix} 3+2 & 0+2 \\ -1+6 & 5+4 \\ -2+5 & 1+(-3) \end{bmatrix} = \begin{bmatrix} 5 & 2 \\ 5 & 9 \\ 3 & -2 \end{bmatrix}$$

Problem 11

$$2A - 3B = 2\begin{bmatrix} 1 & 3 & -1 \\ 0 & 2 & 1 \end{bmatrix} - 3\begin{bmatrix} -2 & 0 & 1 \\ 3 & -3 & 5 \end{bmatrix}$$
$$= \begin{bmatrix} 2 & 6 & -2 \\ 0 & 4 & 2 \end{bmatrix} - \begin{bmatrix} -6 & 0 & 3 \\ 9 & -9 & 15 \end{bmatrix}$$
$$= \begin{bmatrix} 8 & 6 & -5 \\ -9 & 13 & -13 \end{bmatrix}$$

SOLUTIONS TO PRACTICE PROBLEMS

Problem 12
$$\begin{bmatrix} 1 & 3 & 2 \\ 0 & -1 & 1 \end{bmatrix} \begin{bmatrix} -1 \\ 3 \\ 0 \end{bmatrix} = \begin{bmatrix} 1(-1) + 3 \cdot 3 + 2 \cdot 0 \\ 0(-1) + (-1)3 + 1 \cdot 0 \end{bmatrix} = \begin{bmatrix} 8 \\ -3 \end{bmatrix}$$

Problem 13
$$AB = \begin{bmatrix} 1 \cdot 1 + 2 \cdot 2 & 1 \cdot 0 + 2 \cdot 3 & 1(-1) + 2 \cdot 0 \\ 3 \cdot 1 + (-1)2 & 3 \cdot 0 + (-1)3 & 3(-1) + (-1)0 \end{bmatrix} = \begin{bmatrix} 5 & 6 & -1 \\ 1 & -3 & -3 \end{bmatrix}$$

Problem 14

$$\begin{bmatrix} 0 & 1 & 3 & | & 1 & 0 & 0 \\ 0 & ① & 2 & | & 0 & 1 & 0 \\ 1 & 0 & -1 & | & 0 & 0 & 1 \end{bmatrix} \qquad \begin{bmatrix} 0 & 0 & ① & | & 1 & -1 & 0 \\ 0 & 1 & 2 & | & 0 & 1 & 0 \\ 1 & 0 & -1 & | & 0 & 0 & 1 \end{bmatrix} \quad (R_1 + (-1)R_2)$$

$$\begin{bmatrix} 0 & 0 & 1 & | & 1 & -1 & 0 \\ 0 & 1 & 0 & | & -2 & 3 & 0 \\ 1 & 0 & 0 & | & 1 & -1 & 1 \end{bmatrix} \quad \begin{matrix} (R_2 + (-2)R_1) \\ (R_3 + R_1) \end{matrix} \qquad \begin{bmatrix} 1 & 0 & 0 & | & 1 & -1 & 1 \\ 0 & 1 & 0 & | & -2 & 3 & 0 \\ 0 & 0 & 1 & | & 1 & -1 & 0 \end{bmatrix} \quad \begin{matrix} (R_3) \\ (R_1) \end{matrix}$$

We see that the inverse is

$$\begin{bmatrix} 1 & -1 & 1 \\ -2 & 3 & 0 \\ 1 & -1 & 0 \end{bmatrix}.$$

Problem 15

$$X = A^{-1}B = \begin{bmatrix} 1 & -1 & 1 \\ -2 & 3 & 0 \\ 1 & -1 & 0 \end{bmatrix} \begin{bmatrix} 3 \\ -1 \\ 5 \end{bmatrix}$$

$$= \begin{bmatrix} 1 \cdot 3 + (-1)(-1) + 1 \cdot 5 \\ (-2)3 + 3(-1) + 0 \cdot 5 \\ 1 \cdot 3 + (-1)(-1) + 0 \cdot 5 \end{bmatrix} = \begin{bmatrix} 9 \\ -9 \\ 4 \end{bmatrix}$$

Problem 16 (a)

$$\begin{bmatrix} 2 & 1 & 0 & | & 1 & 0 & 0 \\ 0 & 1 & -1 & | & 0 & 1 & 0 \\ ① & 0 & 1 & | & 0 & 0 & 1 \end{bmatrix} \qquad \begin{bmatrix} 0 & 1 & -2 & | & 1 & 0 & -2 \\ 0 & ① & -1 & | & 0 & 1 & 0 \\ 1 & 0 & 1 & | & 0 & 0 & 1 \end{bmatrix}$$

$$\begin{bmatrix} 0 & 0 & -1 & | & 1 & -1 & -2 \\ 0 & 1 & -1 & | & 0 & 1 & 0 \\ 1 & 0 & 1 & | & 0 & 0 & 1 \end{bmatrix} \qquad \begin{bmatrix} 0 & 0 & ① & | & -1 & 1 & 2 \\ 0 & 1 & -1 & | & 0 & 1 & 0 \\ 1 & 0 & 1 & | & 0 & 0 & 1 \end{bmatrix}$$

$$\begin{bmatrix} 0 & 0 & 1 & | & -1 & 1 & 2 \\ 0 & 1 & 0 & | & -1 & 2 & 2 \\ 1 & 0 & 0 & | & 1 & -1 & -1 \end{bmatrix} \qquad \begin{bmatrix} 1 & 0 & 0 & | & 1 & -1 & -1 \\ 0 & 1 & 0 & | & -1 & 2 & 2 \\ 0 & 0 & 1 & | & -1 & 1 & 2 \end{bmatrix}$$

The inverse is

$$\begin{bmatrix} 1 & -1 & -1 \\ -1 & 2 & 2 \\ -1 & 1 & 2 \end{bmatrix}.$$

(b)
$$\begin{bmatrix} 2 & 1 & 0 & | & 1 & 0 & 0 \\ ① & 2 & 1 & | & 0 & 1 & 0 \\ 4 & 5 & 2 & | & 0 & 0 & 1 \end{bmatrix} \quad \begin{bmatrix} 0 & ㋐ & -2 & | & 1 & -2 & 0 \\ 1 & 2 & 1 & | & 0 & 1 & 0 \\ 0 & -3 & -2 & | & 0 & -4 & 1 \end{bmatrix}$$

$$\begin{bmatrix} 0 & ① & \frac{2}{3} & | & -\frac{1}{3} & \frac{2}{3} & 0 \\ 1 & 2 & 1 & | & 0 & 1 & 0 \\ 0 & -3 & -2 & | & 0 & -4 & 1 \end{bmatrix} \quad \begin{bmatrix} 0 & 1 & \frac{2}{3} & | & -\frac{1}{3} & \frac{2}{3} & 0 \\ 1 & 0 & -\frac{1}{3} & | & \frac{2}{3} & -\frac{1}{3} & 0 \\ 0 & 0 & 0 & | & -1 & -2 & 1 \end{bmatrix}$$

From the row of zeros on the left of the last tableau, we see that the matrix has no inverse.

Problem 17
$$U = AX = \begin{bmatrix} .2 & .1 & .3 \\ .4 & .2 & .1 \\ .3 & .5 & .5 \end{bmatrix} \begin{bmatrix} 1000 \\ 1200 \\ 2000 \end{bmatrix} = \begin{bmatrix} 920 \\ 840 \\ 1900 \end{bmatrix}$$

$$S = X - U = \begin{bmatrix} 1000 \\ 1200 \\ 2000 \end{bmatrix} - \begin{bmatrix} 920 \\ 840 \\ 1900 \end{bmatrix} = \begin{bmatrix} 80 \\ 360 \\ 100 \end{bmatrix}$$

Problem 18
$$X = (I - A)^{-1}D = \begin{bmatrix} 2.80 & 1.60 & 2.0 \\ 1.84 & 2.48 & 1.6 \\ 3.52 & 3.44 & 4.8 \end{bmatrix} \begin{bmatrix} 200 \\ 100 \\ 300 \end{bmatrix} = \begin{bmatrix} 1320 \\ 1096 \\ 2488 \end{bmatrix}$$

Thus 1320 units of agricultural products, 1096 units of manufactured goods, and 2488 units of energy are needed.

Chapter 3

Problem 1 (a) Since $8(50) + 3(30) = 490 > 480$, the point $(50, 30)$ is not feasible. Since $8(30) + 3(50) = 390 \leq 480$, $4(30) + 9(50) = 570 \leq 720$, $30 \geq 0$, and $50 \geq 0$, the point $(30, 50)$ is feasible. Since $8(45) + 3(40) = 480 \leq 480$, $4(45) + 9(40) = 540 \leq 720$, $45 \geq 0$, and $40 \geq 0$, the point $(45, 40)$ is feasible.
(b) For $(30, 50)$ we have $R = 16(30) + 12(50) = 1080$, and for $(45, 40)$ we have $R = 16(45) + 12(40) = 1200$. Thus the point $(45, 40)$ is better.

Problem 2 If
$$x = \text{the number of TV sets carried, and}$$
$$y = \text{the number of VCRs carried,}$$

then the information can be summarized in the following table.

	TV	VCR	Available
Weight	40 lbs	25 lbs	4000 lbs
Volume	3 cu ft	2 cu ft	2000 cu ft
Revenue	$5	$4	
Number	x	y	

The linear program is:

$$\text{Maximize } 5x + 4y = R$$
$$\text{subject to } 40x + 25y \leq 4000$$
$$3x + 2y \leq 2000$$
$$x \geq 0$$
$$y \geq 0.$$

Problem 3 The line $2x + 3y = 12$ has the intercepts $(6, 0)$ and $(0, 4)$, while the line $x = 2$ is vertical and goes through $(2, 0)$. Note that $(0, 0)$ satisfies $2x + 3y \leq 12$ but not $x \geq 2$, so the arrows indicate the desired region.

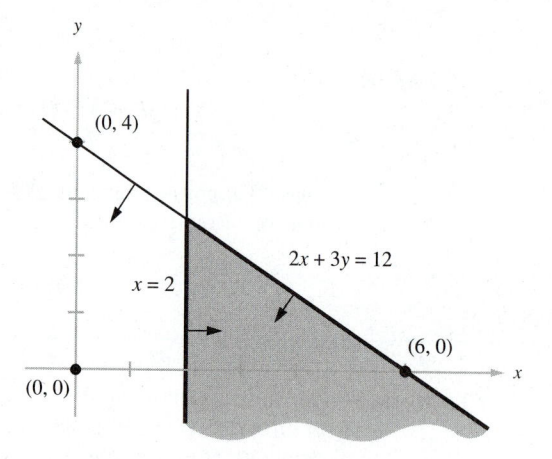

Problem 4 Notice that $3x + 3y = 75$ has intercepts $(25, 0)$ and $(0, 25)$, $2x + 4y = 60$ has intercepts $(30, 0)$ and $(0, 15)$, and $4x + 2y = 60$ has intercepts $(15, 0)$ and $(0, 30)$. The point $(0, 0)$ satisfies none of the inequalities $3x + 3y \geq 75$, $2x + 4y \geq 60$, $4x + 2y \geq 60$.

Problem 5 The four corner points are at the intersections of the line (3) and the y-axis, lines (3) and (1), lines (1) and (2), and line (2) and the x-axis.

SOLUTIONS TO PRACTICE PROBLEMS 557

Problem 6

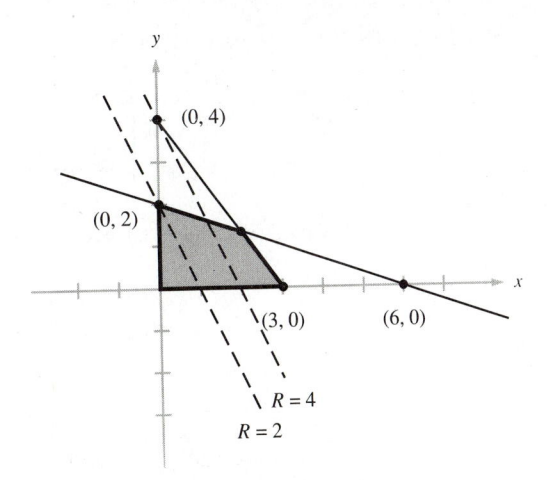

From the sketch the optimal point is at the intersection of the lines $4x + 3y = 12$ and $y = 0$. This is the point $(3, 0)$. Then $R = 2(3) + 0 = 6$.

Problem 7 Let $x + 2y + s = 320$, $5x + 4y + t = 1000$, $s \geq 0$, $t \geq 0$.

$$\begin{array}{c} \\ s \\ t \end{array} \left[\begin{array}{cccc|c} x & y & s & t & \\ 1 & 2 & \boxed{1} & 0 & 320 \\ \boxed{5} & 4 & 0 & \boxed{1} & 1000 \\ \hline 4 & 3 & 0 & 0 & R \end{array} \right]$$

$$\begin{array}{c} \\ s \\ x \end{array} \left[\begin{array}{cccc|c} x & y & s & t & \\ 0 & 1.2 & \boxed{1} & -.2 & 120 \\ \boxed{1} & .8 & 0 & .2 & 200 \\ \hline 0 & -.2 & 0 & -.8 & R - 800 \end{array} \right]$$

The last row corresponds to

$$R = 800 - .2y - .8t.$$

Thus the maximum is 800 for $y = t = 0$. Then $x = 200$ and $y = 0$.

Problem 8 We choose column 3 because 7 is the largest positive indicator. Dividing the other entries in column 3 into the corresponding rightmost entries gives $34/5 = 6.8$, $16/2 = 8$, and $70/8 = 8.75$. Since 6.8 is the smallest of these, we choose row 1.

Problem 9 The row labels of the tableau give the basic variables x, z, and s. The last row corresponds to

$$R = 22 - 3y - 5t,$$

so $R = 22$ when $y = t = 0$. Then $x = 10$ from row 3, and $z = 4$ from row 1.

Problem 10   The compact tableau is

$$\begin{bmatrix} x & y & z & \\ 1 & 2 & 4 & (\geq) \ 12 \\ 3 & 1 & 2 & (\geq) \ 15 \\ \hline 2 & 5 & 3 & \text{(minimize) } C \end{bmatrix}$$

with dual compact tableau

$$\begin{bmatrix} X & Y & \\ 1 & 3 & (\leq) \ 2 \\ 2 & 1 & (\leq) \ 5 \\ 4 & 2 & (\leq) \ 3 \\ \hline 12 & 15 & \text{(maximize) } R \end{bmatrix}.$$

Thus we should maximize $12X + 15Y = R$, subject to

$$X + 3Y \leq 2$$
$$2X + Y \leq 5$$
$$4X + 2Y \leq 3$$
$$X \geq 0, Y \geq 0.$$

Problem 11   In order to satisfy (a) we replace

$$\text{minimize } -3x + y = C$$

with

$$\text{maximize } 3x - y = R.$$

In order to satisfy (d) we replace

$$-x + y \geq -2 \quad \text{and} \quad -2x - 3y \geq -6$$

with

$$x - y \leq 2 \quad \text{and} \quad 2x + 3y \leq 6.$$

Problem 12   The linear program is

Maximize $x + 2y - 3z = R$
subject to
$$x \quad\quad + z + s \quad\quad\quad\quad\quad\quad = 30$$
$$2x + y + z \quad\quad - t \quad + a \quad\quad = 10$$
$$x + 3y + 2z \quad\quad\quad\quad - u \quad\quad + b = 15$$
$$x, y, z, s, t, u, a, b \geq 0.$$

We want to maximize $-a - b = Q$, which leads to the tableau

$$\begin{bmatrix} 1 & 0 & 1 & 1 & 0 & 0 & 0 & 0 & 30 \\ 2 & 1 & 1 & 0 & -1 & 0 & 1 & 0 & 10 \\ 1 & 3 & 2 & 0 & 0 & -1 & 0 & 1 & 15 \\ \hline 1 & 2 & -3 & 0 & 0 & 0 & 0 & 0 & R \\ 0 & 0 & 0 & 0 & 0 & 0 & -1 & -1 & Q \end{bmatrix}.$$

Since rows 2 and 3 correspond to artificial variables, we will add them to the bottom row.

Chapter 4

Problem 1 The first term in the sequence is the initial value x_0, which is 4. The next term x_1 is found by taking $n = 0$ in the defining equation $x_{n+1} = -3x_n + 10$; this yields

$$x_1 = -3x_0 + 10 = -3(4) + 10 = -12 + 10 = -2.$$

Similarly, by taking n to be 1, 2, 3, and 4 in the defining equation, we obtain

$$x_2 = -3x_1 + 10 = -3(-2) + 10 = 6 + 10 = 16,$$
$$x_3 = -3x_2 + 10 = -3(16) + 10 = -48 + 10 = -38,$$
$$x_4 = -3x_3 + 10 = -3(-38) + 10 = 114 + 10 = 124, \text{ and}$$
$$x_5 = -3x_4 + 10 = -3(124) + 10 = -372 + 10 = -362.$$

Problem 2 Applying Theorem 4.1 with $b = -3$ and $x_0 = 36$, we obtain

$$x_n = 36 - 3n.$$

Problem 3 Applying Theorem 4.2 with $a = 2$ and $x_0 = 5$, we obtain

$$x_n = 2^n(5).$$

Problem 4 The amount of interest due is

$$I = Prt = \$8000(0.16)(2.5) = \$3200.$$

Problem 5 The amount owed after 18 months is

$$A = P(1 + rt) = \$10,000[1 + (0.15)(1.5)] = \$12,250.$$

The amount of interest paid is

$$A - P = \$12,250 - \$10,000 = \$2250.$$

Problem 6 We must solve for r in formula (4.3) given that $A = \$22,650$, $P = \$20,000$, and $t = \frac{15}{12} = \frac{5}{4}$.

$$A = P(1 + rt)$$
$$\$22,650 = \$20,000[1 + r(\tfrac{5}{4})]$$
$$1.1325 = 1 + r(\tfrac{5}{4})$$
$$0.1325 = r(\tfrac{5}{4})$$
$$r = 0.1325(\tfrac{4}{5}) = 0.106$$

Thus the bond is paying 10.6% simple interest.

Problem 7 Using formula (4.4), we obtain

$$A = P(1 + i)^{20} = \$6000(1.02)^{20} \approx \$8915.68.$$

560 SOLUTIONS TO PRACTICE PROBLEMS

Problem 8   If we invest $100 at 6% interest compounded monthly, then after 1 year we will have
$$A = P(1 + i)^{12} = \$100(1.005)^{12} \approx \$106.17.$$

This is an increase of approximately 6.17%, and hence the effective rate of interest is about 6.17%.

Problem 9   (a) Using formula (4.5) with $A = \$8000$, $r = 0.15$, and $t = 6$, we obtain
$$P = \frac{A}{1 + rt} = \frac{\$8000}{1 + (0.15)(6)} = \frac{\$8000}{1.9} \approx \$4210.53.$$

(b) Using formula (4.6) with $A = \$8000$, $i = 0.10/4 = 0.025$, and $n = 6(4) = 24$, we obtain
$$P = \frac{A}{(1 + i)^n} = \frac{\$8000}{(1.025)^{24}} \approx \$4423.00.$$

Problem 10   Taking $a = 3$, $b = -4$, and $x_0 = 8$ in Theorem 4.3 gives
$$c = \frac{b}{1 - a} = \frac{-4}{1 - 3} = 2, \quad \text{and}$$
$$x_n = a^n(x_0 - c) + c = 3^n(8 - 2) + 2 = 6(3^n) + 2.$$

Problem 11   Using formula (4.8) with $p = \$150$, $i = 0.005$, and $n = 60$ gives
$$F = \$150 \left[\frac{(1.005)^{60} - 1}{0.005}\right] \approx \$10{,}465.50.$$

Problem 12   Using formula (4.9) with $p = \$50{,}000$, $i = 0.06$, and $n = 20$ gives
$$P = \$50{,}000 \left[\frac{1 - (1.06)^{-20}}{0.06}\right] \approx \$573{,}496.06.$$

Problem 13   Using formula (4.10) with $F = \$120{,}000{,}000$, $i = 0.0225$, and $n = 32$ gives
$$P = \frac{0.0225(\$120{,}000{,}000)}{(1.0225)^{32} - 1} \approx \$2{,}600{,}897.92.$$

Problem 14   The amount to be financed is $9946 - \$1500 = \$8{,}446$. Using formula (4.11) with $A = \$8446$, $i = 0.00825$, and $n = 60$ gives
$$P = \frac{0.00825(\$8446)}{1 - (1.00825)^{-60}} \approx \$179.04.$$

Chapter 5

Problem 1 (a) Since $x^2 - 2x + 1 = (x-1)^2 \geq 0$ for all real numbers x,
$$\{x | x^2 - 2x + 1 < 0\} = \emptyset.$$

(b) Since $x^2 = 1$ is equivalent to $0 = x^2 - 1 = (x+1)(x-1)$,
$$\{x | x^2 = 1\} = \{-1, 1\}.$$

(c) Since $(x-2)^2 \geq 0$ for all real numbers x,
$$\{x | (x-2)^2 \geq 0\} = U.$$

Problem 2 Since $A \cup B$ consists of the elements in A or B or both, we have
$$A \cup B = \{1, 2, 3, 4, 6, 8\}.$$

Since $A \cap B$ consists of the elements in both A and B, we have
$$A \cap B = \{2\}.$$

Since $\overline{B}$ consists of the elements in U but not in B, we have
$$\overline{B} = \{1, 5, 7, 8\}.$$

Since $A \cap C$ consists of the elements in both A and C, we have
$$A \cap C = \emptyset.$$

Since $B \cup C$ consists of the elements in B or C or both, we have
$$B \cup C = \{2, 3, 4, 6, 7\}.$$

Using the set $B \cup C$ computed above, we see that
$$\overline{(B \cup C)} = \{1, 5, 8\}.$$

The set $A \cup B \cup C$ consists of the elements in one or more of the sets A, B, or C; so
$$A \cup B \cup C = \{1, 2, 3, 4, 6, 7, 8\}.$$

Problem 3 (a) Region III lies in B but not in A or C; so region III denotes the set
$$\overline{A} \cap B \cap \overline{C}.$$

Region IV lies in both A and C but not in B; so region IV denotes the set
$$A \cap \overline{B} \cap C.$$

(b) Region II consists of athletes in A and B but not in C; region IV consists of athletes in B and C but not in A; and region VI consists of athletes in A and C but not in B. So the Venn diagram showing the athletes who take exactly two of the three vitamins is shown below.

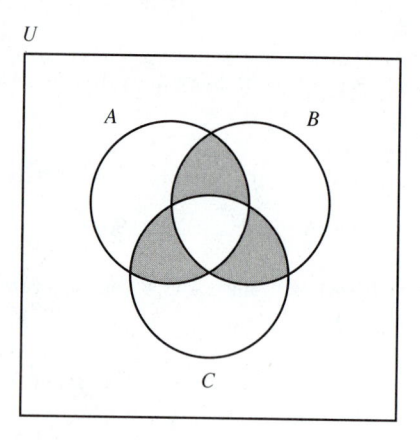

Problem 4 (a) Since 36 executives read *Forbes*,

$$100 - 36 = 64$$

do not read *Forbes*.

(b) By Theorem 5.1(b), there are

$$36 + 28 - 7 = 57$$

executives who read *Forbes* or *Business Week*.

(c) Since 57 executives read *Forbes* or *Business Week*,

$$100 - 57 = 43$$

read neither.

Problem 5 Let B, M, and E denote the sets of students who are taking courses in business, mathematics, and economics, respectively. Proceeding as in Figures 5.14–5.17, we obtain the Venn diagram below.

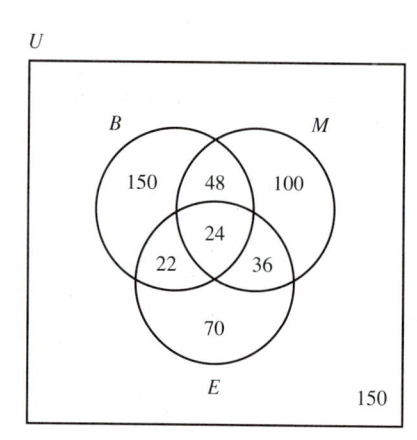

(a) The number of students taking a business course but not taking a course in mathematics or economics is 150.

(b) The number of students taking courses in exactly two of the three disciplines is

$$48 + 36 + 22 = 106.$$

(c) The number of students who were surveyed is the total of all the numbers in the Venn diagram, which is 600.

Problem 6   We can create a tree diagram by successively deciding whether to include each element of S. One possibility is shown in the figure below. The last column contains the eight possible subsets:

$$\emptyset, \{a\}, \{b\}, \{c\}, \{a, b\}, \{a, c\}, \{b, c\}, \{a, b, c\}.$$

Include a?	Include b?	Include c?	Subset
yes	yes	yes	$\{a, b, c\}$
		no	$\{a, b\}$
	no	yes	$\{a, c\}$
		no	$\{a\}$
no	yes	yes	$\{b, c\}$
		no	$\{b\}$
	no	yes	$\{c\}$
		no	$\emptyset$

Problem 7   The multiplication principle shows that the number of possible tests equals the product of the numbers of variations for each question, which is

$$8 \cdot 7 \cdot 10 \cdot 5 \cdot 6 \cdot 8 = 134,400.$$

Problem 8   Since each question can be answered in two different ways, there are

$$2^{10} = 1024$$

ways in which all ten questions can be answered.

Problem 9   (a) Any one of the eight people may sit in the first seat, but only one person (the spouse of the person in the first seat) may sit in the second seat. Any one of the remaining six people may sit in the third seat, but only one person may sit in the fourth seat. Any one of the remaining four people may sit in the fifth seat, but only one person may sit in the sixth seat. Finally either of the two remaining people may sit in the seventh seat, leaving only one person to sit in the eighth seat. Thus by the multiplication principle the number of possible seatings is

$$8 \cdot 1 \cdot 6 \cdot 1 \cdot 4 \cdot 1 \cdot 2 \cdot 1 = 384.$$

(b) Choose a man to sit in the first seat, and then have his wife sit in the second seat. Then choose one of the three remaining men to sit in the eighth seat, and have his wife sit in the seventh seat. Any of the four remaining people may sit in the third seat, but only that person's spouse may sit in the fourth seat. Finally either of the two remaining people may sit in the fifth seat, leaving only one person to sit in the sixth seat. Thus the number of possible seatings is

$$4 \cdot 1 \cdot 4 \cdot 1 \cdot 2 \cdot 1 \cdot 1 \cdot 3 = 96.$$

Problem 10 By the addition principle, the number of courses that can be selected is the sum of the number of business courses, the number of physical education courses, and the number of economics courses. This number is

$$4 + 7 + 3 = 14.$$

Problem 11 With each of the three meat entrees, there are four choices of salad dressing and five choices of vegetable. With each of the two seafood dishes there is a choice of four salad dressings. The number of possible choices (which equals the number of possible ways to order dinner) is

$$3 \cdot 4 \cdot 5 + 2 \cdot 4 = 60 + 8 = 68.$$

Problem 12 The number of ways in which the medals can be awarded equals the number of ways to choose three winners (in order) from among the nine contestants; this number is

$$P(9, 3) = 504.$$

Problem 13 The number of clinks equals the number of different pairs of people, which is

$$C(8, 2) = 28.$$

Problem 14 To form a list of five letters containing three different consonants and two different vowels, we must choose 3 different consonants (from among 21) and 2 different vowels (from among 5). Then we must arrange the five chosen letters in order. These operations can be performed in

$$C(21, 3) \cdot C(5, 2) \cdot P(5, 5) = 1{,}596{,}000$$

different ways.

Problem 15 Row 8 of Pascal's triangle begins and ends with 1. The other entries are obtained by adding pairs of adjacent entries in row 7, which is listed in Example 5.27. The resulting numbers are shown below.

$$1 \quad 8 \quad 28 \quad 56 \quad 70 \quad 56 \quad 28 \quad 8 \quad 1$$

Problem 16 Apply the binomial theorem with $n = 5$ and y replaced by $-2y$. The result is shown below.

$$(x - 2y)^5 = x^5 - 10x^4y + 40x^3y^2 - 80x^2y^3 + 80xy^4 - 32y^5$$

Problem 17 The term involving p^5q^{11} in the expansion of $(2p - q)^{16}$ is

$$C(16, 11)(2p)^5(-q)^{11} = 4368(32p^5)(-q^{11}) = -139{,}776p^5q^{11}.$$

Chapter 6

Problem 1 (a) The tree diagram is shown below.

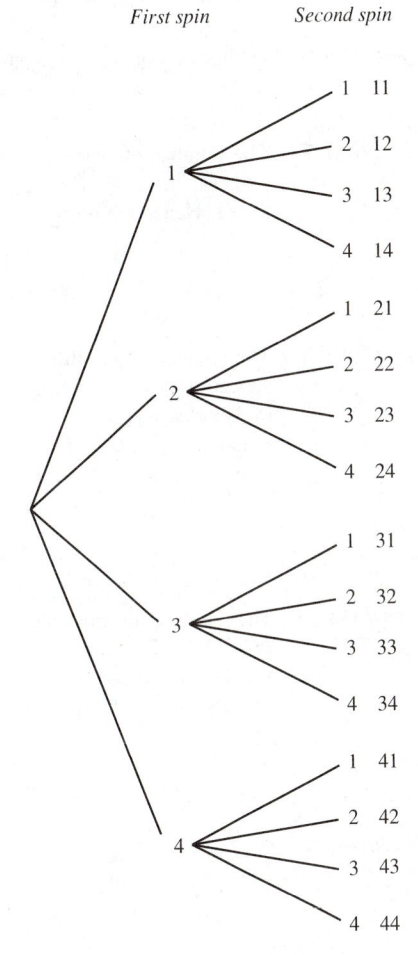

```
          First spin      Second spin

                            1   11
                            2   12
                  1 <
                            3   13
                            4   14

                            1   21
                            2   22
                  2 <
                            3   23
                            4   24

                            1   31
                            2   32
                  3 <
                            3   33
                            4   34

                            1   41
                            2   42
                  4 <
                            3   43
                            4   44
```

(b) Since the 16 outcomes in the tree diagram are equally likely and the outcome 22 occurs once, the probability of obtaining this outcome is 1/16.

566 SOLUTIONS TO PRACTICE PROBLEMS

Problem 2   Since the number of calls is a nonnegative integer not exceeding 15, a sample space for this experiment is

$$\{0, 1, 2, \ldots, 15\}.$$

Problem 3   (a) The event of receiving exactly 10 calls is the set $\{10\}$.
(b) The event of receiving at least 12 calls is the set $\{12, 13, 14, 15\}$.
(c) The event of receiving fewer than five calls is the set $\{0, 1, 2, 3, 4\}$.
(d) The event of receiving 0 calls is the set $\{0\}$.
(e) Since it is impossible to receive 16 calls, this set is $\emptyset$.

Problem 4   Of the five doorways leading from the starting room, two lead to room C. Thus the probability that the mouse will next enter room C is 2/5.

Problem 5   The number of possible outcomes when two dice are rolled is $6 \cdot 6 = 36$. Of this number, there are $P(6, 2) = 6 \cdot 5 = 30$ outcomes in which different numbers are rolled. Hence the probability of obtaining two different numbers is

$$\frac{30}{36} = \frac{5}{6}.$$

Problem 6   The number of possible subcommittees is $C(22, 5) = 26{,}334$. Of these, the number of subcommittees containing exactly 3 representatives of management and 2 representatives of labor is $C(12, 3) \cdot C(10, 2) = 220 \cdot 45 = 9900$. Hence the probability that a randomly selected subcommittee contains exactly 3 representatives of management and 2 representatives of labor is

$$\frac{9900}{26{,}334} = \frac{50}{133}.$$

Problem 7   The number of possible sequences in which the interviews can be conducted is $P(6, 6) = 720$. The number of possible sequences in which the four men are interviewed before the two women is $P(4, 4) \cdot P(2, 2) = 24 \cdot 2 = 48$. Therefore the probability that the four men are interviewed before the two women is

$$\frac{48}{720} = \frac{1}{15}.$$

Problem 8   (a) The probability distribution is formed by dividing each transaction's frequency by 4000, the total number of transactions. It is shown below.

Type of Transaction	Probability
Cashed a check	.398
Made a deposit	.286
Made a withdrawal	.234
Paid a bill	.064
Requested change	.011
Bought a money order	.007
	1.000 Total

(b) The probability that a transaction was either a deposit or withdrawal is the sum of the probabilities of these two types of transactions, which is

$$.286 + .234 = .520.$$

Problem 9 (a) If the odds in favor of E are 7 to 4, then the probability of E is

$$\frac{7}{7+4} = \frac{7}{11}.$$

(b) If the odds against E are 9 to 5, then the probability of E is

$$\frac{5}{5+9} = \frac{5}{14}.$$

Problem 10 (a) If $P(E) = .75 = \frac{3}{4}$, then the odds in favor of E are 3 to $4 - 3$, that is,

$$3 \text{ to } 1.$$

(b) If $P(E) = \frac{2}{9}$, then the odds in favor of E are 2 to $9 - 2$. Hence the odds against E are

$$7 \text{ to } 2.$$

Problem 11 In Figure 6.10 let the outcomes be denoted (w, b), where w is the number rolled on the white die and b is the number rolled on the black die. Then the outcomes corresponding to a sum of 6 are

$$(1, 5), (2, 4), (3, 3), (4, 2) \text{ and } (5, 1).$$

Thus of the 36 possible outcomes there are 5 corresponding to a sum of 6; so

$$P(A) = \frac{5}{36}.$$

The outcomes in which at least one die comes up greater than 4 are

$$(1, 5), (1, 6), (2, 5), (2, 6), (3, 5), (3, 6), (4, 5), (4, 6),$$
$$(5, 1), (5, 2), (5, 3), (5, 4), (5, 5), (5, 6),$$
$$(6, 1), (6, 2), (6, 3), (6, 4), (6, 5), \text{ and } (6, 6).$$

Hence

$$P(B) = \frac{20}{36} = \frac{5}{9}.$$

By similar reasoning, we obtain

$$P(A \cup B) = \frac{23}{36}, \quad P(A \cap B) = \frac{2}{36} = \frac{1}{18},$$
$$P(\overline{A}) = \frac{31}{36}, \quad \text{and} \quad P(\overline{B}) = \frac{16}{36} = \frac{4}{9}.$$

Problem 12 (a) Let E be the event "at most 8 calls are received." Then $\overline{E}$ is the event "9 or 10 calls are received." Hence

$$P(\overline{E}) = P(9) + P(10) = .008 + .004 = .012.$$

Therefore

$$P(E) = 1 - P(\overline{E}) = 1 - .012 = .988.$$

(b) If F is the event "no more than 9 calls are received," then $\overline{F}$ is the event "10 calls are received." Hence $P(\overline{F}) = .004$, and so

$$P(F) = 1 - .004 = .996.$$

Problem 13 If a fair coin is flipped five times, the number of possible sequences of heads and tails that can occur is $2^5 = 32$. Of these, each sequence but one contains at least one head. Hence the probability of obtaining at least one head is

$$\frac{31}{32}.$$

Problem 14 Let A and S be the events of being short of aluminum and steel, respectively. Then the event of having an adequate supply of both metals is $\overline{(A \cup S)}$. Now $P(A) = .06$, $P(S) = .05$, and $P(A \cap S) = .01$, and so

$$P(A \cup S) = P(A) + P(S) - P(A \cap S) = .06 + .05 - .01 = .10,$$

by the union rule. Thus by the complement rule

$$P(\overline{A \cup S}) = 1 - P(A \cup S) = 1 - .10 = .90.$$

Problem 15 Let A, R, and K denote the events that a house was built by the Armstrong Company, Rave Brothers Construction, and Kaisner Construction, respectively. Since the events A, R, and K are mutually exclusive, the probability that a house was built by a member of the Ark Corporation is

$$P(A \cup R \cup K) = P(A) + P(R) + P(K) = .35 + .20 + .19 = .74.$$

Problem 16 Of the 250 patients who received a placebo, 133 felt no better. Hence the conditional probability that a patient who received a placebo felt no better is

$$\frac{133}{250} = .532.$$

Problem 17 We see from Figure 6.2 that the reduced sample space consisting of outcomes in which exactly two of the three flips are heads is $\{HHT, HTH, THH\}$. Hence the conditional probability that the first flip is heads given that exactly two of the three flips are heads is 2/3.

Problem 18 Let A and L denote the events of having purchased automobile and life insurance, respectively, from the agent. Then $P(A) = .60$, $P(L) = .40$, and $P(A \cap L) = .15$.

Hence the conditional probability that someone who has purchased life insurance from this agent will also have purchased automobile insurance from her is

$$P(A|L) = \frac{P(A \cap L)}{P(L)} = \frac{.15}{.40} = .375.$$

Problem 19 Let A and B denote the events that a defective engine passes the first and second inspections, respectively. Then

$$P(A) = 1 - .70 = .30 \quad \text{and} \quad P(B|A) = 1 - .80 = .20.$$

Hence the probability that a defective engine passes both inspections is

$$P(A \cap B) = P(A) \cdot P(B|A) = .30(.20) = .06.$$

Problem 20 Let A and B denote the events "the first transistor is defective" and "the second transistor is defective," respectively. If we select two transistors in sequence from a box containing three defective and seven nondefective transistors, the probability that both of the selected transistors are defective is

$$P(A \cap B) = P(A) \cdot P(B|A) = \left(\tfrac{3}{10}\right)\left(\tfrac{2}{9}\right) = \tfrac{6}{90} = \tfrac{1}{15}.$$

Problem 21 Let A, B, and C denote the events "the turntable fails," "the amplifier fails," and "the speakers fail," respectively. Now

$$P(A) = .05, \quad P(B) = .02, \quad \text{and} \quad P(C) = .03.$$

We are interested in determining the probability that none of these components fail. Because the turntables, amplifiers, and speakers are manufactured independently, we will assume that $\bar{A}$, $\bar{B}$, and $\bar{C}$ are independent events. Hence

$$P(\text{system does not fail}) = P(\text{no component fails}) = P(\bar{A} \cap \bar{B} \cap \bar{C})$$
$$= P(\bar{A}) \cdot P(\bar{B}) \cdot P(\bar{C}) = (.95)(.98)(.97)$$
$$\approx .903.$$

Problem 22 The stochastic diagram for this situation is shown at the left. From this diagram we see that

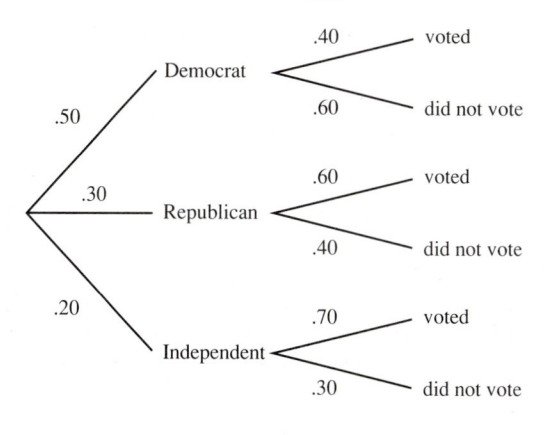

$$P(\text{Democrat}|\text{voted}) = \frac{(.5)(.4)}{(.5)(.4) + (.3)(.6) + (.2)(.7)}$$
$$= \frac{.20}{.52} = \frac{5}{13},$$

$$P(\text{Republican}|\text{voted}) = \frac{(.3)(.6)}{(.5)(.4) + (.3)(.6) + (.2)(.7)}$$
$$= \frac{.18}{.52} = \frac{9}{26}, \quad \text{and}$$

$$P(\text{independent}|\text{voted}) = \frac{(.2)(.7)}{(.5)(.4) + (.3)(.6) + (.2)(.7)}$$
$$= \frac{.14}{.52} = \frac{7}{26}.$$

570 SOLUTIONS TO PRACTICE PROBLEMS

Problem 23 For unacquainted married couples the events of getting divorced can be assumed to be independent. Hence by Bernoulli's formula the probability of having exactly one divorce among 10 couples during the next year is

$$C(10, 1)(.02)^1(.98)^9 = 10(.02)(.98)^9 \approx .167.$$

Problem 24 Since the calls are made at random, we can assume that the events of making a sale are independent. We will use Bernoulli's formula to compute the probability of *not* making at least one sale, that is, the probability of making no sales. This probability is

$$C(8, 0)(.10)^0(.90)^8 = (.90)^8 \approx .430.$$

Hence the probability of making at least one sale is approximately

$$1 - .430 = .570.$$

■ Chapter 7

Problem 1 The mean is

$$\bar{x} = \frac{\Sigma x}{n} = \frac{26 + 38 + 17 + 32 + 41}{5} = \frac{154}{5} = 30.8.$$

Problem 2 The sum of the 25 given numbers is 19,884. Hence the mean number of tornadoes reported in the United States during the years 1960–84 is

$$\frac{19,884}{25} = 795.36.$$

Problem 3 The numbers in Practice Problem 2 listed in increasing order are:

461, 570, 604, 618, 649, 658, 660, 683, 713, 741, 783, 788, 835,
852, 852, 866, 888, 898, 907, 912, 920, 931, 947, 1046, 1102.

The thirteenth number in this list, 835, is the median.

Problem 4 In Example 7.1 we found that the mean of the examination scores is 86. Thus the variance of the four scores is

$$s^2 = \frac{\Sigma(x - \bar{x})^2}{n} = \frac{(84 - 86)^2 + (94 - 86)^2 + (87 - 86)^2 + (79 - 86)^2}{4}$$

$$= \frac{4 + 64 + 1 + 49}{4} = \frac{118}{4} = 29.5.$$

Hence the standard deviation of the scores is $s = \sqrt{29.5} \approx 5.43$.

Problem 5 (a) There are many acceptable choices of intervals; we will use 8 intervals of width 100: 399.5–499.5, 499.5–599.5, 599.5–699.5, 699.5–799.5, 799.5–899.5, 899.5–999.5, 999.5–1099.5, and 1099.5–1199.5.

(b) The frequency distribution is formed by tallying the number of years in which the number of tornadoes lies within each of the intervals above. The results are contained in the table below.

Number of tornadoes	Frequency
399.5–499.5	1
499.5–599.5	1
599.5–699.5	6
699.5–799.5	4
799.5–899.5	6
899.5–999.5	5
999.5–1099.5	1
1099.5–1199.5	1
	25 Total

The histogram using the intervals in part (a) is shown in the figure below.

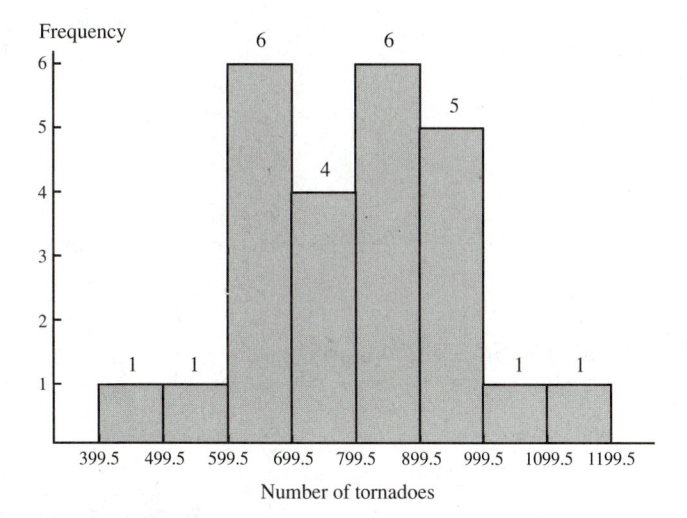

Problem 6 Using the frequency distribution in the answer to Practice Problem 5, we obtain the following table.

Number of tornadoes	Midpoint m	Frequency f	mf
399.5–499.5	449.5	1	449.5
449.5–599.5	549.5	1	549.5
599.5–699.5	649.5	6	3897.0
699.5–799.5	749.5	4	2998.0
799.5–899.5	849.5	6	5097.0
899.5–999.5	949.5	5	4747.5
999.5–1099.5	1049.5	1	1049.5
1099.5–1199.5	1149.5	1	1149.5
		25	19,937.5 Totals

Thus the mean number of tornadoes reported in the United States during the years 1960–1984 is approximately

$$\frac{19{,}937.5}{25} = 797.5.$$

Problem 7 From Practice Problem 6 we see that the mean of these data is approximately 797.5. To approximate the standard deviation, we construct the following table.

Number of tornadoes	Midpoint m	Frequency f	$m - \bar{x}$	$(m - \bar{x})^2$	$(m - \bar{x})^2 f$
399.5–499.5	449.5	1	−348	121,104	121,104
499.5–599.5	549.5	1	−248	61,504	61,504
599.5–699.5	649.5	6	−148	21,904	131,424
699.5–799.5	749.5	4	−48	2,304	9,216
799.5–899.5	849.5	6	52	2,704	16,224
899.5–999.5	949.5	5	152	23,104	115,520
999.5–1099.5	1049.5	1	252	63,504	63,504
1099.5–1199.5	1149.5	1	352	123,904	123,904
		25			642,400 Totals

Thus we see that

$$s^2 \approx \frac{642{,}400}{25} = 25{,}696,$$

and so the standard deviation is approximately

$$\sqrt{25{,}696} \approx 160.$$

Problem 8 When a fair die is rolled, each of the outcomes 1, 2, 3, 4, 5, and 6 is equally likely to occur. Thus the probability histogram is shown below.

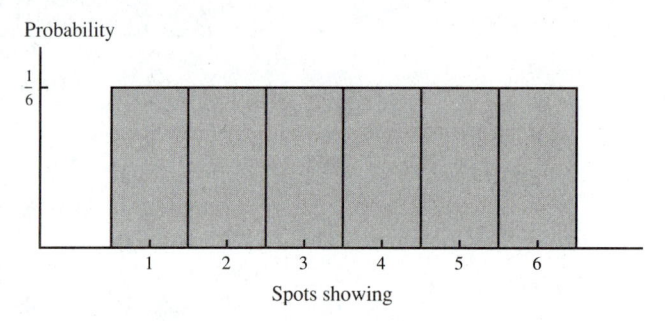

By formula (7.5) the mean is

$$\mu = \Sigma\, xp = 1\left(\frac{1}{6}\right) + 2\left(\frac{1}{6}\right) + 3\left(\frac{1}{6}\right) + 4\left(\frac{1}{6}\right) + 5\left(\frac{1}{6}\right) + 6\left(\frac{1}{6}\right)$$

$$= \frac{21}{6} = 3.5.$$

By formula (7.6) the variance is

$$\sigma^2 = (1 - 3.5)^2 \left(\frac{1}{6}\right) + (2 - 3.5)^2 \left(\frac{1}{6}\right) + \cdots + (6 - 3.5)^2 \left(\frac{1}{6}\right)$$

$$= (6.25 + 2.25 + 0.25 + 0.25 + 2.25 + 6.25)\left(\frac{1}{6}\right)$$

$$= \frac{17.5}{6} = \frac{35}{12};$$

so

$$\sigma = \sqrt{\frac{35}{12}} \approx 1.71.$$

Problem 9 If severe winter weather occurs (and the probability of this happening is .4), the contractor's profit will be $30,000. Otherwise (with probability $1 - .4 = .6$), the contractor's profit will be $50,000. Hence the contractor's expected profit is

$$.4(\$30,000) + .6(\$50,000) = \$12,000 + \$30,000 = \$42,000.$$

Problem 10 (a) There are four sensible actions for the farmer—sending 2, 3, 4, or 5 truckloads of eggs to the city. There are also four possible states of nature, according to whether the demand for eggs is 2, 3, 4, or 5 truckloads. Consider the action of sending 4 truckloads of eggs to town. If the demand for eggs is only 2 truckloads, then the farmer will sell 2 truckloads and lose 2 truckloads to spoilage; his profit in this case is

$$2(\$150) + 2(-\$50) = \$300 - \$100 = \$200.$$

If the demand for eggs is 3 truckloads, then the farmer will sell 3 truckloads and lose 1 truckload to spoilage; his profit in this case is

$$3(\$150) + 1(-\$50) = \$450 - \$50 = \$400.$$

Finally, if the demand for eggs is at least 4 truckloads, then the farmer will sell all 4 of the truckloads that he sent, with a resulting profit of

$$4(\$150) + 0(-\$50) = \$600 + \$0 = \$600.$$

These numbers are the entries of the third row in the payoff table below; the other entries of the table are computed similarly.

Action Number of truckloads sent	State of nature Demand for eggs (number of truckloads)			
	2	3	4	5
2	300	300	300	300
3	250	450	450	450
4	200	400	600	600
5	150	350	550	750

(b) To use the maximin criterion, we choose the action corresponding to the row of the payoff table in which the smallest entry is as large as possible. This is row 1, and so the maximin criterion chooses the action of sending 2 truckloads.

(c) To use the maximax criterion, we choose the action corresponding to the row of the payoff table containing the largest entry. This is row 5, and so the maximin criterion chooses the action of sending 5 truckloads.

(d) To use Bayes's criterion, we compute the expected profit for each action. These expected profits are $300, $410, $440, and $410, respectively. Hence Bayes's criterion chooses the action of sending 4 truckloads.

Problem 11

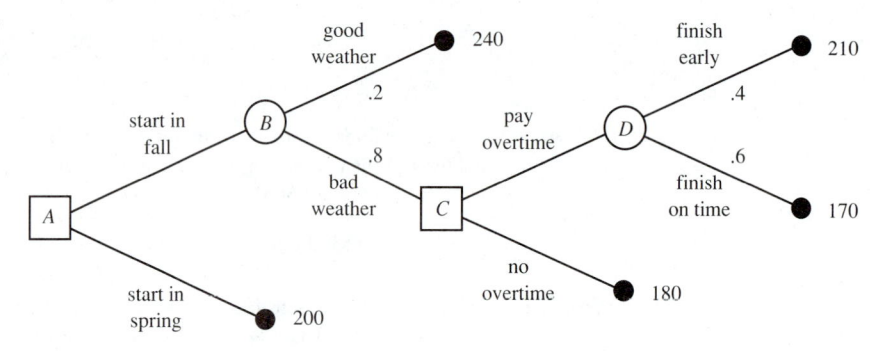

(a) The decision tree is shown in the figure above, where profits are denoted in thousands of dollars. Note that the payoffs are reduced by $20,000 in the event of bad weather, and by an additional $10,000 if overtime is paid.

(b) Working backwards from the end-nodes, we assign to chance-node D the expected value of the nodes to which it branches, which is

$$.4(210) + .6(170) = 186.$$

To choice-node C, we assign the larger of 186 and 180, which is 186. To chance-node B we assign the expected value of the nodes to which it branches, which is

$$.2(240) + .8(186) = 196.80.$$

To choice-node A, we assign the larger of 196.80 and 200, which is 200.

(c) Thus the maximum expected payoff for the contractor is $200,000 when he starts construction in the spring.

Problem 12   (a) The probability of receiving between 3 and 4 service calls, inclusive, is the area under the probability distribution in Figure 7.14 between 2.5 and 4.5. This area is

$$.20 + .10 = .30.$$

(b) The probability of receiving at most 1 service call is the area under the probability distribution in Figure 7.14 between -0.5 and 1.5. This area is

$$.10 + .25 = .35.$$

Problem 13 (a) To determine the distribution of the random variable that counts the number of sixes rolled when a fair die is tossed five times, we apply Bernoulli's formula with "success" meaning that a six is rolled. Then the probability of success is $p = 1/6$, and the probability of failure is $q = 5/6$. Thus the probability of obtaining exactly one six in five rolls is

$$C(5, 1)\left(\frac{1}{6}\right)^1\left(\frac{5}{6}\right)^4 = 5\left(\frac{1}{6}\right)\left(\frac{625}{1296}\right) = \frac{3125}{7776}.$$

Similarly we can compute the other probabilities in the probability density function, which is shown below.

Number of sixes	Probability
0	$\frac{3125}{7776}$
1	$\frac{3125}{7776}$
2	$\frac{1250}{7776}$
3	$\frac{250}{7776}$
4	$\frac{25}{7776}$
5	$\frac{1}{7776}$

(b) The probability of obtaining more than one six in five tosses is

$$P(2, 3, 4, \text{ or } 5 \text{ sixes}) = 1 - P(0 \text{ or } 1 \text{ six})$$
$$= 1 - [P(0 \text{ sixes}) + P(1 \text{ six})]$$
$$= 1 - \left(\frac{3125}{7776} + \frac{3125}{7776}\right)$$
$$= \frac{1526}{7776}.$$

Problem 14 We must determine the mean and standard deviation of the random variable that counts the number of contracts obtained. Since this random variable is binomially distributed, its mean and standard deviation can be computed by using Theorem 7.2 with $p = .2$, $q = .8$, and $n = 30$. Then

$$\mu = np = 30(.2) = 6$$

and

$$\sigma = \sqrt{npq} = \sqrt{30(.2)(.8)} = \sqrt{4.8} \approx 2.19.$$

Problem 15 (a) The probability is ½ of identifying the psychotic person in each pair by chance alone. Hence the probability of identifying the psychotic person in exactly six of the ten pairs by chance alone is given by Bernoulli's formula as

$$C(10, 6)\left(\frac{1}{2}\right)^6\left(\frac{1}{2}\right)^4 = 210\left(\frac{1}{64}\right)\left(\frac{1}{16}\right) = \frac{210}{1024}.$$

(b) As in (a) above we see that the probability of identifying the psychotic person in exactly seven of the ten pairs by chance alone is

$$C(10, 7)\left(\frac{1}{2}\right)^7\left(\frac{1}{2}\right)^3 = 210\left(\frac{1}{128}\right)\left(\frac{1}{8}\right) = \frac{120}{1024}.$$

(c) By calculations similar to those in (a) and (b), we see that the probabilities of identifying the psychotic person in exactly eight, nine, and ten of the pairs by chance alone are 45/1024, 10/1024, and 1/1024, respectively. Hence the probability of identifying the psychotic person in at least six of the ten pairs by chance alone is

$$\frac{210}{1024} + \frac{120}{1024} + \frac{45}{1024} + \frac{10}{1024} + \frac{1}{1024} = \frac{386}{1024} \approx .377.$$

(d) Since the probability in (c) is so large, the graphologist's claim should not be accepted.

Problem 16 (a) In standard units a score of 48 becomes

$$z = \frac{48 - 54}{8} = -\frac{6}{8} = -.75.$$

Hence, as in Example 7.22(a), the area to the right of 48 is

$$1 - A(-.75) = 1 - .2266 = .7734.$$

(b) In standard units scores of 68 and 72 become $z = 1.75$ and $z = 2.25$, respectively. Hence, as in Example 7.22(b), we see that the area between 68 and 72 is

$$A(2.25) - A(1.75) = 0.9878 - 0.9599 = .0279.$$

Problem 17 The random variable that counts the number of defective bottles is binomially distributed with a mean of

$$\mu = np = 3000(.005) = 15$$

and a standard deviation of

$$\sigma = \sqrt{npq} = \sqrt{3000(.005)(.995)} = \sqrt{14.925} \approx 3.86.$$

Since both np and nq exceed 5, we can approximate this binomial distribution by a normal distribution having the same mean and standard deviation. Now the probability of having 20 or more defective bottles is approximately equal to the area under the normal curve to the right of 19.5. Because 19.5 converts to about 1.17 standard units, we see that

$$P(\text{at least 20 defectives}) \approx 1 - A(1.17) = 1 - .8790 = .1210.$$

Chapter 8

Problem 1 In this process there are three states.

State 1: belonging to the upper class
State 2: belonging to the middle class
State 3: belonging to the lower class

The time period between trials is one generation, and the entries in each row of the transition matrix are the probabilities that the children of parents in a particular class will belong to each of the three classes. Hence the transition matrix for this process is

$$\text{Current state} \begin{array}{c} 1 \\ 2 \\ 3 \end{array} \begin{array}{c} \overset{\text{Next state}}{123} \\ \begin{bmatrix} .5 & .5 & 0 \\ .1 & .7 & .2 \\ 0 & .5 & .5 \end{bmatrix} \end{array}.$$

Problem 2 Multiplying the initial state vector times the transition matrix, we obtain the state vector after one generation:

$$[.20 \quad .40 \quad .40] \begin{bmatrix} .5 & .5 & 0 \\ .1 & .7 & .2 \\ 0 & .5 & .5 \end{bmatrix} = [.14 \quad .58 \quad .28].$$

Multiplying the state vector computed above times the transition matrix, we obtain the state vector after two generations:

$$[.14 \quad .58 \quad .28] \begin{bmatrix} .5 & .5 & 0 \\ .1 & .7 & .2 \\ 0 & .5 & .5 \end{bmatrix} = [.128 \quad .616 \quad .256].$$

Hence after two generations the distribution of people among the classes will be 12.8% upper class, 61.6% middle class, and 25.6% lower class.

Problem 3 We must check to see if any of the first 5 powers of A has all positive entries. First we compute

$$A^2 = AA = \begin{bmatrix} 0 & 1 & 0 \\ 0 & 0 & 1 \\ \frac{1}{2} & \frac{1}{2} & 0 \end{bmatrix} \begin{bmatrix} 0 & 1 & 0 \\ 0 & 0 & 1 \\ \frac{1}{2} & \frac{1}{2} & 0 \end{bmatrix}$$

$$= \begin{bmatrix} 0 & 0 & 1 \\ \frac{1}{2} & \frac{1}{2} & 0 \\ 0 & \frac{1}{2} & \frac{1}{2} \end{bmatrix}.$$

We can use A^2 to compute A^4 as follows:

$$A^4 = A^2A^2 = \begin{bmatrix} 0 & 0 & 1 \\ \frac{1}{2} & \frac{1}{2} & 0 \\ 0 & \frac{1}{2} & \frac{1}{2} \end{bmatrix} \begin{bmatrix} 0 & 0 & 1 \\ \frac{1}{2} & \frac{1}{2} & 0 \\ 0 & \frac{1}{2} & \frac{1}{2} \end{bmatrix}$$

$$= \begin{bmatrix} 0 & \frac{1}{2} & \frac{1}{2} \\ \frac{1}{4} & \frac{1}{4} & \frac{1}{2} \\ \frac{1}{4} & \frac{1}{2} & \frac{1}{4} \end{bmatrix}.$$

Therefore

$$A^5 = AA^4 = \begin{bmatrix} 0 & 1 & 0 \\ 0 & 0 & 1 \\ \frac{1}{2} & \frac{1}{2} & 0 \end{bmatrix} \begin{bmatrix} 0 & \frac{1}{2} & \frac{1}{2} \\ \frac{1}{4} & \frac{1}{4} & \frac{1}{2} \\ \frac{1}{4} & \frac{1}{2} & \frac{1}{4} \end{bmatrix}$$

$$= \begin{bmatrix} \frac{1}{4} & \frac{1}{4} & \frac{1}{2} \\ \frac{1}{4} & \frac{1}{2} & \frac{1}{4} \\ \frac{1}{8} & \frac{3}{8} & \frac{1}{2} \end{bmatrix}.$$

Because A^5 contains only positive entries, the Markov chain is regular.

Problem 4 We will proceed as in Example 8.2. Let the equilibrium vector be

$$X = [x_1 \quad x_2 \quad x_3].$$

We must solve the matrix equation $X(I_3 - P) = O$, where P is the transition matrix of the Markov chain, which was obtained in Practice Problem 1.

$$[x_1 \quad x_2 \quad x_3] \begin{bmatrix} .5 & -.5 & 0 \\ -.1 & .3 & -.2 \\ 0 & -.5 & .5 \end{bmatrix} = [0 \quad 0 \quad 0]$$

Multiplying out the left side leads to the system of equations below.

$$.5x_1 - .1x_2 \qquad\qquad = 0$$
$$-.5x_1 + .3x_2 - .5x_3 = 0$$
$$\qquad\quad -.2x_2 + .5x_3 = 0$$

By including the condition that the sum of the unknowns is 1, we obtain the following system.

$$.5x_1 - .1x_2 \qquad\qquad = 0$$
$$-.5x_1 + .3x_2 - .5x_3 = 0$$
$$\qquad\quad -.2x_2 + .5x_3 = 0$$
$$x_1 + x_2 + x_3 = 1$$

We now solve this sytem by the techniques of Chapter 2, as follows.

$$\begin{bmatrix} .5 & -.1 & 0 & | & 0 \\ -.5 & .3 & -.5 & | & 0 \\ 0 & -.2 & .5 & | & 0 \\ \hline 1 & 1 & 1 & | & 1 \end{bmatrix} \quad \begin{bmatrix} 0 & -.6 & -.5 & | & -.5 \\ 0 & .8 & 0 & | & .5 \\ 0 & -.2 & .5 & | & 0 \\ \hline 1 & 1 & 1 & | & 1 \end{bmatrix}$$

$$\begin{bmatrix} 0 & -.6 & -.5 & | & -.5 \\ 0 & .8 & 0 & | & .5 \\ 0 & 1 & -2.5 & | & 0 \\ \hline 1 & 1 & 1 & | & 1 \end{bmatrix} \quad \begin{bmatrix} 0 & 0 & -2 & | & -.5 \\ 0 & 0 & 2 & | & .5 \\ 0 & 1 & -2.5 & | & 0 \\ \hline 1 & 0 & 3.5 & | & 1 \end{bmatrix}$$

$$\begin{bmatrix} 0 & 0 & 1 & | & .25 \\ 0 & 0 & 2 & | & .5 \\ 0 & 1 & -2.5 & | & 0 \\ \hline 1 & 0 & 3.5 & | & 1 \end{bmatrix} \quad \begin{bmatrix} 0 & 0 & 1 & | & .250 \\ 0 & 0 & 0 & | & 0 \\ 0 & 1 & 0 & | & .625 \\ \hline 1 & 0 & 0 & | & .125 \end{bmatrix}$$

Hence in the long run 12.5% of the people will be in the upper class, 62.5% in the middle class, and 25% in the lower class.

Problem 5 (a) To determine the canonical form of the transition matrix, we rearrange the states so that the absorbing states (states B and D) come first. The resulting matrix is

$$\begin{array}{c} \\ B \\ D \\ A \\ C \end{array} \begin{array}{cccc} B & D & A & C \\ \begin{bmatrix} 1 & 0 & 0 & 0 \\ 0 & 1 & 0 & 0 \\ .6 & .2 & .1 & .1 \\ 0 & .2 & .4 & .4 \end{bmatrix} \end{array}.$$

(b) In the canonical form obtained in part (a), the lower right submatrix is

$$Q = \begin{bmatrix} .1 & .1 \\ .4 & .4 \end{bmatrix}.$$

Hence

$$I_2 - Q = \begin{bmatrix} 1 & 0 \\ 0 & 1 \end{bmatrix} - \begin{bmatrix} .1 & .1 \\ .4 & .4 \end{bmatrix}$$

$$= \begin{bmatrix} .9 & -.1 \\ -.4 & .6 \end{bmatrix}.$$

Proceeding as in Section 2.6, we see that the fundamental matrix of the Markov chain is

$$M = (I_2 - Q)^{-1} = \begin{bmatrix} 1.2 & 0.2 \\ 0.8 & 1.8 \end{bmatrix}.$$

580 SOLUTIONS TO PRACTICE PROBLEMS

Problem 6 Label the rows and columns of the fundamental matrix in the solution to Practice Problem 5(b) exactly as the rows and columns are labeled in the canonical form obtained in the solution to Practice Problem 5(a). The resulting matrix is shown below.

$$M = (I_2 - Q)^{-1} = \begin{array}{c} \\ A \\ C \end{array} \begin{array}{cc} A & C \\ \begin{bmatrix} 1.2 & 0.2 \\ 0.8 & 1.8 \end{bmatrix} \end{array}$$

(a) Theorem 8.3(a) states that the number of trials that the system can be expected to be in state A if it begins in state C is 0.8, the entry of M in the row labeled C and the column labeled A.

(b) Theorem 8.3(b) states that the expected number of trials before absorption if the system begins in state A is

$$1.2 + 0.2 = 1.4,$$

the sum of the entries of M in the row labeled A.

(c) To determine the probability of absorption into an absorbing state, we must compute the matrix MR, where M is the fundamental matrix and R is the lower left submatrix in the canonical form:

$$MR = \begin{bmatrix} 1.2 & 0.2 \\ 0.8 & 1.8 \end{bmatrix} \begin{bmatrix} .6 & .2 \\ 0 & .2 \end{bmatrix}$$

$$= \begin{bmatrix} .72 & .28 \\ .48 & .52 \end{bmatrix}.$$

Now we label the rows and columns of MR exactly as the rows and columns of R are labeled in the canonical form of the transition matrix obtained in the solution to Practice Problem 5(a). The resulting matrix is shown below.

$$MR = \begin{array}{c} \\ A \\ C \end{array} \begin{array}{cc} B & D \\ \begin{bmatrix} .72 & .28 \\ .48 & .52 \end{bmatrix} \end{array}$$

By Theorem 8.3(c) the probability that the system will eventually end in state D if it begins in state C is .52, the entry of MR in the row labeled C and the column labeled D.

Problem 7 (a) The canonical form of the transition matrix is obtained by rearranging the order of the states so that the absorbing states precede the nonabsorbing states:

$$\begin{array}{c} \\ DD \\ RR \\ DH \\ DR \\ HH \\ HR \end{array} \begin{array}{cccccc} DD & RR & DH & DR & HH & HR \\ \begin{bmatrix} 1 & 0 & 0 & 0 & 0 & 0 \\ 0 & 1 & 0 & 0 & 0 & 0 \\ \frac{1}{4} & 0 & \frac{1}{2} & 0 & \frac{1}{4} & 0 \\ 0 & 0 & 0 & 0 & 1 & 0 \\ \frac{1}{16} & \frac{1}{16} & \frac{1}{4} & \frac{1}{8} & \frac{1}{4} & \frac{1}{4} \\ 0 & \frac{1}{4} & 0 & 0 & \frac{1}{4} & \frac{1}{2} \end{bmatrix} \end{array}.$$

(b) The matrix Q is the lower right submatrix of the canonical form above; hence

$$I_4 - Q = \begin{bmatrix} 1 & 0 & 0 & 0 \\ 0 & 1 & 0 & 0 \\ 0 & 0 & 1 & 0 \\ 0 & 0 & 0 & 1 \end{bmatrix} - \begin{bmatrix} \frac{1}{2} & 0 & \frac{1}{4} & 0 \\ 0 & 0 & 1 & 0 \\ \frac{1}{4} & \frac{1}{8} & \frac{1}{4} & \frac{1}{4} \\ 0 & 0 & \frac{1}{4} & \frac{1}{2} \end{bmatrix}$$

$$= \begin{bmatrix} \frac{1}{2} & 0 & -\frac{1}{4} & 0 \\ 0 & 1 & -1 & 0 \\ -\frac{1}{4} & -\frac{1}{8} & \frac{3}{4} & -\frac{1}{4} \\ 0 & 0 & -\frac{1}{4} & \frac{1}{2} \end{bmatrix}.$$

Therefore the fundamental matrix is

$$M = (I_2 - Q)^{-1} = \begin{array}{c} \\ \text{DH} \\ \text{DR} \\ \text{HH} \\ \text{HR} \end{array} \begin{array}{c} \text{DH} \quad \text{DR} \quad \text{HH} \quad \text{HR} \\ \begin{bmatrix} \frac{8}{3} & \frac{1}{6} & \frac{4}{3} & \frac{2}{3} \\ \frac{4}{3} & \frac{4}{3} & \frac{8}{3} & \frac{4}{3} \\ \frac{4}{3} & \frac{1}{3} & \frac{8}{3} & \frac{4}{3} \\ \frac{2}{3} & \frac{1}{6} & \frac{4}{3} & \frac{8}{3} \end{bmatrix} \end{array}.$$

(c) If the original parents are in state DR, the average number of generations before absorption is $6\frac{2}{3}$, the sum of the entries in the second row of the fundamental matrix M.
(d) If the original parents are in state DH, the probability that the offspring will eventually become pure dominant is the entry of MR in the row labeled DH and the column labeled DD. Now

$$MR = \begin{bmatrix} \frac{8}{3} & \frac{1}{6} & \frac{4}{3} & \frac{2}{3} \\ \frac{4}{3} & \frac{4}{3} & \frac{8}{3} & \frac{4}{3} \\ \frac{4}{3} & \frac{1}{3} & \frac{8}{3} & \frac{4}{3} \\ \frac{2}{3} & \frac{1}{6} & \frac{4}{3} & \frac{8}{3} \end{bmatrix} \begin{bmatrix} \frac{1}{4} & 0 \\ 0 & 0 \\ \frac{1}{16} & \frac{1}{16} \\ 0 & \frac{1}{4} \end{bmatrix}$$

$$= \begin{array}{c} \\ \text{DH} \\ \text{DR} \\ \text{HH} \\ \text{HR} \end{array} \begin{array}{c} \text{DD} \quad \text{RR} \\ \begin{bmatrix} \frac{3}{4} & \frac{1}{4} \\ \frac{1}{2} & \frac{1}{2} \\ \frac{1}{2} & \frac{1}{2} \\ \frac{1}{4} & \frac{3}{4} \end{bmatrix} \end{array}.$$

Thus the desired probability is $\frac{3}{4}$.

Problem 8 We will proceed as in Example 8.11 with $p_0 = .51$, $q_0 = .38$, and $r_0 = .11$. Now

$$u_0 = p_0 + \tfrac{1}{2} q_0 = .51 + \tfrac{1}{2}(.38) = .70 \quad \text{and}$$
$$v_0 = \tfrac{1}{2} q_0 + r_0 = \tfrac{1}{2}(.38) + .11 = .30.$$

Thus

$$p_1 = u_0^2 = (.70)^2 = .49,$$
$$q_1 = 2u_0 v_0 = 2(.70)(.30) = .42, \quad \text{and}$$
$$r_1 = v_0^2 = (.30)^2 = .09.$$

Therefore the expected genotype distribution will be 49% pure dominant, 42% hybrid, and 9% pure recessive.

Problem 9 We will proceed as in Example 8.12. The proportion of pure recessives is

$$r_1 = \frac{1}{2000} = .0005.$$

Hence

$$v_0 = \sqrt{r_1} \approx 0.02236,$$

and so

$$u_0 = 1 - v_0 \approx 0.97764.$$

Thus

$$q_1 = 2u_0v_0 \approx 2(0.97764)(0.02236) \approx 0.04372.$$

Therefore the disease is carried by about 4.4% of the population.

Chapter 9

Problem 1 (a) Since the payoff matrix is 4×3, player R has 4 moves, and player C has 3 moves.
(b) If R chooses row 2 and C chooses column 2, then the payoff to player R is 4 units, the 2,2 entry of the payoff matrix.
(c) The payoff to player R is -2 units if R chooses row 1 and C chooses column 2. Thus C wins 2 units from R if R chooses row 1 and C chooses column 2.

Problem 2 From the viewpoint of player C, column 2 is preferable to column 3 because the payoffs in column 2 never exceed those in column 3. Hence we delete column 3.

$$\begin{array}{c} \\ 1 \\ 2 \\ 3 \\ 4 \end{array} \begin{bmatrix} 1 & 2 & & 4 \\ -1 & 4 & \vdots & 3 \\ 3 & -2 & \vdots & -3 \\ -1 & 0 & \vdots & 2 \\ 2 & 1 & \vdots & -4 \end{bmatrix}$$

Now player R will prefer row 1 to row 3 because the payoffs in row 1 are always at least as great as those in row 3. Thus we delete row 3.

$$\begin{array}{c} \\ 1 \\ 2 \\ 4 \end{array} \begin{bmatrix} 1 & 2 & & 4 \\ -1 & 4 & \vdots & 3 \\ 3 & -2 & \vdots & -3 \\ 2 & 1 & \vdots & -4 \end{bmatrix}$$

Now column 4 is preferable to column 2 for player C because the payoffs in column 4 never exceed those in column 2. Therefore we delete column 2.

$$\begin{array}{c c}& \begin{array}{c c} 1 & 4 \end{array}\\ \begin{array}{c} 1 \\ 2 \\ 4 \end{array} & \left[\begin{array}{c c c c} -1 & \vdots & \vdots & 3 \\ 3 & \vdots & \vdots & -3 \\ \cdots & \cdots & \cdots & \cdots \\ 2 & \vdots & \vdots & -4 \end{array} \right] \end{array}$$

Player R will now prefer row 2 to row 4 because the payoffs in row 2 are always at least as great as those in row 4. Consequently we delete row 4.

$$\begin{array}{c c}& \begin{array}{c c} 1 & 4 \end{array}\\ \begin{array}{c} 1 \\ 2 \end{array} & \left[\begin{array}{c c c c} -1 & \vdots & \vdots & 3 \\ 3 & \vdots & \vdots & -3 \end{array} \right] \end{array}$$

For player R row 1 is preferable to row 2 if player C chooses column 4, but row 2 is preferable to row 1 if player C chooses column 1. Thus neither row 1 nor row 2 is always preferable to the other. For player C column 1 is preferable to column 4 if player R chooses row 1, but column 4 is preferable to column 1 if player R chooses row 2. Thus neither column in the matrix above is always preferable to the other. Hence we cannot reduce the matrix above any further, and so this matrix is the reduced form of the original game.

Problem 3   The row minima and column maxima for the given matrix are shown below.

$$\begin{array}{c c c c c}& 1 & 2 & 3 & \text{Row Minimum}\\ \begin{array}{c} 1 \\ 2 \\ 3 \\ 4 \end{array} & \left[\begin{array}{c c c} 1 & -4 & 3 \\ -3 & 0 & 4 \\ -1 & 5 & -2 \\ 2 & -3 & 6 \end{array} \right] & & & \begin{array}{l} -4 \\ -3 \\ -2 \leftarrow \text{largest row minimum} \\ -3 \end{array}\\ \text{Column Maximum} & 2 & 5 & 6 & \\ & \uparrow & & & \\ & \text{smallest column maximum} & & & \end{array}$$

Thus the optimal moves are row 3 for Player R and column 1 for player C.

Problem 4   The row minima and column maxima for the given matrix are shown below.

$$\begin{array}{c c c c c c}& 1 & 2 & 3 & 4 & \text{Row Minimum}\\ \begin{array}{c} 1 \\ 2 \\ 3 \end{array} & \left[\begin{array}{c c c c} 0 & -1 & 3 & -2 \\ 1 & -7 & 2 & -3 \\ -5 & 8 & -6 & -4 \end{array} \right] & & & & \begin{array}{l} -2 \leftarrow \text{largest row minimum} \\ -7 \\ -6 \end{array}\\ \text{Column Maximum} & 1 & 8 & 3 & -2 & \\ & & & & \uparrow & \\ & & & \text{smallest column maximum} & & \end{array}$$

(a) Thus the optimal moves for players R and C are row 1 and column 4.
(b) The entry -2 in row 1, column 4 is both the largest entry in its column and the smallest entry in its row. Thus this entry is a saddle point, and so the value of this game is -2.

584 SOLUTIONS TO PRACTICE PROBLEMS

Problem 5   The expected payoff when strategies P and Q are used is the single entry of the matrix PAQ. Since

$$(PA)Q = [-1 \quad -1.3 \quad -.2 \quad -.9] \begin{bmatrix} .1 \\ .2 \\ .3 \\ .4 \end{bmatrix} = [-.78],$$

the expected payoff from player R to player C is .78 unit per play. Thus player C has the advantage in this game.

Problem 6   To find the optimal strategies, we apply Theorem 9.2 to the given payoff matrix. Thus we take $a = 2$, $b = -3$, $c = -4$, and $d = 6$, and so

$$s = a - b - c + d = 2 + 3 + 4 + 6 = 15.$$

Now

$$p_1^\star = \frac{d-c}{s} = \frac{6+4}{15} = \frac{10}{15} = \frac{2}{3},$$

$$p_2^\star = \frac{a-b}{s} = \frac{2+3}{15} = \frac{5}{15} = \frac{1}{3},$$

$$q_1^\star = \frac{d-b}{s} = \frac{6+3}{15} = \frac{9}{15} = \frac{3}{5}, \quad \text{and}$$

$$q_2^\star = \frac{a-c}{s} = \frac{2+4}{15} = \frac{6}{15} = \frac{2}{5}.$$

Thus the optimal strategies for players R and C are

$$P^\star = [\tfrac{2}{3} \quad \tfrac{1}{3}] \quad \text{and} \quad Q^\star = \begin{bmatrix} \tfrac{3}{5} \\ \tfrac{2}{5} \end{bmatrix}.$$

The value of this game is

$$\frac{ad - bc}{s} = \frac{2(6) - (-3)(-4)}{15} = \frac{12 - 12}{15} = 0.$$

Since the value of the game is 0, this game is fair; hence neither player has an advantage.

Problem 7   (a) For player R, row 2 is preferable to row 3 because the payoffs in row 2 are always at least as great as those in row 3. Therefore we can delete row 3 from the payoff matrix to obtain the matrix below.

$$\begin{array}{c} \\ 1 \\ 2 \end{array} \begin{bmatrix} 1 & 2 & 3 \\ 6 & 8 & -2 \\ -1 & 2 & 3 \end{bmatrix}$$

Now player C will prefer column 1 to column 2 because the payoffs in column 1 never exceed those in column 2. Therefore we delete column 2 to obtain the 2×2 matrix game below.

$$\begin{array}{c} & 1 & 3 \\ 1 & \begin{bmatrix} 6 & \vdots & -2 \\ -1 & \vdots & 3 \end{bmatrix} \end{array}$$

In this matrix neither row is always preferable to the other, and neither column is always preferable to the other. Hence this matrix is the reduced form of the original game.

(b) To find the optimal strategies for players R and C in the reduced game above, we apply Theorem 9.2 with $a = 6$, $b = -2$, $c = -1$, and $d = 3$. Now

$$s = a - b - c + d = 6 + 2 + 1 + 3 = 12,$$

$$p_1^\star = \frac{d-c}{s} = \frac{3+1}{12} = \frac{4}{12} = \frac{1}{3},$$

$$p_2^\star = \frac{a-b}{s} = \frac{6+2}{12} = \frac{8}{12} = \frac{2}{3},$$

$$q_1^\star = \frac{d-b}{s} = \frac{3+2}{12} = \frac{5}{12}, \quad \text{and} \quad q_2^\star = \frac{a-c}{s} = \frac{6+1}{12} = \frac{7}{12}.$$

Furthermore, the value of this game is

$$\frac{ad-bc}{s} = \frac{6(3)-(-2)(-1)}{12} = \frac{16}{12} = \frac{4}{3}.$$

Since in the original 3×3 game player R will not use row 3 and player C will not use column 2, we see that the optimal strategies for the row and column players in the original game are

$$P^\star = \begin{bmatrix} \frac{1}{3} & \frac{2}{3} & 0 \end{bmatrix} \quad \text{and} \quad Q^\star = \begin{bmatrix} \frac{5}{12} \\ 0 \\ \frac{7}{12} \end{bmatrix},$$

respectively. Moreover, the value of the original game is the same as the value of the reduced game, namely, $\frac{4}{3}$.

Problem 8 By adding $k = 2$ to each entry of A, we will make each entry of the first row (and also the third row) positive. The resulting matrix B is

$$B = \begin{bmatrix} 6 & 1 & 4 & 3 \\ 0 & 4 & 3 & 2 \\ 8 & 2 & 1 & 3 \end{bmatrix}.$$

Problem 9 We will proceed as in Example 9.14. Note first that the given payoff matrix has no saddle point and is in reduced form. In order to make all the entries of row 2 positive, we add $k = 2$ to each entry of the given matrix. This yields the matrix below.

$$A = \begin{bmatrix} 4 & 0 & 1 \\ 1 & 2 & 3 \end{bmatrix}$$

Thus the linear program to be solved is:

$$\text{Maximize } F = y_1 + y_2 + y_3$$
$$\text{subject to } 4y_1 + y_3 \leq 1$$
$$y_1 + 2y_2 + 3y_3 \leq 1$$
$$y_1 \geq 0, y_2 \geq 0, y_3 \geq 0.$$

Applying the simplex method to this linear program, we obtain the following sequence of tableaus.

$$\begin{array}{c} \\ s \\ t \\ \end{array} \begin{array}{c} \begin{array}{ccccc} y_1 & y_2 & y_3 & s & t \end{array} \\ \left[\begin{array}{ccccc|c} \textcircled{4} & 0 & 1 & 1 & 0 & 1 \\ 1 & 2 & 3 & 0 & 1 & 1 \\ \hline 1^* & 1 & 1 & 0 & 0 & F \end{array} \right] \begin{array}{c} 1/4^* \\ 1/1 \\ \end{array} \end{array}$$

$$\begin{array}{c} \\ y_1 \\ t \\ \end{array} \begin{array}{c} \begin{array}{ccccc} y_1 & y_2 & y_3 & s & t \end{array} \\ \left[\begin{array}{ccccc|c} 1 & 0 & 1/4 & 1/4 & 0 & 1/4 \\ 0 & \textcircled{2} & 11/4 & -1/4 & 1 & 3/4 \\ \hline 0 & 1^* & 3/4 & -1/4 & 0 & F - \tfrac{1}{4} \end{array} \right] \begin{array}{c} \\ \tfrac{3}{4}/2 = 3/8^* \\ \end{array} \end{array}$$

$$\begin{array}{c} \\ y_1 \\ y_2 \\ \end{array} \begin{array}{c} \begin{array}{ccccc} y_1 & y_2 & y_3 & s & t \end{array} \\ \left[\begin{array}{ccccc|c} 1 & 0 & 1/4 & 1/4 & 0 & 1/4 \\ 0 & 1 & 11/8 & -1/8 & 1/2 & 3/8 \\ \hline 0 & 0 & -5/8 & -1/8 & -1/2 & F - 5/8 \end{array} \right] \\ \uparrow \uparrow \\ -x_1 -x_2 \end{array}$$

Since none of the indicators are positive, the final tableau has been obtained. From the final tableau we read the optimal solution

$$y_1 = \frac{1}{4}, \quad y_2 = \frac{3}{8}, \quad y_3 = 0, \quad F = \frac{5}{8}$$

and the negatives of the indicators corresponding to the slack variables

$$x_1 = \frac{1}{8} \quad \text{and} \quad x_2 = \frac{1}{2}.$$

Letting $v = 1/F = 8/5$, we obtain the optimal strategies for players R and C:

$$P^\star = [vx_1 \quad vx_2] = [\tfrac{1}{5} \quad \tfrac{4}{5}]$$

and

$$Q^\star = \begin{bmatrix} vy_1 \\ vy_2 \\ vy_3 \end{bmatrix} = \begin{bmatrix} \tfrac{2}{5} \\ \tfrac{3}{5} \\ 0 \end{bmatrix}.$$

The value of the original game is

$$v - k = \frac{8}{5} - 2 = -\frac{2}{5}.$$

INDEX

A

Absorbing Markov chain, 445–55
Addition principle, 246–48
Allele, 460–62
Amortization, 213–16
Annuity, 202–16
   future value, 205–6
   present value, 206–9
Arithmetic progression, 190
Artificial variable, 172
Augmented matrix, 78
Axis, 3

B

Balance of a loan, 192
Bar graph, 351–52
Basic variable, 62
Bayes's criterion, 383–87
Bayes's formula, 329–34
Bayes, Thomas, 329
Bernoulli, Jacob, 338
Bernoulli process, 338–43, 398–403, 416–17
Bernoulli's formula, 340
Binomial coefficients, 263
Binomial distribution, 399–403, 414–17
Binomial theorem, 265–66
Blood type, 228–29, 240, 325
Break-even analysis, 26
Break-even point, 26
Business
   break-even analysis, 26
   decision making, 382–92
   model for allocating service charges, 69–71

C

Cartesian coordinate system, 3–4
Cayley, Arthur, 78
Center for Disease Control, 337
Circular permutation, 263
Column vector, 80
Combinations, 256–60
Compact tableau, 159
Complement of a set, 225
Compound interest, 196–200
Conditional probability, 311–21, 327
Constant-sum game, 484
Constraint, 116
   equality, 181
Coordinates, 3–4
Corner point, 122
Cost
   fixed, 26
   variable, 26
Cost line, 26
Counting rule for conditional probabilities, 277–83, 312–13
Counting techniques, 231–60
   combinations, 256–60
   multiplication principle, 242–45
   permutations, 252–55
   principle of inclusion-exclusion, 232
   use in probability, 279–83

D

Dantzig, George B., 137
Decision theory, 382–92
Decision tree, 387–92
Demand, 35
   equilibrium, 37
   vector, 106

Dependent system, 21, 59
Dependent variable, 10
Diagnostic tests, 333–34
Difference equation, 189, 202–5
Disjoint sets, 225
Divorce, 342
Dual program, 159
   variable interpretation, 166–67
Duality theorem, 161
Dyson, Freeman, 345

E

Economics
   Leontief input-output model, 102–9
   market equilibrium, 35–40
   supply and demand, 35–40
Ecosystem, 445–46, 457
Elementary operation
   on equations, 45
   on matrices, 48
Elementary row operations, 48
Elimination
   Gauss-Jordan, 50
   method of, 18
Elmira, New York, 323
Empty set, 224
Entry
   i, j, 79
   of a matrix, 78
Equality constraint, 181
Equilibrium demand, 37
Equilibrium price, 36
Equilibrium supply, 37
Equilibrium vector, 436–42, 447–48
Events in an experiment, 276–79, 297–307

Events in an experiment, *continued*
  certain, 274
  impossible, 276–78
  independent, 318–21
  mutually exclusive, 304–5, 326
  odds in favor of, 291–93
  probability of, 277
Expected value, 373–78
Experiment, 274–75
  Bernoulli, 337–43, 398–403, 416–17
  outcomes, 274–75
  sample space, 274–75

F
Factorial notation, 254–55
Fair game, 492
Feasible point, 116
Feasible region, 127
Fermat, Pierre de, 272
First-order linear difference equation, 189, 202–5
Framingham study, 322–23
Free variable, 62
Frequency distribution, 362–66
Functions
  linear, 10
  objective, 116

G
Game theory, 475–510
  constant-sum game, 484
  expected payoff, 489–91
  fair game, 492
  linear programming, 501–5
  matrix game, 475
  moves, 475
  nonzero-sum game, 507–8
  n-person game, 507–8
  optimal moves, 479–81
  optimal strategy, 491–99
  payoff matrix, 475
  players, 475
  prisoner's dilemma, 508
  reduced form, 477–79
  saddle point, 481–83
  strategy, 487–92
  strictly determined game, 484
  two-person zero-sum game, 475
  value of a game, 482–83
Gauss-Jordan elimination, 50
Gene, 460
Genetics, 460–71
Genotype, 460–68
Geometric progression, 191
Graph of a function, 4
Graphical method for linear programs, 129
Gross production vector, 105
Grouped data, 361–66

H
Half-plane, 120
Hardy-Weinberg law, 466
Harris, Louis, 342
Heredity, 460–62
Histogram, 351–52

I
Identity matrix, 92
i, j entry of a matrix, 79
Inconsistent systems, 20, 57
Independent events, 318–26
Independent variable, 10
Indicator, 150
Inductive reasoning, 326
Initial value of a sequence, 189
Input-output matrix, 103
Input-output model, 102–9
Input-output ratio, 103
Input-output theory, 103
Installment loans, 213–16
Intercepts, 6
Interest, 191–200
  compound, 196–200
  effective rate, 198
  nominal rate, 198
  principal, 192
  rate of, 192
  simple, 192–94
Internal consumption vector, 105
Intersection of sets, 225
Inverse matrix, 96
Invertible matrix, 96

L
Land usage, 442
Law of complete probability, 329
Leontief input-output model, 102–9
Leontief, Wassily, 102
Linear equation, 6, 45
Linear function, 10–15
Linear inequalities, 119–23
Linear programming, 116–83, 501–5
  duality, 159–67
  game theory, 501–5
  graphical method, 127–35
  simplex method, 137–83, 501–5
Lines,
  equation of, 10–15
  graphing of, 6–8
  horizontal, 8
  vertical, 8

M
$m \times n$ matrix, 78
Marginal cost, 13
Market equilibrium, 35–40
Markov, A. A., 426
Markov chain, 426–73
  absorbing, 445–55
  absorbing state, 445–55
  canonical transition matrix, 449–50
  equilibrium vector, 436–42, 447–48
  fundamental matrix, 450
  initial state vector, 429
  regular, 438–42
  states of, 426
  state vectors, 429–32
  transition diagram, 426
  transition matrix, 427–29, 449–50
  transition probability, 426
Mathematics of finance, 189–216
Matrix, 78
  augmented, 78

canonical transition, 449–50
of coefficients, 93
difference, 81
entry, 78–79
equality, 79
form of linear system, 93
fundamental, 450
game, 475
identity, 92
input-output, 103
inverse, 96
invertible, 96
payoff, 475
product, 86–89
product with scalar, 81
size, 78
square, 92
sum, 80
technology, 103
transition, 427–29, 449–50
Maximax criterion, 383–87
Maximin criterion, 383–87
Mean, 352, 363, 372
Measure of central tendency, 352
    mean, 352, 363, 372
    median, 354
Measure of dispersion, 355–59
    standard deviation, 356–59
    variance, 355–60, 365–66, 370
Median, 354
Mendel, Gregor, 460
Method of elimination, 18
Méré, Chevalier de, 272, 345
Multiplication principle, 242–45
Mutually exclusive events, 304–305, 326

N

Net production vector, 106
New Zealand, 442
Nodes of a decision tree, 387–88
Nonzero-sum game, 507–8
Normal distribution, 407–18
    approximation to a binomial distribution, 414–18
    equation of, 409

standard units, 410–12
z-values, 410–12
n-person game, 507–8

O

Objective function, 116
Odds, 291–93
Optimal point, 117
Optimization, 115
Ordered pair, 3
Origin, 3

P

Partition of a sample space, 328
Pascal, Blaise, 272
Pascal's triangle, 264–65
Pasture ecosystem, 445–46, 457
Payoff matrix, 475
Payoff table, 382
Pepys, Samuel, 345
Permutation, 252–55
    circular, 263
Phases of the simplex method, 171
Pivot
    column, 49
    operation, 49
    row, 49
Point-slope form of the equation of a line, 12
Pregnancy test, 337
Present value, 199–200, 206–9
Price, equilibrium, 36
Principal, 192
Principle of inclusion-exclusion, 232
Prisoner's dilemma, 508
Probability, 272–341
    of the complement of an event, 300
    conditional, 311–21, 327
    distribution, 287–88, 397–418
    of an intersection of events, 317–20, 326
    of a simple event, 258
    of a union of events, 302–7, 311

Probability density function, 397–418
    binomial, 399–403
    normal, 407–18
Producing sector, 103
Product-mix problem, 67
Profit, 27
Profit function, 27

R

Random variable, 370–78, 401–3
    continuous, 407–8
    discrete, 370–78
    expected value, 373–78
    mean, 373, 381, 401–3
    standard deviation, 374, 401–3
    variance, 373, 381
Real number line, 3
Rectangular coordinate system, 3–4
Reduced form of a matrix game, 477–79
Regular Markov chain, 438–42
Relative frequency, 272, 337
Revenue, 26
Revenue line, 26
Road network, 71–74
Row label, 140
Row vector, 79

S

Saddle point, 481–83
Sales tax, 38
Sample space, 274–75
Scalar, 81
Scalar product, 81
Scissors-paper-stone, 510
Sequence, 189
Service charge allocation, 69–71
Set braces, 223
Set of feasible points, 127
Sets, 223–36, 297–307
    complement of, 225, 298, 300
    disjoint, 225
    elements of, 223
    empty, 224
    equality of, 223

Sets, *continued*
   intersection of, 225, 298
   number of elements in the union of, 232
   subset, 225, 267
   union of, 225, 297–98, 302–5, 311
   universal, 224
   Venn diagrams, 226–36
Simple interest, 192–94
Simplex method, 137–83, 501–5
   algorithm for, 154, 173
   artificial variable, 170–83
   basic variable, 139
   compact tableau, 159
   duality, 159–67
   free variable, 139
   indicators, 150–51
   minimization, 159–74
   phase one, 173
   pivot, 141, 151–53
   pivot column, 151
   pivot row, 151
   simplex tableau, 141
   slack variable, 138–40
   solving matrix games, 501–5
Sinking fund, 211–13
Size of a matrix, 78
Slack variable, 138
Slope-intercept form of the equation of a line, 11
Slope of a line, 11
Snapdragons, 468–69
Square matrix, 92
Stamp tax, 39
Standard deviation, 356–59
Standard maximum program, 149
Standard units, 410–12
States of nature, 382
State vectors, 429–32
Statistics, 351–78, 396–418
   descriptive, 351–39
   inferential, 351
Stochastic process, 326–27

Strictly determined game, 484
Subset, 225, 267
Subsidy, 41
Success runs, 454–55
Summation sign, 353
Supply, 35
   equilibrium, 37
Surplus production vector, 106
System of linear equations
   dependent, 21, 59
   inconsistent, 20, 57
   m by n, 45
   of two equations in two unknowns, 17

T
Tableau form, 47
Tax,
   sales, 38
   stamp, 39
Technology matrix, 103
Terms of a sequence, 169
Three-finger morra, 510
Toronto, Canada, 442
Traffic flow, 71–74
Tree diagram, 236–37, 327–34
Two-finger morra, 476, 494
Two-person zero-sum game, 475
Two-phase method, 171

U
Union of sets, 225, 297–98, 302–5, 311
Universal set, 224
Utility of money, 386

V
Value of a game, 482–83
Value of perfect information, 395

Variable
   artificial, 172
   basic, 62
   dependent, 10
   free, 62
   independent, 10
   slack, 138
Variance, 355–60, 365–66, 370
Vector
   column, 80
   demand, 106
   gross production, 105
   internal consumption, 105
   net production, 106
   of right-hand constants, 93
   row, 79
   state, 429–32
   surplus production, 106
   of unknowns, 93
Venn diagram, 226–36
von Neumann, John, 161, 475

W
World War II, 345

X
x-axis, 3
x-intercept, 6

Y
y-axis, 3
y-intercept, 6

Z
Zero matrix, 83
z-values, 410–12